INCOSE NEEDS AND REQUIREMENTS MANUAL

INCOSE NEEDS AND REQUIREMENTS MANUAL

NEEDS, REQUIREMENTS, VERIFICATION, VALIDATION ACROSS THE LIFECYCLE

INCOSE-TP-2021-002-01
2024

Prepared by:

Requirements Working Group
International Council on Systems Engineering (INCOSE)
7670 Opportunity Road, Suite 220
San Diego, California 92111-2222 USA

Written by:

Louis S. Wheatcraft
Michael J. Ryan
Tami Edner Katz

Published by John Wiley & Sons, Inc., Hoboken, New Jersey.
Published simultaneously in Canada.

Library of Congress Cataloging-in-Publication Data Applied for:

Hardback ISBN: 9781394152742

Cover Design: Wiley
Cover Image: Courtesy of Michael J. Ryan and Louis S. Wheatcraft

Set in 10/12pt TimesLTStd by Straive, Chennai, India

SKY10087261_100924

CONTENTS

INCOSE NOTICES

ADDITIONAL COPIES/GENERAL INFORMATION

Copies of the *Needs and Requirements Manual,* as well as any other INCOSE document can be obtained from the INCOSE Store. General information on INCOSE, the Requirements Working Group, any other INCOSE working group, or membership may also be obtained from the INCOSE Central Office at:

International Council on Systems Engineering
7670 Opportunity Road, Suite 220
San Diego, California 92111-2222 | USA
E-mail: info@incose.org

Telephone: +1 858-541-1725
Toll Free Phone (US): 800-366-1164
Fax: +1 858-541-1728
Web Site: http://www.incose.org

PREFACE

This Manual has been prepared and produced by a volunteer group of authors and contributors within the Requirements Working Group (RWG) of the International Council on Systems Engineering (INCOSE).

AUTHORS

The principal authors of this Manual are:

Lou Wheatcraft, Wheatland Consulting, LLC, USA

Michael Ryan, Capability Associates Pty Ltd, AU

Tami Katz, BAE Systems, USA

MAJOR CONTRIBUTORS

Those who made a significant contribution to the generation of this Manual are:

Mark Abernathy, Retired, USA

James R. Armstrong, Stevens Institute of Technology, USA

Brian Berenbach, Georgia Institute of Technology, USA

Simone Bergamo, Thales, FR

Ronald S. Carson, PhD, Seattle Pacific University, USA

Jeremy Dick, Retired, UK

Celeste Drewien, Sandia National Labs, USA

James R. van Gaasbeek, Retired, USA

Rick D. Hefner, Caltech, USA

Henrick Mattfolk, Whirlpool, USA

Lamont McAliley, Veracity Engineering, USA

Donald McNally, Woodward, USA

Kevin E. Orr, Eaton Corporation, USA

Michael E. Pafford, Retired, USA

Susan E. Ronning, ADCOMM Engineering LLC, USA

Raymond B. Wolfgang, Sandia National Labs, USA

Gordon Woods, East West Railway Company, UK

Richard Zinni, Harris Corporation, USA

REVIEWERS

In addition to the authors and contributors, below are the names of those who submitted review comments that were included during the development of this Manual:

Angel Agrawal, Northrop Grumman, USA

Ben Canty, Ball Aerospace and Technologies Corporation, USA

James B. Burns, Johns Hopkins University, USA

Ken Eastman, Ball Aerospace and Technologies Corporation, USA

Raymond Joseph, Odfjell Terminal, Houston, USA

Jenn Molloy, Ball Aerospace and Technologies Corporation, USA

Michele Reed, Shell Products and Technology, USA

Andreas Vollerthun, Dr., KAIAO-Consulting, DE

Beth Wilson, retired, USA

REVISION HISTORY

Revision	Revision Date	Change Description and Rationale
1.0	January 2022	First release
1.1	May 2022	Editorial and alignment with other RWG products
2.0	November 2024	Major update based on use and comments and alignment with other RWG products.

LIST OF FIGURES

LIST OF TABLES

1

INTRODUCTION

1.1 PURPOSE

This Needs and Requirements Manual (NRM) presents systems engineering (SE) from the perspective of the definition and management across the system lifecycle of needs, requirements, verification, and validation (NRVV). NRVV are common threads that tie together all lifecycle activities and processes.

As presented in this Manual, for acceptance, certification, and qualification, the system or product being developed is verified against design input requirements and validated against its integrated set of needs. To successfully complete system verification and system validation, the needs and requirements of the system as well as the system verification and validation artifacts must be managed throughout the entire system lifecycle. This Manual provides practical guidance on the concepts and activities required to achieve those outcomes.

As shown in Figure 1.1, this Manual supplements and elaborates the INCOSE Systems Engineering Handbook (SE HB) [1] and the Systems Engineering Body of Knowledge (SEBoK) [2], providing more detailed guidance on the "what," "how," and "why" concerning NRVV across the system lifecycle. The NRM also addresses ambiguity and inconsistencies in NRVV terminology and ontology.

Figure 1.1 shows this Manual is further elaborated by several supporting guides. The *Guide to Needs and Requirements* (GtNR) [3] and the *Guide to Verification and Validation* (GtVV) [4] focus further on the "what" and "how" of the specific processes being implemented within an organization. The level of detail is similar in content to an organization's Work Instructions (WIs) or Standard Operating Procedures (SOPs). These guides reference this Manual for specific guidance on the "why" and underlying concepts, maintaining consistency in approach and ontology defined in this Manual.

INCOSE Needs and Requirements Manual: Needs, Requirements, Verification, Validation Across the Lifecycle,
First Edition. Louis S. Wheatcraft, Michael J. Ryan, and Tami Edner Katz.

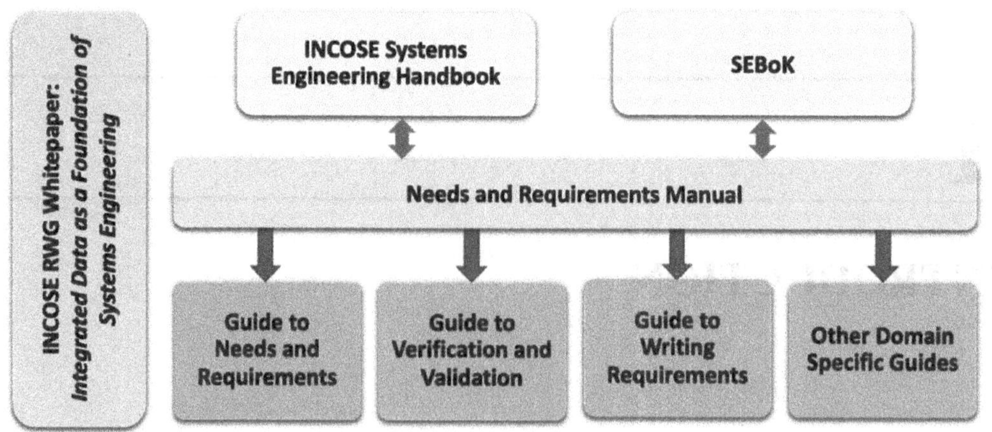

FIGURE 1.1 Relationships Between INCOSE Requirements Working Group (RWG) Products and the INCOSE SE HB and SEBoK.

Although this Manual addresses the activities and underlying analysis associated with defining individual and sets of needs and design input requirements, the actual writing of the need and requirement statements is covered in the INCOSE *Guide to Writing Requirements* (GtWR) [5]. The GtWR includes a list of key characteristics of well-formed needs and requirements and sets of needs and requirements, as well as a set of rules that can help achieve those characteristics. Throughout this Manual, when the activities being discussed contribute to a given characteristic defined in the GtWR, a trace to that characteristic is included.

Figure 1.1 also shows that this Manual and the associated guides advocate a data-centric approach to Project Management (PM) and SE as defined in the INCOSE RWG Whitepaper *Integrated Data as a Foundation of Systems Engineering* [6], as discussed in Chapter 3 of this Manual.

1.2 SCOPE

To support PM and SE from an NRVV perspective, this Manual:

- Provides PM and SE practitioners with an understanding of the best practices for effective NRVV definition and management throughout the system lifecycle.
- Helps organizations understand that NRVV are key elements of SE activities.
- Provides guidance to the successful implementation of NRVV activities as part of PM and SE, in any domain.
- Reinforces the idea that adequate definition of lifecycle concepts and a well-formed set of needs is a prerequisite to the definition of a well-formed set of system design input requirements.
- Provides practical, cross-domain guidance to enable organizations to integrate best practices and concepts within their PM and SE processes, activities, WIs, and procedures.
- Provides a clear description of how the terms verification and validation are applied to the artifacts generated across the system lifecycle.
- Describes the importance of planning early for verification and validation activities across the lifecycle and the inclusion of verification and validation artifacts in system models.

- Provides thorough guidance to readers on planning, definition, execution, and reporting of verification and validation activities across the system lifecycle.
- Provides guidance and best practices that will help customers avoid accumulating technical debt and enable projects and suppliers to deliver a winning product.
- Presents a data-centric approach to NRVV definition and management.
- Provides guidance for organization- and enterprise-wide sharing of data and information associated with developing and managing an integrated set of needs, the resulting design input requirements, and design output specifications, as well as verification and validation artifacts throughout the system lifecycle.

1.3 AUDIENCE

This Manual is intended for those whose role is to perform NRVV activities throughout the system lifecycle. This includes those who *verify* that the design and realized System of Interest (SOI) meet the requirements and those who *validate* that the requirements, design, and realized SOI meet the needs in the intended operational environment when used by the intended users and mitigate risk of any misuse of the SOI and losses because of misuse.

This Manual is addressed to practitioners of all levels of experience. Someone new to PM and SE should find the specific guidance useful, and those more experienced should be able to find new insights concerning NRVV across all stages of the system lifecycle, which is often absent from other texts, guides, or standards, particularly in terms of a data-centric perspective.

Major user groups who will benefit from the use of this Manual include systems engineers, requirements engineers, business analysts, product developers, system architects, configuration managers, designers, testers, verifiers, validators, manufacturers, coders, operators, users, disposers, course developers, trainers, tool vendors, project managers, acquisition personnel, lawyers, regulators, and standards organizations. Specific use cases for various classes of readers are shown in Table 1.1.

1.4 APPROACH

While this Manual addresses the specific application of the activities and concepts associated with NRVV, the specifics of "how" this information is applied are not prescribed. For example, while the use of models and a data-centric approach is advocated, the specifics concerning how to implement these concepts within the project's toolset are not addressed; while the use of Requirement Management Tools (RMTs) is advocated, the specifics concerning any particular RMT are not discussed. In this regard, this Manual is structured to enable other INCOSE Working Groups (WGs) and tool vendors to develop domain or tool-specific guides that tailor these contents to best fit the needs of the organization.

There are many use cases for how organizations practice SE to develop systems and products. This Manual presents a generic set of concepts and activities that can be applied. It is not intended that organizations adopt all the activities presented, but rather use the best practices presented to tailor their product development activities and processes appropriate to their domain, product line, workforce, and culture in such a way that provides the most value. For additional guidance concerning tailoring, refer to Chapter 4 of the INCOSE SE HB [1].

TABLE 1.1 NRVV Use Cases.

Reader	Use Cases
Novice practitioners: those new to SE and NRVV	• Learn and understand NRVV terminology, concepts, and best practices • Access a structured, unambiguous, and comprehensive source of information and knowledge to help learn NRVV from a data-centric perspective • Learn a consistent and unambiguous ontology (meta-model) for NRVV
Seasoned practitioners: those experienced in SE and NRVV but not from a data-centric perspective	• Gain more in-depth understanding of NRVV from a data-centric perspective • Reinforce, refresh, build upon, and renew their NRVV knowledge • Tailor the concepts in the NRM to their organization's product line, processes, and culture • Adopt the NRVV best practices presented in this Manual • Use this Manual to help mentor novice practitioners in the realm of NRVV
Course developers/ educators/trainers: individuals or organizations that specialize in training practitioners and other stakeholders in NRVV processes and tools	• Use a structured, unambiguous, and comprehensive source of information and knowledge to help teach NRVV from a data-centric perspective • Suggest relevant NRVV topics to trainers for their course content • Present a consistent and unambiguous ontology for NRVV. • Lead SE curricula development and revision inside their own organizations based on best practices and knowledge presented in this Manual
Tool vendors: organizations that provide applications that enable the data-centric practice of SE	• Implement recommended features in a toolset to enable practitioners to develop and manage NRVV across the system lifecycle from a data-centric perspective • Align their PM and SE toolset products with a comprehensive set of NRVV activities and artifacts and underlying data and information • Apply Artificial Intelligence (AI) as a "digital assistant," helpful in the performance of the activities defined in this Manual
Project managers: those who manage product development projects	• Understand overall product development lifecycle processes from the perspective of SE and NRVV • Understand what a data-centric practice of SE means and its advantages • Understand the value and importance of SE and NRVV activities to project success—and the importance of budgeting for and scheduling such activities • Understand how measures and metrics managed within the SE toolset can help better manage product development projects • Provide an accurate and comprehensive SE and NRVV reference, for both training and practitioner use • Provide more accurate cost and schedule estimates for complex systems engineering projects • Suggest cost and schedule savings to complex systems engineering projects, even without an engineering background

TABLE 1.1 ***(Continued)***

Reader	Use Cases
Non-SE stakeholders: those involved in non-SE project activities	• Understand basic terminologies, scope, best practices, artifacts, and value associated with SE and NRVV from a data-centric perspective • Understand how various NRVV activities and artifacts relate to other SE and PM activities and artifacts • Understand NRVV activities and the various opportunities for cross-functional collaboration between non-SE stakeholders and SE practitioners
Customers: those who request a work product or outsource the development of an SOI to a supplier.	• Understand overall product development lifecycle processes from the perspective of SE and NRVV activities. • Understand the role that both the customer and supplier have in these activities.
Suppliers: those who supply an SOI to a customer.	• Understand the importance of clearly defining in the supplier Statement of Work (SOW), Statement of Objectives (SOO), Performance Work Statement (PWS), or Supplier Agreement (SA) processes and deliverables and the relationships of the customer/supplier roles and responsibilities concerning the development, manufacturing, coding, verification, and validation of an SOI. • Understand the relationship between needs and a customer's strategic objectives driving them. • Understand why the integrated system must be managed across all lifecycle activities. • Understand why an integrated set of needs must be defined, against which the integrated system will be validated prior to defining the set of design input requirements. • Understand the value of providing suppliers with the underlying analysis and resulting data and information used to define the integrated set of needs and set of design input requirements.

Note: The end item being developed can be referred to as either a system or a product. Either term implies the integrated system, which comprises subsystems and lower-level system elements (assemblies, sub-assemblies, and components) that are defined within the system architecture. An SOI is a specific entity (system, subsystem, or system element) that is to be defined, developed, verified, validated, and delivered to either internal or external customers and then used, sustained, and retired. The concepts discussed here apply to any SOI, no matter where it exists in the physical architecture, so the term SOI will be used to stand for both systems and products.

1.5 MAPPING OF NRVV ACROSS STANDARDS

Various PM and SE organizations and standards may name and group the NRVV activities discussed in this Manual differently, combining several of these activities and assigning various names to the resulting technical processes as shown in Table 1.2.

TABLE 1.2 NRVV Activities as Addressed by Various Organizations and Standards.

	PMI®	SEI®/CMMI®	ISO/IEC/IEEE 15288/29148	INCOSE SE HB	NASA NPR 7123 & SE HB
Requirements Development	X	X			
Requirements Management	X	X			X
Business or Mission Analysis Process			X	X	
Stakeholder Needs and Requirements Definition			X	X	
Systems Requirements Definition			X	X	
Stakeholder Expectations Definition					X
Technical requirements definition					X
System verification	X	X	X	X	X
System validation	X	X	X	X	X

The Project Management Institute (PMI®) and Carnegie Mellon's Software Engineering Institute (SEI®) Capability Maturity Model Integrated (CMMI®) separate requirements development and management into two separate processes, "*Requirements Development*" and "*Requirements Management.*" Verification and validation of the lifecycle concepts and needs and of the resulting design input requirements are divided between the two main processes. Lifecycle concepts analysis and maturation and needs definition are included as part of *Requirements Development*, and management of lifecycle concepts and needs is included as part of *Requirements Management.*

ISO/IEC/IEEE 15288 [7] defines needs and requirements in terms of organizational levels: organizational (enterprise), strategic, operational, system, and system elements. Lifecycle concepts, needs, and requirements (CNR) are defined for each level at a level of abstraction appropriate to that level. The INCOSE SE HB divides requirement development and management into three technical process areas in accordance with ISO/IEC/IEEE 15288: *Business or Mission Analysis Process* for lifecycle CNR at the strategic level, *Stakeholder Needs and Requirements Definition Process* for lifecycle CNR at the operational level, and *Systems Requirements Definition Process* for lifecycle CNR at the system level and below. Chapter 2 of this Manual provides a more detailed discussion of levels of lifecycle CNR.

In the context of the CMMI, *Requirement Development* process, ISO/IEC/IEEE 15288, ISO/IEC/IEEE 29148 [8], and INCOSE SE HB [1] separate *Requirements Development* activities into the above mentioned three areas such that *Business or Mission Analysis Process* and *Stakeholder Needs and Requirements Process* correspond to the development of stakeholder needs, and *System Requirements Definition Process* corresponds to the development of the system-level requirements.

Within ISO/IEC/IEEE 15288, ISO/IEC/ IEEE 29148, and the INCOSE SE HB technical processes, system-level lifecycle CNR are combined into one process area: *Systems Requirements Definition.* In addition, in these documents, there is no *separate* needs and requirement management process to address the activities associated with needs and requirements management. Instead, needs and requirements management is included as part of the *Systems Requirements Definition Process* activities, and other technical process.

While including the lifecycle processes contained in ISO/IEC/IEEE 15288, ISO/IEC/ IEEE 29148 also includes clause 6.6 concerning requirements management as a distinct activity.

The National Aeronautics and Space Administration (NASA) NASA Procedural Requirements (NPR) 7123.1C, NASA SE Processes and Requirements Document [9], and the companion NASA SE HB NASA/SP-2016-6105 [10] define three process areas that address needs and requirements: *Stakeholder Expectations Definition*, *Technical Requirements Definition*, and *Requirement Management*. The *Stakeholder Expectations Definition* and *Requirements Definition* processes are similar to those defined in ISO/IEC/IEEE 15288, ISO/IEC/IEEE 29148, and INCOSE SE HB; however, NASA includes a separate crosscutting *Requirement Management* process.

Another difference between the NASA SE HB and the INCOSE SE HB is that the NASA SE HB *Stakeholder Expectations Definition Process* does not use the terms stakeholder needs and stakeholder requirements. However, the form of communicating stakeholder expectations is not specifically stated, implying that the stakeholder expectations are communicated via Operations Concept (OpsCon) and Concept of Operations (ConOps) documents. *See Chapter 4 for a more detailed discussion concerning OpsCon versus ConOps.*

All organizations include distinct verification and validation processes; however, the organizational element within an organization responsible for these activities tends to vary. PMI/CMMI places verification and validation activities as part of Quality Management (QM), while INCOSE and NASA place verification and validation activities as part of SE.

1.6 THE FIVE NRVV ACTIVITY AREAS DISCUSSED IN THIS MANUAL

As shown in Figure 1.2, the five NRVV activity areas defined below trace to, and are an elaboration of, the *Business or Mission Analysis*, *Stakeholder Needs and Requirements Definition*, *System Requirements Definition*, *Verification*, and *Validation* technical process activities defined in ISO/IEC/IEEE 15288 and INCOSE SE HB. *Note: These activity areas are highly integrated*

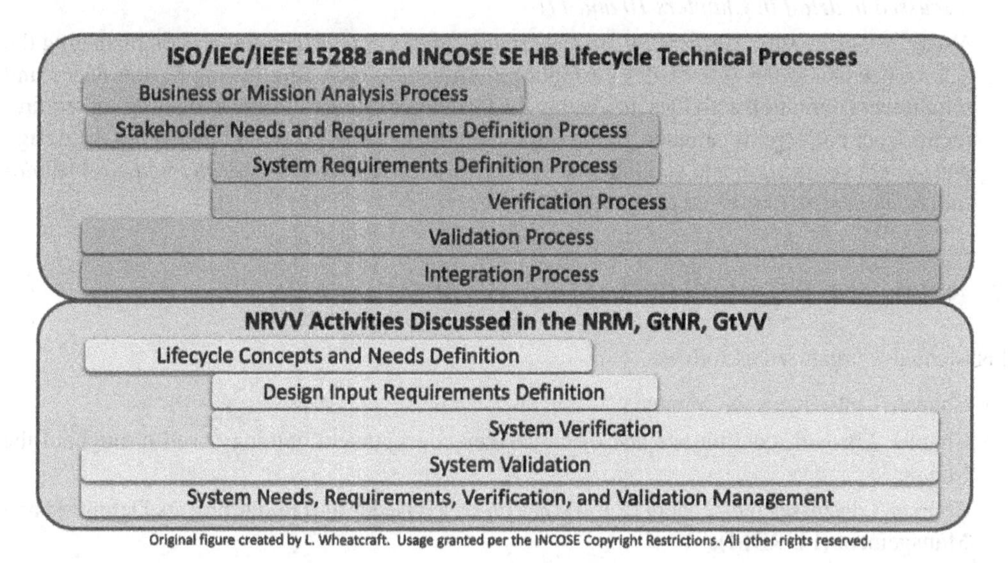

FIGURE 1.2 Relationship of INCOSE SE HB Technical Processes and NRVV Activities Discussed in this Manual.

with and dependent on the other technical processes defined in ISO/IEC/IEEE 15288 and INCOSE SE HB, including the Architectural Definition Process, Design Definition Process, and Integration Process as well as the crosscutting Interface Management and the technical management processes. Because of the dependencies, the activities and processes shown in Figure 1.2, are best performed concurrently rather than serially.

The focus of this Manual is on the NRVV activities required for an SOI to be developed, managed, delivered, used, sustained, and retired. This Manual divides these activities into the following five activity areas:

- *Lifecycle Concepts and Needs Definition:* The focus is on defining a feasible set of lifecycle concepts and a well-formed integrated set of needs that will result in an SOI that will meet the defined problem or opportunity; mission, goals, and objectives (MGOs); measures; needs; and requirements defined at the operational and strategic levels of the organization as well as by an external customer. *Lifecycle Concepts and Needs Definition activities are discussed in detail in Chapters 4 and 5.*
- *Design Input Requirements Definition:* The focus is on transforming the baselined integrated set of needs for an SOI into well-formed design input requirements expressed as "shall" statements that are inputs for defining the SOI physical architecture, flowing requirements down the architecture, and realization of the SOI via a design solution. The design input requirements address what the system, subsystem, or system element must do to satisfy their integrated set of needs from which they were transformed without stating how they are to be implemented (design outputs). *Design Input Requirements Definition is discussed in detail in Chapters 6 and 7.*
- *System Verification.* The focus is on planning for system verification, executing those plans, recording the results, and reporting the results to an Approval Authority. *System Verification is discussed in detail in Chapters 10 and 11.*
- *System Validation.* The focus is on planning for system validation, executing those plans, recording the results, and reporting the results to an Approval Authority. *System Validation is discussed in detail in Chapters 10 and 11.*
- *Needs, Requirements, Verification, and Validation Management.* The focus is on managing the sets of needs and the sets of design input requirements, including managing the needs and requirement definition activities, managing the flow down (allocation and budgeting) of requirements from one level to another, validating bidirectional traceability, and managing the design and system verification and validation artifacts. *Needs, Requirements, Verification, and Validation Management are discussed in detail in Chapter 14.*

1.7 NEEDS AND REQUIREMENTS MANUAL ORGANIZATION

This Manual is organized as follows:

- Chapter 1 introduces the Manual.
- Chapter 2 provides definitions and discusses key concepts and ontology used throughout the Manual.
- Chapter 3 discusses the concept of Information-based Needs and Requirements Definition and Management (I-NRDM).
- Chapter 4 discusses Lifecycle Concepts and Needs Definition activities.
- Chapter 5 discusses Need Verification and Validation activities.

- Chapter 6 discusses Design Input Requirements Definition activities.
- Chapter 7 discusses Design Input Requirement Verification and Validation activities.
- Chapter 8 discusses Design Verification and Design Validation activities.
- Chapter 9 discusses Production Verification activities.
- Chapter 10 discusses system verification and validation common principles.
- Chapter 11 discusses System Verification and Validation activities.
- Chapter 12 discusses the use of Off-the-Shelf (OTS) system elements.
- Chapter 13 discusses design and system verification and design and system validation considerations when using supplier-developed system elements.
- Chapter 14 discusses Needs, Requirements, Verification, and Validation Management activities.
- Chapter 15 discusses the use of attributes.
- Chapter 16 discusses the desirable features of an SE toolset.
- Appendix A lists references for sources of information in this Manual.
- Appendix B lists acronyms and abbreviations used in this Manual.
- Appendix C provides a glossary defining key terms used in this Manual.
- Appendix D provides a comment form and address to which comments on this Manual can be sent.

2

DEFINITIONS AND CONCEPTS

The Needs, Requirements, Verification, and Validation (NRVV) concepts discussed in this Manual are better understood if based on consistent definitions of key terms and an understanding of basic concepts.

2.1 ONTOLOGY USED IN THIS MANUAL

Various authoritative sources provide a wide range of different views of NRVV, which often leads to confusion regarding the concepts themselves as well as the appropriate authoritative source. Consequently, a major challenge in developing this Manual was in the provision of an ontology, in particular the use of various terms and their relationships, especially with respect to the context of their usage.

To establish a framework for NRVV, this Manual presents a specific ontology. The reader is cautioned to be aware of the specific use of terms in context as they read this Manual and the associated INCOSE Requirement Working Group (RWG) guides, particularly if they are familiar with the concepts, processes, activities, and terminology referred to in the PMI® documents, Carnegie Mellon's SEI CMMI®, the International Organization for Standardization (ISO)/International Electrotechnical Commission (IEC)/Institute of Electrical and Electronics Engineers (IEEE) standards, INCOSE's SE HB, or National Aeronautics and Space Administration (NASA) procedure requirements standards and handbooks, and other organizations' standards associated with product and system development. Because of the differences in the use of terms in these various organizational standards, discussion of the differences and challenges is appropriate, so the following paragraphs provide examples of ontology challenges and how they are addressed in this Manual. (*Section 2.2 and the Glossary provide definitions of key terms used in this Manual and its supporting Guides.*)

INCOSE Needs and Requirements Manual: Needs, Requirements, Verification, Validation Across the Lifecycle, First Edition. Louis S. Wheatcraft, Michael J. Ryan, and Tami Edner Katz.

2.1.1 Stakeholder Expectations, Needs, and Requirements

Various guides, textbooks, and standards refer to stakeholder "expectations, needs, and requirements," "stakeholder needs," or "user needs and requirements" as if they are the same, often combining them in a single document, such as a User Requirements Document (URD), Program [or Project] Requirements Document (PRD), Stakeholder Requirements Document (StRD), or Stakeholders Needs Document (StND)—communicating both the needs and requirements in the form of "shall" statements.

This can result in confusion as to the scope of each of the terms, in terms of what is an "expectation" versus what is a "need" as opposed to what is a "requirement." All three terms can have different contractual meanings, and the resultant System of Interest (SOI) may not be what the customer expected if all three terms are not clarified and agreed-to. In addition, stakeholder expectations can be communicated in various forms, such as user stories, use cases, user scenarios, operational scenarios, Concepts of Operation (ConOps), Operational Concepts (OpsCon), or as textual statements.

For some practitioners, as described in ISO/IEC/IEEE 29148 [8], "stakeholder expectations" are communicated in terms of stakeholder needs and then transformed into stakeholder requirements. The purpose of both stakeholder needs and requirements is to represent a user-oriented view of the system as viewed externally: what stakeholders need from the system, what they need the system to do, how they plan to interact with the system, and what interactions the system has with its external environment.

To minimize confusion, this Manual refers to *stakeholder needs*, which is inclusive of stakeholder requirements.

Section 1.8 of the GtWR [5] provides a more detailed discussion concerning stakeholder expectations, stakeholder needs, and stakeholder requirements as opposed to system requirements.

2.1.2 Needs Versus Requirements

To avoid ambiguity in the use of the terms "needs" and "requirements," this Manual follows the conventions for each described below.

2.1.2.1 Integrated Set of Needs The *integrated set of needs* represents the integrated and baselined set of needs that were transformed from the set of lifecycle concepts for the SOI. This concept is highlighted below and further discussed in Chapters 4 and 5.

To establish a complete understanding of what is needed from the SOI, maturation and analysis of the lifecycle concepts are undertaken, which is an input into defining a fully integrated set of needs at the project/system level (shown in Figure 2.1).

Stakeholder expectations, needs, and requirements are treated as inputs into this process, which can be in the form of stakeholder needs and requirements (or other forms of expression), the problem statement, the mission statement, goals, objectives, and measures (or some combination of those forms). These elements are established using the processes of *Business or Mission Analysis* and *Stakeholder Needs and Requirements Definition* as described in ISO/IEC/IEEE15288 and the INCOSE SE HB. Note that any customer-supplied system requirements are treated as higher-level requirements that are inputs into the lifecycle concepts analysis and maturation and needs definition activities.

The use of the word "integrated" with the "set of needs" is important. As shown in Figure 2.1, there are many possible inputs into the lifecycle concepts and needs definition activities. These inputs can be poorly communicated and may be inconsistent, incomplete, incorrect, or not feasible. As part of the lifecycle concepts analysis and maturation activities and the needs analysis activities, these

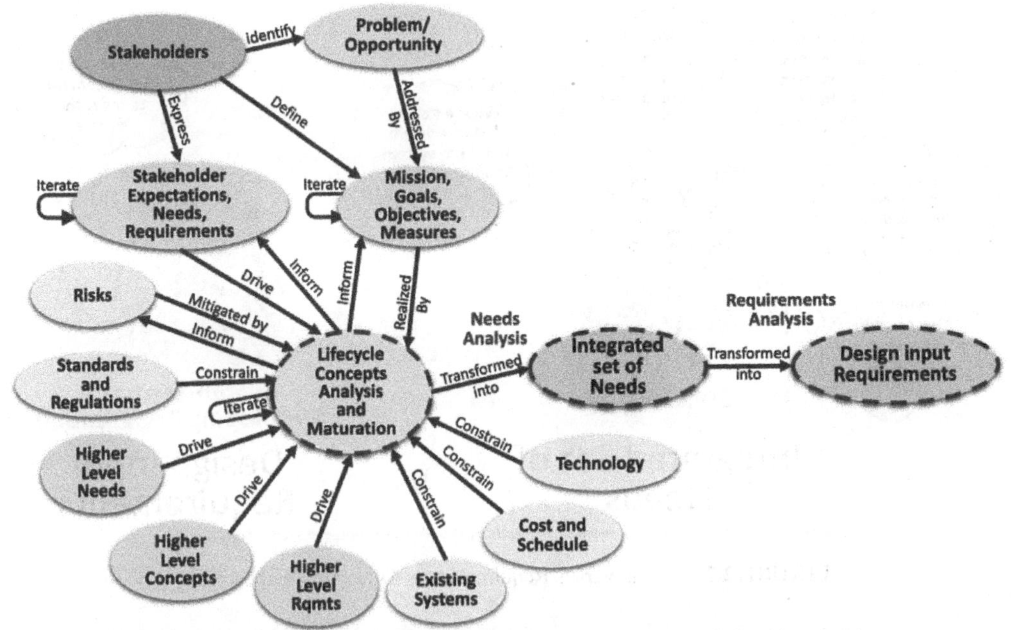

FIGURE 2.1 Lifecycle Concepts, Integrated Set of Needs, and Design Input Requirements.

issues are addressed such that the resulting integrated set of needs is consistent, complete, correct, and feasible. The integrated set of needs represents the scope of the project and is what the SOI design input requirements, design, and the SOI are validated against, as discussed in Chapters 7, 8, 10, and 11.

The need statements for an SOI are written in a structured, natural language and have the characteristics defined in the GtWR for well-formed need statements and sets of needs. Because needs are not requirements, they do not contain the word "shall." *As discussed later in this Manual, a common error is to communicate the stakeholder, user, or customer needs as "shall" statements, adding confusion as to what are needs as opposed to what are requirements.*

2.1.2.2 *Design Input Requirements* The *design input requirements* represent the *technical* design-to set of system requirements for the SOI. This concept is discussed in Chapters 6 and 7.

Design input requirements are established through a transformation from the baselined integrated set of needs as shown in Figure 2.1 and are inputs to the *Architecture Definition* and *Design Definition* processes of ISO/IEC/IEEE 15288. The design input requirements are written in a structured, natural language as textual "shall" statements that have the characteristics defined in the GtWR for well-formed requirement statements and sets of requirements. The set of design input requirements is the focus of both design verification discussed in Chapter 8 and system verification discussed in Chapters 10 and 11.

The term "structured, natural language" refers to the textual form of need and requirement statements such that the sentence is not treated as an atomic entity [11] but has a grammatical structure appropriate for communicating needs and requirements; for example, "When <condition clause>, the <subject clause> shall <action verb clause> <object clause> <optional qualifying clause>." This allows specific templates and a set of rules to be defined, such as those in the INCOSE GtWR

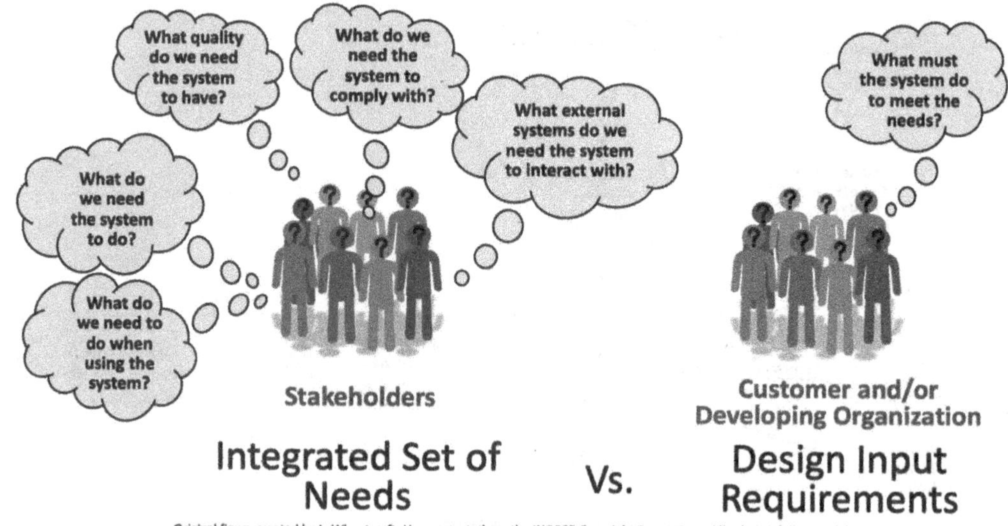

FIGURE 2.2 Needs Versus Requirements—Different Perspectives.

that will result in the needs and requirements having the characteristics of well-formed need statements and requirement statements.

2.1.2.3 Different Perspectives Communicated by Needs and Requirements As illustrated in Figure 2.2, the approach taken in the Guide to Needs and Requirements (GtNR), GtWR, and this Manual, which is consistent with ISO/IEC/IEEE 29148, is that the integrated set of needs communicates the stakeholder's perspective concerning their expectations of what they need the SOI to do, while the design input requirements communicate the customer/developer perspective concerning what the SOI must do to meet the set of needs.

2.1.3 Stakeholder Requirements Versus System Requirements in the Context of System Verification and System Validation

Use of the term "stakeholder requirements" to reflect needs defined at the operational level may lead to confusion with the SOI requirements, which are defined at the system, subsystem, and system element levels (refer to Section 2.3 for a detailed discussion on organizational levels). It can be confusing if the word "requirement" is used in both the context of *system verification* (meeting the system requirements) and *system validation* (meeting the operational-level stakeholder requirements), which is advocated in ISO/IEC/IEEE 15288 and the INCOSE SE HB. To avoid confusion, rather than defining system validation with respect to meeting stakeholder requirements, the following definitions are used in this Manual:

- *System verification* addresses whether the SOI meets its design input requirements.
- *System validation* addresses whether the SOI can do what is intended when operated in the operational environment when used by the intended users, does not enable unintended users to negatively impact the intended use of the system, or use the system in an unintended way, *as defined by the integrated set of needs.*

In this context, the operational-level stakeholder needs, along with any customer-supplied system requirements, are integrated with other sets of stakeholder needs that were obtained during elicitation activities. These are identified as drivers for the SOI and are addressed during the activities associated with SOI lifecycle concepts analysis and maturation and the definition of the integrated set of needs.

Note that the integrated set of needs will include references (via traceability) to the operational-level stakeholder needs and requirements as appropriate. Similarly, the resultant set of design input requirements will contain child requirements that trace to the applicable higher-level allocated stakeholder requirements and customer-supplied system requirements, as well as trace to the set of needs or other sources from which they had been transformed.

2.1.4 Requirements Versus Specifications and Documents Versus Specifications

Other sources of confusion are the use of "document" versus "specification" and the use of "requirements" versus "specifications." These terms are often used interchangeably: requirement document or requirement specification, for example, System Requirements Document (SRD), Software Requirements Specification (SRS), or Software Requirements Document (SRD). In this context, the words "document" and "specification" when used in their singular form, represent containers of design input requirements. However, the terms document, specification, and requirements represent different concepts, which are described further below.

2.1.4.1 Design Inputs Versus Outputs A distinction is made between the terms for requirements and specifications, where requirements refer to inputs into the design process and specifications refer to design outputs as part of the design definition process. It is also common practice to refer to documentation concerning as-built system characteristics as specifications.

As shown in Figure 2.3, the integrated set of needs and resulting set of design input requirements are considered inputs into the *Architecture Definition* and *Design Definition* processes of

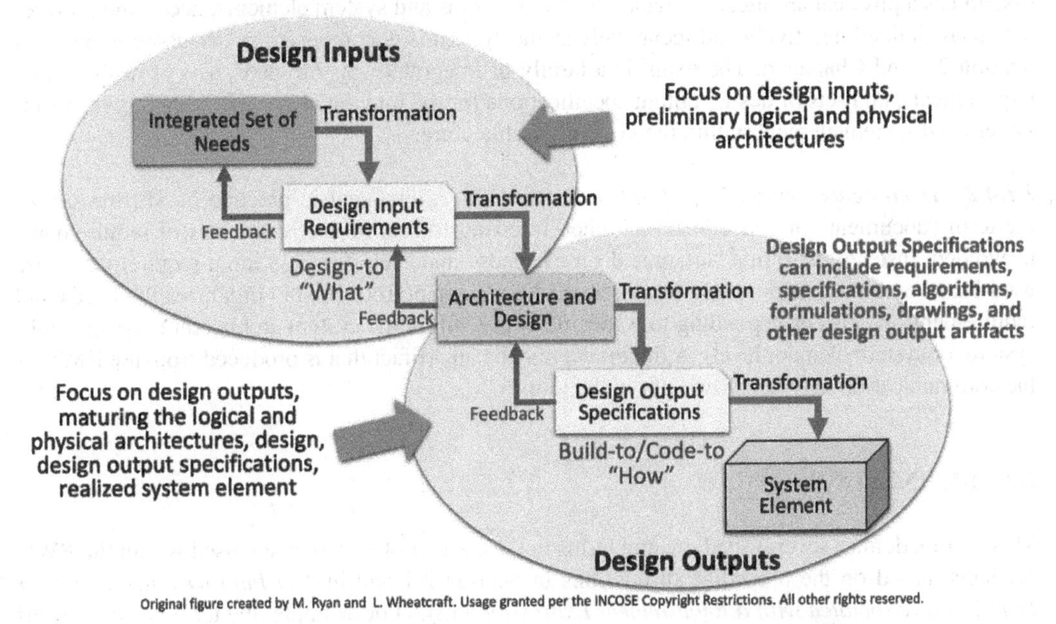

FIGURE 2.3 NRVV Activity Relationships to the INCOSE SE HB.

ISO/IEC/IEEE 15288, which transform the design input requirements into sets of design output specifications (outputs of the architecture and design definition processes activities) to which the system element is realized.

The use of design input requirements is appropriate in that they are *inputs* to the architecture and design definition processes as opposed to *outputs* of these processes.

The INCOSE SE HB [1] refers to outputs of the design definition process in terms of design definition records, artifacts, reports, strategy/approach as well as system design characteristics and design descriptions. NASA's SE HB refers to design outputs as "design descriptions." Design outputs are also often referred to as end-item specifications or the Technical Data Package (TDP), which include parts lists [Bill of Materials (BOM)], drawings, wiring diagrams, plumbing diagrams, labeling diagrams, requirements, logic diagrams, algorithms, computer-aided design (CAD) files, or Standard Tessellation Language (STL) files (for 3D printing), and represent criteria to which the SOI will be manufactured or coded. As such, they can be thought of as the build-to/code-to requirements no matter their form. The form used must have the characteristic, verifiable, in that the built or coded system will be verified against the design output specifications as discussed in Chapter 9.

To avoid confusion and ambiguity, this Manual refers to these *design output* artifacts as *design output specifications* that communicate the *design* to those who will build or code the SOI.

The realized SOI will be verified to have met its *design output* specifications, which is commonly stated as verifying that the SOI was "built to spec." This verification is often done by the organization's quality function as part of *production verification* discussed in Chapter 9. *A major concept that is advocated in this Manual is that design output specifications (how) are not to be included within the sets of design input requirements (what) – or as some phrase it, the "design input requirements should be design agnostic."*

While Figure 2.3 could be viewed as waterfall with the transformations occurring serially, that is not the intent. SE is a knowledge-based discipline; as more knowledge is gained, there is feedback to previous activities that often result in change. In addition, as shown in Figures 1-2 and 2-15 the NRVV activities described in this Manual are intended to be performed concurrently. Because the system has a physical architecture made up of subsystems and system elements, needs and requirements are defined iteratively and recursively as the system is decomposed, as discussed in detail in Section 2.3 and Chapter 6. The result is a family of integrated sets of needs, sets of design input requirements, and sets of design output specifications for the integrated system as well as each subsystem and system element within the system's architecture.

2.1.4.2 Documents Versus Data-Centric Sets With a data-centric practice of SE, the use of the term "document" or "specification" when referring to sets of needs and sets of requirements is avoided. Instead, the terms "integrated set of needs" and "set of design input requirements" are used based on how a data-centric Requirement Management Tool (RMT) organizes the needs and requirements in sets corresponding to a specific entity within the system architecture (system, subsystem, or system element level). A document may be an artifact that is produced from the RMT, or the communication may occur directly in the tool.

2.2 DEFINITIONS

This section defines several fundamental terms in the context of how they are used within the RWG products, based on the preceding discussions in Section 2.1 and in "*An Improved Taxonomy for Definitions Associated with a Requirement Expression* [12]." For example, the term *needs* by itself in the RWG products refers to the needs included in a well-formed integrated set of needs and the

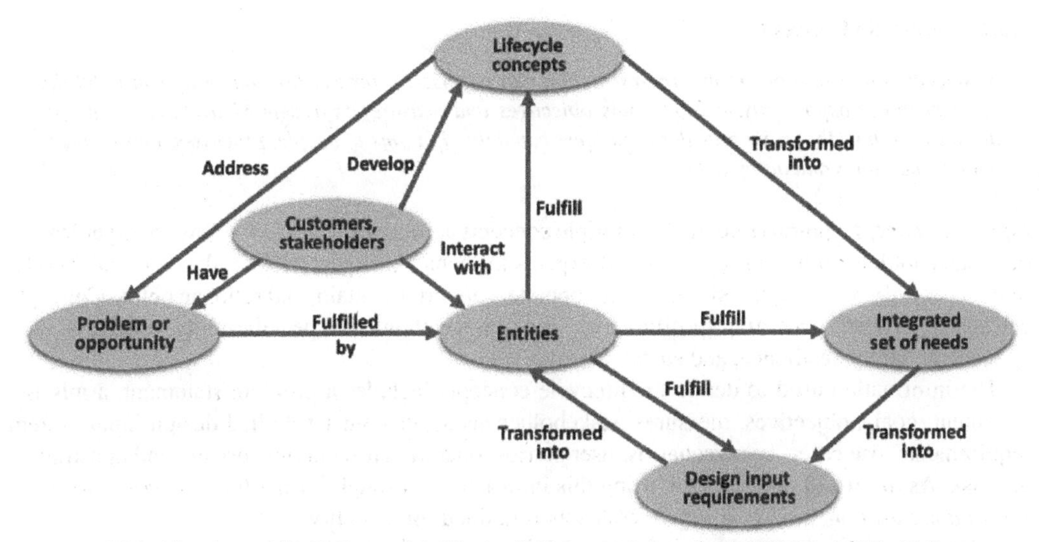

FIGURE 2.4 Entity–Relationship Diagram for Customers, Concepts, Entities, Needs, and Requirements Terms.

term *requirements* refers to the requirements contained within a well-formed set of design input requirements having the characteristics defined in the GtWR.

When describing system development, some form of distinction is commonly made between lifecycle concepts, needs, and requirements (CNR), as shown in Figure 2.4. Lifecycle CNR should be developed for entities at all levels of the organization and system architecture.

Note: In the definitions, the word "customer" is used to refer to the organizations or persons (internal or external) requesting or procuring a work product and/or will be the recipient of the work product when delivered. Customers are key stakeholders that exist at multiple levels of an organization and may be internal or external to the enterprise. As such, there are multiple customers.

2.2.1 Entities

Needs and *requirements* apply to an *entity*, which could exist at any level of the organization or the system architecture in the context of a problem or opportunity fulfilled by the entity. Since terms such as product, SOI, system, subsystem, and system element are level-specific, a general term is needed that can apply at any level of the organization or architecture and to any single item at that level. For this, the term *entity* is used, which has lifecycle CNR which the entity fulfills.

> *An **entity** is a single item to which a concept, need, or requirement applies: an organization, business unit, project, supplier, service, procedure, hardware SOI (system, subsystem, system element), software SOI (application, package, module, feature), product, process, or stakeholder class (such as user, operator, tester, maintainer.)*

There are three general types of entities— physical or software entities, such as the engineered systems to be developed, process entities, such as procedures or work instructions (WIs), and business or human entities, such as business units, users, customers, developers, suppliers, and other stakeholders.

2.2.2 Lifecycle Concepts

A ***concept*** *is a textual or graphic representation that concisely expresses how an entity can fulfill the problem, threat, opportunity, mission, goals, objectives, and measures it was defined to address. Concepts demonstrate how the entity provides a business capability in terms of people, processes, and products within constraints and acceptable risk.*

Lifecycle concepts (plural) refer to the multiple concepts across the lifecycle for how the organization (and stakeholders within an organization) expects to manage, acquire, define, develop, build/code, integrate, verify, validate, transition, install, operate, support, maintain, and retire an entity. Concepts can also be defined concerning specific capabilities the stakeholders expect the entity to address, such as security, safety, resilience, and sustainability.

The information used to define the lifecycle concepts includes a problem statement, a mission statement, goals, objectives, measures, stakeholder needs, customer-supplied design input system requirements, use cases, user scenarios, user stories, operational scenarios, drivers and constraints, and risk. As discussed in Chapter 4, using this information, through formal *lifecycle concepts analysis and maturation*, a set of lifecycle concepts is defined for an entity.

Lifecycle concepts can be communicated in various forms, including textual (OpsCon or ConOps), graphical representations (diagrams and models), and/or electronic (databases). There are multiple lifecycle concepts that apply to each entity, so it is useful to develop a necessary and sufficient set of lifecycle concepts from which an integrated set of needs can be derived.

Lifecycle concepts can be defined from the perspective of the organization, the SOI, or the macrosystem in which the organization and SOI exist.

As discussed in the INCOSE SE HB, lifecycle stages for an SOI include concept, development, production, utilization, support, and retirement. From a holistic perspective, concepts concerning the approaches the project team will implement in these lifecycle stages must defined early in the project both from a project management perspective as well as a systems engineering (SE) perspective.

2.2.3 Needs

Based on the set of lifecycle concepts, through formal *needs analysis*, an integrated set of needs is derived using a formal transformation process involving decomposition, derivation, elaboration, diagrams, and architectural and analytical/behavioral models. The realization of this integrated set of needs will result in the set of lifecycle concepts for the entity to be realized.

Needs are well-formed textual statements of expectations for an entity stated in a structured, natural language from the perspective of what the stakeholders need the entity to do in a specific operational environment communicated at a level of abstraction appropriate to the level at which the entity exists.

A ***need statement*** *is the result of a formal transformation of one or more sources or lifecycle concepts into an agreed-to-expectation for an entity to perform some function or possess some quality within specified constraints with acceptable risk.*

2.2.4 Requirements

As with defining an integrated set of needs, defining requirements is an engineering activity which, through formal *requirements analysis*, determines specifically what the entity must do to meet the needs they are being transformed from using a formal transformation process involving decomposition, derivation, diagrams, and architectural and analytical/behavioral models. A deeper exploration and elaboration of the lifecycle concepts, including a thorough examination of interactions between the entity and other entities, are part of the transformation from needs into requirements. As a result of the *requirements analysis*, there may be more than one requirement defined for each need; conversely, there may be cases where a single requirement addresses more than one need.

Requirements are well-formed textual "shall" statements that communicate in a structured, natural language what an entity must do to realize the intent of the needs from which they were *transformed.*

*A **requirement statement** is the result of a formal transformation of one or more sources, needs, or higher-level requirements into an agreed-to-obligation for an entity to perform some function or possess some quality within specified constraints with acceptable risk.*

The analysis used to transform lifecycle concepts into needs and to transform needs into requirements is frequently referred to as *business analysis* or *mission analysis* at the enterprise and strategic levels of the organization and *needs analysis* and *requirements analysis* at the business operations, system, subsystem, and system element levels. Diagrams and models are important tools used as part of this analysis to help achieve correctness, consistency, completeness, and feasibility of the transformations. For completeness, it is expected that lifecycle CNR are developed for entities at all levels.

The reader is urged to be diligent with the lifecycle CNR approach advocated in this Manual and resist the temptation to jump from the lifecycle concepts directly to defining requirements or, even worse, going directly from lifecycle concepts to candidate design solutions, skipping needs and requirements definition entirely—doing so will result in the realized SOI failing to meet stakeholder expectations (failing to achieve system validation).

2.2.5 Attributes

Need and requirement statements are supported by associated attributes that aid in the definition and management of a need, sets of needs, a requirement, and sets of requirements. These attributes also aid in achieving many of the characteristics defined in the GtWR.

*An **attribute** is additional information associated with an entity which is used to aid in its definition and management.*

Well-chosen attributes, when properly defined and tracked, can make the difference between being able to correctly interpret and manage needs and requirements definition throughout the system lifecycle and to adjust accordingly or finding out late in the program that the needs or requirements were defective in the first place. When errors in the needs and requirements—or errors in their interpretation—are discovered late in the program, they can be expensive and time consuming to fix.

Figure 2.5 illustrates the relationship of attributes with the other activities in support of the definition and management of needs and requirements (and their sets).

Chapter 15 provides a more detailed discussion on attributes as applied to needs and requirements.

2.2.6 Needs and Requirements Expressions

Although each need statement and requirement statement are individually important, needs expressions and requirements expressions are more than just well-formed textual statements that are written succinctly in a standard format having the characteristics defined in the GtWR. The full expression comprises a statement and its associated supporting attributes.

*A **need expression** includes a need statement and a set of associated attributes.*
*A **requirement expression** includes a requirement statement and a set of associated attributes.*

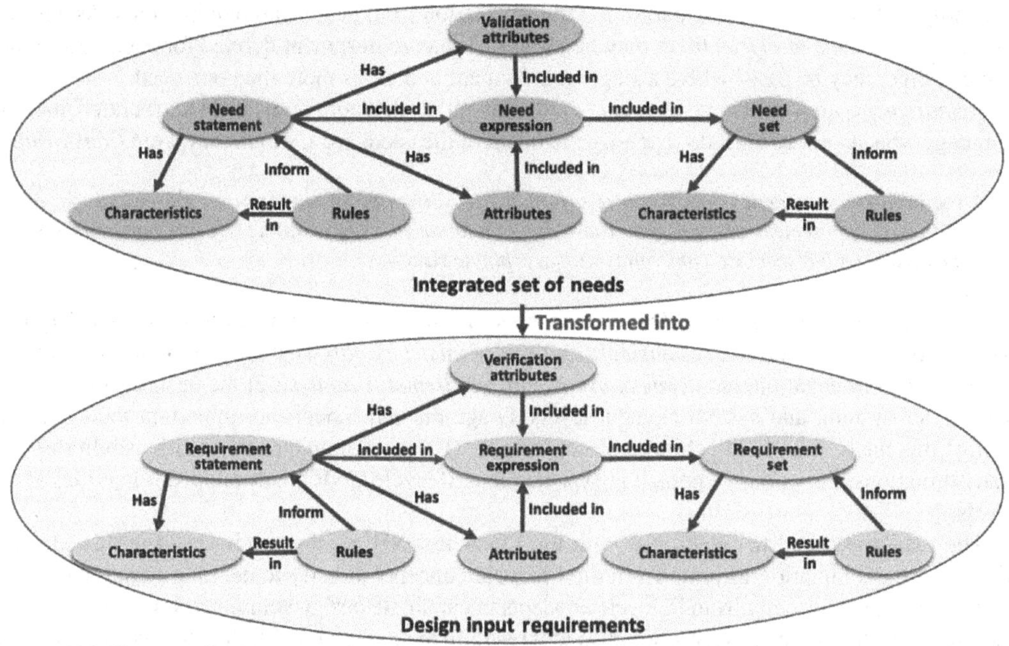

FIGURE 2.5 Entity–Relationship Diagram for Needs and Requirements Terms. Source: Adapted from Ryan et al. [12].

2.2.7 Sets of Needs and Sets of Requirements

Although each individual need and requirement expression is important, it is ultimately the set of needs and resulting set of design input requirements that will describe what the entity must do and be, and it is the set of needs and/or set of requirements that most often will be agreed-to as a contractual obligation. *See the GtWR for a description of the characteristics of well-formed sets of needs and sets of requirements and rules that help result in those characteristics.*

> *A **need set** is a well-formed set of agreed-to-need expressions for the entity and its external interfaces.*
> *A **requirement set** is a well-formed set of agreed-to requirement expressions for an entity and its external interfaces.*

2.3 BASIC CONCEPTS

2.3.1 Needs, Requirements, and the Entity to Which They Apply

When defining needs and requirements, it is important to understand the significance of the use of the word "entity" in the above-mentioned definitions.

In terms of need and requirement statements, the entity is the *object* of a need statement ("The <stakeholders> need the <entity> to … …") which, in turn, will be the *subject* of a subsequent requirement statement ("The <entity> shall … … ") that has been transformed from that need.

For example, the need for an entity called a bicycle may incorporate the constraint, "The customers need the bicycle to transport the rider using non-motorized power." This would be transformed into one or more requirements on the bicycle system that would result in the need being met; one such requirement could be, "The bicycle shall transport a rider using the propulsion force supplied by the rider."

Systems engineers must make clear the applicability of needs and requirements. Types of requirements are often defined in terms of the entity to which they apply, usually following the people, process, and product model. For example:

- There will be *project needs and requirements* on a project or organizational elements within the enterprise dealing with activities and deliverables the project is responsible for that will be recorded in a Project Authorization Document (PAD), Project Management Plan (PMP), and other plans and procedures that may take the form:

  The <stakeholders> need the <project> to … ….

  The <stakeholders> need the <xxxx team> to … ….

  The <project> shall … ….

  The <xxxxx team> shall … ….

- There will be *supplier needs and requirements* on a supplier, vendor, or contractor addressing activities and deliverables that will be recorded in SOW, Supplier Agreements (SA), or Purchase Order (PO) that may take the form:

  The <stakeholders> need the <supplier/contractor> to … ….

  The <supplier/contractor> shall … ….

- There will be procedural needs and requirements, as discussed in Chapter 10, addressing actions the person or organization responsible for conducting steps within a procedure resulting in those actions; for example, a system verification or system validation procedure, such as the following:

  The <stakeholders> need the <operator, technician, engineer> to verify the SOI meets <this requirement> per the defined success criteria using the defined system verification approach using the defined system verification method.

  The <operator, technician, engineer> shall <stimulate the SOI in some manner>.

  The <operator, technician, engineer> shall <record the results of the stimulation>.

- There will be system needs and requirements on the SOI being developed as defined earlier that are recorded within the SOI design input requirements or design output specifications as shown in Figure 1.2, which provide expectations concerning the production of the SOI:

  The <stakeholder> needs the <SOI> to provide the capability to <do something in a given operational environment with a required performance of xxxxx>. (Need)

  The <SOI> shall <perform some function with the required performance under some operating condition>. (Design input requirement)

  The <SOI component> shall be manufactured to <the physical dimensions shown in drawing xyz>. (Design output requirement)

Needs and requirements for the different entities must not be combined within a single set of needs or set of requirements. Each set of needs and requirements must only include statements that relate to the single entity to which the set applies. This is important because, for each entity, there is an expectation that objective evidence can be obtained, which can be used to assess, with some level of confidence that the entity has met the requirements (system verification) or needs (system validation).

Additionally, combining requirements that apply to different entities within the same set can create ambiguity as to the requirement's intent (especially if written in passive voice) as well as creating the risk of missing requirements or struggling to determine which entity needs or requirements are applicable, which can make system verification and validation planning much harder and can lead to confusion among both the customer and supplier.

Unfortunately, it is common to see nontechnical SOW/SA/PO supplier (people/process) requirements mixed in with the technical design input requirements for the SOI the supplier is developing in the same document, many of which are written in a passive voice such that it is not clear whether the requirement is a nontechnical requirement that applies to the supplier or a technical design input requirement that applies to the SOI.

The approach as to how a project verifies that an SOI meets its design input requirements is different from the approach used by a customer to verify that a supplier meets its SOW/SA/PO requirements. A more formal approach is applied concerning how the project or supplier will verify an SOI meets both its design input requirements and design output specifications.

Separating requirements based on the entity to which they apply allows the verification approaches and resulting activities and artifacts to be recorded, managed, and implemented separately.

In some cases, such as an SA or PO that is a combination of a contract, SOW, and technical requirements all in one document; it is important to separate the requirements based on the entity to which they apply (different sections or appendices) as well as to ensure that the requirements are well-formed in accordance with the GtWR.

It is also common to see design input requirements (what) and design output specifications (how) included in the same set of requirements without being clearly marked. For successful projects, this mixing must be avoided.

2.3.2 Levels of Lifecycle Concepts, Needs, and Requirements

Caution should be used to understand the context and intent in which the word "level" is used: levels of organization, levels of architecture, levels of detail, and levels of abstraction.

It is common to hear statements labeling requirements as high-level requirements or low-level requirements. What is really meant by "high-level" and "low-level"?

- In some cases, use of high- and low-level refers to levels of abstraction or levels of detail—strategic and operational-level lifecycle CNR are communicated at a higher level of abstraction than system-level lifecycle CNR (likewise, system-level lifecycle CNR are communicated at a higher level of abstraction than system element-level lifecycle CNR).
- For others, higher-level requirements may refer to "what" or "design-to" design input requirements, while lower-level refers to "how" or "build-to" design output specifications.

2.3.2.1 Levels in the Context of an Organization As illustrated in Figure 2.6 lifecycle, CNR exist at several levels of an organization and from different points of view [13].

At each level, there are real-world stakeholder expectations (as opposed to "as communicated" or "as understood" expectations), which are addressed by lifecycle CNR. In the end, it is the real-world expectations that must be realized throughout the operational life of the system. The lifecycle CNR

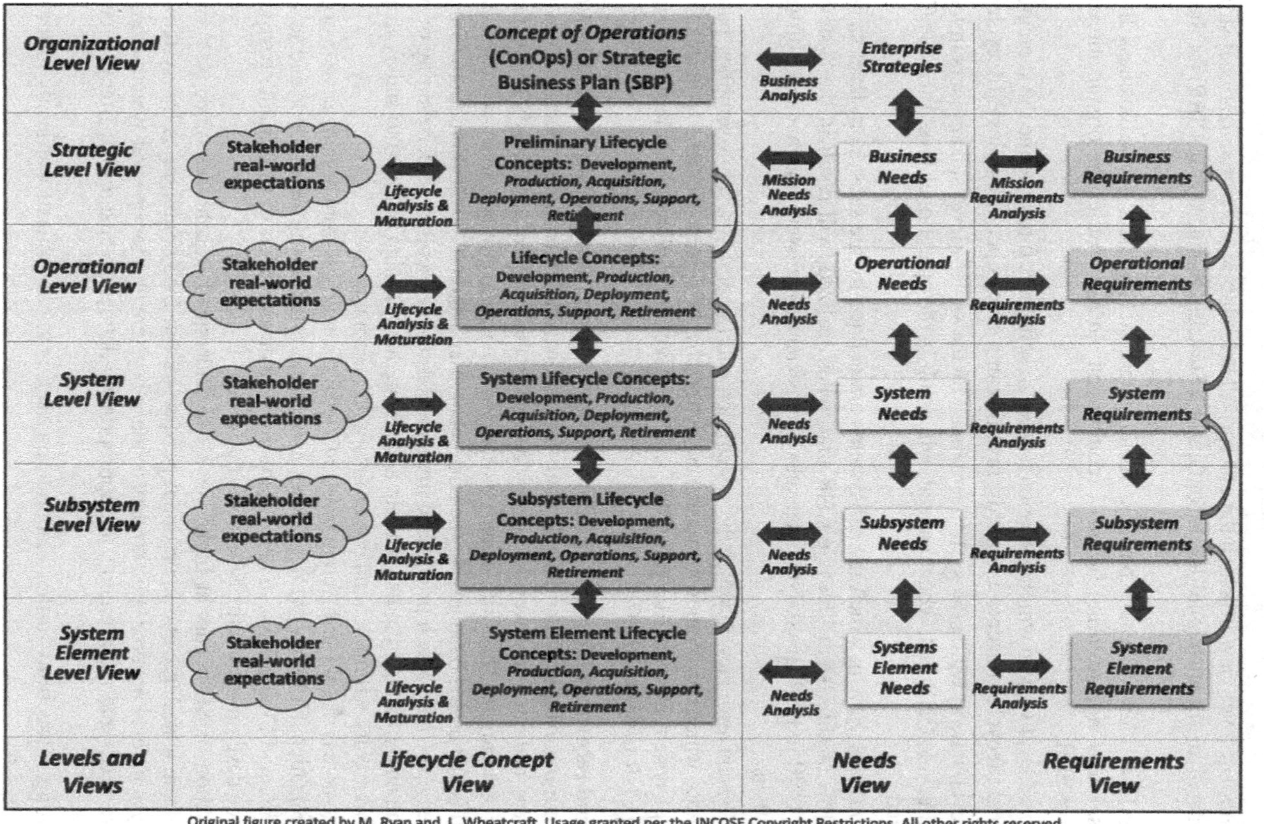

FIGURE 2.6 Levels in the Context of an Organization. Source: Adapted from Ryan [13].

at one level flow down and are drivers for the next level of lifecycle CNR. At this next level, the lifecycle CNR must be consistent with and compliant with the higher-level lifecycle CNR.

The stakeholders have a viewpoint appropriate to the level they exist. Each level elaborates on the lifecycle CNR defined at the previous level, as shown in Figure 2.6. There are a number of views:

- *Organizational* level view in which the enterprise leadership sets the organizational strategies in the form of an organizational-level ConOps or a Strategic Business Plan (SBP) for the enterprise.
- *Strategic* level view in which business management stakeholders derive business lifecycle concepts, needs, and resulting business requirements in response to the organizational-level ConOps or SBP.
- *Operational* level view in which operational-level stakeholders further refine the strategic-level operational lifecycle concepts and needs, resulting in a set of operational-level lifecycle CNR.
- *System* level view in which the system is defined by the system-level stakeholders in terms of system lifecycle CNR that, when realized, will result in the operational-level lifecycle CNR to be realized.
- *Subsystem* level view in which the subsystems are defined by the subsystem-level stakeholders in terms of subsystem lifecycle CNR that, when realized, will result in the system-level lifecycle CNR to be realized. Note that subsystems can be hardware, mechanical, software, or combinations.
- *System Element* level view in which the system elements (assemblies, sub-assemblies, parts, and components, which can include firmware) are defined by the system-element-level stakeholders in terms of system-element lifecycle CNR, that when realized, will result in the subsystem-level lifecycle CNR to be realized.

Defining and recording lifecycle CNR for an entity is more than just an exercise in writing; it is also an engineering activity that, through formal analysis, determines specifically what the customers, users, and other real-world stakeholders expect the entity to do to satisfy a specific problem, threat, or opportunity. This formal analysis starts at the organizational and strategic levels of the enterprise and is elaborated by engaging the customers, users, and other stakeholders to formalize a number of lifecycle CNR, which provide an implementation-independent understanding of what is expected of the entity (design inputs) without addressing how (design outputs) to satisfy a specific problem or opportunity within defined constraints with acceptable risk.

One thing to note is that these different levels of views are not rigid from organization to organization. The organizational level and strategic level in one enterprise may differ in scope and scale from another. The intent of Figure 2.6 is to help establish the scale and levels of stakeholder perspectives, which influence development at the SOI level (the system-level view).

The application of SE occurs at higher levels to understand stakeholder objectives as stakeholder needs from a strategic and operational perspective but then continues as the systems engineers at the system level become stakeholders to the subsystem and lower-level system elements in helping ensure needs and requirements are appropriately defined to ensure SOI lifecycle concepts are met.

The following subsections further elaborate on each of the organizational levels in the model shown in Figure 2.6.

2.3.2.1.1 Organizational Level At the *Organizational* level, the enterprise has several strategies that will guide its future. A system has its genesis in the enterprise-level ConOps or SBP, which communicates the leadership's intentions regarding the management and operation of the organization—in terms of existing systems, systems to be developed or procured, services, domains, and product line(s). At this level, the ConOps (or SBP) defines the organization in terms of "brand" (distinctive identity) and establishes a vision or mission statement and corresponding goals, objectives (MGOs), and measures that clearly state the reason for the enterprise, its strategy for moving forward and expected outcomes. The focus of the enterprise-level ConOps/SBP is how the organization will conduct business within the domain, market, industry, and regulatory environment in which it exists.

2.3.2.1.2 Strategic Level At the *Strategic* level, business management stakeholders use the guidance in the organizational-level ConOps/SBP to define a set of strategic-level lifecycle concepts, which capture the business management stakeholder expectations for principal capabilities and product line(s), branding, quality, safety, security, sustainability, supply chain, human resources, data governance, regulations and standards compliance, and high-level processes. (for example, product definition and development, project management, SE, information technology (IT), marketing, sales, procurement, manufacturing, operations, support, maintenance, and retirement.) These lifecycle concepts are captured as preliminary concepts whose focus is on defining the lifecycle concepts at the strategic level in terms of the enterprise's internal capabilities, performance, activities, standards, and processes, which will be used to realize the vision or mission statement, goals, and objectives for the enterprise.

The preliminary lifecycle concepts are elaborated, formalized, and transformed into an integrated set of strategic-level business needs, which are transformed into a set of strategic-level business requirements using *Mission or Business Analysis* activities defined in ISO/IEC/IEEE 15288. The lifecycle CNR at this level are about strategic-level capabilities provided by the people, processes, and products within the enterprise.

Strategic-level lifecycle CNR often focus on governance of the organization's business management and operations. At this level, the lifecycle CNR can apply to various entities, including the organization, programs, projects, processes, and people responsible for supplying servicers or developing systems (products) as well as the services provided and systems the enterprise procures, delivers, or develops. Once the strategic-level stakeholders are satisfied that their lifecycle CNR are complete at the appropriate level of abstraction, they are passed down to the operational-level for implementation.

At the strategic level, the focus is on classes of potential solutions satisfying the organizational-level ConOps/SBP rather than on a specific service, product, or system within a class.

2.3.2.1.3 Operational Level At the *Operational* level, lifecycle CNR are defined at a lower, tactical level of abstraction using the organization-level ConOps/SBP and the strategic-level preliminary lifecycle CNR as guidance. The operational-level stakeholders elicit operational-level stakeholder needs and develop lifecycle concepts appropriate to the operational level, which are recorded in the operational-level set of lifecycle concepts.

Needs analysis activities are used to extract operational-level needs from the operational-level lifecycle concepts and transform them into a formal integrated set of operational-level needs, which are transformed into a formal integrated set of operational-level requirements for the organizational entities responsible for developing systems or services.

The operational-level lifecycle CNR are recorded in the organization's toolset for storing the organization's data and information. Hereafter, this is called the organization's *integrated dataset*.

Business units, programs, and projects exist at the operational level. At this level, a key area of focus is on the infrastructure: specific programs/projects, people, processes, and software applications used within the organization that will be involved in operations and in supplying services or developing systems (products) for both internal and external customers.

The operational-level needs concerning the people, processes, and tools are implemented via the organization's plans, SOPs, WIs, or Process Definition Documents (PDDs). A key point is that the focus of these lifecycle CNR is on the organization, programs/projects, and people involved in business operations activities.

A second focus at the operational level is on the lifecycle CNR concerning the services provided by the enterprise and systems (products) that the enterprise develops or procures, to satisfy the needs of both internal as well as external customers.

Lifecycle concepts for the services supplied and the systems and product lines supplied by the organization units are developed at a more detailed level of abstraction. Operational-level needs are extracted from the operational-level lifecycle concepts and transformed into operational-level requirements at a level of abstraction appropriate to the operational level. The resulting lifecycle CNR are recorded in the organization's integrated dataset. The transformation is guided by formal *lifecycle concepts analysis and maturation*, *needs analysis*, and *requirements analysis* activities.

A key point is that this second set of operational-level lifecycle CNR is on the services supplied (or products being developed or procured) rather than on the organizational units supplying the service or developing or procuring the product. These sets of operational-level lifecycle CNR flow down and constrain the system-level lifecycle CNR (which must be consistent and compliant with the operational-level lifecycle CNR). The projects are responsible for verification that the operational-level requirements for a product or service have been met and validation that the operational-level needs have been met.

2.3.2.1.4 System Level At the *system* level, the focus is on a specific service being supplied or an SOI to be developed/procured for a customer (internal or external).

At this level, a system concept is defined to meet operational-level needs and requirements. A major advantage of postponing consideration of a specific solution until the system level is to ensure that the organizational-, strategic-, and operational-level lifecycle CNR can be considered when assessing candidate solutions. With this big picture in mind, the higher-level lifecycle CNR are considered when defining the system-level lifecycle CNR.

Caution: Proposing a solution without understanding the higher-level organizational lifecycle CNR is a significant cause of project failure. This approach adds risk that a candidate solution will gain momentum within the organization; once higher-level needs and requirements are discovered, there is often a temptation to fit those needs and requirements to the proposed solution (instead of the other way around). This approach often leads to either a failed project or higher-level needs and requirements that are not met and a failure of the realized system to be successfully validated against its integrated set of needs.

Using the approach discussed in this Manual, the operational-level lifecycle CNR are drivers for the specific SOI being procured/developed or service to be supplied. The project team uses a structured engineering process to work with the various system-level stakeholders associated with the product or service, both internal and external, to elicit system-level stakeholder needs and identify risks specific to the SOI or service being developed.

With this information, the project team defines and matures a feasible set of system-level lifecycle concepts that will result in the operational-level needs and requirements to be met with risk acceptable for this lifecycle stage.

Based on this set of lifecycle concepts, the project team defines an integrated set of needs for the system, which is transformed into the set of system-level design input requirements. The system-level lifecycle concepts, integrated set of needs, and set of design input requirements are recorded in the SOI's integrated data set. The system will be verified against the system-level design input requirements and validated against the system-level integrated set of needs.

As part of the lifecycle concepts analysis and maturation activities, a logical architecture for the system is defined consisting of subsystems and systems elements. If the organization decides the system needs further elaboration, this set of system-level design input requirements will be allocated to the subsystems at the next level of the architecture.

2.3.2.1.5 Subsystem Level At the *subsystem* level, the focus is on specific subsystems that are part of the system logical architecture.

Included within the system-level needs and requirements allocated to each subsystem are MGOs, measures, and constraints that clearly communicate the real-world expectations of the system-level stakeholders for each subsystem. Together, the system-level lifecycle CNR allocated to the subsystems are drivers for the specific subsystems being procured or developed.

The project team responsible for each subsystem works with the various subsystem-level stakeholders, both internal and external, to elicit subsystem-level stakeholder needs and to identify risk and constraints specific to each subsystem. With this information, the project team defines and matures a feasible set of subsystem-level lifecycle concepts for each subsystem that will result in the system-level allocated needs and requirements to be met with risk acceptable for this lifecycle stage.

Based on this set of subsystem-level lifecycle concepts, an integrated set of subsystem-level needs are derived for each subsystem, which are transformed into subsystem design input requirements. The subsystem-level lifecycle concepts, integrated set of needs, and the set of subsystem-level design input requirements are recorded in the SOI's integrated dataset. Each of the subsystems will be verified against their design input requirements and validated against their integrated set of needs.

If the organization decides a subsystem needs further elaboration, this set of subsystem-level design input requirements will be allocated to the system elements at the next level of the architecture.

2.3.2.1.6 System Element Level At the *system element* level, the focus is on specific elements (assemblies, sub-assemblies, parts, and components) that are part of their parent subsystem.

Included within the subsystem-level needs and requirements allocated to each system element are MGOs, measures, and constraints that clearly communicate the real-world expectations of the system-level stakeholder for each subsystem. Together, the subsystem-level lifecycle CNR allocated to the system elements are drivers for the specific system elements being procured or developed.

The project team responsible for each system element works with the various system-element-level stakeholders, both internal and external, to elicit system-element-level stakeholder needs, identify risk, and identify drivers and constraints specific to each system element. With this information, the project team defines and matures a feasible set of system-element-level lifecycle concepts for each system element that will result in the allocated subsystem-level needs and requirements to be met with risk acceptable for this lifecycle stage.

Based on this set of system-element-level lifecycle concepts, the project defines an integrated set of needs for each system element, which are transformed into the system-element design input requirements. The system-element-level lifecycle concepts, integrated set of needs, and the system-element-level set of design input requirements are recorded in the SOI's integrated dataset. Each of the system elements will be verified against their design input requirements and validated against their integrated set of needs.

2.3.2.2 Use of Systems Thinking—The SOI in Context of the Macrosystem It is critical that systems engineers adequately consider (or have access to) higher-level organizational and architectural lifecycle CNR both internal to their organization as well an external customer's organization. Failure to recognize this top-down process is a common cause of project failure.

When a project team is responsible for developing a SOI, it needs to use systems thinking to address the macro macrosystem of which the SOI is a part. This macro system includes external systems as well as processes or organizations which interact with the SOI.

If the SOI is a subsystem or system element, both the customer and developing organizations must never lose sight of the higher-level system of which it is a part, of the interactions between their subsystem and other subsystems, or of the interactions of their system element with other system elements. They also must acknowledge that their SOI must respond to the higher-level allocated lifecycle CNR to ensure the integrated system or macro system is optimized, even if it means the SOI is not optimized.

2.3.2.2.1 Example of Organizational Levels of Lifecycle Concepts, Needs, and Requirements Consideration of the organizational, strategic, and operational levels is necessary because they represent significantly different points of view that must be defined and implemented in a top-down manner. The following is an example of why organizations must address the levels of lifecycle CNR above the system level to be successful.

Consider a wood-felling company whose management has just returned from a 1930s trade show where they witnessed a demonstration of the revolutionary wood-felling technological device called a "chainsaw" that was introduced to the market by Andreas Stihl's new company.

Although the ability to cut down trees at a greater rate is extremely attractive, the company cannot simply nominate one of the stakeholders to define a set of requirements for the procurement of chainsaws. From a systems thinking perspective, how will this new, transforming technology be phased into operations? Further, management cannot even ask stakeholders at the operational level to describe what they want from the introduction of the new chainsaw capability in the form of needs and requirements. The current business operations managers are experienced in managing axe men who are not able (and probably not willing since they are going to be retrained at best and, at worst, let go) to describe in any manner how the new device is to be operated since they have no familiarity with the operating procedures, training, safety, or the maintenance necessary for the new device. Similarly, the logistics and procurement staff at the operational level will know, in considerable detail, current logistic information, such as the number of axe handles broken per linear meter of hardwood cut in support of current operations, but will know nothing of the support for a device that will need a different maintenance methodology, significantly different support materials such as fuel and lubrication, different storage, and many more parts of much greater variety (such as starter ropes, chains, sparkplugs, and pistons).

Therefore, before the operational-level stakeholders can begin to define their needs and requirements, the stakeholders at the strategic level above them must define lifecycle concepts for how the new capability is to be introduced into the organization. Is the reason to procure chainsaws to fell trees faster, or to cut down the same amount of wood more efficiently (with fewer operators, for example)? In either case, will the current tradesmen be retained and retrained, or will new operators be required? With axes, the tradesmen own their own axes and are responsible for maintaining the axes (keeping them sharp and replacing broken handles supplied by the company.) With the company procuring the more expensive chainsaws, what new logistics and procurement support will be required, and how will it be acquired? How will the company transport and store materials—such as fuel, lubrication, and spare parts—that are not currently supported? How will the new devices be maintained (engine and chains), and by whom? How will the organizational structure need to be changed? How will these new devices be deployed—across all operators at once, or team by team, or region by region? If wood is to be produced faster, will additional transport vehicles be required to extract the product, and will additional sawmill capacity be required? Ultimately, of course, business management must also consider whether they want to produce more wood at the risk of flooding their own market, resulting in a falling price and reduced profits, or will the new chainsaws result in lower production costs, allowing the enterprise to offer products to the market cheaper, thus making them more profitable?

If the company simply buys 200+ of these new chainsaws before thinking through at least some of these questions, they could cause a lot of churn in the supply chain at best or seriously injure or kill someone at worst. Before an acquisition can be considered, therefore, the stakeholders at the strategic level must draft the initial versions of the acquisition concept, deployment concept, operating concept, maintenance concept, and retirement concept. Those concepts can then be elaborated by the selected stakeholders at the operational level. Once at least one feasible set of lifecycle concepts has been defined, system-level lifecycle concepts and an integrated set of needs for the chainsaws, supporting infrastructure, and materials can be recorded and transformed into a set of system-level design input requirements. These can then be passed on to systems designers and procurement to develop the design output artifacts and SOW, SA, or PO. At the very least, this will help the company buy the "right" chainsaws and supporting parts, equipment, and enabling systems. In addition, procurement plans and maintenance plans, including lubrication and sharpening schedules, operating instructions, safety plans, and training, can be developed.

Looking at the system development top-down then, the need for the three distinct levels views above the system level is evident, and it is easier to understand why the lack of attention to the strategic and operational-level lifecycle CNR is a principal reason for project failure—especially for systems with custom-built hardware and/or software. Buying chainsaws is one endeavor; building a sophisticated military system is quite another.

This same systems-thinking mindset is needed when an organization wants to adopt any new technology, methodology, or process. To successfully adopt concepts such as Agile, Lean, Six-Sigma, and Model-based System Engineering (MBSE), each of these levels needs to first be considered.

For example, it could be an expensive mistake to invest heavily in a new MBSE tool without considering how it will be used, by whom, the supporting infrastructure, the value, return on investment (ROI), and how such a tool will be used and maintained.

2.3.2.3 Portfolio Management Within the Operational Level In the above-mentioned discussion, the point was made that there are distinct lifecycle CNR at the strategic and operational levels. The strategic level addresses the organizational units, and the operational level addresses the products

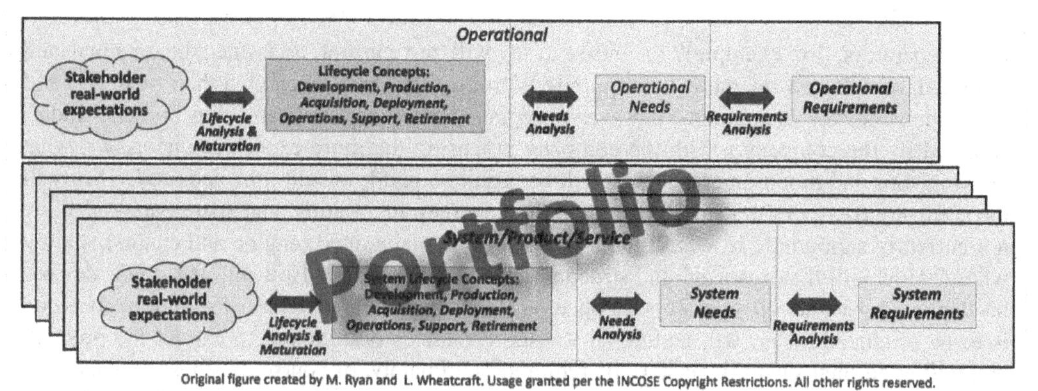

FIGURE 2.7 Portfolio of Projects Within the Operational Level.

and services provided by the organizational units. The stakeholders at each level will identify problems, opportunities, threats, and potential solutions in the form of products and services to address the problems, opportunities, and threats.

This could yield a solution set of multiple products and services, resulting in the need to develop individual operational-level lifecycle CNR for each product and service. Together, the organization manages the set of projects, which realize products or services as a portfolio, as shown in Figure 2.7.

For each product or service in the portfolio, the stakeholders at the strategic or operational level of the organization define lifecycle CNR for both the project teams as well as the products to be produced or the service being supplied by each project team.

Each product or service can be in various stages of planning and maturity. Some may be at the idea stage, while some will have progressed to the concept stage, where key technologies are identified, the maturity levels of key technologies are assessed, and operational-level lifecycle CNR are defined. Value to the organization is assessed in terms of how the potential products fit within the strategic-level business lifecycle CNR and ROI. These activities would form the basis of managing the portfolio of products and services.

2.3.2.4 SOI Customer/Supplier Relationship Considerations Projects are formed at the operational level to produce a product or provide a service. In some cases, the product/service will be developed within the organization. In other cases, they will be acquired from an outside supplier. In this case, the organization level is the customer of the product/service.

For the cases where there is a customer/supplier relationship, the customer will have a project team within their organization's operational level responsible to their strategic- and operational-level stakeholders to develop a product/service consistent with the strategic- and operational-level lifecycle CNR. Before issuing a Request for Proposal (RFP), the project team will determine which level of needs or requirements they will include in the contract: the operational level or the system level, as indicated in Figure 2.8.

When the acquisition occurs at the operational level, the customer's project team will include their organization's operational-level needs in the RFP (in some domains, the integrated set of needs may be in a Statement of Objectives (SOO)). For the winning bidder, acceptance will be when the supplier supplies objective evidence that the realized system or provided service meets the customer-owned integrated set of needs or SOO (system validation) as well as the requirements in the SOW or SA (contract verification).

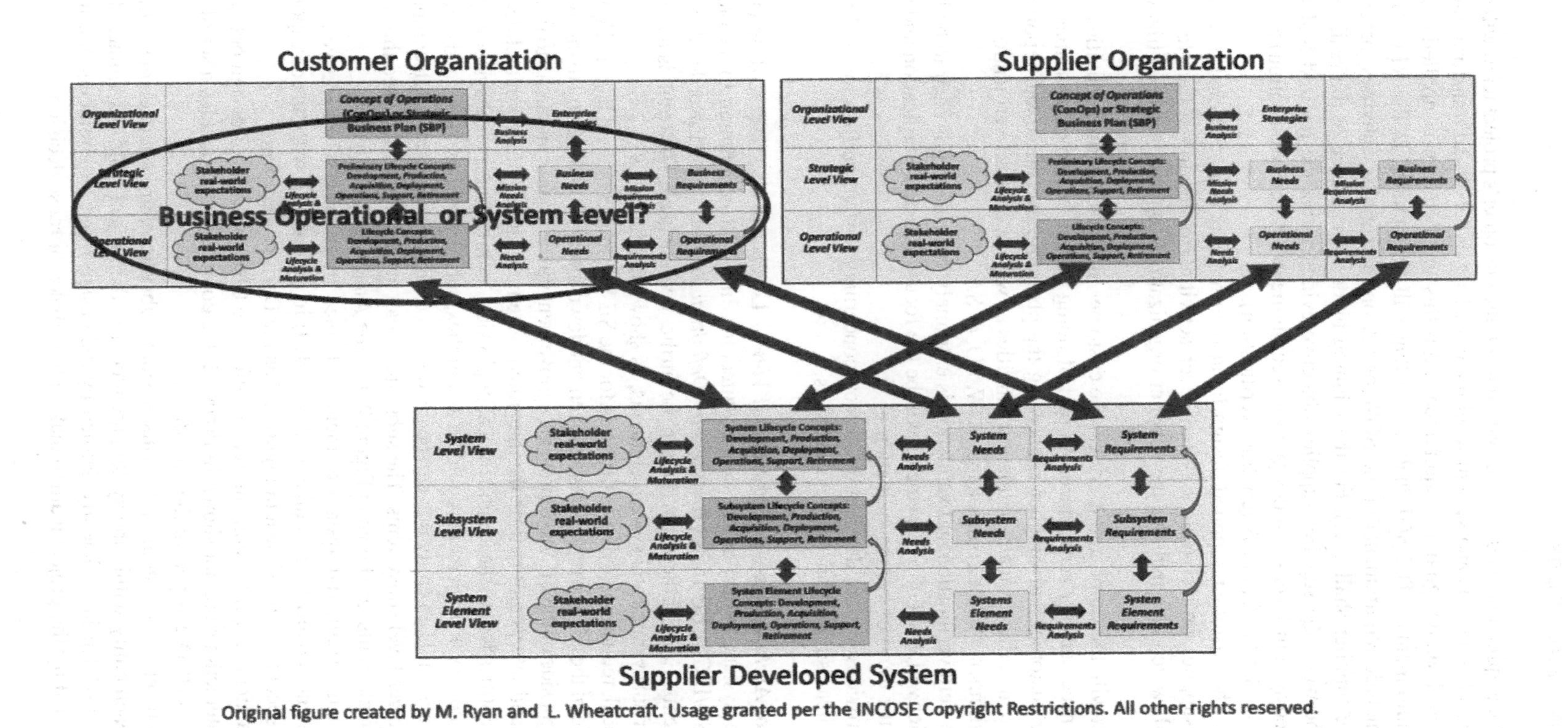

FIGURE 2.8 Supplier-Developed System.

However, if the acquisition occurs at the system level, the customer's project team will define the system-level lifecycle concepts and system-level integrated set of needs and transform those needs into a set of system-level design input requirements. It is this set of customer-supplied system-level design input requirements ("The SOI shall ....") that will be included in the RFP. The project team will also include requirements in the SOW or SA on the supplier activities, processes, and deliverables (The supplier shall ....). For the winning bidder, acceptance of the realized product or service will be when the supplier supplies objective evidence that the realized SOI meets the customer-supplied system-level set of design input requirements (system verification) as well as the requirements in the SOW, PO, or SA (contract verification).

In either case, the customer-supplied design input requirements or operational-level needs are inputs into the supplier project team's system-level activities to develop system-level lifecycle CNR for the SOI, which also includes the supplier organization's operational-level lifecycle CNR (Figure 2.8).

In the contract with the supplier, it is critical the customer make clear which activities and which deliverables are necessary for acceptance. As part of doing so, the customer must make clear the roles of both the customer and supplier concerning the planning, execution, and reporting of the results of the system verification and validation activities as well as contract verification concerning the supplier requirements in the contract and SOW, PO, or SA.

Addressing system verification and validation is extremely critical, so it is clear to both the customer and supplier which set of requirements will be the focus of system verification and which set of needs will be the focus of system validation.

Chapter 13 provides a more detailed discussion concerning system verification and validation considerations when using supplier-developed SOIs.

2.3.2.5 Levels of Architecture—a Hierarchical View Levels can also refer to different tiers or layers within the system logical or physical architecture. In SE, it is common to decompose the system via the ISO/IEC/IEEE 15288 *Architecture Definition Process* in the lower-level subsystems and system elements. Each of these can, in turn, be further decomposed. The result is a hierarchy of multiple levels of subsystems and system elements, as shown in Figure 2.9.

This hierarchical, reductionist view is the basis for the SE Vee model (shown later in Figure 2.15), where the left side of the SE Vee represents a movement down the system hierarchy to the system elements, and the right side of the SE Vee represents a movement up the system hierarchy to integrate the system elements and subsystems to form the integrated system. In terms of levels of architecture, this Manual refers to entities within the system logical or physical architecture as "subsystems" and "system elements," where system elements could be assemblies, sub-assemblies, parts, and components. In this context, a system is an entity that can be decomposed through elaboration (decomposition or derivation) into lower-level subsystems. The subsystems can be further decomposed through elaboration into system elements, as shown in Figure 2.9 as the Analysis/Decomp flow.

Each system, subsystem, and system element is defined by its own set of lifecycle CNR and each system element is realized (manufactured or coded) in accordance with its corresponding set of design output specifications, shown in Figure 2.9 as the Design/Spec flow.

The result is a hierarchical "document tree" of lifecycle CNR and sets of design output specifications. While interactions are not shown in Figure 2.9, a complete tree includes a set of Interface Control Documents (ICDs) as part of the flow down.

An SOI could be either a system-of-systems, system, subsystem, or system element, depending on context. For each entity within the system architecture, the project team determines if it needs further elaboration or if it is sufficiently defined by the design input requirements that no further elaboration is needed for the project team to make a buy, make, or reuse decision.

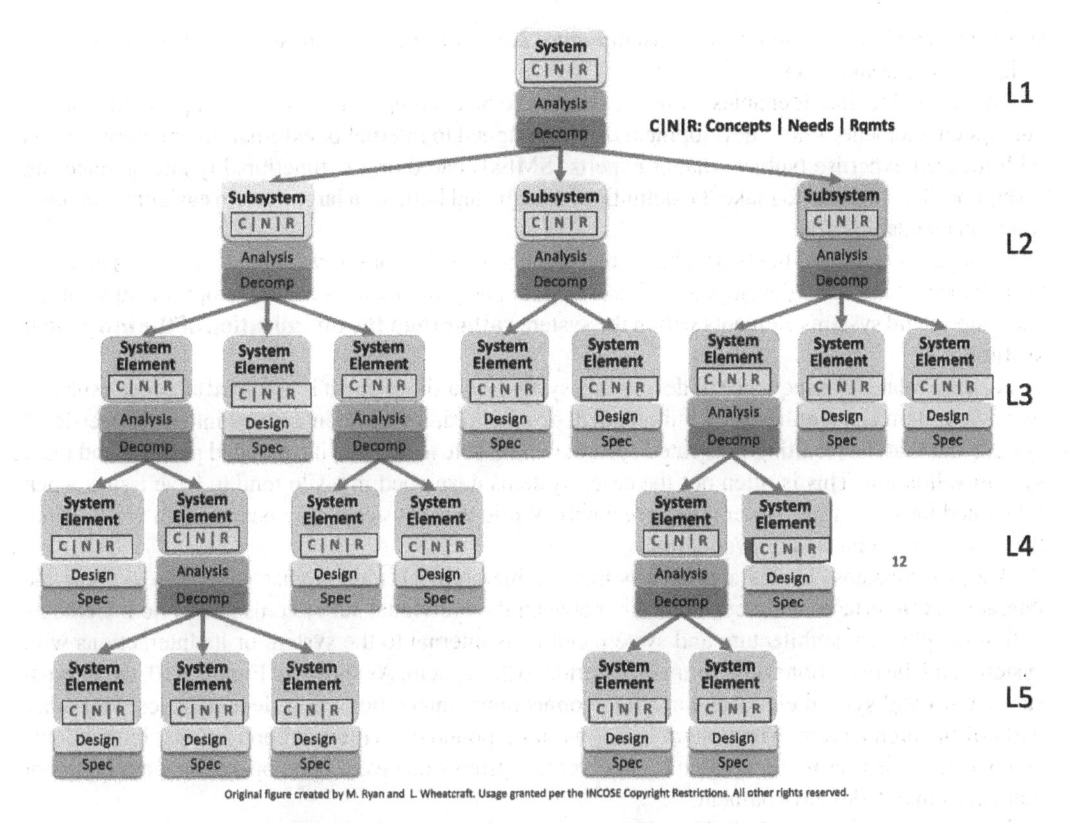

FIGURE 2.9 Levels of a System—Hierarchical View.

Buy could mean the project team will procure a system, subsystem, or system element commercial off-the-shelf (COTS) that has already been manufactured or coded. "Buy" could also mean the project team will contract out further elaboration and development to a supplier, as discussed earlier. From a supplier's perspective, that system, subsystem, or system element defined in the contract is their SOI.

"Make" refers to the case where the SOI will be developed (designed, assembled, manufactured, or coded) in-house. "Reuse" refers to cases where an SOI already exists within the organization and can either be use as-is off-the-shelf (OTS) or modified OTS (MOTS) to meet the lifecycle CNR defined for that SOI. *Chapter 12 provides more detail concerning the use of OTS and MOTS SOIs. Chapter 13 provides more detail for supplier-developed SOIs.*

If the SOI is to be developed in-house, using the *Design Definition Process*, the set of design input requirements is transformed into a design, which will result in a physical realization of the SOI. As part of the *Design Definition Process*, the design team will decompose the SOI into physical entities referred to as assemblies, sub-assemblies, parts, and components. The design of these physical entities is communicated via a set of design output specifications that are used to manufacture or code the physical entities that, when integrated together, will result in the physical realization of the SOI, which can then be integrated with other realized SOIs into the next higher-level SOI in the architecture.

2.3.2.6 Levels of Architecture—a Holistic View Developing systems per the hierarchical view (also referred to as a reductionist view since we are decomposing the system to its fundamental subsystems and system elements shown in Figure 2.9) is one of the greatest strengths of SE and one of its greatest weaknesses.

The strength is that it enables complex systems to be decomposed into less complex subsystems and system elements whose development can be assigned to internal or external organizational units with focused expertise (subject matter experts [SMEs]) based on the functionality and engineering discipline. The intent is to make the definition, design, and build of a large system easier by breaking it into manageable parts.

A major weakness of the hierarchical, reductionist view is that the resulting subsystems and system elements tend to be developed in silos. Consequently, the focus is often the optimization of the subsystems and systems elements within the system **rather than the optimization of the integrated system.**

Another related consequence of developing systems in a silo is that it is assumed after each subsystem and system element has been built, coded, or procured, it will be integrated into its higher-level system and that the resulting integrated system will be able to achieve its intended purpose and meet system validation. This is often not the case. Systems developed in a silo tend to have issues when integrated into the larger system they are a part. While these systems may pass system verification, they are likely to fail system validation.

Another weakness in this approach is that the hierarchical, reductionist view fails to show the interactions (interfaces and dependencies) between the individual subsystems and system elements within the physical architecture and system elements internal to the system or its interactions with systems and the operational environment external to the system. As shown in Figure 2.10, the internal subsystems and system elements have interconnections, interactions, and dependencies with other parts of the architecture. The system exists within a boundary with interconnections, interactions, and dependencies across that boundary to external systems that exist in an operational environment that exists in a wider environment.

In this context, interconnections and interactions refer to what is commonly referred to as an interface boundary across which two systems, subsystems, or system elements interact. These interactions

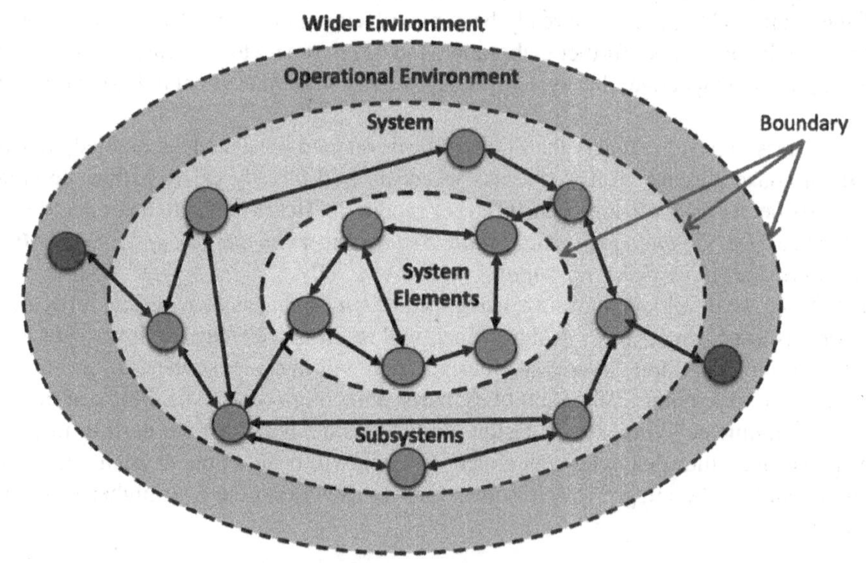

FIGURE 2.10 Interactions and Dependences Internal and External to a System.

can involve the flow of information, energy, or matter. These interactions can include interactions with the natural environment or induced environments (such as acoustics, vibrations, electromagnetic, or thermal).

The interactions can also be in the form of systems, subsystems, or system elements competing for the use of common resources (bandwidth, CPU time, memory, power, weight, or mass [for space systems]). Dependencies can be in the form of a system, subsystem, or system element that provides something as an enabling system to another system, subsystem, or system element or in cases where subsystems or system elements share an allocated or budgeted resource or performance measure (such as mass, power, time, total allowable error, precision, accuracy, jitter, or quality characteristic).

Because the interconnections, interactions, and dependencies are related to the behavior of the system elements, the subsystem they are a part of, and the integrated system, a more holistic view of the system, such as shown in Figure 2.11, is needed. (*Figure* 2.11 *contains the same system elements as shown in Figure* 2.10, *only arranged in a more holistic view with an emphasis on interconnections, interactions, and dependencies between the subsystems and system elements within the system as well as with external systems in the operational environment.*)

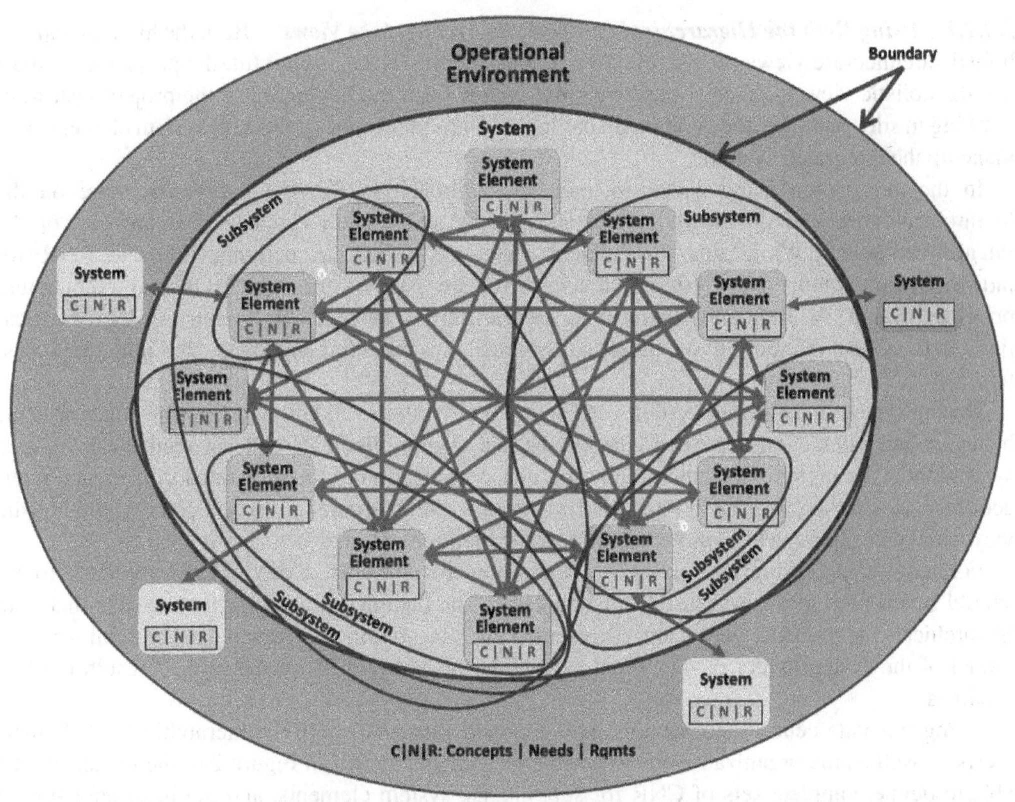

FIGURE 2.11 Holistic View of the SOI.

While the hierarchical, reductionist view is useful when defining a system, subsystem, and system elements in terms of lifecycle CNR, the project team must also have a holistic view of the system architecture and resulting behavior throughout the system lifecycle. This holistic view requires the project team to understand the basic tenets of systems thinking.

- A system is more than a sum of its parts.
- A system's behavior is a function of the interaction of its parts as well as the interactions of the integrated system with the external systems and environment of which it is a part.
- A system's parts are often interdependent.
- A system will have emergent behaviors (good and bad) beyond what was defined within the needs, requirements, and specifications.
- Optimizing a system's performance almost always means that some parts that make up the system may have to be suboptimized.

The approach described in this Manual is meant to drive the project team toward systems thinking and have a holistic view of system development across all aspects of the system lifecycle.

Viewing the system holistically, the focus is on the behavior of the integrated system as a function of the interaction of the subsystems and system elements that make up the system as well as the behavior of the system in the context of its interaction with the systems that make up the macro system, the operational environment, users, operators, and maintainers.

2.3.2.7 Using Both the Hierarchical and Holistic Architecture Views Both the hierarchical and holistic architecture views are useful in the practice of SE. To be successful, the project team must use the holistic view to manage the *integrated system* from the beginning of the project instead of working in silos and only focusing on managing the individual subsystems and system elements that make up the integrated system.

In the past, when using a document-centric approach to SE, the focus was more on the hierarchical view and on the individual subsystems and system elements that are part of the integrated system. While this hierarchical view was useful in defining lifecycle CNR for individual subsystems and system elements that are part of the system logical architecture prior to design, it frequently results in issues during system integration, system verification, and system validation of those subsystems, system elements, and the integrated system.

During system integration, system verification, and system validation, a primary concern is the behavior and interactions between subsystems and system elements and between the integrated system and external systems, emerging properties, optimization of the integrated system, and interactions between the integrated system and its operational environment. In this context, the holistic view shown in Figure 2.11 is more useful than the hierarchical view.

For today's increasingly complex, software-intensive systems, a data-centric approach to SE should be adopted, discussed further in Chapter 3. The data-centric approach advocates that both hierarchical and holistic views are represented by a common integrated data and information model of the system under development as well as the PM and SE processes and resulting work products.

Using this data-centric approach, blending considerations of both the hierarchical and holistic views as well as the organizational/ system/element model shown in Figure 2.6, the reader will be able to define complete sets of CNR for subsystems, system elements, and the integrated system that, when designed and built/coded, will yield an integrated system that can be verified to meet its set of design input requirements and validated to meet its integrated set of needs. The result is an integrated system that can be integrated into the customer's technical environment as well as their strategic and organizational ones—this is the ultimate goal of SE.

2.3.3 Verification and Validation in Context

The concepts associated with the terms verification and validation are distinctly different depending on the context. The intended meaning of the concepts represented by each term is often misunderstood, and the terms are often used interchangeably without making clear the context in which they are used, resulting in ambiguity and a failure to communicate.

To avoid this ambiguity, each of these terms should be preceded by a modifier (i.e., the subject), which clearly denotes the proper intended context in which the term is being used, as shown in Figures 2.12 and 2.13:

- *Needs verification* or *needs validation* for verification and validation of the needs,
- *Requirements verification* or *requirements validation* for verification and validation of the design input requirements,
- *Design verification* or *design validation* for verification and validation of the design and associated design output specifications, or
- *System verification, system validation*, or *production verification* for verification and validation of the realized SOI.

Not shown in Figure 2.12 are the development of lifecycle concepts and the transformation of those concepts into the integrated set of needs. Figure 2.13 expands on the box on the left of Figure 2.12, *"Integrated Set of Needs,"* illustrating the information that feeds into the lifecycle concepts analysis and maturation activities and the resulting transformation of the approved lifecycle concepts into the integrated set of needs. Figure 2.13 also shows verification and validation in context of the integrated set of needs.

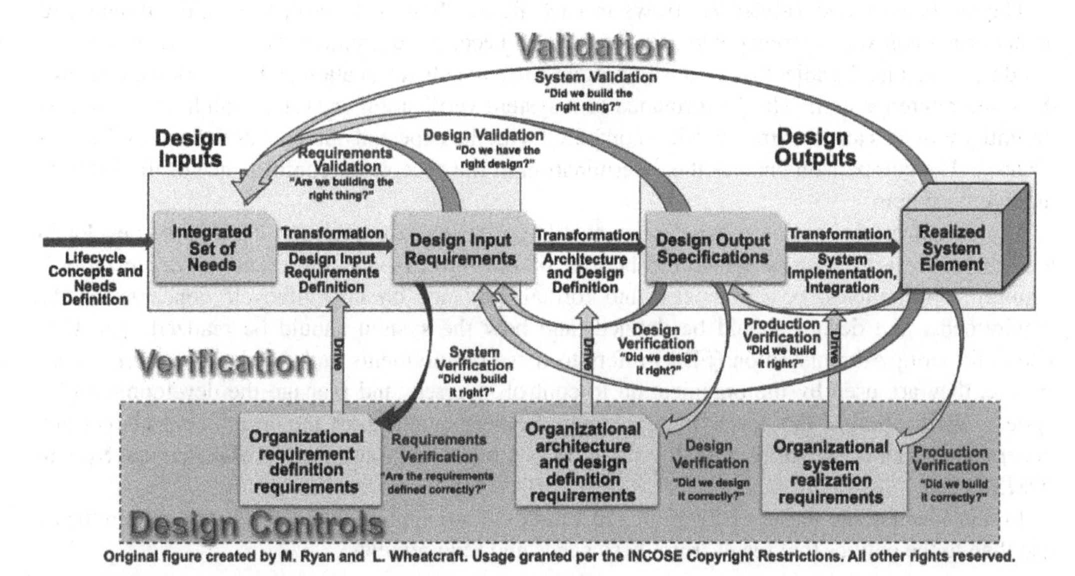

FIGURE 2.12 Verification and Validation Confirm that SE Artifacts Generated During Transformation are Acceptable. Source: Adapted from Ryan and Wheatcraft [14].

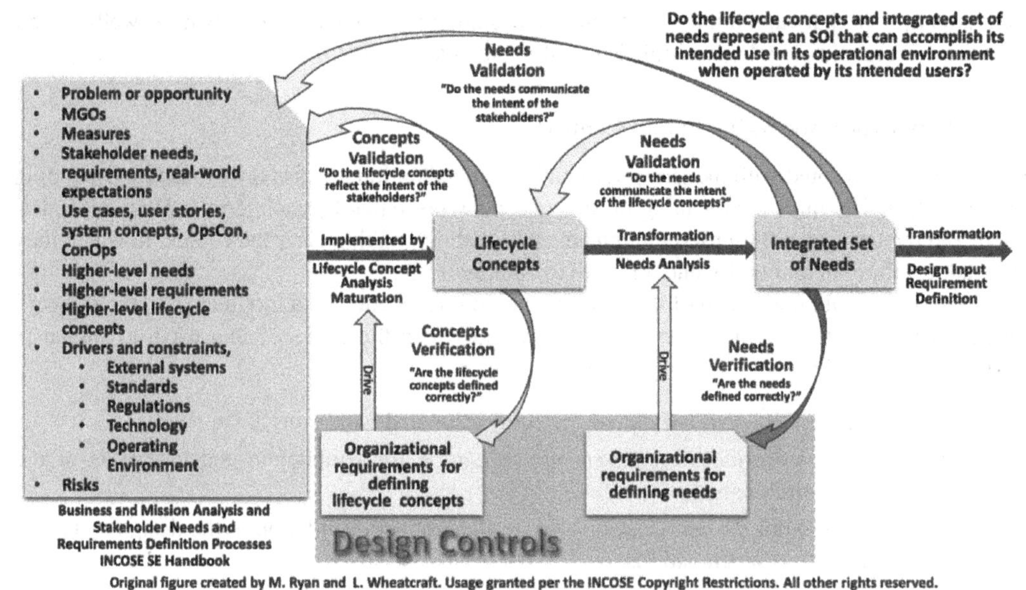

FIGURE 2.13 Needs Verification and Validation.

Figures 2.12 and 2.13 illustrate the importance of making it clear as to the context in which the terms verification and validation are being used. The development activities are represented as the successive transformation of lifecycle concepts into an integrated set of needs, which are transformed into a set of design input requirements, which are transformed into design output specifications, which are transformed into the realized system element. This is represented in the main horizontal path across the center of both figures.

The verification and validation arrows in each figure show that verification and validation of an entity is *against a reference item* (for example, a need, a requirement, design characteristic, or standard)—that is, the adjective used is important since not all verification and validation are against the same reference item. The performance of a system verification or system validation action on the entity provides an outcome, which is compared with the expected outcome as defined by *Success Criteria*. The comparison enables the determination of the acceptable conformance of the entity to the reference item.

The organizational requirements and standards governing how each artifact is to be developed are labeled design controls across the bottom of the figures—and represent the organization's requirements, policies, best practices, and formal guidance on how lifecycle concepts, needs, requirements, the design should be defined, and how the system should be realized. The U.S. Food and Drug Administration (FDA) refers to these requirements and standards as *design controls* as they are used by the organization to control, oversee, and manage the development of a system. Verification of the development activity requirements against the design controls is often undertaken by an organization's quality function as a part of a larger Quality Management System (QMS).

In this Manual, the terms verification and validation are applied to the needs and requirements and then to the entities that result from these needs and requirements:

- *Verification of the needs and requirements.* The GtWR defines characteristics and rules that help to achieve those characteristics for forming need and requirement statements and sets of needs and requirements. In this context, verification involves verifying that the statements and sets of statements have the characteristics of well-formed need and requirement statements resulting from following the rules such as those in the INCOSE GtWR or similar organizational guide or standard as well as completing the activities discussed within this Manual. Verification of needs and requirements is what is commonly referred to as assessing the quality of the need and requirement statements and sets of needs and requirements. *Refer to Chapter 5 for a more detailed discussion on need verification and Chapter 7 for a more detailed discussion on requirement verification.*

- *Validation of the needs and requirements.* Validation of a need statement determines whether a need statement clearly communicates the intent of the lifecycle concepts or source from which it was derived or transformed. Validation of a requirement statement determines whether the requirement statement clearly communicates the intent of the need, source, or parent requirement from which it was derived or transformed. A need or requirement statement can be well-formed in accordance with the GtWR rules (that is, the statement can be verified in terms of its quality)—but can speak to the wrong need or requirement, or it may not communicate the true intent of the concept, need, or parent requirement from which it was derived or transformed. Again, the focus is on the quality of the need and requirement statements and sets of needs and requirements. *Chapter 5 provides a more detailed discussion on need validation, and Chapter 7 provides a more detailed discussion on requirement validation.*
- *Verification of the entity.* Verification of the entity (design, design output specifications, subsystem, system element, or the integrated system) against the verified and validated set of design input requirements. In this context, verification is a formal process that results in objective evidence that can be used to determine that the entity is being/has been formed in the right way as defined by the set of design input requirements with the required level of confidence. *Refer to Chapter 8 for a more detailed discussion on design verification, Chapter 9 for a more detailed discussion on production verification, and Chapters 10 and 11 for a more detailed discussion on system verification.*
- *Validation of the entity.* Validation of the entity (design, design output specifications, subsystem, system element, or the integrated system) against the verified and validated integrated set of needs. In this context, validation is a formal process that results in objective evidence to be used for the acceptance, certification, and qualification of the SOI that can be used to determine the right entity is being (or has been) formed as defined by the integrated set of needs, with the required level of confidence. *Refer to Chapter 8 for a more detailed discussion on design validation and Chapters 10 and 11 for a more detailed discussion on system validation.*

2.3.3.1 Distinctions Between the Concepts of Verification and Validation While there is overlap between these concepts, it is important to understand where the concepts of verification and validation differ in terms of the questions that are being addressed by each of the two activities.

Verification addresses the questions, "Are we designing and building the system *in the right way*?" (i.e., in accordance with its requirements), and "Are we designing and building the system correctly?" (i.e., in accordance with the design controls), and thus verifies two things at each stage of development:

- The SE artifacts generated across the lifecycle are verified against their success criteria, for example, needs and requirements against the organizational best practice for developing needs and requirements, design and design output specifications against the design input requirements, and the manufactured/coded SOI against both its design output specifications (production verification) and its design input requirements (system verification) as shown across the center portion of Figures 2.12 and 2.13.

- The activities performed by the organization developing the SE artifacts are verified against the organization's requirements for development, design, manufacturing, system integration, system verification, and system validation as communicated in the organization coding best practices, standards, procedures, and WIs as shown at the bottom of Figures 2.12 and 2.13 as design controls.

Validation addresses the question, "Are we asking for, designing, and building the right entity?" and, as such, validates:

- The set of design input requirements, design and subsystems, system elements, and integrated system against their respective set of needs as shown in Figure 2.12.
- The set of needs against lifecycle concepts from which the set of needs was transformed, as shown in Figure 2.13 and against the stakeholder real-world expectations, as shown later in Figure 2.14.

In EIA-632, *Processes for Engineering a System*, there is guidance that states: "Validation in a system lifecycle context is the set of actions ensuring and gaining confidence that a system is able to accomplish its intended use, goals, and objectives" [15]. This note supports the view of not waiting until the final integrated system is released for use to determine whether it is acceptable, but rather the intent is to determine the conditions for acceptance at the beginning of the project. In this way the project's modeling and simulation tools can be used during development to detect flaws and missed needs and requirements prior to production and system integration, system verification, and system validation.

Validation is critical because it addresses the risk that the final system will not satisfy the customers, users, and operational- and system-level stakeholder's real-world expectations concerning its intended use in the operational environment when operated by the intended users and does not enable unintended users to negatively impact the intended use of the system. One could argue that this is the ultimate purpose of engineering as a discipline and, specifically, of the discipline of SE [16, 17].

Other perspectives concerning the distinction between the concepts of verification and validation are:

- **The focus of verification is more on form.** Do the sets of needs and sets of design input requirements have the characteristics as defined in the GtWR? Does the design result in a system that can be verified to meet its set of design input requirements? Is the SOI made correctly in accordance with its set of design output specifications? Are the internal relationships, like traceability and dependencies, recorded properly within the subsystems, system elements, and integrated system? Are the interactions between the entity and external entities recorded properly?
- **The focus of validation is more on content and intent concerning what is being communicated.** Do the set of needs or set of design input requirements communicate the right thing? Does the need or requirement statement clearly communicate the correct intent of the parent or source from which it was derived? Is the set of needs or set of design input requirements correct, complete, consistent, and feasible? Will the resulting SOI correctly represent its intended use in the operational environment when operated by the intended users and does not enable unintended users to negatively impact the intended use of the system?

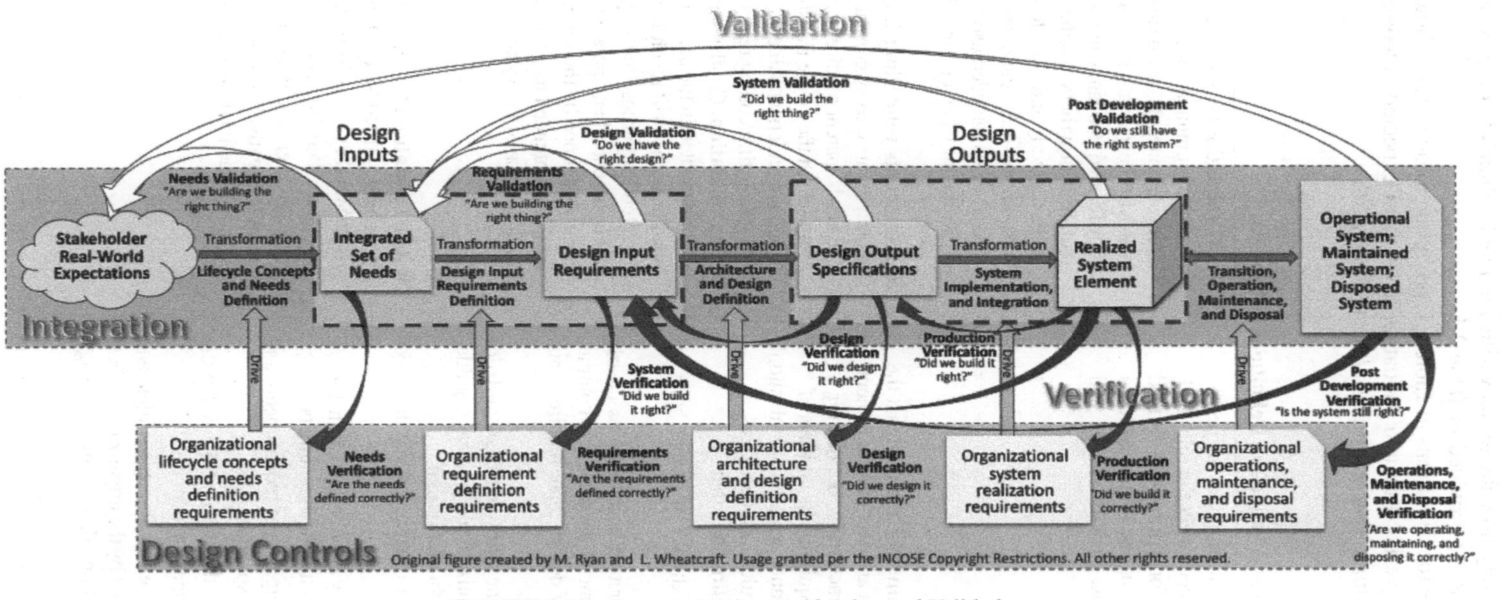

FIGURE 2.14 Post-production Verification and Validation.

To show more precisely how the terms verification and validation are used within this Manual, the following definitions are provided in Table 2.1.

Table 2.2 expands on the definitions in Table 2.1, providing a comparison showing the differences of the various kinds of verification and validation in terms of expected outcomes. *Note: the list of outcomes includes examples and is not intended to be exhaustive.*

TABLE 2.1 Verification and Validation Definitions in Context.

Verification	Validation
Needs verification: Confirmation that the need statements and set of needs meet the rules and characteristics defined for writing well-formed needs and sets of needs in accordance with the organization's standards, guidelines, rules, and checklists	**Needs validation:** Confirmation that the needs clearly communicate the intent of the agreed-to lifecycle concepts, constraints, and stakeholder real-world expectations from which they were transformed in a language understood by the requirement writers. Confirmation that the set of needs correctly and completely captures what the stakeholders need and expect the system to do in context of its intended use in the operational environment when operated by its intended users
Requirements verification: Confirmation that the requirement statements and sets of requirements meet the rules and characteristics defined for writing well-formed requirements and sets of requirements in accordance with the organization's standards, guidelines, rules, and checklists	**Requirements validation:** Confirmation that the requirements clearly communicate the intent of the needs, parent requirements, and other sources from which they were transformed into a language understandable by the design and manufacturing/coding teams
Design verification: Confirmation that: a) The design reflects the set of design input requirements b) The set of design output specifications clearly implements the intent of the design as communicated by the set of design input requirements, and c) The design meets the rules and characteristics defined for the organization's processes, guidelines, and requirements for design	**Design validation:** Confirmation that the design, as communicated in the set of design output specifications, will result in a system that meets its intended purpose in its operational environment when operated by the intended users as defined by the set of needs and does not enable unintended users to negatively impact the intended use of the system
System verification: Confirmation that the designed and built or coded SOI: a) Has been produced by an acceptable transformation of design inputs into design outputs b) Meets its set of design input requirements and set of design output specifications c) No error/defect/fault has been introduced at the time of any transformation, and d) Was produced per the requirements, rules, and characteristics defined by the organization's best practices and guidelines defined in the design controls	**System validation:** Confirmation that the designed, built, and verified SOI will result or has resulted in an SOI that meets its intended purpose in its operational environment when operated by its intended users and does not enable unintended users to negatively impact the intended use of the system as defined by its integrated set of needs

TABLE 2.2 Verification and Validation Comparisons in Terms of Outcomes.

Needs verification: The focus is on the wording and form • Are individual need statements worded or structured correctly in accordance with the organization's requirements, rules, and checklists? o Do the needs have the characteristics of well-formed needs and sets of needs? • Is the set of needs complete, containing needs dealing with form, fit, function, quality, and compliance? • Does each need statement trace to a source? • Does each source have a corresponding need statement?	**Needs validation:** The focus is on the message being communicated • Has an acceptable elicitation process been followed? • Have all relevant stakeholders been consulted? • Are the needs the right needs, i.e., do they accurately represent the agreed-to lifecycle concepts from which they were transformed? • Do the needs correctly and completely capture what the stakeholders need the SOI to do in the operational environment in terms of form, fit, function, compliance, and quality? • Will the set of needs, if met, adequately address the problem, opportunity, or threat?
Requirements verification: The focus is on the wording and form • Are individual requirement statements worded and structured correctly in accordance with the organization's requirements, guidelines, rules, and checklists? o Do the requirements have the characteristics of well-formed requirements and sets of requirements? • Do the requirements trace to the need(s), parent requirement(s), or source requirement(s)? • Have the requirements been allocated to the next level of the architecture? Are there any gaps?	**Requirements validation:** The focus is on the message being communicated • Are the requirements and sets of requirements the right requirements, i.e., do they accurately represent the intent of the needs, parent requirements, or source from which they were transformed? • Do the requirements and sets of requirements address the right need? • Do the resulting child requirements represent a necessary and sufficient set to meet the intent of the parent's need, requirement, or source? • Are the requirements communicating the right things?
Design verification: The focus is on the design. • Has an acceptable transformation of the set of design input requirements into design output specifications been applied? • Does the design satisfy the set of design input requirements? • Do the design output specifications clearly and accurately communicate the agreed-to design to the suppliers/builders/coders? • Can the system be built to the design output specifications? • Has traceability been established between the design input requirements and the design artifacts? • Did the design team follow the organization's requirements for the design activities? • Did we design the thing right?	**Design validation:** The focus is on the message the design is communicating. • Will implementing the agreed-to design output specifications result in an SOI that will meet the needs such that the SOI can be used as intended in the operational environment? • Do we have the right design (as defined by the design output specifications)?

TABLE 2.2 ***(Continued)***

System verification: The focus is on the built or coded SOI and how well it meets the agreed-to design input requirements and design output specifications • Did we correctly follow our organization's requirements for manufacturing or coding? • Is there objective evidence the SOI satisfies the design input requirements and design output specifications? • Did we detect all errors/defects/faults that may have been introduced during the transformation? • Did we build the thing right?	**System validation:** The focus is on the completed SOI and how well it addresses the problem, opportunity, or treat, MGOs, and stakeholder real-world expectations as defined within the agreed integrated set of needs • Does the SOI satisfy the needs for its intended use when operated by the operators/users in the operational environment? • Does the SOI protect against unintended users negatively impacting the intended use of the system? • Did we build the right thing?

Further details for each of these activity areas are included in:

- Chapter 5: Needs Verification and Validation.
- Chapter 7: Design Input Requirements Verification and Validation.
- Chapter 8: Design Verification and Validation.
- Chapter 9: Production Verification.
- Chapters 10 and 11: System Verification and Validation.

2.3.3.2 System Verification and Validation After Product Release Verification and validation do not end with production and release of a SOI for use. As shown in Figure 2.14, there is post-development system verification and validation that must take place.

Post-production validation addresses the question: "Do we **still** have the right system?" i.e., does the system still meet the stakeholders' real-world expectations and post-production verification addresses the question: "Is the system **still** right?", i.e., does the system still meet its design input requirements? Both post-production verification and validation are a continuous set of activities performed across the system operational life.

For some products, there may be post-production verification and validation performed on returned items (fault or root-cause analysis) or items in the field to gather data about operational performance over time (quality analysis). Such efforts can identify defects in the released system that were missed during initial system verification and validation activities—or can detect aging (degradation over time) and end-of-life (EOL) defects before a catastrophic failure occurs.

Additionally, depending on the system and qualities of its operational environment, emergent properties may be identified after an extended period of fielding, which may not have been identified during system development and system validation. These post-production verification and validation activities are often part of an organization's post-release quality surveillance program required by regulatory agencies that feed the organization's warrantee and recall efforts.

This information can also be used when a new version of the SOI is being developed, enabling the project to take advantage of the lessons learned from the current system and incorporate them into the new version.

This is demonstrated with the example of a bathtub curve of a system's life, which is a plot of defects found over time. The curve starts high early in the system's use and then curves down like the first edge of a bathtub in profile. As time progresses to the right, at some point, the number of defects slopes up sharply, representing the other side of the tub—and a warning the system is nearing EOL and merits close observation and possible replacement (this is especially important for systems that are still operational long after their design lifetime). For some systems, based on lessons learned, maintenance history, recalls, and warranty work (all are part of post-production verification and system validation), solutions such as upgrades, modifications, or expanded preventive maintenance procedures are developed to address these issues as well as extend the system's operational lifetime.

2.3.4 Integration, Verification, and Validation and the SE Vee Model

Figure 2.15 shows an enhanced version of the SE Vee model, which is commonly used to show the iterative and recursive nature of SE. The figure also shows the relationship of the lifecycle process activities in relation to the SE Vee model.

The SE Vee model is one of several visualizations to help communicate the SE processes over time. The left side of the SE Vee shows a top-down series of activities where the definition of lifecycle concepts, integrated sets of needs, sets of design input requirements, architecture, and design take place. Correspondingly, the right side of the SE Vee shows a bottom-up series of activities where the physical system elements and subsystems that make up the system's physical architecture are

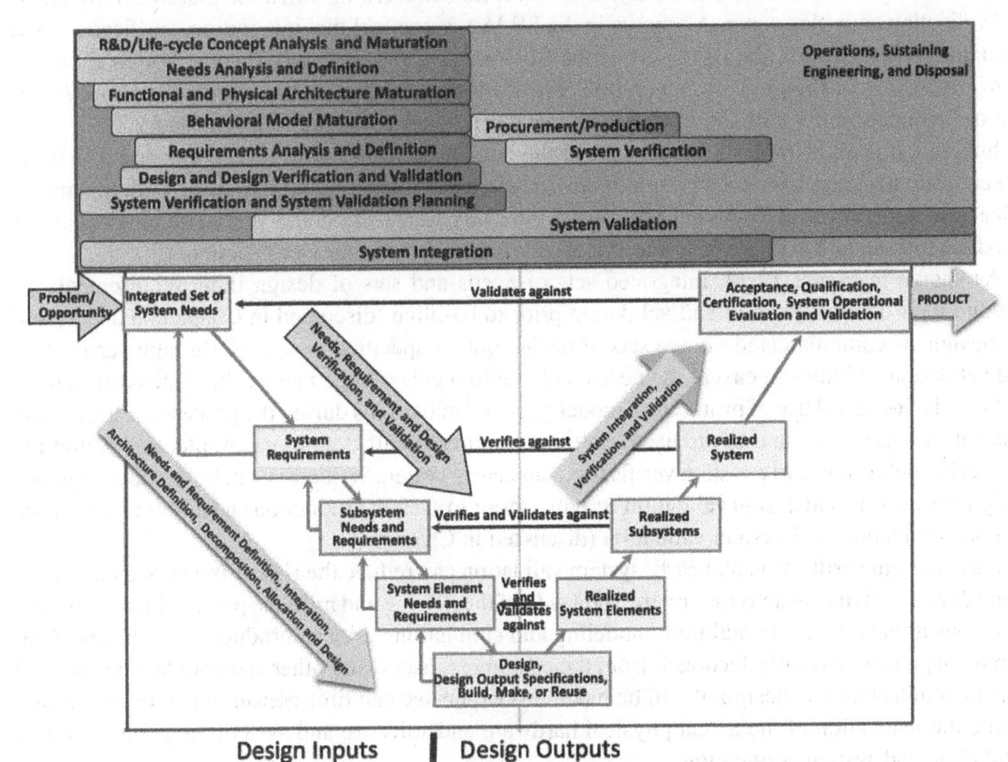

Adapted from M. Ryan, L. Wheatcraft, L.S., "On the Use of the Terms Verification and Validation", February 2017 and INCOSE SE HB, Version 4, Figures 4.15 & 4.19

FIGURE 2.15 Lifecycle Processes in Relation to the SE Vee Model.

verified against their respective requirements and validated against their respective needs that were defined during the SE activities on the left side of the SE Vee. Thus, system integration, system verification, and system validation on the right side of the SE Vee are distinctly separate activities from the activities concerned with needs definition, requirements definition, architecting, design, and production of the parts that occur on the left side of the SE Vee.

This distinction is especially important in terms of budgeting and scheduling. A common issue with many projects is a failure to appreciate the time and effort associated with system integration, system validation, and system verification activities associated with the right side of the SE Vee. Consequently, it is common for projects to not include sufficient budget and schedule to successfully complete the system integration, system verification, and system validation activities—especially when issues are discovered that can lead to expensive and time consuming rework.

Because of this, it is important for projects to understand that the success of the system integration, system validation, and system verification activities on the right side of the SE Vee is highly dependent on the time spent and quality of the SE activities and resulting artifacts developed on the left side of the SE Vee. Additionally, the time invested and quality of the work performed during verification and validation of the SE artifacts generated on the left side of the SE Vee can also influence and improve the quality of the system design by driving requirements and needs refinement.

2.3.4.1 Concurrent Integration, Verification, and Validation A major issue many have with the SE Vee model is there is sometimes a misconception concerning when integration, verification, and validation activities occur. As shown in the SE Vee, some feel that integration, verification, and validation only occur on the right side of the SE Vee, as it is commonly shown. This is false. As shown at the top of Figure 2.12, integration, verification, and validation are three distinct *activities* that occur concurrently with the other SE technical processes across all lifecycle stages.

Integration is an activity that must start at the beginning of the project. It is important that the project team manages the development of any system as an integrated system using both the hierarchical and holistic views discussed previously. Similarly, verification and validation are concurrent activities that occur across all lifecycle stages, as discussed in the previous section.

As shown in Figure 12.14, integrated sets of needs and sets of design input requirements go through their own verification and validation prior to baseline (discussed in Chapters 5 and 7) and the design, as communicated via the sets of design output specifications, goes through verification and validation activities to ensure the needs and requirements will be met by the realized system.

With the increased use of prototypes, modeling, and simulations during the processes for lifecycle concepts analysis and maturation, needs definition, requirements definition, architecture definition, and design definition, early system verification, and early system validation can be conducted during design verification and design validation activities prior to system production and system integration, system verification, and system validation (discussed in Chapter 8).

Early system verification and early system validation can reduce the risk of issues and anomalies being discovered during activities on the right side of the SE Vee and mitigate potential for costly and time consuming rework. In addition, modeling and simulations prior to production enable expectation management and early feedback from the customers, users, and other stakeholders on the final system architecture and design; it will be much less expensive and time consuming to resolve issues before the realization of the actual physical hardware and software and system integration, system validation, and system verification.

Note: In some organizations, design verification is referred to as a Functional Configuration Audit (FCA), where the design for the integrated system is verified to meet its functional baseline approved at the Preliminary Design Review (PDR) and Critical Design Review (CDR) as communicated in the sets of design input requirements.

2.3.4.2 System Verification Versus Requirement Verification System elements, subsystems, and the integrated system are verified against predefined and agreed-to verification success criteria defined for each requirement using an agreed-to verification strategy and method as part of system verification. In terms of a contract, most often, it is the design input requirements against which the project will verify the SOI by providing objective evidence that the SOI meets the agreed-to verification success criteria defined when the requirements were defined. *Refer to Chapter 10 for a more detailed discussion concerning Verification Success Criteria, Strategy, and Method.*

In the past, it was a customary practice that the shorter phrase "requirement verification" was used rather than defining system verification in terms of the verification of a system, subsystem, and system element against its design input requirements. This, in turn, led to the use of phrases and wording, such as "Requirement Verification Matrix (RVM)" and "requirement verification events," rather than "System Verification Matrix (SVM)" and "system verification events." However, it is important to understand that verification of a system, subsystem, or system element against its requirements is not actually requirement verification; it is system verification as discussed previously.

An issue with the focus on "requirement verification," in this context is that often, what is referred to as "requirements" are only the design input requirements. Consequently, verifying the system was successfully produced via the design output specifications is often not included as part of system verification activities. *Note: In many organizations, verifying the built system meets the design output specifications, known as "built per spec," is a quality organization function. This is sometimes referred to as a Physical Configuration Audit (PCA), performed as part of production verification, discussed in Chapter 9, that occurs during manufacturing or coding and not as part of the system verification activities.*

2.3.4.3 System Verification Based on Lower-Level Verifications Using the phrase "requirement verification" rather than "system verification" can result in a focus on the design input requirements and their relationships rather than addressing verification that the system elements, subsystems, and the integrated system each meet their design input requirements. As a result, some systems engineers believe that there is a correlation between the satisfaction relationship between parent/child requirements and the choice of system verification activities against the different levels of requirements as they move up the system's physical architecture during system integration.

With this perspective, some advocate that system verification consists of the verification activities directly associated with a given requirement as well as verification activities concerning the lower-level child requirements that trace directly to it, and state that the successful verification of the child requirements at one level is a prerequisite to system verification of the parent requirements at the next higher level. These systems engineers advocate that satisfaction of the child's requirements may be sufficient objective evidence that the parent requirements have been met, while others state that, in some cases, the satisfaction of a parent requirement can be used as evidence the child's requirements have been met.

However, there is a major flaw in this line of reasoning, which is a lack of understanding of the different types of verification and validation (as shown in Figure 2.12). System verification and system validation are neither exercises in verifying design input requirements nor in validating needs. Rather, they are exercises in verifying the realized physical subsystems, system elements, and integrated system meet their design input requirements (system verification) and the system needs (system validation) defined during development.

As discussed in Chapter 6, parent/child relationships are important concepts during design input requirements definition as they deal with the flow down (allocation) of requirements from the system at one level of the physical architecture to subsystems and system elements at the next lower level of the physical architecture. The flow down of requirements also involves budgeting of resources, quality, and performance to ensure the integrated physical system meets the design input requirements and needs when all the subsystems and system elements that make up the SOI are integrated.

However, the determination of the proper parent/child relationships and whether the child design input requirements are both necessary and sufficient are part of requirement verification, requirement validation, design verification, and design validation activities during development—not system verification and validation activities during integration of the physical subsystems and system elements into the integrated system.

As such, parent/child requirement relationships should not be the focus of system verification during system integration. These relationships are important to ensure the system elements are supportive of the system requirements, but verifying child requirements have been met does not inherently verify the parent requirements have been met; it ensures the intent of the parent requirement(s) have been met by the system element, subsystem, or integrated system where they apply.

This is especially true when addressing budgeted quantities. Consider when a parent requirement has a quantity (for example, performance, physical characteristic, or quality attribute) that was allocated by a budgeting process to the subsystems and system elements at the next lower level of architecture. Because the requirement is associated with a budgeted quantity, there is a dependency between *each* child requirement as well as with the parent such that if a budgeted child requirement is not met, one of the other dependent child requirements might not be met or the parent requirement itself might not be satisfied.

As an example, if the parent system has a weight constraint and its required weight was apportioned (budgeted) to the subsystems and system elements that are part of its physical architecture, and one or more of these lower-level subsystems or system elements exceed their budgeted weight, it is possible that the parent system, subsystem, or system element will fail to meet its weight requirement during its verification activities even if all of the other children are satisfied. It is for this reason that using a holistic approach for system verification is more appropriate than a hierarchical one.

2.3.4.4 Verification and Validation in Context As highlighted in Table 2.1, when this Manual refers to needs verification, requirement verification, needs validation, or requirement validation, what is really meant is verification and validation of the needs and requirements themselves and not verification and validation of the realized SOI. When this Manual refers to "system verification and system validation," what is meant is verification and validation of a realized SOI against their respective sets of needs and requirements. The same is true for design verification and design validation.

To help correct the misuse of the phrases "requirement verification" and "requirement validation" when "system verification" and "system validation" is the real intent, common artifacts have been renamed throughout this Manual. For example, rather than an RVM or Requirement Verification Compliance Matrix (RVCM), the names have been changed to System Verification Matrix (SVM) and System Verification Compliance Matrix (SVCM). In addition to the verification matrices, there are similar validation matrices: System Validation Matrix (SVaM) and System Validation Compliance Matrix (SVaCM).

2.3.4.5 The Intent of System Verification and System Validation A key outcome of system verification and validation is to collect objective evidence to be used to show that the physical realized subsystems and system elements at the lower levels of the architecture meet their needs and requirements prior to being accepted and approved for integration into the next higher level of the physical architecture.

Once the system elements and subsystems are integrated, they represent a higher-level system element, subsystem, or the integrated system whose behavior is a function of the interactions of each of the physical parts (hardware, mechanical, and software) as well as interactions between the integrated system and the macro system of which it is a part.

A larger question concerning these subsystems and system elements is whether the architecting, allocation, flow down, and budgeting were done properly, such that the physical subsystems and system elements, when integrated into their parent system, will result in the integrated parent physical system being able to meet its requirements (system verification) and needs (system validation) and be accepted for use by the customer or regulatory agency.

2.3.4.5.1 Internal Operational Environment of a System Below the top level of the SOI architecture, validation of subsystems and system elements takes on a slightly different meaning than validation of the integrated system. The operational environment for the subsystems and system elements includes the interactions between each other across interface boundaries (internal and external) and the associated induced environments (for example, vibrations, acoustics, thermal dynamics, or EMI/EMC) within the integrated system, which may or may not be well-defined and modeled).

A subsystem or system element may have been verified that it meets its design input requirements; however, after being integrated into the next higher level of the architecture and exposed to the resulting operational environment, it may fail to do so, failing validation that it has met its needs concerning its performance within the integrated system.

2.3.4.5.2 System Verification and Validation of the Physical System Versus Modeled System Because the behavior of a system is a function of the interaction of its parts, a major goal of system validation is to assess the behavior of the integrated physical system and to identify emergent properties not specifically addressed by the needs or requirements nor discovered during modeling and simulations. Emerging properties may be beneficial (in that they are the properties we wish the system to possess when its elements are integrated) or detrimental (in that the emergent property is an undesirable consequence of integration—for example, cascading failures across multiple interface boundaries between the system elements that are part of the system's architecture). Relying on models and simulations of the system and its operational environment may not uncover all the emergent properties and issues that occur in the physical realm.

While system validation using models and simulations allows a theoretical determination that the modeled SOI will meet its needs in the operational environment by the intended users once realized, the assessment of the actual behavior (system validation) must be undertaken, whenever possible, in the physical realm with the actual hardware and software integrated into the SOI in the actual operational environment and operated by the intended users.

In some cases, such a failure of the integrated system to pass system verification and validation could be the result of an interface or operational environment definition issue or a failure to develop a well-formed and complete set of design input requirements against which to verify the system or develop a well-formed and complete integrated set of needs against which to validate the system.

There are cases when it may not be practical in terms of the intended use and actual operational environment to conduct system validation of the physical subsystems, system elements, or the integrated system. However, the reader is cautioned not to substitute validation of the actual hardware and software with the design and early system validation results using models and simulations unless necessary. Doing so adds risk to the project and reduces the confidence level (as compared to system validation against the actual integrated physical system in its actual operational environment when operated by the intended users) and adds risk of the realized physical system failing system validation. As long as the physical system is not completely integrated and/or has not been validated to operate in the actual operational environment by the intended users, no result should be regarded as definitive until the acceptable degree of confidence is realized.

2.3.4.6 Reducing Risk by Addressing System Verification and Validation Early in the Lifecycle

At the end of a project, all project teams want to be able to say they have delivered a winning system that delivers what is needed within budget and schedule and with the desired quality. In this context, all project teams should focus on the risk that could result in a failure to deliver a winning product and failing to meet the stakeholders' real-world expectations as defined by their integrated set of needs and resulting set of design input requirements.

There are several reasons why system verification and validation should be planned early:

- Helps improve the quality of the sets of needs and sets of design input requirements by ensuring they are well-formed having the characteristics defined in the GtWR. Asking, "How will we verify the system to meet this requirement or validate the system to meet this need?" is an effective approach to identify poorly formed needs and requirements.
- Enables the system to be designed to enable system verification and validation activities and to gather the data required to provide objective evidence that the system has met a need or requirement as defined by the *Success Criteria*. If two design options are equally viable, why not choose the one that makes it easier to obtain the data?
- Ensures the additional resources and enabling systems needed to conduct the system verification and validation activities are available when needed (for example, construction of facilities, test stands, support equipment, or other enabling systems). These facilities and equipment cost money and may have to be reserved months in advance (if existing) or, if not existing or need modifications, will take time to define, develop, build, and pass their own system verification and validation activities. Project planning for these resources is key to making sure they are available when needed.
- In some cases, the customer's presence may be required at the verification event. Planning verification early in the system life allows for collaborative scheduling across stakeholders. Internally, the attendance of quality management stakeholders may also be required.
- For enabling systems which also serve as systems within a System-of-Systems (SoS), planning verification in advance also accommodates scheduling of verification activities for other systems within the SoS. This expands the perspective of verification beyond the SOI to the SoS scope.
- Helps ensure all needed resources are included in the project master schedule, budget, and work breakdown structure (WBS). When contracting out the development of the system, it is important to include system verification and validation activities in the supplier's contract. If this is not planned early, it can be expensive to add later.

SE best practices include planning for system verification and validation activities from the beginning of the project and continuing to mature the planning artifacts for system verification and validation activities throughout the lifecycle. Failing to do so can result in risk of massive cost overruns and schedule slips as well as a system that does not meet its requirements and fails to meet its needs.

Given these impacts, all project managers need to mitigate this risk from the beginning of their project. Recognizing the need to mitigate this risk is critical to being able to deliver a winning product. The amount (budget, resources, and time) allocated to system verification and validation activities should be proportional to the project risk associated with the system not meeting a particular requirement or need. In assessing risk, the likelihood and consequences of failure should be considered in terms of project size, complexity, and visibility.

2.3.5 Project and Systems Engineering Management Concepts

2.3.5.1 Importance of an Integrated, Multidiscipline, Collaborative Project Team The concepts discussed in this Manual are enabled by adoption of an integrated multidiscipline, collaborative project team—minimizing the silos between specializations. Project success is achieved when the various disciplines work together collaboratively across the lifecycle, completing the various PM and SE activities and recording the results within an integrated/federated set of data. As shown in Figure 2.16, this team includes stakeholders from all lifecycle stages and associated SMEs who have a role in the activities discussed in this Manual.

The integrated, multidiscipline, and collaborative project team includes stakeholders responsible for SE activities as well as stakeholders responsible for traditional PM activities: defining the WBS and Product Breakdown Structures (PBS), planning, budgeting, scheduling, procurement, contracting, risk management, quality, monitoring, controlling, configuration management (CM), and other activities associated with both project and SE management.

The SE processes, activities, work products, and artifacts are highly interdependent and together form a system that represents the project team members, activities, work products, artifacts, and interactions between team members and external organizations. For the project to be successful, it is critical that team members have a systems-thinking perspective. The interactions between project team members involve the communications and flow of data and information in many forms and media types as well as between external organizations that make up the macrosystem of which the project team is a part.

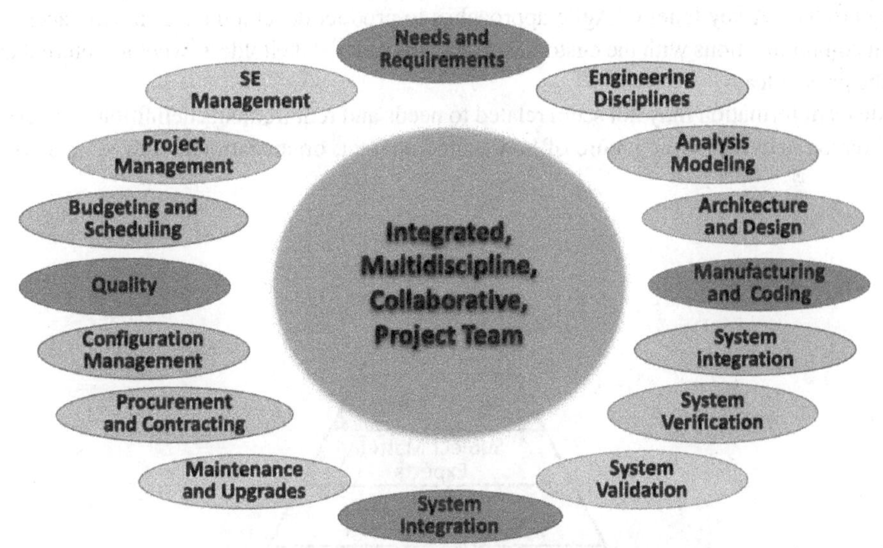

FIGURE 2.16 Integrated, Multidiscipline, Collaborative Project Team.

An advantage of strong collaboration (aka "removing silos") and practicing PM and SE from a data-centric perspective is that it allows the system under development to be managed from the top as an integrated system using an integrated data and information model of the SOI, as opposed to a silo approach where each of the parts is managed independently. Managing the integrated system enables the project team to optimize system performance and use of resources, helps to ensure consistency between artifacts generated across the lifecycle, and enables different activities (such as lifecycle concepts analysis and maturation, needs definition, developing candidate architectures, defining requirements, and design) to take place concurrently as shown in Figure 2.15.

This approach is preferred for its back-and-forth feedback flow between project team members as well as stakeholders, both internal and external to the project. Feedback across the lifecycle is a key consideration as part of the continuous validation activities to ensure stakeholder's real-world expectations will be realized.

Figure 2.17 presents one example of how a project team structure may look in an organization. In this example, the project team would be organized based on roles, levels of responsibilities, and expertise. There would be a core team responsible for the overall project and SE management, planning, monitoring, and controlling. The core team is supported by a PM and Integration Working Group (PM&I WG) whose focus is on the day-to-day PM activities and a SE and Integration Working Group (SE&I WG) whose focus is on the day-to-day SE activities.

While the day-to-day focus of each WG is different, they would have frequent joint status and planning meetings co-chaired by the leads of each WG. The leads of each WG would also be members of the core team. The WGs would be supported by SMEs, who in turn, are supported by other lifecycle stage support personnel. Representatives of each subject matter area of expertise would be a member of the applicable WG.

Internal to the project, the architecture and design teams as well as those stakeholders involved in manufacturing, integration, verification, validation, maintenance, and disposal phases, are key in assessing feasibility of the lifecycle CNR in context of the mission, goals, objectives, measures, drivers and constraints, stakeholder expectations, and risk. Frequent communications between team members as well as with external stakeholders are critical to the project being able to deliver a winning product. A key tenet of Agile approaches to product development is having access to and frequent communications with the customer, users, and other stakeholders, whether internal or external to the project team.

While team formation may not seem related to needs and requirements definition and verification and validation activities—the nature of how teams interact, organizational culture, and where and

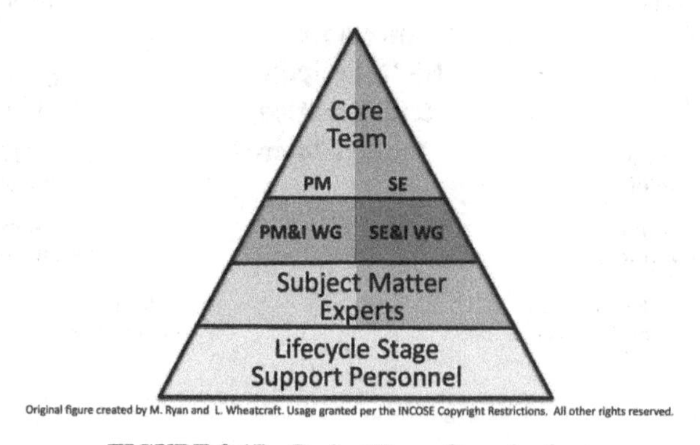

FIGURE 2.17 Project Team Organization.

how to access subject matter expertise *will* determine how easily and effectively PM and SE activities take place throughout the development lifecycle.

2.3.5.2 Importance of Effective Communications While an in-depth discussion of communication theory is outside the scope of this Manual, a brief mention is merited since miscommunication between team members and stakeholders can be disastrous to projects, even to small endeavors.

Fundamentally, the ultimate purpose of defining well-formed needs and requirements is effective communication, which is only successful when the received message or content is the same as that which the originator intended to communicate. Just because communication appears to have taken place does not mean it was successful. To be effective, the sender must use the form and media most appropriate for what is being communicated and the audience to which the message is intended.

In Figure 2.18 below, the goal is for the two boxes labeled "message as intended" and "message as understood," to represent the same message. If there is an intended action associated with the message, the resulting action should also match.

When the messages differ, problems are going to exist such that the system under development may fail either, or both, system verification or fail system validation. The reason for failure is often a "failure to communicate."

When communicating lifecycle CNR, increasingly, the debate is about which means (form and medium) of communication is best. The debate is usually among those who practice traditional SE, those who are adopting the use of language-based models, and those who are following Agile principles. Depending on the specific information item or concept being communicated and the domain, culture, people, and processes within a specific enterprise, one means of communication is often advocated (with a lot of passion in many cases) over the others.

To address the debate concerning the best means of effective communication of lifecycle CNR, architecture, and design, it is helpful to understand the basic communication model shown in Figure 2.18. There is information to be communicated in the form of a message; there is a sender of the message, the message sent (as intended), the form of the message and the media used to

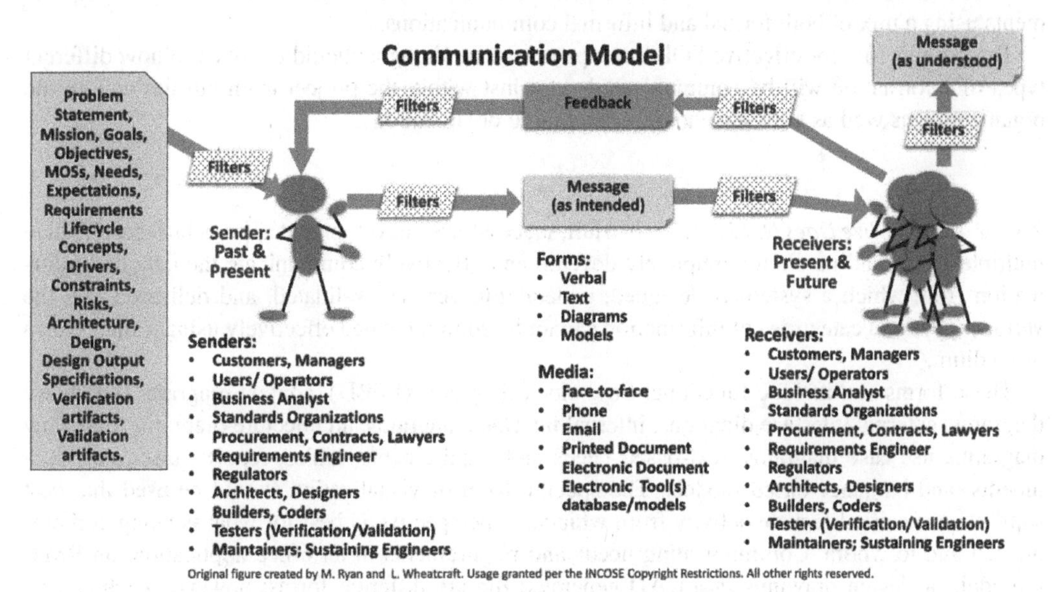

FIGURE 2.18 Communications Model [18, 19].

communicate the message, the message sent, the receiver(s) of the message, feedback, and the message received (as understood).

People often have built-in biases which are used to filter the information sent and received. *Refer to the INCOSE SE HB v5* [1], *Section 1.4, Cognitive Bias, for more details.* Although this information comes from many sources, the sender receives this information through the senders' personal filters and biases. The sender then uses these to encode the information into a message, which is transmitted via some means (form and media) to the intended receiver(s). They, in turn, decode the message via their own personal filter(s) and biases.

The encoding of the message by the sender and the decoding of the message by the receiver(s) is based on their understanding of language, but also training in product development methodologies, processes, tools, culture, domain, education, experience, and work environment of the organization they are employed. An individual may have worked in a given domain (consumer products, government procured systems, and standalone software applications as examples) all their career and may therefore assume that everyone does product development the same way and uses key basic terms in the same way with the same meaning.

If an individual has only worked in an SE environment that is formal, document-based, and uses the traditional serial product development processes, that individual may have a built-in bias to follow that approach to encode and decode the messages. If an individual's experience base is solely related to standalone software application development, they may have a built-in bias to use the approaches and terminology based on that experience. If an individual is used to developing products using Capability Maturity Model Integration (CMMI), Project Management Institute (PMI), or other domain-specific approaches and standards, they might assume everyone else develops products using those same approaches and standards and associated terminology.

In an organization that is implementing SE from a data-centric perspective, including the use of language-based models, and project team is trained in one or more of the modeling languages (for example, UML/SysML), they are likely to encode and decode needs and requirements via various diagrams and visualizations that make up an overall model of the SOI.

In an Agile software environment, the project team is likely to communicate needs and requirements using a mix of both formal and informal communications.

The point is that for effective SOI development, some thought should be given to how different types of information will be communicated—not just within the project team but throughout the organization as well as with stakeholders external to organizations.

2.3.5.2.1 One Size Does Not Fit All In truth, successful SE development approaches must include multiple forms and media to completely develop and effectively communicate the data and information from which a system is designed, coded/built, verified, validated, and delivered. All the various types and categories of information cannot be communicated effectively using a single form or medium.

These forms can include functional flow block diagrams (FFBD), context diagrams, boundary diagrams, external interface diagrams, internal interface diagrams, architecture diagrams, data flow diagrams, use case diagrams, text-based needs and requirements, tables, reports, electronic documents, and language-based models. The specific form of visualization should be used that best supports a specific lifecycle activity from whichever perspective is best for what is being communicated and to whom. Communicating needs and requirements in an office application, an RMT, a model, or design drawings in a CAD-generated file are different forms; however, each can be appropriate depending on the intent of the sender and the intended audience.

To effectively communicate data and information, project managers, business analysts, and systems engineers need to recognize the need to use whichever form and medium is the most appropriate based on both what they are communicating and the audience—it is important to know the audience.

It is important to realize that the responsibility for ensuring that communication has taken place and that the message was transmitted effectively rests with the sender, as only the sender knows the original message and intent. Therefore, it is the responsibility of the sender that the message being communicated is in a form that will be understood as intended by the receiver(s). To communicate project information effectively, the sender must acknowledge the various filters and biases used to encode and decode the message that is being sent such that the sender's meaning is interpreted and understood as intended, no matter the form or medium used.

If the message is needs and requirements that need to be captured and communicated for features and functions that will supply those features, then an FFBD may be the most effective form of communication. If the customer can meet face-to-face frequently with the project team, then user stories, use cases, and agreed-to success, evaluation, and acceptance criteria can be used to communicate the information. However, once the communication takes place, the content needs to be recorded and archived so there is a record of the communication that is retrievable and understandable.

If an agency is formally communicating standards and regulations to present and future developers, a textual set of requirements in a printed or electronic form may be the most effective means of communication. If a customer is developing an RFP to be released to multiple geographically separated potential bidders, then both the technical requirements for the system as well as the SOW, PO, or SA need to be communicated formally as well-formed textual need or requirement statements.

The key is that not one single form and media type will be best for all the types and categories of data and information that must be communicated, recorded, and archived.

Referring to Figure 2.18, identify the content to be communicated (the input), pick one type of sender, formulate the message, decide who the recipient(s) is/are, whether current or future recipient(s) is/are who the message is being communicated to, and then pick the means of communication that is the most effective to communicate the message and will meet the needs of the recipient(s). Following this process will result in different approaches for communicating different types of the inputs listed, the specific message and intent, and the intended recipients. *(Refer also to Chapter 3 for a detailed discussion concerning text-based needs and requirements versus models and diagrams.)*

2.3.5.3 Avoiding Technical Debt Failing to spend time at the beginning of the project to understand business strategic level, operational-level, and system-level needs and requirements, define and mature the lifecycle concepts, and establish completeness, consistency, correctness, and feasibility leads to an accumulation of *technical debt.*

The point of this section is to urge both SE and PM teams to consider program decisions from a risk-based, technical debt avoidance perspective—especially those that are tempted to "cut corners" or implement temporary "band-aid fixes." Management should always be aware of the old saying: "*Pay now or pay later; if later, you will be paying a lot more!.*"

2.3.5.3.1 What is Technical Debt? Technical debt [20] is a metaphor coined by Ward Cunningham, coauthor of the Manifesto for Agile Software Development, to describe what occurs when a project team uses a quick short-term solution that will require additional development work later to meet the needs of the stakeholders. From a project perspective, technical debt refers to the eventual consequences of poor SE and PM practices.

Technical debt is closely related to project and technical risk—not performing or delaying key activities early in the development lifecycle adds risk to the project due to the consequences later in the lifecycle of not doing those activities when they should have been done and addressing the consequences of "kicking the can down the road."

A key part of the concept of technical debt is that, like financial debt, it must be repaid at some point in the future, along with the accumulated interest. Moreover, like financial debt, the interest on technical debt compounds over time. From a risk perspective, if risk is accumulated—even for seemingly valid reasons—the probability of the risk becoming an issue and cascading into worse issues becomes high, making delivery of a winning product less likely.

This interest represents the increased cost and time along with cost and time associated with rework that could have been avoided, or at least minimized, if the work was done at an earlier point in the program when the cost and schedule impacts of change are less before hardware is built and software coded. This increase in cost across the lifecycle is illustrated in the INCOSE SE HB v5 [1], Section 1.2, Figure 1.4, which shows how the lifecycle cost and defect cost increase over time. These increased costs represent technical debt.

Also, much like actual debt, there is potential that accumulated debt and interest grow to the point where the contracting organization defaults on their "loan." This is represented by the potential for the supplier to develop a system, which does not meet stakeholder or customer needs. To mitigate that technical debt, contracting organizations may attempt to reconcile their debt in the form of renegotiating customer needs and system requirements. The worst-case scenario is default/bankruptcy, where the supplier receives a stop work order, and the creditor loses their investment.

The analogy continues with the concept of credibility. A default indicates a company or program lacks the capacity to manage the work and is likely to incur technical debt in future work. Therefore, their credibility is impacted due to poor performance and inability to manage technical risk and pay back technical debt.

2.3.5.3.2 Why Does Technical Debt Accumulate? Major reasons for technical debt accumulation include:

- The project team does not always have a complete understanding of the big picture and what is necessary for acceptance at the outset of the project.
- The project team may be provided with, or go directly to defining, a set of requirements without first defining and maturing lifecycle concepts, establishing feasibility, and defining an integrated set of needs.
- Developers providing a proposed solution (design) without first clearly understanding the problem, stakeholder needs, mission, goals, objectives, drivers and constraints, risk, defining and maturing lifecycle concepts, establishing feasibility, and defining an integrated set of needs and transforming them into a set of well-formed design input requirements.
- Leadership of projects that are cost-constrained or schedule-constrained may be tempted to reduce verification and validation activities across the lifecycle, which can make up a sizable portion of the overall development budget, without clearly understanding the big picture, associated risks, and what is necessary for acceptance before beginning design.

In each case, the result is a large amount of technical debt resulting in cost and schedule overruns and, ultimately, project failure.

2.3.5.3.3 Technical Debt and Risk Technical debt represents risk that the project will not be able to deliver a winning product. The project team needs to take steps to prevent technical debt and avoid the high-interest consequences later in the project.

For customers that will outsource the development of an SOI to a supplier, failing to complete the work necessary to develop well-formed sets of system requirements, SOWs, POs, and SAs accumulate technical debt in the form of expensive contract changes.

Toward the end of the project, a high amount of technical debt puts a project at risk of failure of system verification and system validation and exposes the project to the risk of potential costly rework to meet the needs and requirements.

Understanding the concepts and performing the activities advocated in this Manual will help projects avoid technical dept and resulting risk. While the activities discussed in this Manual may lead some readers to feel that the number of activities is overwhelming, they need to understand that failing to undertake these activities can result in the accumulation of technical debt that, unfortunately, is all too often repaid in the form of a failed project and associated consequences to both the enterprise and the project team.

As stated previously, the intent of this Manual is not that organizations adopt all the activities discussed, but rather adopt and tailor the activities based on the value to the organization that will reduce technical debt and associated risk and thus increase their ability to deliver winning products and stay in business.

2.3.5.4 Greenfield Versus Brownfield Systems When developing a system, it is important to understand the concepts of "greenfield" and "brownfield" systems.

- *Greenfield* system development is when a new system is being developed where there is not an adequate predecessor system. There may be other similar systems, but the organization or customer has decided to start with a "blank piece of paper." For example, an organization is building a new medical diagnostic device. There may be other similar devices in the market, but the organization is implementing an innovative approach or technology in this new device.
- *Brownfield* systems, on the other hand, involve legacy or heritage systems where there is an existing predecessor system that can be evolved or transformed into the desired system. This is often the case in consumer products, where the organization periodically produces revisions or updates to existing products, new product versions for the upcoming year, or derivatives of existing products.

There can also be a combination. A medical device company may have a general-purpose diagnostic instrument that can support multiple types of inputs and analyses. In this case, the instrument is existing (brownfield), but the biological sample, assay, and analysis software are new (greenfield) projects. Or a new medical device may have the same hardware and functionality but use different software to improve performance and quality.

One of the major advantages of adopting a data-centric approach to SE as described in Chapter 3, is that there will be a data and information model of the current "as-is" brownfield system. With this data and information model, the organization can identify what changes need to be made to the existing data and information model that will result in the desired future state/to-be system.

Changes to the needs will result in an updated set of needs. Changes to this set of needs will identify which of the system's technical design input requirements will need to be changed. Traces from these requirements to the design artifacts will indicate what design changes will need to be made. This information will also help to develop a more accurate budget and schedule as well as decide what needs to be verified and validated and what does not (existing data may be sufficient). This capability can result in an organization being able to release an updated system in a much shorter time at a reduced development cost.

An important point is that the concepts and activities discussed in this Manual need to be tailored, as appropriate, depending on whether the system to be developed is brownfield or greenfield. If brownfield, the focus is on which activities are needed to define the "deltas" between the predecessor and the system to be developed. If greenfield, much more rigor is needed for all activities across the lifecycle.

Also refer to the INCOSE SE HB [1] *Sections 4.3.1 and 4.3.2 for additional information concerning greenfield and brownfield systems.*

2.3.6 Overview of System Development Activities Covered in This Manual

Figure 2.19 provides an overview of the system development activities discussed in this Manual. This section uses this figure to provide the reader with an overview of these activities across the lifecycle.

At the beginning of any project, stakeholders have real-world expectations for a SOI that will address their problems, opportunities, or threats.

At the strategic and operational levels of the organization, a problem/opportunity is identified that a project team is to address, along with the definition of lifecycle concepts, stakeholder needs, and stakeholder requirements appropriate to those levels of the organization.

As discussed in Chapter 4, at the system level, the project team responsible for the development of the SOI elaborates on these higher-level lifecycle CNR, identifies the relevant system-level stakeholders, elicits system-level needs and requirements, identifies drivers and constraints, and identifies risk. This information is used to define, analyze, and mature a set of lifecycle concepts that will address the problem or opportunity, stakeholder needs, and stakeholder requirements within the defined drivers and constraints.

From this set of lifecycle concepts, an integrated set of needs for the SOI is derived. This set of needs is verified and validated (Chapter 5) and baselined as part of a gate review; for example, Scope Review (SR) or Concept Review (CR) (Chapter 14); the baselined integrated set of needs represents the scope of the project.

These baselined integrated sets of needs are then transformed into the set of design input requirements (Chapter 6) go through requirement verification and requirement validation (Chapter 7) and are baselined as part of a gate review; for example, Systems Requirements Review (SRR). The resulting design input requirements represent the system "functional" baseline.

At this point, the organization can decide to either buy, make, or reuse the SOI. If build/code, they will proceed with defining subsystems and system elements at the next level of the system physical architecture. If they make a buy decision (for example, to contract out the development of the system), the supplier will proceed with defining the subsystems and system elements at the next level of the system physical architecture.

Based on the analytical/behavioral and functional architecture models and preliminary architecture developed concurrently with the lifecycle CNR, the architecture is refined via the *Architectural Definition Process,* and the system-level design input requirements are allocated to subsystems and system elements at the next lower level of the physical architecture (discussed in Chapter 6). These allocated requirements represent drivers to the subsystems and system elements at the next level of the architecture. For each subsystem and system element, the lifecycle concepts analysis and maturation, needs analysis/definition, design input requirements analysis/definition, and allocation cycle are repeated as the developing organization moves down the levels of the architecture.

These activities repeat until the developing organization can decide to either buy, make, or reuse each subsystem and system element. This is not a serial, one-way process; rather it is iterative and

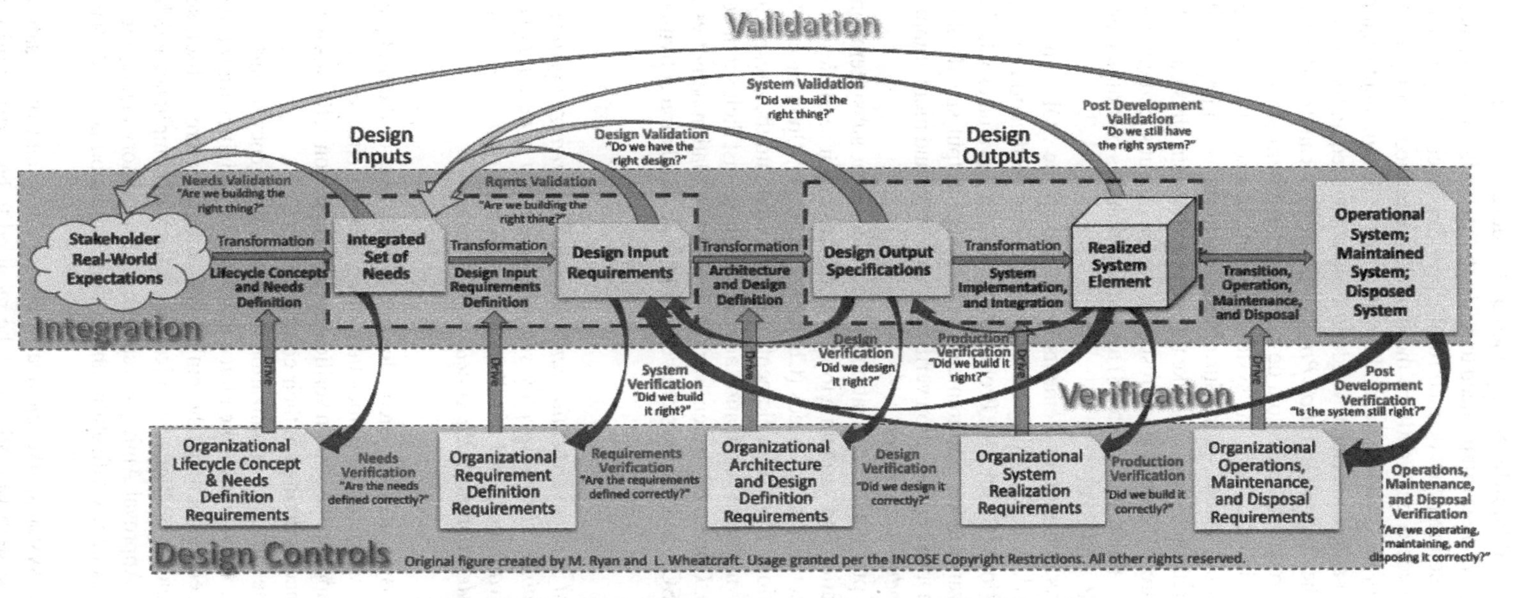

FIGURE 2.19 System Development Activities Across the Lifecycle.

recursive. As more detailed knowledge is gained as the developing organization moves down the levels of the architecture, there is often a need to refine the lifecycle concepts, integrated set of needs, and set of design input requirements defined for the previous level. Using a model-based approach, these refinements are made concurrently as the models are refined as the developing organization moves down the levels of the architecture, and updated sets of design input requirements go through requirement verification and requirement validation and are subsequently baselined.

In accordance with the *Design Definition Process*, the design team transforms the allocated baseline set of design input requirements into a design for each system element and communicates the design to those that will build or code the system element via a set of design output specifications. Realizing the design output specifications during manufacturing or coding results in the creation of the subsystems and system elements that make up integrated system physical architecture (hardware, mechanical, and software). The design and resulting design output specifications go through design verification and design validation (Chapter 8) and are baselined as part of one or more gate reviews; for example, SDR, PDR, and CDR.

Once the subsystems and system elements that make up the SOI architecture are developed (procured, manufactured, or coded), they undergo production verification to verify they meet their design output specifications (Chapter 9). Production verification evaluates whether a resultant product (hardware or software) meets requirements—the design output specifications in this case. In some domains, performance of production verification for larger and more complex system elements is sometimes called *product acceptance*.

Upon completion of production verification and acceptance, the subsystems and system elements undergo system verification and validation to verify they meet their set of design input requirements and to validate they meet their set of needs (Chapters 10 and 11). Upon successful system verification and system validation, the system elements are integrated into their respective systems, which then go through their own system verification and validation against their set of design input requirements and set of needs. Once their system verification and validation activities are complete, they are integrated into the SOI, and then system verification and validation activities are performed on the integrated system.

Note in Figure 2.19 *that system verification and validation of a realized system, subsystem, system element, or SOI is "against" their respective set of needs and set of design input requirements that were defined, verified, validated, and baselined as design inputs for each system, subsystem, and system element at each level of the SOI physical architecture.*

For each system, subsystem, and system element, at each level of the SOI physical architecture, system verification and validation activities take place. For systems, subsystems, and system elements developed by a supplier, the system verification and validation activities for each are completed as part of their acceptance as defined in their contract.

It is important that the customer organization (or regulatory agency) clearly define both their role and the developer's role in system verification and validation activities for the system as well as all subsystems and system elements in the system's physical architecture as discussed in Chapter 13.

At the end of the development lifecycle, objective evidence of compliance will be documented that can be evaluated by the Approval Authority (customer or regulatory agency) to determine whether the system integration, system verification, and system validation activities have been completed successfully.

For highly regulated systems, like medical, transportation, and other safety-critical systems, system verification and system validation may be required to be certified prior to a governmental regulatory agency authorizing the system to be released for use. Following the customer or regulatory agency evaluation, qualification, and certification, the system can be accepted, and ownership transferred to the customer, or released into the marketplace.

This Chapter has provided much of the foundation for the concepts and activities to be presented in the rest of this Manual. The next Chapter presents a discussion of information-based requirements development, completing the foundation before each of the various activities highlighted in this Chapter will be described in more depth.

3

INFORMATION-BASED NEEDS AND REQUIREMENT DEVELOPMENT AND MANAGEMENT

Today's system development environment presents many key challenges because of increases in:

- Complexity.
- The role software has in the system architecture (software-intensive systems are the norm).
- Dependencies and number of interactions among parts of the system.
- The interactions between a system and the macro system it is a part.
- The number of threats across interface boundaries and vulnerabilities to those threats.
- Dependencies between PM and SE.
- Dependencies among development lifecycle process activities and artifacts.
- Oversight.
- Competition.
- The pressure (and need) to reduce development time and time to market.
- Risks (program/project, development, manufacturing, integration, system verification, system validation, and operational).
- The number of projects that are over budget and experiencing schedule slippage.

To address these challenges and successfully develop increasingly complex software-intensive systems, PM and SE practitioners must move to a data-centric approach for system development using a combination of text-based and model-based communications, depending on what is being communicated and to whom.

Using a data-centric approach as discussed within the INCOSE RWG whitepaper *Integrated Data as a Foundation of Systems Engineering* [6, 21], systems engineering is practiced from the perspective that system-level needs and requirements, along with all SE artifacts (such as models, designs, design output specifications, documents, diagrams, or drawings) generated during the performance

INCOSE Needs and Requirements Manual: Needs, Requirements, Verification, Validation Across the Lifecycle, First Edition. Louis S. Wheatcraft, Michael J. Ryan, and Tami Edner Katz.

of system lifecycle process activities, are visualizations of the underlying integrated/federated data and information model of the System of Interest (SOI).

Similarly, PM should be practiced from a data-centric perspective that recognizes the work products developed (such as plans, budgets, schedules, WBS, PBS, or contracts) generated in the performance of all PM phases are also visualizations represented by underlying sets of data and information.

PM and SE need to recognize that both the PM and SE activities and resulting work products and artifacts are interrelated and highly dependent. There are risks, costs, schedules, activities, and resources associated with each SE artifact generated across all system lifecycle stages that must be planned for, monitored, and controlled.

In a document-centric approach to SE, it is difficult to identify an Authoritative Source of Truth (ASoT) that is accurate, consistent, current, and trustworthy. As the project progresses through the development lifecycle activities and changes to the baseline, the various PM and SE documents and artifacts often become out-of-date and lose synchronization. In a data-centric approach to SE, with the ASoT maintained within the SOI's integrated/federated dataset, changes will automatically be reflected in artifacts across all lifecycle stages, minimizing this risk; resulting in an ASoT that is accurate, consistent, current, and trustworthy.

Note: Some larger organizations with multiple systems are using the concept "Federated Source of Truth" (FSoT) instead of ASoT at the Operational level discussed in Chapter 2. In this context, while each individual system may have its own ASoT, there may be inconsistencies between systems. An FSoT enables the organization to ensure all systems have access to the same data and information, which helps to eliminate inconsistencies and errors that can arise when different systems use different data and information sources. With an FSoT, data and information are indexed, retrieved, and presented in a user interface layer where it is created and can be accessed by all stakeholders. This approach is used to ensure that the information used by different systems is consistent and current. The concept of FSoT would apply also when forming a System of Systems (SoS). Refer to the INCOSE SE HB v5 [1] Section 4.3.6 for additional information concerning SoS.

3.1 INFORMATION-BASED NEEDS AND REQUIREMENTS DEFINITION AND MANAGEMENT

The approach described in this Manual is referred to as Information-based Needs and Requirements Definition and Management (I-NRDM) [22, 23]. Using the I-NRDM approach, the focus of NRDM activities is on the design inputs as shown in Figure 3.1, defining the system-level integrated set of needs and transforming these needs into a well-formed set of design input requirements, which are linked to the functional and logical/behavioral models and physical architecture.

This I-NRDM approach is a practical implementation of PM and SE from a data-centric perspective as defined in the INCOSE RWG whitepaper *Integrated Data as a Foundation of Systems Engineering*:

"*SE, from a data-centric perspective, involves the formalized application of shareable sets of data to represent the SE work products and underlying data and information generated to support concept maturation, needs and requirements development, design, analysis, integration, system verification, and validation activities throughout the system lifecycle, from conceptual design to retirement.*"

Some organizations practicing Model-Based Systems Engineering (MBSE) are actually focused on model-based design (MBD), as shown in Figure 3.1. For this effort, the focus is on transforming

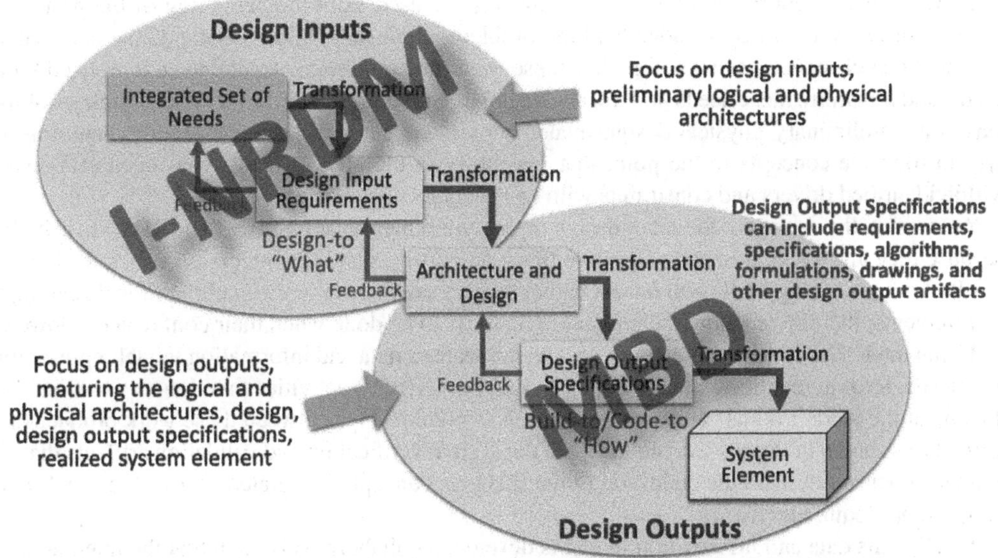

FIGURE 3.1 I-NRDM + MBD = MBSE.

design inputs into design outputs. In this context, MBD activities often begin with an already established and baselined set of design input requirements. This approach is common, especially in cases when a customer outsources the design, manufacturing, and coding of the SOI to a supplier.

The MBD team develops diagrams and models to analyze the set of design input requirements received from the customer (internal or external). As part of this analysis, the MBD team iteratively and recursively develops functional architecture and analytical/behavioral models of the SOI per the activities included in the *Architectural Definition Process* and *Design Definition Process* (described in the ISO/IEC/IEEE 15288). Based on this analysis, the MBD team addresses and corrects any defects that may exist within the requirements set. For systems being outsourced to a supplier, at this point in the development, these corrections can result in expensive contract changes.

With approval of the customer, the MBD team then transforms the resulting functional architecture and analytical/behavioral models into a design and physical architecture, which is communicated via a set of design output specifications for each of the system elements that are part of the physical architecture. The design output specifications are provided to the organization (internal or external) responsible for manufacturing/coding each of the system elements.

A major issue with the MBD approach is that the MBD team often works in a silo separate from the lifecycle Concepts, Needs, and Requirements (CNR) definition activities. As a result, they may lack the knowledge of, or access to, the underlying analysis that resulted in the definition of the system-level lifecycle concepts and integrated set of needs from which the set of design input requirements was transformed. Without this knowledge and understanding, the MBD team is at risk of delivering a system that can be verified to meet the set of customer-supplied design input requirements but fails system validation (fails to meet the integrated set of needs) and is not accepted for use by the customer or regulatory agency.

As shown in Figure 3.1, if the concept of I-NRDM (design inputs) is combined with the concept of MBD (design outputs), i.e., I-NRDM + MBD, the result is the real intent of MBSE. Adopting this perspective will help organizations move closer to INCOSE's Vision 2035 [24].

In the I-NRDM approach, diagrams and models are used from the beginning of the project as a means of analysis to better understand the problem, elicit stakeholder needs, define and mature feasible lifecycle concepts that will realize these needs and requirements, define an integrated set of needs, and transform these needs into a set of design input requirements for the SOI. To help establish feasibility, preliminary physical design-related activities are conducted concurrently, maturing the system lifecycle concepts to the point that feasibility in the physical realm has been established within identified drivers and constraints with an acceptable level of risk.

As a result, the quality of the set of design input requirements supplied to the MBD team is higher because they are based on mature, feasible lifecycle concepts and the integrated set of needs. As a result, those practicing MBD will have a higher level of confidence in the quality of the design input requirements, thereby reducing the amount of analysis to be done when their confidence is lower.

Using the I-NRDM approach, the project team creates a data and information model representing the Needs, Requirements, Verification, and Validation (NRVV) activities that define, establish, and document the work products and artifacts as well as relationships between these work products and artifacts, as shown in Figure 3.2. Included are the system verification and validation artifacts that are defined concurrently with the definition of the lifecycle concepts, integrated set of needs, and set of design input requirements.

Ideally, this data and information model is developed collaboratively such that the integrated set of needs, design input requirements, functional architectural and logical/behavioral models, preliminary physical architecture, and PM work products are developed concurrently.

When the development of the SOI is contracted out to a supplier, applicable parts of the data and information model developed by the customer project team as part of the lifecycle CNR definition activities can be shared with the supplier's project team. If the toolset used by both the customer and supplier has the capability of sharing data, the supplier can import the needs, requirements, functional architectural and logical/behavioral models, and the preliminary physical architecture directly into their toolset. *Refer to Chapter 16 for a detailed discussion on features the project SE toolset should have.*

In either case, the organization responsible for the design of the SOI does not have to start the design activities from scratch. Instead, they can focus their efforts on maturing and finalizing the functional architectural and logical/behavioral models and preliminary physical architecture that were defined during the lifecycle CNR definition activities, significantly reducing the time and cost of design.

The advantage of practicing PM and SE based on the I-NRDM approach is that their work products and artifacts developed across all system development lifecycle stages will be more consistent, complete, and correct. With this concurrent approach, the time to define feasible system lifecycle concepts, baseline an integrated set of needs representing those concepts, and transform these needs into an agreed-to set of design input requirements will be shortened.

Since correctness, consistency, completeness, and feasibility will have been addressed prior to issuing the contract, the risk of expensive contract changes and schedule slips will be reduced. This will help PM and SE practitioners develop winning products that deliver what is needed, within cost and schedule, with the desired level of quality.

3.1.1 Key Characteristics of the I-NRDM Approach

The I-NRDM approach has several key characteristics, as described in the following sections.

3.1.1.1 Implementing Development Lifecycle Phase Activities Concurrently Rather Than Serially Rather than a serial approach to performing lifecycle activities, using the I-NRDM

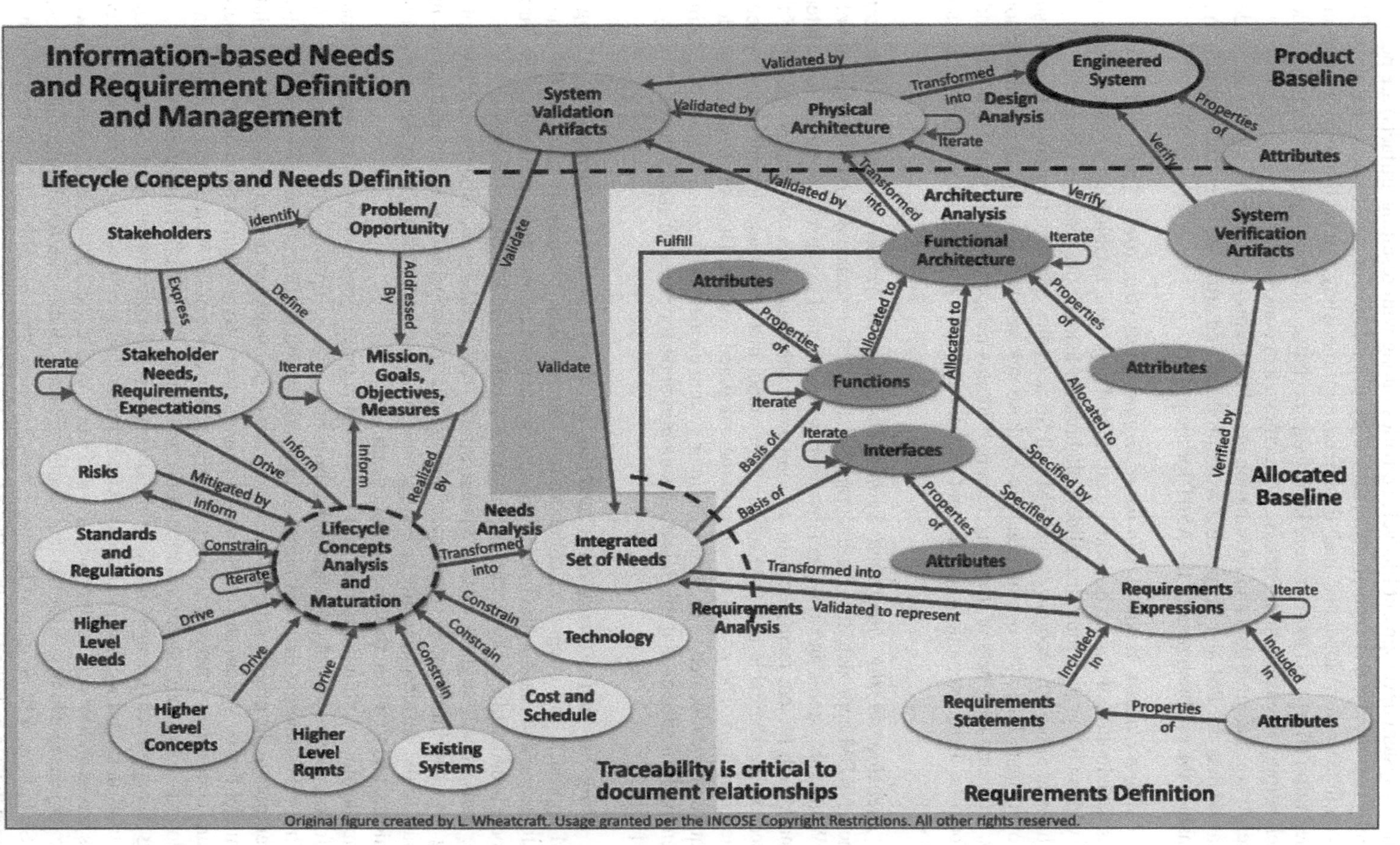

FIGURE 3.2 Information-based Requirement Development and Management Model. Source: Adapted from Ryan et al. [12].

approach, the system development lifecycle activities (along with resulting artifacts) are performed concurrently, iteratively, and recursively, enabled by the establishment of connected data. These activities include the functional and analytical/behavioral modeling and preliminary physical architecture definition efforts. As more mature knowledge becomes available, updates can be included in the model, helping to ensure completeness, correctness, and consistency.

The result is a shareable data and information model that represents both the system under development as well as the PM and SE process work products and artifacts. This data and information model enables the establishment of the ASoT for the project.

3.1.1.2 Using a Holistic Versus A Silo-based Organization Structure The I-NRDM approach advocates the adoption of a more holistic organizational approach, establishing a collaborative environment with an integrated, multidiscipline, collaborative team as discussed in Chapter 2.

This holistic approach recognizes that PM and SE activities and resulting artifacts are systems in their own right. Thus, the overall behavior of these systems is a function of the interaction of the parts, including members of the project team. With this perspective, the practice of PM and SE should focus on these interactions and not put each lifecycle activity and the resulting artifacts into separate silos.

3.1.1.3 Focusing on the Integrated System Optimization Versus Subsystem Optimization Using the I-NRDM approach, it is the integrated system that is managed from the beginning of the project. While subsystems and system elements may be assigned to separate organizations (internal or external), each project team needs to develop a shareable data and information model of their SOI based on the shareable data and information model that represents the integrated system of which they are a part. The goal is to optimize the integrated system. Thus, the system-level design input requirements allocated to each subsystem or system element will contain values that optimize the system performance even if this results in suboptimal performance of each subsystem or system element [25].

3.1.1.4 Using SE Tools Having the Capability to Share Data and Information Using the I-NRDM approach, the integrated, collaborative, and multidiscipline team will use PM and SE toolsets to define and manage the underlying sets of data and information representing the various work products and artifacts developed as part of the development lifecycle processes.

These toolsets will support various visualizations (tables, diagrams, models, and text) of the resulting data and information. Team members use the tool capabilities and visualizations most appropriate to the development lifecycle stage activities they are working, what they are trying to communicate, and to whom they are communicating.

The applications within the toolset will use a schema that is consistent with the project's master schema for organizing and storing data and information while complying with interoperability standards. This will allow the data and information model resulting from these activities to be shared with other tools in the project's digital ecosystem and to be integrated into a common data and information model for the system and its constituent components. By using a tool that uses a compatible schema and conforms to interoperability standards, those responsible for architecture and design will be able to import the data and information generated during the I-NRDM activities.

Using this approach, changes made to the data and information in one tool can be propagated to other tools in the project's digital ecosystem, helping ensure consistency, correctness, and completeness across all lifecycle activity artifacts and work products and maintain the ASoT. This would allow the project team to have a toolset that would support both the traditional capabilities of existing PM and SE tools as well as support language-based models allowing artifacts generated within special purpose tools to be linked together. *Refer to Chapter 16 for a detailed discussion of the desirable features of the project SE toolset.*

3.1.1.5 Establishing Well-Formed Sets of Needs and Design Input Requirements with an Underlying Data and Information Model Because all artifacts are represented by a sharable data and information model, correctness, completeness, consistency, and trustworthiness should be achievable with more ease than in a non-data-centric approach. The design input requirements resulting from use of a data and information model can readily be assessed to show they have the characteristics of a well-formed set of requirements, as defined in the Guide for Writing Requirements (GtWR). Using this approach mitigates time-consuming and expensive rework to discover and correct a defective set of needs or set of design input requirements by those responsible for architecture and design definition.

3.1.1.6 Defining a Project Ontology and Master Database Schema at the Beginning of the Project Using the I-NRDM approach, an ontology and master schema is developed from the beginning of the project resulting in both consistent use of terms [26, 27] within the needs and requirements sets as well as all other artifacts represented by the data and information model. Having a defined ontology and master schema is critical to the sharing of data between SE tools. This is enabled by use of a digital ecosystem that uses tools where the tool supplier schemas are consistent with the project's ontology and master schema and the tool supplier fully supports interoperability standards. *Refer to Chapter 16 for a detailed discussion of the desirable features of the project SE toolset.*

3.1.1.7 Capturing Dependencies and Relationships A key benefit of the data-centric approach is that traceability between all the artifacts, including dependencies and relationships, is captured as part of the data and information model. This means that the data established among the lifecycle CNR is traceable to each other, to the stakeholders, and to the elements in the architecture.

3.1.1.8 Using Tools That Allow Requirement Data to Be Referenced and Modeled Within the needs and design input requirement statements, meaningful terms and entities may be referenced, such as parts of the system architectural components, states, conditions, functions, data elements, interfaces, and other entities drawn from an integrated system data and information model. These entities typically appear in other modeling contexts, such as functional block diagrams, context diagrams, boundary diagrams, external interface diagrams, and state-transition diagrams, to name a few [11]. With this capability, completeness, consistency, and correctness are much easier to assess and manage. From an NRVV perspective, this gives the project team a powerful capability. In addition, change impact assessment is a built-in capability no matter where a change occurs in the development lifecycle.

3.1.1.9 Using RMT and Modeling Tools That Help Ensure Need and Requirement Statements are Well-Formed Using the I-NRDM approach, the SE tools used will have the capability to define and maintain the project ontology as well as include the structured, natural language processing/artificial intelligence (NLP/AI) capability [27] to aid the authors in writing need and requirement statements that have the characteristics defined in the GtWR, enabling ability to assess the overall quality of the set of needs and set of design input requirements. NLP/AI tools can also be used to ensure consistency in the use of terms across all lifecycle artifacts based on the project's official ontology.

These capabilities improve not only the quality of the need and requirement statements but also the quality of all the artifacts represented by the data and information model. With this capability, needs and requirement verification (i.e., verification that the need and requirement statements have the characteristics of well-formed needs and requirements and follow the organization's rules for writing text-based needs and requirements) is enhanced enabling the definition of higher-quality needs and design input requirement statements within the model.

3.1.1.10 PM Work Products are Developed Concurrently with the Development of the SE Artifacts Using the I-NRDM approach, PM work products will be consistent with the SE artifacts, resulting in budgets and schedules that have a higher degree of confidence. Another benefit is that concurrent PM and SE processes can be integrated, such as Configuration Management (CM); risk management; problem definition; definition of the project mission, goals, and objects; identifying drivers and constraints; and lifecycle concepts definition, analysis, and maturation.

3.1.2 Implementing the I-NRDM Approach

To successfully implement the proposed I-NRDM approach, the following actions must have been completed prior to the start of the project [12].

- There is an enterprise-level "champion" advocating the use of the I-NRDM approach.
- Senior management has agreed to implement SE from a data-centric perspective.
- Data governance and information management policies have been defined and are enforced.
- The level of data-centric SE capability consistent with the needs of the project has been agreed to.
- An IT infrastructure has been put into place that meets the needs of the enterprise and projects.
- A PM and SE toolset consistent with the needs of the projects has been procured, and licenses put in place that support both I-NRDM and MBD.
- The project has a defined ontology and master schema for the SOI's integrated dataset.
- An integrated, collaborative, multidiscipline PM and SE team has been formed.
- The project team members are trained in practicing PM and SE from a data-centric perspective, using the proposed I-NRDM approach, the PM and SE tools, sharing of data and information between tools, defined schema, plans, processes, procedures, and work instructions.

Organizations that have achieved SE Capability Level (SCL) 3 as defined in the INCOSE RWG whitepaper *Integrated Data as a Foundation of Systems Engineering* will have addressed the abovementioned actions as part of their normal practice of SE from a data-centric perspective.

3.2 EXPRESSION OF TEXT-BASED NEEDS AND REQUIREMENTS WITHIN RMTS VERSUS MODELS AND DIAGRAMS

The I-NRDM approach is based on the concept of duality [28] as applied to the text-based expression of needs and requirements. Depending on what is being done and what is being communicated, text-based needs and requirements expressed within a Requirement Management Tool (RMT) or expressed within diagrams and models are two sides of the same SE coin. Neither is solely sufficient—both are needed. The following provides some considerations when establishing the means of expression of needs and requirements for an SOI development effort.

A major critique of text-based need and requirement statements is the inherent ambiguity in the use of an unstructured, natural language. To help avoid this ambiguity, the language advocated in the GtWR is a structured, natural language defined by a set of rules. Performing the activities defined in this Manual, and following the rules defined in the GtWR, will result in individual and sets of needs and requirements that have the characteristics defined in the GtWR. In this Manual, when the phrase "well-formed" is applied to text-based need and requirement statements and sets of needs and requirements, it refers to need and requirement statements that are written using a structured, natural language having the characteristics defined in the GtWR no matter whether they are expressed within an RMT or a diagram or model.

3.2.1 Expression of Text-based Needs and Requirements Within RMTs

For many ideas and concepts that are to be communicated, well-formed, text-based needs and requirements expressed within an RMT have proven to be a more effective form of communication (particularly in formal contracting) as compared to their expression within a diagram or model. The strengths of expressing requirements within an RMT include:

3.2.1.1 Communication and Comprehension Despite the increased emphasis on the use of language-based models in systems engineering, there remains a sizable audience who cannot interpret, do not understand, or who are not willing to work with diagrammatic or model representations of needs or requirement statements, especially when the formal technical aspects of such representations are not intuitively obvious to the reader.

Some managers, customers, regulators, and users of the system or other nontechnical stakeholders may not have been trained in language-based models or find the terminology used in some diagrams and models confusing and nonintuitive, making them difficult to understand. Even engineers who create visual models (such as System Modeling Language (SysML) [29]) need to be trained on the tool and technique, which may not be a trivial expense.

Forcing stakeholders to learn a specific detailed technical language to describe their needs and requirements may well result in them losing interest in the critical definition activities. Consequently, diagrammatic or model representation of needs and requirements must be supported by well-formed textual statements and descriptions for the representations to be understood unambiguously by all stakeholders.

The combination of well-formed need and requirement statements within an RMT along with visual models can be extremely powerful during analysis and in communicating abstract or hard-to-describe concepts and features. The use of both is highly encouraged.

More than anything else, text-based needs and requirements are a form of communication no matter where they are expressed. As such, it is vital that the intended message is clearly and unambiguously communicated to those for whom the message is intended over time as discussed in Chapter 2.

Even when stakeholders are willing to spend the time to learn modeling languages (such as SysML), these SE modeling tools have limited capabilities of displaying or exporting requirements in a tabular or document form as compared to the more robust capabilities in many modern RMTs.

Being able to provide text-based needs and requirements in an electronic document format (such as pdf, or common office application formats) allows the stakeholders to view the needs and requirements in common office applications that have been installed on their computers without further training or expense. In addition, there are stakeholders who still prefer, and demand, printed, text-based documents, and will continue to do so for the foreseeable future.

This concept supports the GtWR characteristics: C3—Unambiguous, C7—Verifiable, C13—Comprehensible, C14—Able to be Validated.

3.2.1.2 Power of Expression There is a wide variety of types of needs and requirements that must be expressed. Use cases, user stories, scenarios, diagrams, and models tend to focus on the functional architecture and behaviors expressing functions, performance, and interactions. However, these forms of expression are not presently well suited to expressing nonfunctional needs and requirements that deal with the physical system elements associated with quality (-ilities), regulations, standards, environments, and physical characteristics. Expressing well-formed needs and requirements within an RMT enable the expression for all types of needs and requirements outside of physical "build-to-print" drawings and other design representations included in the set of design output specifications.

Other points that illustrate the enduring importance of well-formed text-based need and requirement statements expressed within an RMT include:

1. **Problem statements, operational scenarios, use cases, and user stories should be written from the perspective of the user's (actor's) interaction with other actors in the context of the system under development**. In contrast, many models describe the system from the developer perspective of what the system under development must do in order for the users to interact with the SOI in the way they expect. While those forms of expression are excellent conceptual tools for stakeholder expectation and needs elicitation and management, they do not always effectively replace well-formed, text-based needs and requirements expressed within an RMT for the various ideas and concepts that must be communicated, especially what is sometimes referred to as nonfunctional needs and requirements as discussed above.
2. **Use cases, diagrams, models, and other alternate forms may not be able to communicate stakeholder needs as effectively as can be done using a text-based, structured language expressed within an RMT that can be clearly understood by all stakeholders *over time*.** While these alternate forms are useful, they are not sufficient by themselves without also defining a resultant set of text-based need and requirement statements.
3. **Often the model-based entity referred to as a "requirement" is not expressed as a well-formed requirement statement within a model,** rather the requirement is expressed as a short phrase, e.g., "receive data", "send data", or "acceleration = 200" resulting in requirements that do not have the characteristics defined within the GtWR.

This concept supports the characteristics defined in the GtWR: C3—Unambiguous, C4—Complete, C7—Verifiable, C13—Comprehensible, C14—Able to be Validated.

3.2.1.3 Managing Sets of Needs and Requirements Text-based needs and requirements expressed within an RMT lend themselves to the presentation of large numbers of different types of needs and requirements in an easily digestible form. In contrast, SysML [29] requirement diagrams can present individual requirement statements but are not well suited to representing multiple or large sets of requirements associated with all the parts of the system architecture.

Note: While most SysML modeling tools allow the entity type "requirements" to be exported in the form of a table displayable in common office applications, the features to produce a document or table visualization of sets of requirements are limited as compared to most of the (RMTs available in the market.

Note: Tool vendors are now providing tools that allow the linking of requirements contained within language-based models with the same requirements defined and managed within a RMT. This enables the requirements to remain consistent between tools and allows practitioners to view needs and requirements in whichever form is needed for what they are doing. This also enables the requirements contained within the language-based models to be well-formed.

Note: Currently, SysML does not include an entity type "needs," needs relationships, nor corresponding needs diagrams.

This concept supports the GtWR characteristics: C4—Complete, C13—Comprehensible.

3.2.1.4 Use of Attributes Both the need and requirement expressions include a set of attributes. Attributes are a powerful tools to produce meaningful reports and dashboards for use by management stakeholders.

Modeling tools do allow users to define an entity having the name *attribute* and link that entity to a need or requirement statement, however, few practitioners do so, especially when there are multiple attributes that the project team has decided to use and define.

While several modeling tools are making progress in enabling users to define a number of attributes for each need or requirement statement, RMT expression of attributes are often more straightforward. In addition, RMTs enable attributes to be defined and managed for not only needs and requirements, but any entity contained within the application.

The use of attributes within an RMT also enables dashboards and reports to be generated using the information communicated withing the attributes, helping the project team to better manage the project by keeping track of the status of the project and quickly identifying issues.

Chapter 15 provides a detailed list, definition, and discussion concerning the use of attributes.

This concept supports the GtWR characteristics: C1—Necessary and C13—Comprehensible.

3.2.1.5 The Need for a Formal, Binding Agreement The legal aspect of system development represents perhaps the greatest barrier to adoption of models and non-textual tools for expressing needs and requirements. Text-based need and requirement statements expressed within an RMT or a report generated by an RMT are more easily understood in a formal agreement or contract-based system development effort by a wider, and often, nontechnical group of stakeholders, including business management, PM, CM, contract administrators, and legal practitioners.

To be part of a binding agreement, especially in a legal contract, the sets of needs and requirements must be expressed formally and configuration managed in a form that 1) makes it clear the statements are binding and 2) has the characteristics of well-formed need and requirement statements and sets of needs and requirements as defined in standards and guides such as the GtWR.

Use of "shall" in requirement statements, or another term defined to have the same meaning, makes it clear that what is being communicated is formal, the requirement statement is contractually binding, and the system must be verified to meet the requirements.

While drawings and figures can certainly be included within a contract, it is currently difficult to include a model file as a contractual obligation. Well-formed textual requirements statements expressed within an RMT or a report from the RMT are far easier to enforce legally. Perhaps one day, this will evolve; however, until then, the reader is recommended to continue defining well-formed, text-based needs and requirements within an RMT, particularly in support of contract-based developments.

This concept supports the GtWR characteristics: C1—Necessary, C3—Unambiguous, C4—Complete, C6—Feasible, C7—Verifiable, C13—Comprehensible, C14—Able to be Validated.

3.2.1.6 Necessary for Acceptance, Qualification, Certification, and Approval for Use Formal, contract-based product development and management processes, as well as highly regulated products, define what is "necessary for acceptance" and require objective evidence that these criteria have been met. Acquiring this objective evidence is the purpose of system verification and system validation. The resulting objective evidence is used as part of acceptance, qualification, certification, and approval for use. *Refer to Section 4.1.1 for a more detailed discussion concerning what is necessary for acceptance.*

In highly regulated safety-critical industries (such as the medical device industry, ground transportation, aviation, and consumer products), objective evidence that the design outputs and final product meet the design inputs (integrated set of needs [system validation] and set of design input requirements [system verification]) is required prior to the product being approved for its use.

Historically, well-formed text-based needs and requirements expressed within an RMT and reports generated by the RMT have been the focus of defining what is necessary for acceptance, and many will continue to do so. However, with the move toward MBSE, the increased use of models to express requirements, and the inclusion of the "verified by" relationship with models, some acquiring organizations are issuing contracts allowing requirements expressed within a model to be the focus of verification of the system to be against these requirements. This should be an acceptable alternative if the requirements within the models have the characteristics of well-formed requirements as defined in the GtWR and the above-mentioned issues have been satisfactorily addressed.

This concept supports the GtWR characteristics: C7—Verifiable, C14—Able to be Validated.

3.2.2 Expressing Needs and Requirements Within Models and Diagrams

On the other side of the SE coin, for many ideas and concepts that need to be communicated, models and diagrams have been proven to be an effective form of communications. Advantages [28] of models and diagrams to discover and express needs and requirements are described in the following sections.

3.2.2.1 Analysis from Which Needs and Requirements are Derived Models and diagrams are excellent analysis tools for defining and maturing feasible lifecycle concepts by providing a context for needs and requirements. As part of lifecycle concepts maturation, functions are defined, and relationships between those functions (interactions and interfaces) are identified. From this knowledge, FFBDs can be developed as well as context diagrams, boundary diagrams, and external interface diagrams.

These artifacts can then be transformed into functional architecture and analytical/behavioral models, which can, in turn, be transformed into a physical architecture. These models are excellent sources of needs and requirements dealing with functions, performance, and interactions between the subsystems and system elements within the system physical architecture as well as between the system and external systems in its operational environment. *Refer to Chapter 6 for a more detailed discussion concerning assessing interactions and defining interface requirements.*

This concept supports the GtWR characteristics: C1—Necessary, C6—Feasible, C8—Correct, C10—Complete, C11—Consistent, C12—Feasible, C13—Comprehensible, C14—Able to be Validated.

3.2.2.2 Completeness A key issue when defining needs and requirements is completeness. Models and diagrams provide the capability to address completeness in terms of functions, inputs to those functions, sources of those inputs, outputs, and customers (destinations/users) for those outputs. When developing these models or diagrams, missing sources for inputs or missing customers for the outputs become apparent and enable the project team to address these issues. In the model, functions can be decomposed to reveal subfunctions that must be addressed, along with their inputs, sources of inputs, outputs, and customers for those outputs.

This concept supports the GtWR characteristics: C10—Complete, C13—Comprehensible, C14—Able to be Validated.

3.2.2.3 Consistency Another key issue when defining needs and requirements is consistency. As the number of needs and requirements grows for today's increasingly complex systems and the number of subsystems and system elements within the system architecture grows, it becomes increasingly difficult to comprehend and manage all the associated data and information as well as the artifacts represented by the data and information. With each subsystem and system element within the system physical architecture defined by their own lifecycle CNR, consistency can be an issue, not only in the use and definition of terms (ontology) but also in what the needs and requirements are communicating.

Models and diagrams provide the capability to address consistency not only within sets of needs and requirements but also between needs and requirements in other sets associated with other subsystems and system elements within the system physical architecture as well as external systems with which the system interacts.

This concept supports the GtWR characteristics: C11—Consistent, C13—Comprehensible, C14—Able to be Validated.

3.2.2.4 Identify and Manage Interdependencies A key tenet of SE is that system behavior is a function of the interactions between the parts of the system as well as interactions with the external systems and the operational environment of which the system is a part. A key area where there are dependencies is the budgeting of performance, quality, and physical attribute values contained in design input requirements for systems at one level of the architecture to subsystems and system elements at the next lower level of the architecture.

When this is done, the budgeted values allocated to the subsystems and system elements are interdependent—a change in one will affect the others. Tying these interdependencies together can result in an equation of dependent variables. Managing these interdependencies within a model is much easier than in document-based approaches to SE, where these interdependencies are often not managed as dependent variables.

Note: Once these dependencies have been identified within a model, traceability of these dependencies can be established within an RMT. Without the analysis within the model, the traceability of dependent requirements within an RMT are often not established.

Refer to Section 6.4 for a more detailed discussion on allocation and budgeting.

This concept supports the GtWR characteristics: C8—Correct, C10—Complete, C11—Consistent, C13—Comprehensible, C14—Able to be Validated.

3.2.2.5 Support Simulations Language-based analytical/behavioral models can be used to develop higher fidelity models that allow simulations of the SOI. These simulations can be a significant part of design verification and design validation. With a simulation capability, design issues can be identified and corrected before baselining the design output specifications and building, coding, and integrating the realized parts that make up the SOI—saving both time and money by avoiding expensive and time-consuming rework that often occurs during system integration, system verification, and system validation. *Refer to Chapter 8 for a more detailed discussion of using models and simulations as part of design and early system verification and system validation.*

This concept supports the GtWR characteristics: C7—Verifiable, C13—Comprehensible, C14—Able to be Validated.

3.2.2.6 Comprehensive System View In many cases, models and diagrams help facilitate communication by making complex systems and processes easier to understand. As the old saying goes: "A picture is worth a thousand words."

This concept supports the GtWR characteristic: C13—Comprehensible, C14—Able to be Validated.

4

LIFECYCLE CONCEPTS AND NEEDS DEFINITION

The focus of this Chapter is an elaboration of the ISO/IEC/IEEE 15288 *Stakeholder Needs and Requirements Definition Process* showing application through the definition of System of Interest (SOI) lifecycle concepts and needs. Because there are multiple stakeholders involved, each having their own sets of needs, the results of the activities in this section are referred to as an "integrated set of needs" as discussed in Chapter 2 which will be transformed into a set of design input requirements for the SOI.

Note: The Stakeholder Needs and Requirements Definition Process (defined in ISO/IEC/IEEE 15288 and the INCOSE SE HB [1]*) applies to the operational-level lifecycle Concepts, Needs and Requirements (CNR) for a system, representing the sets of operational-level stakeholder needs for the SOI (Chapter 2). These activities can also be applied to each level within the architecture (e.g., the development of system, subsystem, and system element level CNR).*

4.1 INTRODUCTION

For an SOI under development, project success depends on the project team understanding the source of concern, problem/opportunity, higher-level lifecycle concepts, higher-level needs, and higher-level requirements that constitute acceptability or desirability of a solution (what?), measures (how well?), and the conditions in which the SOI must operate (in what operational environment?) as defined at the previous organizational or architectural level. No matter the level, the project team responsible for the SOI must understand and comply with the higher-level CNR when defining the CNR for their SOI.

When there is a customer/supplier relationship, the supplier project team must address both the customer's as well as their own organization's needs and requirements, as discussed in Chapter 2. These inputs enable lifecycle concepts for the SOI to be defined, analyzed, and matured, and a formal integrated set of needs defined, agree-to, and baselined. These activities enable the definition of

INCOSE Needs and Requirements Manual: Needs, Requirements, Verification, Validation Across the Lifecycle, First Edition. Louis S. Wheatcraft, Michael J. Ryan, and Tami Edner Katz.

well-formed need and design input requirement statements with the characteristics defined in the Guide for Writing Requirements (GtWR).

It is this integrated set of the needs that communicate agreed-to capabilities, functions, performance, interactions with external systems, quality, and compliance the stakeholders expect from the SOI to address the stated problem or opportunity. This set of needs (and resulting set of design input requirements) is also the form of communication used to inform those responsible for designing, building/coding, integrating, verifying, validating, and delivering the SOI.

4.1.1 Necessary for Acceptance

The focus of *Lifecycle Concepts and Needs Definition* activities addressed in this section and *Design Input Requirement Definition* activities addressed in Chapter 6 is to clearly define what is *necessary for acceptance*.

The goal of all development projects is that an *Approval Authority* accept their output. The *Approval Authority* (which can be the customer(s) or some regulatory agency) is responsible for qualification, certification, acceptance, or approval for use of the SOI. In the end, it is the *Approval Authority* that 1) *decides* what *constitutes necessary for acceptance*, and 2) *determines* what *is necessary for acceptance*. While the *Approval Authority* has the last vote, other stakeholders can be ranked as to their say in what is necessary for the system to be acceptable. When there is an inconsistency or a disagreement, the rank of a given stakeholder will be taken into consideration.

During all lifecycle stages, the project team must be continuously focused on what is *necessary for acceptance* to ensure they are defining, designing, making an SOI that will meet these criteria. Failure to do so will result in system validation failure and a failed project.

Successfully completing the *Lifecycle Concepts and Needs Definition* and *Design Input Requirements Definition* activities as defined in this Manual ensures what is *necessary for acceptance* have been clearly defined and recorded early in the project. Doing so helps ensure the customers have defined success for the developers, and the developers understand what they must deliver for the SOI to be accepted by the *Approval Authority* and approved for its intended use by its intended users in its intended operational environment (system validation). Therefore, it is critical that system verification and validation *Success Criteria, Method*, and *Strategy* are stated that clearly defines what is *necessary for acceptance* for each need and requirement statement. *Refer to Chapter 10 for a more detailed discussion on Success Criteria, Method, and Strategy.*

Acceptance can be done in stages. Where there is an acquirer/supplier relationship, an SOI may be first accepted contractually as a result of passing system verification and then accepted for use in its operational environment when operated by its intended users through customer managed system validation (Reference Chapter 13). In this case, for each stage of acceptance, there are different criteria that define what is *necessary for acceptance*. As an example, for US government funded systems, USC Title 9 requires system verification, referred to as developmental test and evaluation (DT&E), and system validation, referred to as operational test and evaluation (OT&E), to be done by different organizations; each stage having different acceptance criteria.

For many highly regulated systems, acceptance is based on the SOI passing system validation – that is, confirmation that the system can be used as intended safely and securely, in its intended operational environment, when operated by its intended users.

4.1.2 Lifecycle Concepts and Needs Definition Activities

Lifecycle Concepts and Needs Definition involves a number of activities, as shown in Figures 4.1a and 4.1b. Each activity results in data and information that will be used to define the integrated set of needs for the SOI as well as elements within the SOI architecture. Figures 4.1a and 4.1b also highlight the specific Manual section number that contains descriptions of the activity shown.

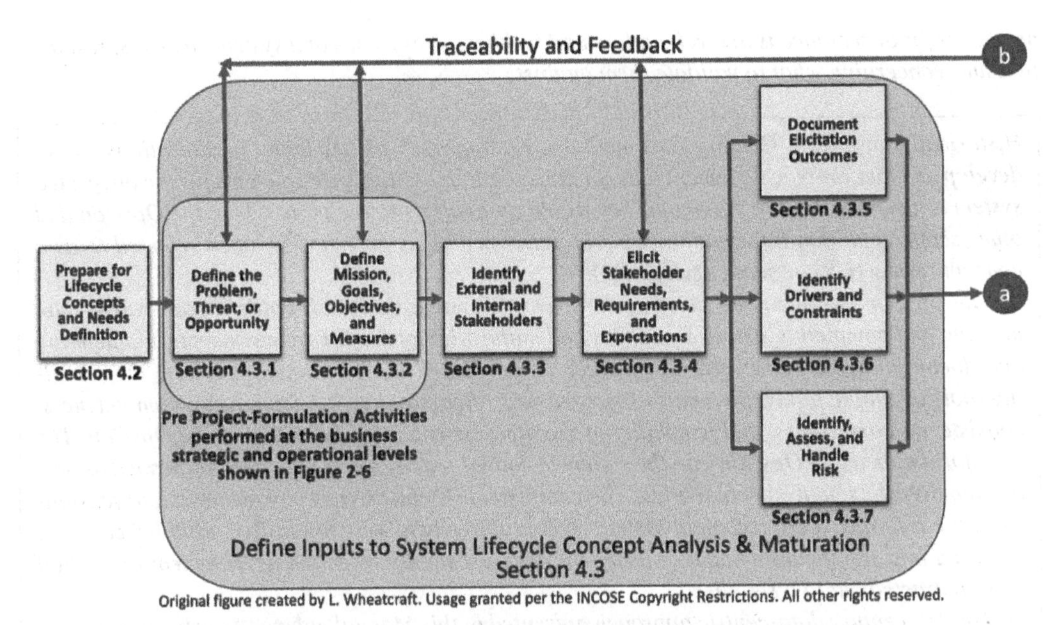

FIGURE 4.1a Lifecycle and Needs Definition Activities Part 1.

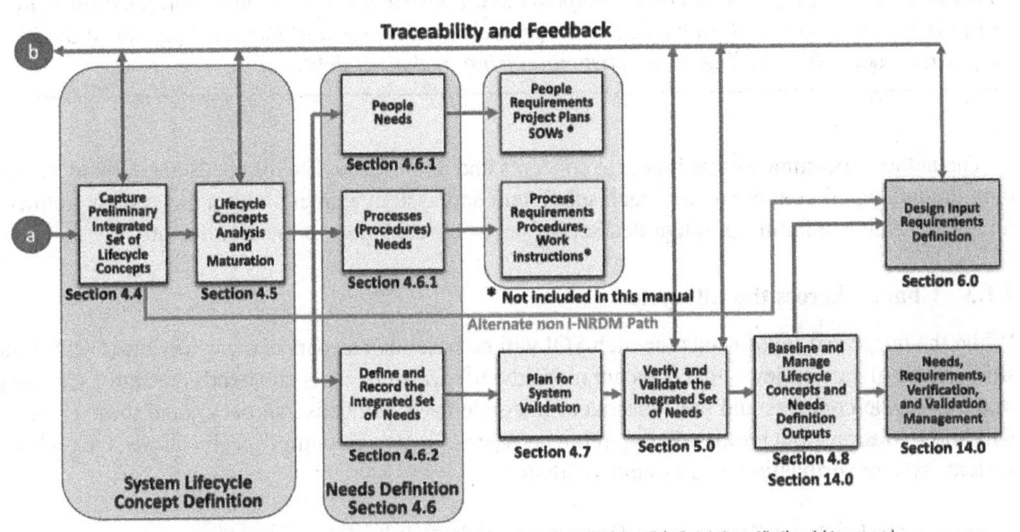

FIGURE 4.1b Lifecycle and Needs Definition Activities Part 2.

Note: While theoretically, lifecycle concepts and an integrated set of needs should be defined for subsystems and system elements at all architectural levels for the integrated system, some organizations may not do so. Some may only formally define an integrated set of needs at the system level, but not for subsystems and system elements at the lower levels of the architecture. Without doing so, what is necessary for acceptance is not well understood for these subsystems and system elements, resulting in issues concerning what to validate them against.

Historically, in a 20th Century document-centric approach to SE, many organizations would develop an OpsCon-type document that contains preliminary lifecycle concepts for the integrated system as discussed in this Manual. They would do analysis to the point where the OpsCon and other scope-definition information could be baselined via a gate review, often referred to as a concept review (CR) or scope review (SR).

The OpsCon has often been used as the source of system-level stakeholder needs, rather than defining and completing detailed analysis and maturation of the lifecycle concepts and defining and documenting an integrated set of needs as advocated in this Manual. Lacking the modeling tools available today, the analysis needed was often not done to help ensure completeness, consistency, correctness, and feasibility of the lifecycle concepts described in the OpsCon. The transformation of the OpsCon into the system technical requirements would be performed as part of "requirements analysis" activities. The completeness, consistency, correctness, and feasibility of the resulting set of technical system-level requirements were not always established, often resulting in a significant amount of technical debt, especially when the development of the SOI was contracted out to a supplier.

The 21st century data-centric approach presented in this Manual, advocates using models and diagrams to perform the system-level lifecycle concepts analysis and maturation activities before defining an integrated set of needs and transforming them into a set of design input requirements. This approach is preferred because completeness, consistency, correctness, and feasibility are established early before baselining the integrated set of needs and the resulting set of design input requirements, avoiding, or at least minimizing, technical debt.

The authors recommend that lifecycle concepts and an integrated set of needs are defined for not only the integrated system but also each subsystem and system element within the system architecture. The result is a family of integrated sets of needs for each part of the architecture.

4.1.3 Change Across the Lifecycle

While the integrated set of needs for each SOI will be baselined as part of a gate review (CR, SR, or similar type of gate review), that does not mean the lifecycle concepts and needs are static. Changes to the lifecycle concepts and the subsequent integrated sets of needs can occur, and these changes will impact design input requirements, architecture, design, design output specifications, the realized system, system verification, and system validation.

During *Design Input Requirement Definition activities*, it is common to discover issues with the lifecycle concepts and integrated sets of needs, requiring them to be updated when the set of design input requirements is baselined. During maturation of the physical architecture, design verification, design validation, early system verification, early system validation, and development of the design output specifications (reference Chapter 8), it is also common to discover issues with the lifecycle concepts and integrated sets of needs (and resulting design input requirements), requiring them to be updated.

The ability to assess and manage change across the lifecycle is enabled by traceability between SE and PM artifacts and work products across the lifecycle (reference Chapter 14). This traceability is made possible by practicing SE from a data-centric perspective as discussed in Chapter 3.

4.1.4 Benefits of Lifecycle Concepts and Needs Definition Activities

Focusing on defining lifecycle concepts and a feasible set of needs for the SOI before transforming them into design input requirements has many benefits:

- Clearly defines what is *necessary for acceptance* at the beginning of the project.
- Ensures what the customers and operational-level stakeholders expect and need from the SOI is defined and recorded.
- Clearly sets the boundaries for the SOI – makes clear what is in scope and what is not.
- Helps avoid scope battles later in the development lifecycle.
- Helps to identify and resolve issues early in the development lifecycle.
- Helps prevent poorly formed design input requirements in terms of completeness, correctness, consistency, and feasibility.
- Reduces the time to define and baseline the sets of design input requirements.
- Reduces costly rework.
- Helps to manage change.
- Results in an integrated set of needs that are the focus of design input requirements definition, design, and system validation.

However, while there are benefits, there are also additional costs associated with doing this extra work. It is important to weigh benefits against these costs; organizations can tailor the activities presented in this Manual to adopt those that reduce technical debt and provide the most value.

The following sections describe each of areas of focus shown in Figures 4.1a and 4.1b that are part of *Lifecycle Concepts and Needs Definition* activities:

Section 4.2: Prepare for Lifecycle Concepts and Needs Definition.
Section 4.3: Define Inputs to Lifecycle Concepts Analysis and Maturation.
Section 4.4: Document Preliminary, Integrated Set of Lifecycle Concepts.
Section 4.5: Lifecycle Concepts Analysis and Maturation.
Section 4.6: Define and Record the Integrated Set of Needs.
Section 4.7: Plan for System Validation.
Section 4.8: Baseline and Manage Lifecycle Concepts and Needs Definition Outputs.

4.2 PREPARE FOR LIFECYCLE CONCEPTS AND NEEDS DEFINITION

A prerequisite to the successful definition of lifecycle concepts and needs are the enablers shown in Figure 4.2. These enablers are part of NRVV activities discussed in Chapter 14. Enablers include an enterprise-tailored GtWR and processes and work instructions for the development and management of needs and requirements.

The project's plans, guides, processes, and work instructions should be consistent and in compliance with organizational, strategic, and operational-level product development concepts and processes. Project team members involved in the development of the integrated set of needs should be trained in, and knowledgeable of, how to perform *Lifecycle Concepts and Needs Definition* activities based on organization application guidance.

This section assumes the project will record and manage the inputs and outputs of the *Lifecycle Concepts and Needs Definition* activities, as well as all the output artifacts using a project toolset

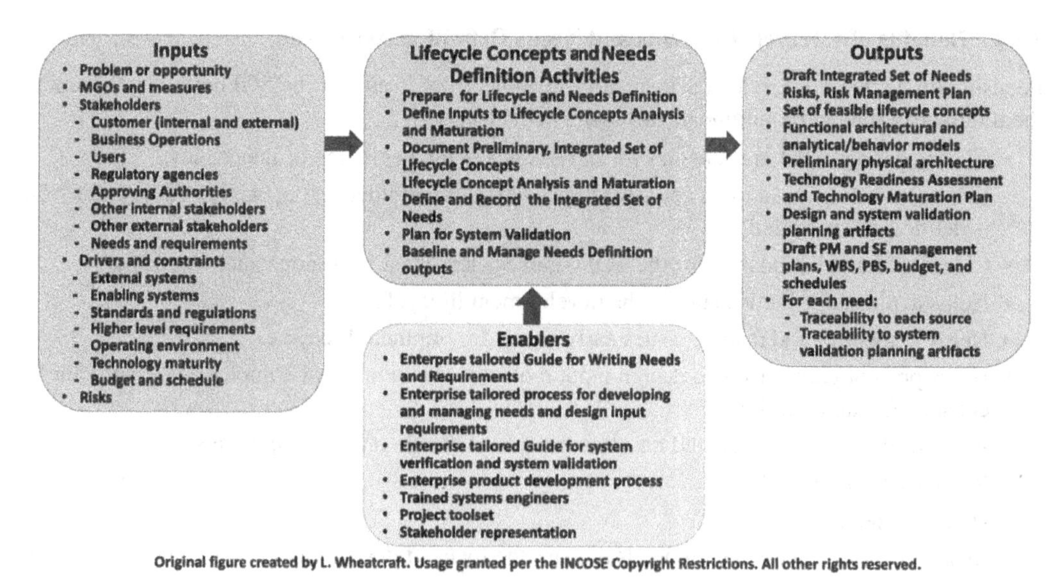

FIGURE 4.2 *Lifecycle Concepts and Needs Definition* IPO Diagram.

that supports the data-centric I-NRDM approach discussed in Chapter 3. Provisions must be made for project team members to have access to tools that meet their needs, including the RMTs and modeling/diagramming tools used to produce the artifacts, along with the ability to link and share data and information between tools.

Preparing for the *Lifecycle Concepts and Needs Definition* activities consists of gathering the required input artifacts shown in Figure 4.2. Ideally, standards, regulations, and higher-level lifecycle CNR will already be managed within the organization's PM and SE toolsets, such that the applicable information can be imported into the project's toolset and traceability can be established.

A key preparation activity is the creation of a needs inspection checklist tailored to the project. If the organization has a generic checklist as part of their design controls, the project can tailor the checklist to the SOI being developed and the project's specific processes. This checklist serves as a standard to guide the project team's *Lifecycle Concepts and Needs Definition* activities, as well as a standard that will be the basis of the needs verification activities and validation activities defined in Chapter 5. Addressing the areas and questions within the checklist will aid in the successful completion of the needs definition activities defined in this section. (*An example of a need verification checklist is contained in the GtNR Appendix D.*)

4.3 DEFINE INPUTS TO LIFECYCLE CONCEPTS ANALYSIS AND MATURATION

The focus of this section is defining the inputs that feed into the lifecycle concepts analysis and maturation activities. Some organizations may classify these activities under the heading "requirements elicitation." In this Manual, this activity is described as elicitation of stakeholder needs, which is inclusive of their needs, requirements, and real-world expectations for lifecycle concepts. The result of the activities in this section is the data and information used to define a preliminary integrated set of lifecycle concepts.

Note: The activities in Section 4.3.1, Define the problem, threat, or opportunity, and 4.3.2, Define mission, goals, objectives, and measures, are activities that normally occur at the organization's strategic and operational levels prior to project formulation. These two sections are elaborations of the ISO/IEC/IEEE 15288 Business or Mission Analysis process activities. For lower-level systems and systems elements within the system physical architecture, the project team responsible for the parent system will define the problem or opportunity, MGOs, and measures for the lower-level subsystems and system elements.

4.3.1 Define the Problem, Threat, or Opportunity

"One of the most dangerous forms of human error is forgetting what one is trying to achieve."
Paul Nitze (1907–2006)

Expanding on this quote, an even more dangerous form of human error is to proceed with the project without knowing what the project is trying to achieve—that is, not understanding why the SOI is needed and a lack of clearly defined expected outcomes at the beginning of the project. As part of project formulation, a project champion collaborates with key stakeholders at the organization's strategic and operational levels to clearly define the problem, threat, or opportunity that the project team is to address. This will enable the project team to understand why the project is worth doing, why the system is needed, and what capabilities, functions, performance, and features are important to the customers, users, and operators of the SOI.

The steps to defining the problem or opportunity include:

1. Identify the organization's strategic and operational-level stakeholders that are impacted by the problem or threat or those who will benefit from pursuing the opportunity.
2. Work with these stakeholders to understand how they are impacted by the problem or threat or those that will benefit from pursuing the opportunity.
3. Clearly define a statement of the problem, threat, or opportunity.
4. Obtain stakeholder agreement on the problem, threat, or opportunity statement.

When identifying and defining the problem, threat, or opportunity, it is important to ensure that it is the right problem, threat, or opportunity that is being addressed. For example, is the problem or threat as stated the real issue, or is the stated problem or threat really one of several associated with an underlying issue that is the real problem or threat that needs to be addressed?

If the project focuses on the wrong thing, or one symptom of the underlying problem or threat, the delivered solution will not be the correct solution, and the project will have failed to meet the stakeholder's real-world expectations. Ensure the project is addressing the "right" thing.

Likewise, if the project focuses on the wrong opportunity, or perceived source of an opportunity, the delivered solution may not be a sufficient solution, and the enterprise may miss the real opportunity, resulting in lost revenue and enabling competitors to take over a larger part of the intended market. If this is the case, the project will have failed to meet the stakeholder's real-world expectations. Ensure the project is addressing the "right" opportunity.

When formulating the problem statement, it is important that the right questions are asked. Examples include:

- Why is that a problem, threat, or opportunity?
- What is the underlying issue that is the cause of the problem or threat or source of the opportunity?
- Are there other problems, threats, or opportunities that are a result of that issue or source of the opportunity?
- What is the "right" problem, threat, or opportunity that the project should be addressing?

Example Problem Statement Defined by the LIR Project Team

Assume there is a factory whose purpose is to fill jars with a customer-supplied product, label the jars, place a lid on the jars, and package the jars for customer pickup. All parts of the process have been automated except the placement of lids on the jars. This task is currently being done by a human, but the placement of the lids on the jars is slow and is limiting the overall output of the factory. As a result, the owner has decided to automate this function and procure lid installation robots (LIR) to replace the current human lid installers. To make this happen, a LIR Project Team has been formed. One of the first things they did was work with the owner to define the problem statement.

For this example, the problem statement could be: "*Manual installation of lids is severely limiting the number of jars that can be processed, which prevents us from meeting the needs of our customers and maximizing our profit.*"

4.3.2 Define Mission, Goals, Objectives (MGOs), and Measures

Once the problem, threat, or opportunity has been defined, recorded, and agreed to, the project champion will collaborate with the stakeholders that participated in identifying and defining the problem, threat, or opportunity to better understand what they would view as an acceptable outcome.

- How do they define success?
- What measures would the stakeholders use to define success?
- What is the intended use of the SOI in what operational environment?
- What capabilities, features, functions, and performance do they need?
- What are their expectations for safety, security, and resilience?
- What are their expectations for quality?
- What are their expectations for compliance (with standards and regulations)?
- What specific outcome(s) do they expect once the SOI is delivered?

For cases where there is no existing system or product (green-field systems), a common approach is to characterize the as-is or present-state of the organization in terms of the problem, threat, or opportunity and then characterize the to-be or future-state of the organization in terms of the resolution of the problem, threat, or the ability to pursue the opportunity.

For existing systems that need to be updated (brown-field systems), a common approach is to list the problems or issues with the existing as-is SOI and the reasons the SOI needs to be updated.

Key information includes what needs to be updated, why, and what value will result from the update. What can the existing SOI no longer do, what performance needs to be improved, and what changes need to be made concerning interactions with external systems and the current and future operational environment? For the strategic and operational-level stakeholders, a key concern is the ROI if the proposed updates are made and the consequences if they are not.

Key drivers for updates to brown-field systems include the need for new features driven by the market and competitors, new technologies, old technologies that are out of date or no longer supported, new or changed operational environments, changes to external systems the SOI must interact with, changes in the supply chain (key parts or resources needed to produce the parts of the existing system are no longer available), changes to regulations, and changing needs of the customers, users, or consumers.

Assuming the organization has been doing the post-development verification and validation of the existing system as discussed earlier, most of the answers to these questions will have already been determined. Using this information, the project champion will collaborate with the stakeholders concerning their vision for the updated version or new model of the SOI, defining the to-be or future-state real-world expectations of the stakeholders.

For either of the above cases, the project champion will do a gap analysis comparing the as-is current state to the to-be or future-state. The result is the identification of the changes that need to be made to the existing as-is state that will result in the to-be or future-state, the value of making these changes, and the expected ROI.

With this knowledge, the project champion will work with the organization's strategic and operational-level stakeholders to elicit their needs, define preliminary lifecycle concepts, and define an integrated set of operational-level needs and requirements for the SOI. The needs elicitation activities at the operational level include MGOs and measures that address the stakeholder expectations for the project, and SOI to be developed, or the service to be supplied, that clearly communicate the expected outcome(s).

This information is captured within the organization's toolset, such that subsequent artifacts developed at lower levels can trace back to this information.

Note: Stakeholder needs are expressed at various levels of abstraction. The MGOs and measures are at the top of the hierarchy of the set of needs. For some organizations, rather than defining a mission statement, they define a top-level Need statement (with a capital "N" to distinguish this statement from lower-level needs). With this approach, rather than MGOs, the NGOs are at the top of the hierarchy. The MGOs/NGOs and measures defined by the stakeholders at the previous level of the organizations are further elaborated during the SOI lifecycle concepts analysis and maturation activities and are reflected within the SOI's integrated set of needs. Within this Manual the phrase "mission statement" is used to be consistent with ISO/IEC/IEEE 15288 and the INCOSE SE HB.

4.3.2.1 Mission Statement The *mission* statement is the top tier of the hierarchy of needs based on the above analysis of a problem, threat, or opportunity that the project was formed to address. The mission statement defines the "why" – why does the project exist? and the "what" – what does the organization's strategic- and operational-level stakeholders or the customer need the system to accomplish (what is the expected outcome) that will address the defined problem statement?

The mission statement should not be to obtain the SOI (that is, the project's mission); rather, it should articulate the expected outcome that will result from obtaining or developing the SOI. For example, a customer's primary need is not a new coffee maker, but to be able to make a "great" cup of coffee. What "great" means to them will be part of the goals and objectives defined for the SOI.

The mission statement is communicated in a single-thought sentence that encapsulates the integrated set of needs, from which the elements of the set can be elaborated (decomposed and derived).

Multiple sentence or multiple part-mission statements often indicate the organization is not able to agree on a single mission statement. There have been cases when a project has multiple sponsors or funding sources. Unfortunately, each of the sponsors or funding sources may have a different mission in mind. Multiple mission statements or multiple part-mission statements will often result in failure when the different mission statements are in conflict.

Rationale should be defined for the mission statement that clearly communicates the intent and expectations for the outcome. Why is it worded the way it is, and how does it address the problem, threat, or opportunity from which it was derived?

Once a mission statement has been formulated and agreed to, the project champion collaborates with the stakeholders to further elaborate the mission statement in terms of goals and objectives.

Example Mission Statement Defined by the LIR Project Team

For the example case study concerning the development of the LIR, the mission statement for the project from the perspective of the business could be as simple as: "*Maximize profits.*"

From the perspective of the LIR to be developed, the mission statement could be "*Provide a faster and more effective way to install lids on jars.*"

As stated, the mission statement for the LIR can be traced directly to the business mission statement.

4.3.2.2 Goals *Goals* are upper-level needs that form the second level of the hierarchy of the integrated set of needs. Goals are elaborated from the mission statement, communicating what needs to be achieved that will result in achieving the mission.

Goals allow the organization to divide the mission statement into manageable pieces and promote a shared understanding between the project team and the organization's strategic and operational-level stakeholders or customers of the outcomes that will result in the mission statement being met.

Some general guidelines for writing goal statements include:

- Goals are often initially written as a list of phrases, each beginning with an action verb.
- Goals should be written as single-thought statements that are clear, positive, concise, and grammatically correct.
- Goals may be more qualitative than quantitative (for example, a goal could be to "reduce processing time," while a child objective statement would address the actual change in processing time either as a percentage or an actual value).
- The number of goals should focus on what is most important and be limited by the general rule of 7 ± 2. A long list of goals will be harder to achieve and may be confused with more detailed needs and requirements.
- Each goal should address a critical outcome, chart a clear course, reflect primary activities and approach of the project, and address mandates from management or the parent organization.
- Goals must be congruent with the mission statement, be independent, and not conflict with each other or the parent mission statement.
- Achievement of each of the goals and achievement of the goals as a set should result in achievement of the mission statement.
- As for the mission statement, each goal should be accompanied by a rationale.

Example Goals Defined by the LIR Project Team

For the example case study concerning the development of the LIR, the goals for the project from the perspective of the business could be:

BG1: Automate the installation of the lids on jars.

BG2: Reduce per jar processing cost.

BG3: Increase capacity of the facility (number of jars processed).

BG4: Increase performance in terms of the speed of processing jars.

BG5: Retrain as many existing human workers as is practical to fulfill new roles associated with automation and other activities within the facility.

From the perspective of the LIR, the goals could be:

RG1: Automate lid installation (BG1).
RG2: Increase number of lids installed per shift (BG3).
RG3: Increase yield (BG4).
RG4: Minimize cost of implementation (BG2).
RG5: Conduct safe food handling, hazardous materials handling, and robotic operations.

Note: "BGx" stands for a business goal, and "RGx" stands for a LIR goal. The mission statement, goals, and objectives should be identified in such a way as to enable traceability. As can be seen in the set of robot goals, there is traceability to several of the business goals.

4.3.2.3 Objectives *Objectives* are upper-level needs that form the third level of the hierarchy of the integrated set of needs. Objectives are elaborated from the goals, providing more details concerning what must be done to meet the goals that will result in the mission to be achieved, i.e., what the project team and the SOI need to achieve so the SOI can fulfill its intended purpose (mission) in its operational environment when operated by its intended users.

Some general rules for writing objective statements include:

- Objectives are often initially written as a list of phrases, each beginning with an action verb.
- Objectives should be written as single-thought statements that are clear, positive, concise, and grammatically correct.
- Objectives are more quantitative in that the statements that can be validated, i.e., the completed SOI, should be able to be validated to have met each of the objectives. (For example, while the goal could be to "reduce processing time," the associated objective would address the actual change in processing time, either as a percentage or an actual value.)
- The number of objectives should focus on what is important and be limited by the general rule of 7 ± 2 – the objectives are not a list of requirements (however, they will be implemented via need and requirement statements that will trace back to the objectives as their source).
- Objectives must be able to be decomposed/derived from goals—they are children of goals and aggregate to be the goals.
- Objectives address a critical need and address mandates from the customers, management, or parent organization that resulting SOI will be validated against.
- From a project or contract perspective, objectives could include key deliverables.
- Achievement of the sum of the objectives should result in achievement of the parent goal, such that the achievement of the sum of all objectives should result in the achievement of the sum of all goals, which should result in the achievement of the mission (that is, achievement of the set of needs).
- As for the mission and goal statements, include rationale for each objective.

As a set, the MGOs need a time and budget defined by which they will be met. The budget and schedule are drivers and constraints for the project team and are discussed in more detail in Section 4.3.6.5.

Some organizations will include priorities when defining the objectives, referring to the high priority objectives as "primary" and lower priority objectives as "secondary", for example a NASA science mission may define primary science objectives and secondary science objectives. The project can claim project success when all the primary science objectives have been met and claim additional credit when any of the secondary objectives are also met.

From a planning perspective, the project can plan to initially meet both the primary and secondary objectives; however, as determined by the lifecycle concepts analysis and maturation activities discussed later, doing so may not be feasible, and some of the primary objectives may need to be changed and some or all of the secondary objectives may need to be removed from the list of objectives. If initially deemed feasible at this lifecycle stage, when issues occur later in the lifecycle that require the project to de-scope, it is the secondary objectives that would be first removed from the list. In doing so, any downstream artifacts linked to those objectives would also need to be assessed for removal as well.

In the next section, measures are discussed separately from the objectives; however, it is common to include measures within the set of objectives, as is done in this example.

4.3.2.4 Measures It is important that the project champion define and get agreement on key measures that will be used to both validate the objectives as well as manage system development across the lifecycle. When defining objectives, the project team must define what measures the SOI will be validated against.

Measures are referred to by various terms: Measures of Suitability (MOS), Measures of Effectiveness (MOEs), Measures of Performance (MOPs), Key Performance Parameters (KPPs), Technical Performance Measures (TPMs), Leading Indicators (LIs), mission success criteria, primary science objectives, secondary science objectives, acceptance criteria, as examples.

Note: This Manual does not attempt to define the various measures. There are multiple documents available from INCOSE that go into detail concerning the definition of measures and how they can be used to better manage a project. Also refer to the INCOSE SE HB Section 2.3.4.7, Measurement Process.

Achievement of the MGOs and measures should result from the achievement of the integrated set of needs defined for the SOI.

Example Objectives/Measures Defined by the LIR Project Team

For the example case study concerning the development of the LIR, the objectives for the project from the perspective of the business could be:

BO1: Procure automated JPS systems. (BG1)
BO2: Reduce the per jar processing cost by [TBD 50%]. (BG2)
BO3: Increase the daily capacity at each production facility by at least [TBD 600%]. (BG3)
BO4: Increase performance yield to [TBD ≥ 99%] for each production facility. (BG4)
BO5: Retain at least [TBD 35%] of current human workers. (BG5)

From the perspective of the LIR, the objectives could be:

RO1: Procure LIRs to install lids – Off the Shelf (OTS) if possible. (RG1) (BO1)
RO2: Increase daily capacity to install lids to >[960 TBD] lids/shift for each LIR. (RG2) (BO3)

RO3: Increase yield such that the number of jars successfully installed per shift at each production facility is ≥ 99% (RG3) (BO4)
RO4: Limit changes to existing facilities and processes. (RG4) (BO2)
RO5: Stay within the acquisition budget. (TBD) (RG4) (BO2)
RO6: Comply with OSHA robotics regulations. (RG5)
RO7: Comply with FDA food handling regulations. (RG5)
RO8: Comply with EPA hazardous materials regulations. (RG5)

Note: "BOx" stands for a business objective, and "ROx" stands for a LIR objective. Again, the mission statement, goals, and objectives should be identified in such a way as to enable traceability. As can be seen in the set of business objectives, there is traceability to the business goals, and for the robot objectives as well as traceability to both the robot goals as well as the business objectives. Also note the use of brackets and "TBD" for values or measures that need further analysis to define and baseline the objectives.

4.3.2.5 Special Considerations

4.3.2.5.1 Business versus Consumer/User Perspective As illustrated in the above examples, it is important to understand different perspectives from which the MGOs and measures are defined. The consumer/user of the system does not care about the developing organization's profits, time to market, market share, or reuse of resources. The consumer cares about how the resulting product meets their specific needs and addresses their specific problem, threat, or opportunity. Thus, there will be two sets of MGOs and measures that need to be defined and met by the project team from both a business perspective and a consumer product perspective.

In general, the verbs used to define project goals and objectives apply to what the project must achieve, while verbs used to define SOI goals and objectives apply to what the SOI must achieve. Each defines different activities and expected outcomes. This may lead to conflicts, which the project team must address (for example, product price versus profitability and market share).

It is important that the two sets of MGOs and measures are defined at the beginning of the project, as each will have its own set of lifecycle concepts, resulting set of needs, and set of requirements. Each set will have different verification and validation activities. Refer to Section 4.6.1 for a more detailed discussion concerning needs and requirements for organizations and processes as opposed to the needs and requirements being developed for an SOI.

4.3.2.5.2 Dealing with Uncertainty Initial values stated within the objectives and measures are often questionable in terms of feasibility (cost, schedule, technology, legal, ethical, environmental, etc.). The actual feasibility may not be known until several iterations by the project team of the lifecycle concepts analysis and maturation activities to the point where the feasibility likelihood is established with risk acceptable for this lifecycle stage. The lifecycle concepts maturation activities will inform the definition of the MGOs and measures, which may need to be updated based on the knowledge gained during the lifecycle concepts analysis and maturation activities.

As illustrated in the above examples for the objectives and measures, a best practice is to enclose numbers contained in objectives and measures within brackets [...] with "to be determined" (TBD) or "to be resolved" (TBR) to indicate they are preliminary and will be finalized once the lifecycle concepts have matured and needs defined, recorded, and baselined. *Also refer to Section 4.6.3.1 for a more detailed discussion on managing unknowns and the use of TBDs and TBRs.*

4.3.2.5.3 Goals versus Objectives Definitions of goals and objectives are often interchanged. To avoid confusion, some organizations will combine the list of goals and objectives. When this is done, once the list has been formed, often a goal/objective hierarchy is apparent. The important thing is that all the goals and objectives are captured, and they form a complete, non-conflicting set.

4.3.2.5.4 Getting Stakeholder Agreement Getting the organization's strategic- and operational-level stakeholders to agree on the wording of the MGOs and measures can be challenging and may take multiple iterations to reach a set of project and SOI MGOs and measures that the stakeholders will agree to. Even though defining the MGOs can be time-consuming, the effort is well worth it as it ensures everyone is on the same page, sharing a common vision when it comes to what outcomes are expected for the project to be successful.

Many project failures can be traced back to the fact that the organization's strategic- and operational-level stakeholders (or perhaps an external customer) either did not define a set of MGOs and measures for the project and SOI, or the project or SOI did not meet the defined and agreed-upon MGOs and measures. *Without a common vision, each stakeholder will define needs based on their vision for the SOI, which may conflict with other stakeholders' visions, resulting in conflicts and inconsistencies.*

4.3.2.5.5 Assessing Feasibility before Baselining the MGOs and Measures There have been cases where the project champion defined the MGOs and measures, and the specific values appeared to be well-defined; however, the feasibility (cost, schedule, technology, legal, ethical, and environmental) was not adequately assessed before the MGOs and measures were baselined. In these cases, as the project moves through the SE and PM lifecycle process activities, it becomes apparent that the MGOs and measures cannot be met as defined. When this happens, the organization can either go back and update the MGOs and measures based on what is feasible or cancel the project. Failing to act often results in a failed project. For an SOI that will be acquired from a supplier, the failure to establish feasibility within the needs and requirements before a contract is issued is a major cause for expensive and time-consuming contract changes.

4.3.2.6 Recording and Managing the MGOs and Measures The MGOs and measures are recorded within the SOI's integrated dataset and linked to the problem/opportunity statement. It is important that both the organization and the project manage the MGOs and measures so that implementing system-level lifecycle concepts, needs, and design input requirements can be traced to the MGOs and measures they apply.

Once the project champion has obtained agreement on the problem or opportunity statement, operational-level lifecycle CNR, and resulting MGOs and measures for both the project and SOI, a project is formally formed to develop an SOI that will result in the MGOs and measures being achieved, the problem being resolved, or the capability to pursue the opportunity being realized. The project team must develop a solution that is consistent with the operational-level lifecycle CNR.

At the top of the integrated set of needs hierarchy, the agreed-to MGOs and measures are a major focus of the SOI lifecycle concepts analysis and maturation activities, lower-level needs definition, design input requirements definition, architecture, design, and system validation.

Ultimately, meeting the MGOs and measures is the key to delivering a winning product and realizing the stakeholders' real-world expectations. In the end, successfully meeting the MGOs and measures is a major factor in passing system validation.

4.3.3 Identify External and Internal Stakeholders

A critical first step to defining, analyzing, and maturing the lifecycle concepts is to identify the stakeholders, both internal and external, who are relevant to these activities at this level and with whom the project team will interact.

Stakeholders are the primary source of needs and requirements; therefore, for the project to be successful, all relevant stakeholders must be identified and included at the beginning of the project. Stakeholders can be both internal and external to the project team and their organization. Leaving out a relevant stakeholder often results in missing needs and requirements, resulting in a failure of the SOI to pass system validation and not being accepted for its intended use. These stakeholders are not only important in defining the needs and requirements, but they must also be involved in the validation of all SE artifacts across all lifecycle activities to ensure their real-world expectations are being met.

The choice of a relevant set of stakeholders is critical to the determination of completeness (*GtWR Characteristic C10—Complete*) of the integrated set of needs and the resulting design input requirements. A different set of stakeholders would lead to a different set of needs and requirements based on different perspectives on their unique problems, threats, opportunities, needs, requirements, MGOs, and measures [30].

Stakeholders are defined as any individual or organization, who may be affected by the SOI, who will participate in the development of the SOI, are able to influence the definition and development of the SOI, or with whom the project team will interact across the SOI lifecycle.

As such, some organizations classify stakeholders as "VIPs," which is anyone who has a **V**ested interest or "stake" ("interest or concern" in the Oxford dictionary) in the SOI, has **I**nfluence concerning the funding, development, procurement, approval, or acceptance of the SOI, or will **P**articipate in any of the SOI's lifecycle phase activities.

Stakeholders can include, but are not limited to, customers, sponsors, organization decision makers, regulatory organizations, developing organizations, integrators, testers, users, operators, maintainers, support organizations, disposers, the public at large (within the context of the business and proposed solution), and those involved in the disposal or retirement of the SOI. Stakeholders can be both internal and external to the organization.

A key part of stakeholder identification is to determine who the *Approval Authority* is within the group of stakeholders. It cannot be assumed that the only stakeholder that has this authority is the "customer" (the one paying for the SOI or development of the SOI, requesting the SOI, or procuring the SOI). It is critical that these stakeholders be included in the needs definition activities and in the resulting design input requirements definition; making sure what constitutes *necessary for acceptance is clearly defined*. They must agree that the set of needs and resulting set of design input requirements will result in an SOI that can be verified to meet those requirements and validated to meet the baselined set of needs in the operational environment when used by the intended users.

There can be various levels of *Approval Authority*. During development, the *Approval Authority* could be a stakeholder internal to the developing organization. Once the system integration, system verification, and system validation activities have been completed, the *Approval Authority* could be an external stakeholder, either the customer to whom the SOI is being delivered or, for highly regulated systems, a regulatory agency.

For a developer-funded effort (in-house product development), determination of the *Approval Authority* is up to the developer. In this case, the developing organization will determine who the *Approval Authority* is both internal and external to the organization. For highly regulated industries, of particular interest is who is to be the *Approval Authority* is, for example for medical devices the US Food and Drug Administration (FDA) and for aircraft and commercial space the US Federal Aviation Administration (FAA). For consumer products, customers and users in the marketplace will determine the ultimate acceptability of the solution based on sales and social media feedback.

From a customer/acquirer-funded effort (an organization procures the SOI or contracts out the development of the SOI with an external supplier), the task of defining what is necessary for acceptance and identifying the *Approval Authority* becomes a negotiation among the customer/acquirer, supplier/developer, users/operators, and possibly other stakeholders as determined by the customer/acquirer (the one supplying the funding). The results of these negotiations must be reflected in the contract. However, it is still the supplier's/developer's responsibility to alert the customer/acquirer regarding risk of ignoring stakeholders and any concerns or characteristics deemed necessary for acceptance by these stakeholders.

There can be many stakeholders for a SOI over its lifecycle; therefore, considering the stakeholders involvement in each lifecycle stage provides a thorough source for stakeholder identification. Table 4.1 contains the lifecycle stages and examples of potential stakeholders.

TABLE 4.1 Potential Stakeholders over the System Lifecycle.

Lifecycle Stage	Potential Stakeholders
Define	*Paying customer, sponsor, project team, project manager, procurement, research and development, suppliers, Approving Authority, public, marketing, end users, operators, compliance office, regulators, owners of enabling systems, owners of external systems, Approval Authority.*
Develop	Project team, SMEs, system architects, design engineers, suppliers, procurement, etc.
Produce	Production organization, process engineers, quality control, production verification, product acceptance, and supply chain.
Integrate, verify, and validate	Test engineers, system integration engineers, system verification engineers, system validation engineers, operators/users, owners of enabling systems, facility personnel, contracting, *Approval Authority*, regulators, safety personnel, and security personnel.
Operate	Transporters, installers, users, operators, safety engineers, security engineers, owners of external systems, IT, regulators, quality, and mission assurance.
Sustain/maintain	Customer/technical support, replacement part providers, service technicians, trainers, IT, quality engineers, inspectors, CMs, those conducting post-development system verification and validation activities.
Dismantle/dispose	Operators, waste management, regulators, and the public.

Example Stakeholders Identified by the LIR Project Team

For the LIR, the initial set of stakeholders identified included:

S1: Company Owner (Internal)
S2: Customers (External)
S3: Ronald Roundtop, Chief of Operations (Internal)
S4: Tom Torque, Facility Manager (Internal)
S5: Shelia Stores, Chief of Shipping, Receiving, & Storage (Internal)
S6: Sally Lidisan, Sales, Lids Unlimited, Inc. (External)
S7: Bob Jarhead, Sales, Jars Unlimited, Inc. (External)
S8: Mary Safelton, Chief Safety and Compliance (Internal)
S9: "Scotty" Wrench, Chief of Maintenance & Logistics (Internal)
S10: Irene Teichmann, Chief of IT (Internal)

Note: Unique designations "Sx" are assigned to each stakeholder to aid in traceability.

To help determine whether this was a complete list, the following questions were addressed:

What about potential LIR suppliers?
What about the regulators?
What about the knowledge of the lid installers being replaced?

For all the external systems the LIR will interact, have their respective stakeholders been included?

What about other stakeholders within the business – human resources, finance/budgeting, procurement/contracting, community relations, security, etc.?

4.3.3.1 Stakeholder Perspectives As shown in Table 4.2, each stakeholder has a unique perspective concerning the SOI to be developed based on their "stake" in the SOI, their involvement in their interactions with the SOI, their specific roles and responsibilities, and involvement in one or more of the lifecycle stages.

Each of these perspectives is a potential source of needs and requirements. Missing a key stakeholder and their unique perspective, results in an incomplete integrated set of needs and an incomplete set of design input requirements transformed from those needs.

Example LIR Stakeholder Perspectives defined by the LIR Project Team

Below are the perspectives of three of the LIR stakeholders.

S1: Company Owner (Internal)

MGOs achieved within the drivers and constraints.
Meet the customer demand for filled jars.
Can the LIRs meet my MGOs and other needs?
Safe, secure, and sustainable operations.

S2: Customers (External)

Facility can receive and store customer-supplied product in bulk containers.

Facility can properly handle, label, and fill jars with a variety of eatable food stuffs as well as non-eatable cleaning chemicals and compounds.

Jars can be either glass or plastic.

Lids are properly installed on the jars with a positive seal with a torque applicable to the size of lid and jar material.

Meet customer demand for filled jars.

Able to package and stage filled jars for pickup and delivery to customer's customers.

Facility cleans and makes available for pickup empty product bulk containers for reuse.

Bottom line – can the LIR meet the customer lid installation requirements?

S3: Ronald Roundtop, Chief of Operations (Internal)

No changes to existing facility.
Meet owner and customer needs.
Lids installed on jars per customer specs.
Maximize yield per shift.
Provide JPSs and associated LISs with daily quantity of product, jars, labels, and lids.
Workable LIRs are available during operational shift.
Safe, secure, and sustainable operations.

TABLE 4.2 Example Stakeholders' Perspectives and Concerns.

Stakeholders	Example Perspectives/Concerns
Customer/sponsors (those providing resources or have requested the system)	Cost, schedule, and resources to develop. Cost and resources to operate and maintain – lifecycle costs. Whether the system will meet their needs. Safety, Security (misuse and loss). Quality. Risks. ROI.
Customers (those that will buy and use a consumer product)	Features/capabilities (functions). Cost/schedule. User interface / user experience. Whether the system will meet their real-world expectations. Performance: speed, accuracy, precision. Safety, Security (misuse and loss). Quality: reliability, maintainability, updateability, lifecycle costs.
Users/operators (those using or interacting with the operational system)	Features/capabilities, functionality. Whether the system will meet their real-world expectations. User interface, user experience, usability, operability. Performance: speed, accuracy, precision. Quality: availability, reliability, robustness. Safety, Security (misuse and loss). Operational environment.
Marketing	Capabilities, features, performance, and quality better than their competitors. Providing a product that consumers need, want, and will buy. Meet customer real-world expectations, increase market share, increase revenue.
Project management	Understanding the stakeholder needs Feasibility of what is being asked for. Drivers and constraints. Enabling systems (hardware, SE tools, networks). Schedule/budget. Available resources and personnel. Risks (misuse and loss). Compliance with standards and regulations.
Contracting/procurement	Defining the customer's real-world expectations. Defining the deliverables. Defining the measures of performance. Defining what is necessary for acceptance. Schedule and budget.
Contractors, suppliers, vendors	Understanding the customer real-world expectations. Understanding the deliverables. Understanding the customer's measures. Understanding what is necessary for acceptance. Schedule and budget. Profits. Keeping their customers happy.

TABLE 4.2 (*Continued*)

Stakeholders	Example Perspectives/Concerns
Developers/engineering	Understanding the stakeholder needs. Feasibility of what is being asked for. Drivers and constraints. External system interactions. Enabling systems (hardware, SE tools, networks). Schedule/budget. Operational environment. Measures (what will make customers happy and accept the SOI).
Production engineers	Timeliness and quality of the design output specifications. Ability to produce quality products – repeatably, at the needed rate and yield. Changes/updates needed to the existing manufacturing capability. Feasibility of what they are being asked to do. Cost and Schedule.
Integrators, verifiers, validators	Quality of the built or coded system (built/coded) in accordance with the design output specifications. A realized, integrated system that can be verified to meet the design input requirements and validated to meet the set of needs. Testability of the system. Infrastructure to enable integration, system verification, and system validation (hardware/software). Access to test points to get needed data.
Maintainers/technical support	Installation and setup. Maintainability. Tools needed to maintain and repair. Logistics of spare parts. Shipping and storage. Reliability, failure rate, and time to repair. Configuration management.
Facility personnel	Operational environment. Support services. Enabling systems. Safety and security (misuse and loss).
Regulators and government	Compliance with standards and regulations. Consumer safety.
Information technology (IT)	IT hardware (networks, servers, workstations). Application support software. Security (misuse and loss).
Disposers	Use of hazardous materials. Safety and security (deposition of protected or classified data). Environmental impacts. Reusability of parts and materials.
The public at large	Safety and security (misuse and loss). Environmental impacts.

4.3.3.2 Creating a Stakeholder Register The project team must identify and manage the stakeholders, or classes of stakeholders, who will participate with the project team to develop the SOI. One approach that can be used is to develop a stakeholder register that includes a list of stakeholders involved in some way with the SOI across its life along with key information for each stakeholder. This register is often included in the Project Management Plan (PMP) and/or Systems Engineering Management Plan (SEMP) or a project's Communication Plan or Stakeholder Management Plan.

Key information for each stakeholder includes:

- Stakeholder's name, location, mailing address, email, phone, etc.
- Stakeholder's organization and job title.
- Whether the stakeholder is internal or external to the organization.
- Stakeholder's PM or SE role: How are they involved? What is their "stake"? Are they an *Approval Authority*? Are they decision makers, end users, controllers of assets or resources, influencers, interested parties, procurement, legal, or compliance?
- Which part of the system are they primarily involved? In what way?
- What lifecycle(s) are they primarily concerned with or involved in? In what way?
- Which PM or SE discipline do they represent?
- Classification of the stakeholder in terms of "VIP" as discussed earlier.
- What information does the project need from the stakeholder?
- What information does the stakeholder need from the project team?

The stakeholder register can be useful; often there is a large set of stakeholders, and stakeholders often change over the SOI's lifecycle. This means that elicitation activities may need to be revisited if stakeholders change during development. It is recommended that the project team re-evaluate the stakeholder community periodically to ensure successful engagement with stakeholders, keep them engaged, and manage changes in stakeholders and their needs. *Refer to the GtNR for an example stakeholder register.*

4.3.3.3 Stakeholder Considerations

4.3.3.3.1 Position and Role Not all stakeholders are equal. Based on their position and role, some stakeholders have more "power" and influence than others (such as the customer and the *Approval Authority*). In this case, higher-ranked stakeholders needs will have more importance (higher priority) than those of lower-ranked stakeholders. Higher-ranked stakeholders often have a broader perspective and think at a higher level of abstraction than other stakeholders.

The rank of stakeholders is used to resolve any needs or requirements that are conflicting or cannot be met by the proposed solution within the defined constraints. Higher ranking stakeholders are often paying customers, sponsoring agencies, acquirers, and the *Approval Authority* which will have the authority to accept, qualify, certify, or approve the SOI for its intended use.

A discussion should occur between the acquirer(s), developer(s)/project team, and any stakeholders that supply resources or interact with the SOI over its lifecycle to determine roles and relative ranking of the stakeholders.

4.3.3.3.2 Communication Approach The project team's communication approach with the stakeholders is heavily influenced by stakeholder rank, whether the stakeholder is internal or external to their organization, stakeholder availability, and stakeholder accessibility.

The communication approach with stakeholders should be decided and recorded as part of the project's Communication Plan or Stakeholder Management Plan. Communicating with internal stakeholders can vary in form and timing compared with communicating with external stakeholders. Forms of communication may be written documents, electronic communication, verbal communication, a gate review, or a variety of forms as appropriate. A project team point of contact (POC) should be assigned for communication with various stakeholders and to develop a plan for when and how things will be communicated so that communication channels are understood, and messaging is consistent.

Interaction with stakeholders is not limited to the elicitation activities at the beginning of the project. The project team must establish and maintain frequent communications with the stakeholders throughout the lifecycle activities. These communications should not be limited to just formal "gate reviews" but rather enable frequent informal participation and feedback. A fundamental premise of Agile development methods is frequent communication between the project team and key stakeholders, which can enable continuous validation of the various artifacts generated throughout the development lifecycle.

4.3.3.3.3 Availability/Accessibility For the project to proceed on schedule, it is important for the stakeholders to be available and accessible when needed, especially for projects using Agile methodologies. The project will need to consult each stakeholder to make sure they are available when needed or have a designated individual take their place with the authority to represent the specific stakeholder.

There are cases where a stakeholder becomes involved late in the development lifecycle with major needs that were not communicated when the project started. Depending on the rank or importance of the stakeholder, this could result in significant rework and resulting cost and schedule issues.

In some cases, accessibility can be an issue for higher-ranked or external stakeholders. If a key stakeholder is not accessible, it may be necessary for them to name a designated individual or surrogate who can represent their views to the project team.

4.3.3.3.4 Personas/User Classes For products that have a significant variety of users, the users may have different characteristics that must be considered regarding factors like demographics, anthropometrics, education, skills, language, needs, training, job function, and lifecycle stage involvement. In addition, different classes of users may be associated with different sets of use cases associated with their specific role.

Often, each user class is given a name; this named user class is referred to as a "persona." When used within need and requirement statements, the persona represents the characteristics defined for that user class.

4.3.3.3.5 Changing Stakeholders There may be cases where a stakeholder moves on to a different project and is replaced. When this happens, the replacement stakeholder may not agree with the perspective of their predecessor; they may have different priorities, and their needs are different from what was communicated when the project was formed. Depending on the rank/importance of the stakeholder, this could result in significant rework and resulting cost and schedule issues. Highly visible, long-term projects are especially vulnerable to changing stakeholders, especially when the stakeholder is of higher rank, controls the funding, and is politically motivated.

When stakeholder inputs change due to new stakeholders, it should be treated as a change in stakeholder needs.

4.3.3.3.6 Changing Stakeholder Needs During a project, stakeholder needs often change. There are several reasons for this. First, the operational environment (physical as well as cultural) may change over time due to changes in external systems with which the SOI must interact, changes to stakeholders as discussed above, changes in the drivers and constraints, or changes in the problem, threat, or opportunity. Again, these changes are more common for projects with longer development times. There have been cases where a system is developed and delivered five years after the project started, only to find that the original problem, threat, or opportunity no longer exists or has significantly changed.

Another reason for change is the fact that SE is a knowledge-based practice. As stakeholders become more informed and have a better understanding of the problem and appropriate solutions, their needs may change. As an example, in the chemical processing industry, when addressing needs from process operators for the implementation of automated control systems, various scenarios were explored to define a complete and consistent set of needs. After six months of lifecycle concepts maturation activities, a review of the lifecycle concepts and resulting needs with the operators, the project found that the operators had learned significantly new concepts that radically altered their original needs. Just by discussing the issues and exploring alternative concepts, their viewpoints changed, and they began to think more critically.

When it is possible that stakeholder needs may change, such changes represent a risk to the project's success. Part of the elicitation activity is to ask stakeholders about possible changes and their likelihood. These possible changes will be evaluated, and the risk will be quantified. The project can then determine how they plan to manage these risk of the possible changes. *Refer to Section* 4.3.7 *Identify, Assess and Handle Risk).*

The risk of changing stakeholder needs can also be mitigated by involving the key stakeholders, like the operators in the example above, in the lifecycle concepts analysis and maturation activities. They are often best qualified to define value and feasibility when selecting concepts from the alternatives. Once a set of feasible lifecycle concepts has been selected, the resulting integrated set of needs will reflect this concept, and the possibility of changes in the set of needs and resulting design input requirements is reduced. *Refer to Section 14.2.5 for a more detailed discussion concerning managing change and assessing the impacts of a change.*

For needs that have a degree of uncertainty and are thus likely to change, the attribute *A26 – Stability*, can be used. Furthermore, if a stakeholder states a number or quantity but is not certain if that number is correct or is likely to change, a common practice is to but "[....]" around the number and assign a TBD or TBR to that value. *For a more detailed discussion on TBDs and TBRs, refer to Section 4.6.3.1, Managing Unknowns.*

4.3.3.3.7 Stakeholder Representatives Often, there will be multiple members of a stakeholder group, for example, users, operators, marketing, sales, safety, regulators, customers, and the public, who will be buying, using, operating, or maintaining a product or may be affected by the product in some way. It may not be practical to collaborate with every member of the group to elicit their needs and requirements. In this case, a means must be implemented to name a representative of the group concerning their needs and requirements who has the authority to speak for the group. For safety, a representative from the safety office; for users, a representative from a user group. For consumers who will be buying a commercial product, a member of the marketing organization, or a consumer/user group. For standards and regulations, a member of the organization's compliance group. For the public, a representative from the applicable regulatory agency or consumer safety organization.

4.3.4 Elicit Stakeholder Needs

The project team engages the stakeholders associated with the SOI to understand their needs for all lifecycle stages. Common elicitation techniques (*Section 4.3.4.3*) are employed in accordance with the organization's SEMP and PMP. Needs come from multiple sources, so eliciting and capturing the stakeholder needs represents a significant effort for the project team. Because of this, adequate resources and time must be included in the project budget and schedule to enable these activities.

Stakeholder needs originate from concerns, problems, challenges, issues, risks, opportunities, experience, failures, and successes. During elicitation, the project team should collaborate with the stakeholders to understand the source and context of their stated needs and requirements.

The elicitation activities allow the project team to discover and understand what is needed, what processes exist, how stakeholders expect to interact with SOI, what happens over the SOI's lifecycle (good and bad), what modes, states, and transitions the SOI might undergo or experience during use (such as nominal, alternate nominal, and off-nominal operations, as well as misuse and loss scenarios). Misuse addresses cases where unintended users could misuse the system in the wrong way or for the wrong purpose, resulting in the SOI operating in unintended ways that could result in losses of some kind (for example, loss of the SOI, loss of the intended use of the SOI, loss of information, loss of money, and even loss of life).

To help ensure nothing has been left out during an elicitation session, toward the end of each session with a stakeholder, the project team should ask, "What should we have asked that we didn't?," "Is there anything else you would like to add?" "What are the loss scenarios you are concerned?," and "In what ways could the SOI be misused?"

Note: Many organizations limit needs definition to the elicitation of user or customer "requirements." As discussed in Section 4.3.1, *there are other stakeholders directly associated with the SOI, both internal and external to the organization that must be included in the elicitation activities. Failing to include all stakeholders can result in issues concerning completeness, consistency, correctness, and feasibility of the resulting integrated set of needs and design input requirements. The scope of needs definition is much broader than just the customers or users. Rather than focusing on only the voice of the customer (VOC), the voices of all stakeholders (VOX) must be considered.*

4.3.4.1 Plan for Elicitation Before conducting elicitation activities, the project team must plan for each stakeholder elicitation session or activity. Key things to determine include:

- The system-level stakeholders who will be involved, how frequently (once or multiple times), and when.
- The types of information that are needed from the stakeholder(s) (*Section 4.3.4.2*).
- The elicitation techniques that will be used. (*Section 4.3.4.4*).
- Existing documentation that should be available, including existing requirements, specifications, interface definitions, or user documents for a similar product, regulations and standards, customer survey info, problem logs, and consumer complaints as examples.
- How information will be captured during the elicitation sessions.
- The form and media that will be used to document and manage the captured information.

4.3.4.2 Information to Be Elicited from Stakeholders For each stakeholder, there is a common set of information that should be the focus of the elicitation interactions.

4.3.4.2.1 Problem or Opportunity Statement At the beginning of the elicitation activities, present the problem, threat, or opportunity statement defined at the operational level to the SOI level stakeholders and ask if they agree? Also ask:

- Why does the stakeholder think the new SOI is needed?
- Is the stakeholder aware of any problems concerning the existing SOI?
  - Are there needed changes to inputs? Outputs? Functionality? Performance? Quality?
  - What changes would they like to see to address these problems?

4.3.4.2.2 MGOs and Measures Present the MGOs and measures to the stakeholders. Questions to ask:

- Are there any changes they think need to be made to the Mission statement?
- Are there changes they think need to be made to the goals or objectives?
- Are there other measures that need to be defined? Measures changed?
- Do they think the values stated in the MGOs and measures are feasible? If not, why?

These first two items have a dual purpose.

1. It is important to align the stakeholders with a common vision for the project and SOI to be developed. This information will help define the context in which the stakeholders are addressing a specific problem, threat, or opportunity, as well as ensure everyone is "on the same page." Without this understanding, it is common to get conflicting information from the stakeholders based on their individual understanding of the problem, threat, or opportunity and their unique vision of the reason for the SOI and outcomes as a result of developing the SOI.
2. There may have been things overlooked by the organization's strategic and operational-level stakeholders during the formulation of the MGOs and measures that other SOI-level stakeholder(s) can help fill in the gaps.

These activities are validation of the higher-level needs that are communicated by the MGOs and measures. Any issues must be resolved prior to proceeding with stakeholder elicitation, lifecycle concepts analysis and maturation, and definition of the integrated set of needs.

4.3.4.2.3 Lifecycle Stages Identify the lifecycle stages the stakeholder represents or is involved in. Lifecycle stages could be procurement, development, test, verification, validation, manufacturing, transportation, deployment, installation, transition, training, operations, logistics, maintenance, upgrades, or disposal. Often, there is a distinct set of stakeholders and interfaces associated with each lifecycle stage, each having unique needs. Not addressing a lifecycle stage could result in missed needs. Table 4.1, shown earlier, lists various lifecycle stages and typical types of stakeholders.

4.3.4.2.4 Use Cases For each lifecycle stage, ask the stakeholder to describe a "day-in-the-life" of the SOI. This could include multiple use cases. Address both nominal and alternate nominal scenarios. Capture the initial conditions and state of the SOI at the end of the use case.

Note: When users or operators are defined in terms of "personas" discussed earlier, the project team will develop use cases for each defined persona. When doing so, it is important to base the use cases on actual users or operators. This information can be obtained using the elicitation techniques and methods listed in Section 4.3.4.3. Actual observations of actual users and operators in the actual operational environment are highly recommended.

The resulting use cases provide key insights into the stakeholder needs concerning features, capabilities, functionality, performance, interaction with other systems, standards, regulations, physical attributes, quality, safety, security, resilience, sustainability, etc.

4.3.4.2.5 Off-nominal Scenarios In addition to the nominal and alternate nominal scenarios, the stakeholder should also be asked about off-nominal scenarios – what could go wrong, or currently, what often goes wrong? (*The off-nominal scenarios will be a major source of risk discussed later as well as Failure Mode and Effects Analysis [FMEA].*) Included with the off-nominal scenarios, ask about safety and security – what are the threats? Hazards? What needs to be protected? What are the *misuse cases and loss scenarios*?

4.3.4.2.6 Capabilities, Features, Functionality, and Performance Different users/operators often have different needs. What capabilities, features, functionality, and performance do they need from the new SOI? Why? While it is common to address "functional requirements" that focus on the primary purpose and use of the SOI, functionality in terms of safety and security are equally important. What capabilities, functionality, and performance must be addressed to address the stakeholder's safety and security issues identified earlier?

- How many users/operators?
- What capabilities and functionality do the users need? For each function, what is the expected performance? Is there a specific trigger for a given function?
- Within what state or mode of the SOI does a given function apply or need to occur?
- What are the conditions (circumstances, events, or triggers) under which the function is to be performed?
- What constraints apply to performing the function?
- What is the operational environment in which the function occurs, or the performance level apply?
- Does the function occur repeatedly or continuously?
- How should the SOI respond to or behave when an off-nominal condition occurs? This could be an off-nominal input or an operational environment outside nominal parameters.

4.3.4.2.7 Interactions/Interfaces with External Systems During the stakeholder's description of the activities during each lifecycle stage, pay particular attention to any interactions of the SOI with external systems. These interactions could represent interfaces or could be the result of induced environments, or the competition of resources. *Refer to Section 4.3.6.3 for more information concerning what the project team needs to consider during elicitation activities concerning interfaces.*

4.3.4.2.8 Human Interactions During the stakeholder's description of the activities during each lifecycle stage, pay particular attention to any interactions of users, operators, installers, maintainers, and disposers with the SOI. Understanding the human/machine interface (HMI) is critical for acceptance. Often, these types of interactions are managed by human factors experts trained and knowledgeable in Human Systems Engineering (HSE) and Human Systems Integration (HSI). Many of the quality (-ilities) needs and requirements discussed later involve human/machine interactions.

4.3.4.2.9 Priority/Criticality It is common for a specific stakeholder to consider some aspects of the SOI more important than others. During elicitation activities, it is important to ask the stakeholders to prioritize what they are asking for. Some things will be especially important to the stakeholder, while other things may be nice-to-have but not critical to the system being able to achieve the agreed-upon MGOs and measures. There will be some things that the stakeholder may be able to live without given budget or schedule constraints or conflicting needs. Ask them if they could identify which of their needs could be cut in the event of a project de-scope. Establishing priorities is important when having to make tradeoffs as well as when a project must be de-scoped due to budget or schedule issues.

While priority addresses what stakeholders think is the most important, criticality addresses needs from the perspective that their realization is essential in terms of the system's ability to fulfill its primary purpose in the operational environment when operated by its intended users. Criticality also applies to needs associated with risk mitigation. Establishing criticality is important when tradeoffs are being made or the project is de-scoped; critical items should never be deleted from the integrated set of needs, and it is important when accessing and managing change. All critical items are of highest priority.

When there is a difference in opinion as to the relative priority or the criticality, the ranking of the stakeholders must be considered. In some cases, the difference in opinion may be a communication issue, where one party does not have the knowledge or insight of the other party. When recording stakeholder inputs, it is important to record within the SOI's integrated dataset the priority or criticality of their needs and rationale concerning the designation. *(Priority and criticality are key attributes of both needs and requirements expressions, as discussed in Chapter 15.)*

The benefits of establishing priorities and criticality include enabling the project team to better plan and manage the development effort, provide trade space, help improve communications, and often result in fewer needs and design input requirements. As discussed in Chapter 10, organizations may use priority and criticality as criteria to focus system verification and validation activities to reduce costs and shorten delivery times. From a safety or security perspective, regulatory agencies tend to focus more on changes to critical needs and requirements.

4.3.4.2.10 Quality Ask the stakeholders what are their expectations for quality (-ilities) such as:

- *Usability* – With or without training?
- *Reliability* – To what level? Over what length of time? Mean time to failure?
- *Availability* – To how many? For how long? Over what time?
- *Serviceability/Sustainability* – Through user updates, system upgrades, etc. Mean time to repair?
- *Scalability* – How many users? How much data and information?
- *Testability/Inspectability* – Built-in capability? With provided tooling and/or equipment?
- *Agility/Adaptability/Resilience* – How well and how fast can the system to respond or adapt to changes to the mission, threats, and operational environment?

Note: There are many "-ilities". Which applies depends on the SOI and needs of the stakeholders. A search for "non-functional requirements," "quality requirements," or "-ilities" on the Internet will result in a much larger list with definitions and examples.

4.3.4.2.11 Drivers and Constraints Ask the stakeholders what they think the drivers and constraints are. This includes available technologies, standards, regulations, operational environment, production capabilities, cost, and schedule. *Refer to Section* 4.3.6 *for a detailed discussion concerning drivers and constraints.*

4.3.4.2.12 Issues/Risks Ask the stakeholders what issues/risk they think could impact the project team's ability to successfully develop and deliver a winning product. (*All issues must be recorded in the SOI's integrated dataset. They will be a major source of risk discussed later.*) Also, ask about issues/risk associated with user/operator interactions with the SOI. What are the hazards? Safety and security risks? Which risks need to be mitigated? In what way? What is the likelihood? The impact? *See Section* 4.3.7 *for a more detailed discussion on risk identification, assessment, and handling.*

4.3.4.2.13 Assumptions Ask the stakeholders what they are assuming to be true for what they have said to be valid. Assumptions are propositions that are taken for granted as if they were known to be true, whether they are true or not. Assumptions can relate to the business or mission, technology (maturity, feasibility, and performance), resource availability (people, facilities, enabling systems, etc.), safety, security, resilience, cost, schedule, expected outcomes, and stakeholder needs. An assumption must not be considered to be true until there is objective evidence that it is true. If a need or design input requirement is based on an assumption that later turns out to be false, that need may not be valid. All assumptions must be recorded *in the SOI's integrated dataset.*

4.3.4.2.14 Rationale For each need elicited from the stakeholders, it is important to capture the rationale concerning "why." Rationale helps understand intent. To understand the real need, stakeholders may have to be asked "why?" multiple times. If the need or requirement is based on an assumption, that assumption should be included in the rationale attribute. If the need includes a number, the rationale should also include a description from which that number was derived. *If a stakeholder cannot provide rationale, why include the need or requirement in the set?* This information will be used to document the rationale for the resulting needs and design input requirements in the rationale attribute defined in Chapter 15.

The goal of elicitation is to provide an implementation-free understanding of the stakeholders' needs and requirements by defining what is expected (design inputs) without addressing how (design outputs) to satisfy the set of needs.

It is important to always remember that final acceptance of a SOI is based on the system passing system validation. If any of the above information is not well understood and reflected in the lifecycle concepts and integrated set of needs, there is a good chance the SOI will fail system validation.

Example Assumptions Documented by the LIR Project Team

The following are assumptions the project team identified. During elicitation, the stakeholders were asked to validate each of these.

The number of shifts will stay the same.

The hours of facility operation will stay the same.

The number of Jar Processing Systems (JPS) within the facilities will not change.

The operational environment will remain as it currently is.

The size of jars/lids will be the same for a given shift.

Jars are sitting, unrestrained on the conveyor belt; however, the location of jars on the conveyor belt will not change.

During installation of the lids on the jars, the LIRs must not change the location of the jars on the conveyor belt.

Timing of jar deliveries will not change during a shift.

The LIR will be notified when a jar is in place and ready for lid installation.

The daily supply of Lids of the proper size will be supplied to the LIR by a maintenance technician prior to shift start. *The method by which the lids will be delivered and made available to the LIR is TBD.*

The lid supplier will be able to supply the increased number of lids.

The jar supplier will be able to supply the increased number of jars.

The room in which the LIR will operate will have sufficient space for operations.

The LIRs will be monitored and commanded from the existing facility control room.

4.3.4.3 Use of Checklists As part of planning for an elicitation session, it is useful to develop a checklist tailored to the specific stakeholder. Table 4.3 shows an example checklist.

The checklist in Table 4.3 is a generic list of topics to address during elicitation. In practice, the project team will brainstorm what they want to ask the stakeholders and what information they need from them. This effort will result in a generic master list, they will tailor depending on the individual or group of stakeholders with whom they are meeting.

TABLE 4.3 Example Checklist for an Elicitation Session.

Purpose/Function of SOI	Operational environments	Priority and Criticality
Performance levels	Storage	Interfaces
Safety and Security	Transportation	Assumptions
Issues and Risks	Process improvement	Constraints
Maintenance or upgrade	Reporting	Assembly/disassembly
User interfaces	Compliance	Training
Access control	Documentation	Cost/schedule

Example List of Information Needed Prepared by LIR Project Team Prior to Elicitation

Information the LIR Project Team needs from the stakeholders:

Stakeholder perspective on how successful operations will be conducted once the LIRs are in place.

What changes will have to be made to the facility to accommodate the increase rate of lid installation?

What other systems or processes are affected by using the LIRs?

What are the drivers and constraints?

What could go wrong during operations?

What are the risks?

Are there any issues associated with automating the lid installation of the jars?

What are the stakeholders assuming for what they are telling us to be valid?

What does the LIR Project Team want to know more about?

Jars: Size, required torque for lid installation, location of portion of jar where the lid will be installed, best location to grasp the jar to keep the jar from rotating during lid installation.

Frequency and position of jars delivered on the conveyor belt – JPS ICD.

Operating shift. Actual number of hours per day, which hours? Number of days/year.

How do the LIRs know when the shift begins/ends? When a jar is in place?

Lids: When, where, how delivered to LIRs, in what form, accessibility, location within the LIS.

Lids: Number, size, shape, grip locations, and allowable grip force.

How will the LIRs be able to obtain the lids?

The operational environment: Temperature, humidity.

Allowable induced environment by the LIRs: EMI/EMC, noise, thermal.

Space within the LIS room.

Power and quality of the existing power within the Lid Installation System (LIS)?

Location and size of the opening between the LIS and the JPS conveyor belt.

LIR storage environment when not in the LIS room – crated and uncrated.

Transportation/shipping environments.

Expectations for safety, security, reliability, maintainability, lifetime, and other -ilities.

Interfaces/interactions of LIRs with JPS, LIS, Facility Control Room (FCR), operators, and maintenance technicians.

Codes, regulations, and standards the project team and LIRs must be compliant.

4.3.4.4 Elicitation Techniques and Methods There are various techniques and methods used for elicitation. Which is used depends on the specific stakeholders and information to be obtained. Below is a summary of the more popular methods. These methods are quite common and well recorded, so this Manual only provides a brief description.

- *Brainstorming*. Large collections of inputs from a group that must be distilled. Best if structured to consider all lifecycle stages.
- *Workshops or focus groups*. Gain more clarification and detail concerning specific topics, lifecycle stages, stakeholder needs, or design input requirements.
- *Interviews*. One-on-one discussions that can define a use case, scenario, or concept, bound a problem, and/or identify specific stakeholder's needs or requirements.
- *Feedback and document analysis*. Information on existing or similar products can remind the project team of what types of needs or requirements may exist and inform them of problems or improvements desired in a new product.
- *Interface analysis*. Paper study or observation to identify interactions of the SOI with external systems and types of interactions with which the SOI must be compatible—form (physical and mechanical characteristics) and functional (input and output of signals and communication, for example). Includes human interactions with the SOI.
- *Observation of comparable products*. Stakeholders can identify features, functionality, performance, and capabilities that are desired or need to be changed or added based on competing products.
- *Prototyping*. Provide a prototype based on initial understanding of stakeholder's needs or requirements to allow stakeholders to explain how the prototype meets or misses their needs

or requirements. This provides an opportunity for stakeholders to point out what they like and do not like and what they would like to be changed, added to, or removed.

- *Surveys/Questionnaires*. Used to obtain a general feel for stakeholder's needs or requirements; however, it can be challenging to develop a useful survey or set of questions that provides clear information without the need for follow-on discussions.
- *Site visits*. The value of site visits cannot be overstated. Site visits give the project team first-hand knowledge of the operational environment and document the operations from the perspective of the actual users/operators in the actual operational environment. Site visits allow the project team to make observations concerning not only the operational environment but also specific actions and human factor considerations that may be overlooked or not communicated in other elicitation settings. This is an extremely important consideration when the customer may send a procurement representative to the elicitation meetings rather than actual users/operators, which often results in missing needs and requirements.
- *Models and diagrams:* The old saying "a picture is worth a thousand words" applies. Sometimes the stakeholders will gain new insights when seeing a model or a diagram. Upon viewing a model or diagram, the stakeholders may identify missing or incorrect information. Any inconsistencies may be more apparent, enabling the project team to collaborate with the stakeholders to reach a satisfactory resolution. *Refer to Section* 4.5 *for a detailed discussion concerning models and diagrams.*

4.3.5 Document Elicitation Outcomes

It is critical that all the information obtained from the elicitation activities be recorded in the SOI's integrated dataset. There are various forms and media that can be used to document the information gained from the elicitation activities. In some cases, there is a predefined format defined within the SE tools. In other cases, the information is captured as text. Below are common forms used:

- Operational scenarios
- User stories
- Use cases
- Diagrams/drawings
- Models
- Lists
- Tables
- Text

The elicitation medium used is also important. As discussed earlier, from a data-centric approach to SE, this information must be recorded in the SOI's integrated dataset so that it can be accessed quickly and supports traceability so that subsequent artifacts can be traced back to this information. A major benefit of practicing SE from a data-centric perspective is the increased use of models as analysis tools to both establish traceability between artifacts as well as enable the ability to view the information within the SOI's integrated dataset from different perspectives depending on the needs of the project team members.

When defining personas/user classes, it is important to record them within the integrated dataset so that they can be referred to within need and requirement statements. One approach is to include the personas/user classes in the project data dictionary or glossary.

The individual outcomes from the elicitation activities represent a unique perspective of a stakeholder or group of stakeholders. These perspectives will be analyzed and integrated into an integrated set of needs. The form used to record the elicitation outcomes depends on the organization, culture, processes, toolset, and domain.

4.3.5.1 Considerations Concerning the Elicitation Outcomes There are several key challenges that must be considered when evaluating the information resulting from the elicitation activities.

4.3.5.1.1 Terminology, Abbreviations, and Acronyms It is common that stakeholders will use specialized terminology, abbreviations, and acronyms. Critical to being able to define well-formed needs and requirements is for the project to establish a *project glossary* or *data dictionary* to record these terms and their definitions. The project team must ensure consistency in the use of these terms for all needs and requirements, as well as models and other artifacts generated across the lifecycle.

4.3.5.1.2 Ambiguous Terms and Phrases The elicited needs will frequently include ambiguous terms and phrases such as *user-friendly, robust, easy-to-use, works fast, safe, affordable, pleasant, easy to test, cost-effective, and works just like the last one only better*. It is common for stakeholders to state their needs at a high-level of abstraction, resulting in ambiguity as to their exact intent as it applies to the SOI.

For example, the need for the SOI to be "safe" or "secure" is ambiguous. The project team cannot meet this need without more detailed information because there are no measures from which to design, verify or validate. Therefore, further analysis must occur to define "safe" or "secure" in terms against which the system can be verified and validated. What hazards and associated threats could exist that make the system unsafe or not secure? What level of safety or security is needed? Which safety or security standards and regulations need to be imposed on the SOI, which will result in a system that meets the stakeholder needs concerning safety or security? What capabilities, functionality, and performance are needed to address these?

During elicitation activities, the project team must collaborate with the stakeholders to remove these ambiguities. With clarification, the project team may be able to meet the intent of each stakeholder need with additional refinement from the original statement. To remove the ambiguity, the project team may need multiple cycles of elicitation with the stakeholders to resolve ambiguous statements. *When there is ambiguity, ask, "What do you mean by xxxxx" "Please elaborate on what you mean by xxxxx" "What do you need the SOI to do such that the system is xxxxx?"*

Below is a classic example of a need that is not appropriate:

"The marketing stakeholders need the product to perform [in some manner] 10 % better than the competition, three years from now."

Unless the project team can see into the future, there is no way they could meet the intent of this need statement. They would have to work with the marketing department to define needs that are attainable in the present based on an assessment of the maturity of technologies needed to attain the required level of performance. Alternatively, this could drive a research and development activity to identify and mature the technologies necessary to achieve a target value.

4.3.5.1.3 Stakeholders Unsure of the Problem to be Addressed The SOI-level stakeholders do not always know what is needed or what specificproblem, threat, or opportunity the project and resulting SOI are intended to address. Therefore, it is critical that this information is clearly communicated to the stakeholders prior to the elicitation activities (*Section* 4.3.2 *and* 4.3.3) and then presented to the SOI-level stakeholders at the beginning of the elicitation sessions, as discussed earlier.

4.3.5.1.4 Needs Expressed as Requirements It is often the case that the stakeholder needs may first be annunciated as requirements, even if they are not substantiated with analysis and lack the characteristics of well-formed requirements as defined in the GtWR. These must be subject to the same reconciliation and prioritization as other needs communicated during elicitation sessions. Failure to reconcile and prioritize all these needs incurs technical debt and ongoing risk for system development.

4.3.5.1.5 Needs Expressed in Terms of Implementation Stakeholders often state their needs as implementation statements or solutions rather than address the problem and their needs concerning a solution to the problem. The focus should be on the "what," not "how." When a stakeholder states a specific implementation or solution, ask, "Why?" and "What does that implementation or solution allow you to do?" A common approach is to ask "why" multiple times. The answers will help uncover the real needs. In addition, avoiding implementation allows the project team to be more innovative in defining an effective design solution.

4.3.5.1.6 Addressing Needs for Lower-level System Elements Stakeholders will often address needs for lower-level system elements that make up the SOI, rather than focusing on the integrated system. For some, it is hard to address higher levels of abstraction when their level of knowledge and involvement will be associated with a specific subsystem or system element. For example, a scientist who has developed a method for detecting a specific characteristic in a biological sample may find it difficult to think at the integrated system level for an instrument that will enable their method to be implemented.

Some developers find it more interesting to jump to an architectural and design solution, rather than spend the time to understand the problem and develop lifecycle CNR.

For any given SOI, have the stakeholders focus on what would be observable externally for the integrated SOI, rather than diving into the internal architecture and design of the SOI.

4.3.5.1.7 Not Articulating Implicit Needs Stakeholders will often have both explicit and implicit needs. Stakeholders will often focus on functionality, performance, and user interfaces, assuming everything else will be addressed by the developing project team. A common error is that the stakeholders assume that the project team knows about their assumptions. The stakeholder's expectations for their implicit needs are often the same as what they have explicitly stated.

If the implicit needs are not met, the system may fail system validation, even though those needs were not explicitly stated. To avoid this issue, the project team must ask questions concerning areas of interest not explicitly stated by the stakeholders. For example, what quality attributes do they need the SOI to have? What standards or regulations need to be adhered to? What are the drivers and constraints? Do you have any safety or security concerns that need to be addressed by the realized system?

4.3.5.1.8 Limited Coverage of "Non-functional" Needs Stories, scenarios, use cases, system concepts, operational concepts, concepts of operation, etc. often focus on primary functionality, performance, and interactions with other systems during operations, but are often incomplete, not addressing other "non-functional" needs such as enabling functions, quality (-ilities), safety, security, resilience, sustainability, design and construction standards, regulations, and physical characteristics. These other needs must also be included – otherwise, the integrated set of needs and resulting set of design input requirements will be incomplete. *Refer to the INCOSE SE HB* [1] *Section 3.1.10 concerning sustainability; and Section 3.1.9 concerning resilience.*

4.3.5.1.9 Inconsistencies and Conflicts Individual stakeholders often have a narrow perspective and only focus on their specific interests rather than considering the needs of other stakeholders. Their needs may unknowingly be inconsistent or in conflict with other stakeholders. These conflicts make it difficult for the project team to satisfy the conflicting needs without further clarification and understanding of the priority, criticality, and rationale for each of the conflicting needs.

The project team must ask questions to encourage the stakeholders to consider a system view and how one stakeholder's perspective could affect other stakeholders. For example, a stakeholder may demand a need be met, no matter the cost or impact on the project's schedule, or ability to meet other stakeholders' needs. The project team may have to help the stakeholder rethink and update their needs.

A key role of the project team is to identify and resolve these inconsistencies and conflicts. During elicitation, the project team can collaborate with the stakeholders to point out these inconsistencies and conflicts and seek a resolution. In Section 4.4, when the preliminary integrated set of lifecycle concepts is defined, these inconsistencies and conflicts can be flagged and addressed during lifecycle concepts analysis and maturation activities.

In some cases, the conflicts are mutually exclusive, so the project team will have to decide which stakeholder needs to implement. The reason for the conflict could be that one stakeholder did not understand the MGOs and measures defined by management or that the needs of one of the stakeholders were based on false assumptions. In either case, the project team will collaborate with the stakeholders to resolve and get agreement on how to move forward. Depending on stakeholder rank, one of the stakeholders may have to rethink and reset their expectations. If they cannot, the project team may need to seek the involvement of the project champion.

4.3.5.1.10 Feasibility Another issue is that the initial stakeholder needs may not be realistic or feasible given the constraints the project team must work within. Because of this, stakeholder perceptions of their needs, no matter the form, should be considered as tentative or preliminary statements subject to further analysis, refinement, prioritization, and reconciliation during lifecycle concepts definition and maturation. In these cases, some stakeholder expectations will have to be modified by what is realistic and feasible given the drivers, constraints, and risks. Again, depending on rank, some stakeholders may have to rethink and reset their expectations to what is feasible, and the project champion may have to become involved.

What defines the SOI is the integrated "set" of needs and resulting "set" of design input requirements. While individual needs may be feasible, the combined set may not be feasible in terms of cost, schedule, technology, and risk. Every stakeholder need, as initially stated, may not be able to be satisfied. *Refer to Section 4.6.3.4 for a more detailed discussion concerning needs feasibility and risk.*

4.3.5.1.11 Conflicting Priorities In addition, one of the parameters of the triad "faster, better, cheaper" may be out prioritized by one of the other parameters of the same triad. Because of this, it is important to decide at the beginning of the project which of the stakeholders has the authority to decide on the necessity, criticality, and prioritization of the operational-level and system-level stakeholder needs.

Most often, the evaluation will require an analysis of the SOI lifecycle concepts so that the developer, acquirer, and user can understand the tradeoffs available among competing sets of stakeholder needs. The project team will use the feasibility, priority, and criticality assessments to prioritize functions, features, and capabilities to help ensure that they are truly "requirements" and *necessary for acceptance* rather than simply preferences or "desirements." The decision authority can then use this information to make an informed decision. *Refer to Section 4.6.3.4 for a more detailed discussion on Needs Feasibility and risk.*

The project team must understand and take into consideration the above considerations and issues. This may involve several iterations with the stakeholders to resolve the issues and reach agreement. Collaborating with stakeholders often involves soft skills in addition to the hard skills used for SE.

4.3.5.2 Elicitation Completeness Checklist Following the elicitation activities to help ensure completeness, the project team should ensure:

- All relevant stakeholders have been involved in the elicitation activities.
- Al lifecycle stages have been addressed.
- All elicitation outcomes have been captured.
- The elicitation outcomes are communicated at the right level of abstraction. Are they an implementation-free understanding of the needs – defining what is expected (design inputs) without addressing how (design outputs) to satisfy the MGOs and achieve the defined measures?
- Safety, security, resilience, and sustainability expectations have been addressed.
- Expectations concerning quality have been addressed.
- Expectations concerning standards and regulations have been addressed.
- Issues and risks have been recorded, and a plan for mitigation has been established, including misuse cases and loss scenarios.
- Rationale has been captured for each stakeholder need and requirement.
- The priorities and critical needs of the stakeholders have been established and recorded.
- Conflicts and inconsistencies have been captured, and a plan for resolution has been developed.
- Ambiguous terms such as user-friendly, easy-to-use, fast, high quality, good tasting, affordable, robust, etc. have been resolved.
- The feasibility of the recorded outcomes has been assessed, and those that are questionable have been identified for further analysis during lifecycle concepts analysis and maturation activities.
- All interactions with external systems identified during elicitation have been recorded.
- Drivers and constrains from the elicitation activities have been recorded.

Before the project team can proceed to document, analyze, and mature the lifecycle concepts, they must implicitly address the drivers and constraints as well as the risks that the lifecycle concepts must take into consideration.

4.3.6 Identify Drivers and Constraints

Completeness is a major consideration during *Lifecycle Concepts and Needs Definition*. During the elicitation activities, the project team should have recorded drivers and constraints within the SOI's integrated dataset. To help ensure completeness, it is helpful for the project team to explicitly focus on identifying drivers and constraints as an additional concurrent activity. In doing so, drivers and constraints not communicated by the stakeholders during elicitation will be identified, along with the stakeholders associated with those drivers and constraints that may not have been included previously in the elicitation activities. Because of this, the elicitation activities and identification of drivers and constraints activities are complementary.

Drivers and constraints represent a major source of needs and requirements that drive and constrain the lifecycle concepts analysis and maturation activities, as well as the solution space available to the project team. Compliance is mandatory – failing to show compliance will result in the SOI failing validation, certification, acceptance, and approval for use.

Drivers and constraints can include:

- Design constraints (parts, materials, organizational design best practices, etc.).
- Standards and regulations (processes, workmanship, technical, design, testing, qualification, acceptance).
- Regulations (law).
- Supply chain (parts, materials, rare-earth metals, politics, sustainability, etc.).
- Production constraints (existing technology, facilities, equipment, cost, throughput, yield, etc.).
- Human factors (HSI/HMI).
- Operational environment (natural, induced; social, cultural).
- Existing systems (interactions, interfaces, dependencies).
- Technology maturity.
- Cost.
- Schedule.
- Mission drivers – examples for space missions include launch date, launch vehicle performance, orbit, destination, duration, logistics, crewed/un-crewed, orbital mechanics, etc.
- Higher-level requirements allocated to the SOI. At the system level, these will be operational-level stakeholder needs. At lower levels of the architecture, these will be requirements allocated to the SOI from the level above.

Concurrently with the stakeholder elicitation activities, drivers and constraints need to be identified and recorded within the SOI's integrated dataset, which could be a large and time-consuming task.

Once a set of feasible system lifecycle concepts has been defined, traceability between the lifecycle concepts and the drivers and constraints will be established and addressed within the integrated set of needs. Once the integrated set of needs has been transformed into the design input requirements, traceability between the requirements, needs, drivers, and constraints will be established within the project toolset.

The following sections expand on several key drivers and constraints.

4.3.6.1 Standards and Regulations Standards and regulations are a major source of needs and resultant design input requirements. Because of this, compliance represents a significant part of the development activities, cost, and schedule. Showing compliance can represent a substantial portion of design, system verification and validation activities, cost, and schedule.

4.3.6.1.1 Identifying Relevant Standards and Regulations Standards and regulations are documents that contain requirements related to:

- Safety, security, and resilience that address prevention of harm, loss, misuse, and recovery from loss.
- Production processes and workmanship (for example, soldering, crimping, and coding).
- Processes for quality management: CM, PM, SE.
- Design approaches for certain types of systems (e.g., petroleum extraction, distribution, and processing; medical devices; transportation; consumer products).
- Requirements for testing, verification, validation, acceptance, certification, and qualification of certain types of systems (for example, medical devices, pharmaceuticals, safety-critical systems, automotive, aircraft, and human-crewed space systems).

- Compatibility/interoperability (for example, interactions between systems, communication protocols, Application Programming Interface [API], ICD, Interface Definition Document [IDD], connectors).

Standards and regulations can apply to all levels of the system architecture and all lifecycle stages. In some cases, the standards and regulations are written at the design input level of abstraction, stating what needs to be done and why, but not how. The "how" is left up to the architecture definition, design definition, test, and manufacturing organizations.

In other cases, the standards and regulations address specific design "how" implementation-type requirements for the architecture, design, testing, or manufacturing. There are standards and regulations governing specific parts and materials used within a system, as well as their specific configuration and use. Compliance with these requirements will be communicated in the design output specifications.

In the medical device and pharmaceutical world, regulations concerning system validation are of critical importance for approval for use (for example, animal versus human testing, staged approach to human testing, and human trials).

Projects must make sure they include all *relevant* industry standards and standards and regulations mandated by the customer and government regulatory agencies. "Relevant" is highlighted in that it is important that the project only address standards and regulations and portions of those standards and regulations that apply to the project and SOI under development. Rationale must be provided for each standard and regulation, making it clear why that standard or regulation is relevant and why the cost of compliance is necessary.

When identifying the relevant standards and regulations, the project must address the countries where it will be marketed, not just the country where it is produced.

A key mistake is invoking a generic list of standards and regulations on a project without specifically identifying which apply. The project will have to show compliance with each requirement invoked by a standard and regulation – thus, only those requirements within standards and regulations that are relevant to the specific SOI should be invoked by the project to reduce development time and cost.

4.3.6.1.2 Managing Standards and Regulations within the Enterprise Within an organization, business unit, and product line, strategic level and operational-level requirements should identify the applicable standards and regulations that apply to the products and services developed or supplied by the organization. Ideally, the requirements within these standards and regulations will have been imported into the organization's toolset database in a form that will:

- allow the organization to assign applicability of the standards and regulations to specific projects and systems to be developed, and
- allow the projects to establish traces from their implementing requirements to the applicable requirements within a standard or regulation.

For all standards and regulations applicable to an organization, the organization should define the governance and processes at the enterprise, business, and operational levels to be followed by all projects within the organization and show traceability of the processes to standards and regulations to which the processes are compliant.

For each standard or regulation, the organization needs to establish which requirements within the standard or regulation are applicable to the type of products produced or services provided by the organization. It is dangerous to call out a complete standard or regulation when only a portion of the requirements apply. All requirements invoked within the set of design input requirements will have to be implemented in the design and the system must be verified to meet those requirements.

The requirements will then be allocated to the projects and products to which they are applicable. One way to establish this is to develop an applicability matrix [31, 32] for each standard and regulation that is applicable to the organization. In the applicability matrix for a given applicable standard or regulation, sections of requirements or individual requirements are listed as a row heading. Individual projects, products, and services are listed as column headings. For each row, an "X" is placed in any column to which the requirement identified for that row applies. (*Note that in a RMT or similar tool, the "X" represents a link or trace.*)

Need statements would invoke the columns within the organization's applicability matrices for the SOI being developed. For example:

The stakeholders need the SOI to comply with <regulation xyz>requirements as indicated in column F of [regulation xyz <project>applicability matrix.]

Using this approach, the project team would then know which specific requirements in the regulation or standard apply to their SOI and to which they would have to show compliance.

Compliance matrices can also be used to sub-allocate specific regulation or standard requirements applicable to lower-level subsystems and system elements using this same approach.

The number of requirements invoked on a project dealing with standards and regulations can be large, taking considerable time, resources, and money to manage, implement, and show compliance [33]. There are several approaches the organization can take to address this issue.

- One approach is for the organization to derive specific requirements concerning stakeholder expectations for quality, safety, security, etc. rather than calling out all the standards and regulations concerning these topics. The result is often a smaller number of requirements that still clearly communicate the intent and expectations, but can be implemented much more effectively.
- Alternately, the organization could develop its own version of each standard or regulation tailored to the product lines developed by the organization, containing well-formed derived requirements having the characteristics as defined in the INCOSE GtWR (or similar document) that meet the intent of the source requirements within the standards and regulations.

With either approach, the organization would establish traceability between their requirements and the source requirements. They would also need to collaborate with the *Approval Authority* to define the system verification and validation attributes for each requirement such that verification that the SOI meets that requirement would provide objective evidence that the intent of the source requirement within the standards or regulations would be met. Using this information, the *Approval Authority* could accept the organization's tailored standards and regulations as equivalent to those mandated [33].

Another important consideration is design and system verification. Each requirement within an invoked standard or regulation will have to be implemented in the design, and the design and system will have to be verified to meet each of those requirements. The system verification and validation attributes would need to be defined for each of these requirements, and the system verification and validation artifacts discussed in Chapter 10 would have to be developed. This would involve a lot of time and resources if each project did this, especially if the requirements in a standard or regulation are poorly written, and the intent is not clear. If the organization develops their own tailored standards and regulations, they could also include the system verification and validation attributes and artifacts, allowing the projects to reuse this information.

4.3.6.1.3 Managing Changes to Standards and Regulations The organization must have a process that enables it to be knowledgeable when any standard or regulation changes, as well as which projects could be impacted by those changes. This capability is often contained within the organization's CM or compliance office.

Organizations also need a policy, supported by processes, to determine how these changes will be managed. Not all changes will impact products being developed by the organization in the same way. If not impacted by the change, projects will not have to make updates to their needs or requirements. There are some changes that will have an impact, for example, a change to a regulation (law) applicable to the SOI. To be accepted for use, the project will have to show objective evidence that the requirements within the updated standard or regulation have been met. In addition, changes to an interface definitions must be evaluated, or else the SOI may not be able to interact successfully with the other system without invoking the change to the interface requirements within the project's set of design input requirements.

There may also be changes that affect the SOI, but not its intended use in the current operational environment or a critical function. In these cases, there should be provisions to "grandfather" the changes so that the current products do not have to be compliant, but any new versions or models will have to be compliant.

It is common for a standard or regulation to call out other standards or regulations that contain requirements. This can be a serious issue when one document calls out another, that document calls out yet another, and so on. Failing to address this issue, the organization will have lost control of their projects, because they no longer have a say as to which of these lower-tier requirements apply and which they will have to show compliance with.

One way to address this issue is to consider the current integrated set of needs as "Tier 1." Any requirement in a standard or regulation invoked directly by a requirement in the Tier 1 design input requirements is considered a "Tier 2" requirement. If that Tier 2 requirement calls out requirements in another document, those requirements are considered "Tier 3." In cases like this, it is recommended that the organization have a policy that says, "*Any Tier 3 or lower requirements are not applicable to the project – if any requirements within a Tier 3 or lower document do apply, they will be invoked directly within the Tier 1 design input requirements.*"

4.3.6.1.4 Consequences of Failing to Show Compliance Failing to identify relevant standards and regulations early in the lifecycle can result in missing needs and requirements, adding to the risk of non-compliance. How can the SOI be expected to meet security or safety needs and requirements if they were not identified? This leads to missing lifecycle concepts concerning compliance, resulting in missing security related needs and requirements addressing capabilities, functionality, and performance.

Failing to show compliance (failing system verification and validation), will result in the system failing qualification, certification, and approval for use and being rejected by the customer or regulatory agency. If this happens, the result will be expensive rework and schedule slips.

In some cases, failing to meet all relevant standards and regulations could lead to bankruptcy. For example, a concept for a new medical device is defined, and the principal project stakeholders are able to get investors and form a company to mature that concept and build that device. In this case, they must follow the USA's FDA processes for developing and qualifying the device for its intended use, as defined in the FDA's set of regulations (USA Code of Federal Regulations [CFR] Title 21) [34].

Also, assume the intent is to market the medical device worldwide. This company must identify all the standards and regulations not only for the USA but also for each country in which they wish to market the device. If they fail to comply with all the relevant standards and regulations, they will not be allowed to sell their device in the intended market(s). Millions of dollars will have been spent getting to the point where all the paperwork can be submitted to the regulatory agencies. If these agencies fail to approve the device for use, those millions of dollars could be in jeopardy. Either the company will have to seek more funding from investors or it may have to go out of business.

4.3.6.1.5 Compliance When Developing Products for a Customer If a project is developing a SOI for an external customer, that customer will frequently specify which standards and regulations apply, from their perspective, and make it clear in the contract that compliance is necessary for acceptance. The project team will have to define specific needs and design input requirements that invoke the standards and regulations, prompting the generation of objective evidence, through system verification and validation, that their SOI is compliant with the applicable requirements within those standards and regulations. This compliance will be part of the project's customer's or regulatory agency's qualification and acceptance activities.

Even if not explicitly stated by the customer or other stakeholders, showing compliance with the relevant standards and regulations is crucial for the successful acceptance of the SOI.

For design and construction standards invoked by a customer or regulatory agency, organizations may be able to do an equivalency check. Rather than meeting the specific customer or regulatory standard, the supplier may be able to show that their organization's internal design and construction standards meet the intent of the customer or regulatory agency requirements and are thus equivalent. This approach is based on whether the customer or regulatory agency allows an equivalency assessment – in some cases, a customer may accept a supplier's existing methodology to reduce cost or schedule.

An example of this is use of a European soldering standard in a contract that specifies a standard released in the United States, or the other way around. Another example is when the project already has in place a quality management system and design controls that address the standards and regulations applicable to the specific products being developed.

For projects responding to a customer contract, specific project requirements for documentation and contract deliverables concerning compliance with standards and regulations need to be included in the contract, making these deliverables contractually binding. The SOW should also make it clear what role the customer will have in system verification and validation activities. In some cases, the supplier is responsible for the verification that the SOI meets the design input requirements and providing the associated acceptance artifact deliverables. In other cases, the customer assumes responsibility for validating the SOI meets their needs and developing the associated validation artifacts. *Refer to Section 6.2.1, Compliance, for a more detailed discussion concerning strategies concerning implementing requirements contained in standards and regulations. Also refer to Chapter 13 concerning supplier developed SOIs.*

4.3.6.1.6 Compliance Stakeholders Often, the organization will have an internal compliance or quality group responsible for ensuring the relevant standards and regulations are identified, managed, and recorded within the SOI's integrated dataset. This group will work with the project team to ensure lifecycle concepts are defined, which will result in the requirements in those standards and regulations to be addressed. The project team must consult with this group throughout the lifecycle to help ensure compliance.

This compliance group will interact with the external regulatory agencies to make sure the intent of the standards and regulations is being met, as well as to ensure the objective evidence needed to show compliance is obtained and data needed to obtain certification or qualification and acceptance is properly recorded in the Approval Packages and submitted to the regulatory agency's *Approval Authority* in accordance with the requirements defined by the agency. *Refer to Chapter 10 for more information concerning Approval Packages.*

If the system is developed for a customer, the objective evidence and data needed for certification, qualification, and acceptance will be supplied to the customer per the deliverables defined in the SOW, PO, or SA. *Refer to Section 10.4 for a more detailed discussion on the recording of compliance data and information that is part of the Approval Package to be supplied to the Approval Authority.*

4.3.6.2 Technology Maturity Technical maturity is a key driver and constraint. The maturity of a technology relates directly to its feasibility for its intended use in its intended operational environment. If a need is dependent upon the maturity of a critical technology, yet that technology has a low maturity, its feasibility is questionable and represents project risk.

A common tool used to address technology maturity and associated risk is Technology Readiness Assessment (TRA) as defined by the US Government Accountability Office (GAO) [35] and Technology Readiness Levels (TRLs), which are expressed on a scale of 1–9. The lower the number, the less mature the technology is for that specific use and operational environment, and the higher the risk to the project. If a need is not feasible, then the risk of failure of the system to meet that need is more likely, and the risk of failing both system verification and system validation is high.

Many technology-dependent projects fail because they do not understand the concept of technology maturity as it relates to risk and they do not understand and plan for the advancement of the technology maturity in their budget and schedule.

Of particular concern are the resources, budget, and time that will be needed to mature the critical technologies – referred to as the Advancement Degree of Difficulty (AD^2) [36]. The concept of AD^2 is closely related to the concept of TRLs, with a focus on doing an assessment as to the degree of difficulty in terms of resources, cost, and time it will take to advance from one TRL to another. The result of this assessment is communicated in a Technology Maturation Plan (TMP), which is often a subplan to the project's SEMP.

In cases where there may be more than one candidate effort to mature a technology, the project will need to assess the AD^2 associated with advancing each candidate technology to the needed TRL. The result of this assessment is a key driver as to which technology can be considered for inclusion in the SOI.

Refer to Section 4.5.1 *for a more detailed discussion on technology maturity and feasibility.*

4.3.6.3 Existing Systems Another source of drivers and constraints are interfaces with existing external systems. An interface is a boundary at or across which systems interact. The external systems could be engineered systems, human systems, or natural systems. Identifying these interactions facilitates the definition of the SOI's boundaries and clarifies the dependencies the SOI has on other systems (enabling systems) and dependencies other systems have on the SOI.

Identifying interactions and interface boundaries also helps ensure compatibility between the SOI and the external systems in which it interacts. Integrating legacy components into new systems is often overlooked. It forces discussions that lead to understanding the operational needs for new and existing (if applicable) systems, including Systems of Systems (SoS) (if applicable).

During elicitation activities the project team begins the identification of external interfaces. To get a complete understanding of all the interactions between the SOI and external systems, the project team can develop context diagrams, boundary diagrams, or external interface diagrams as part of their elicitation activities. An example context/boundary/external interface diagram is shown in Figure 4.3. *Refer to Section 6.2.3 for additional discussions concerning the identification of interface boundaries and interactions across those boundaries.*

It is helpful to develop these diagrams during the elicitation activities and the development of the preliminary lifecycle concepts to help establish completeness and better understand how the SOI interacts with the other systems that make up the macro system of which it is a part.

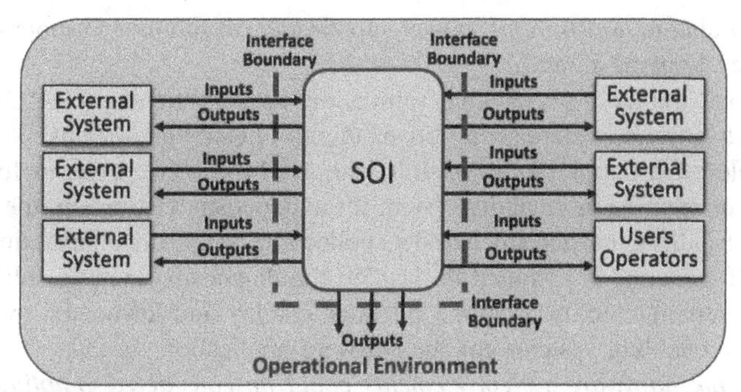

FIGURE 4.3 Example Context Diagram, Boundary Diagram, and External Interface Diagram.

Failure to identify an interface boundary and interactions across that boundary is a significant risk to the project, especially during system integration, system verification, system validation, and operations.

Example Context Diagram Developed by the LIR Project Team

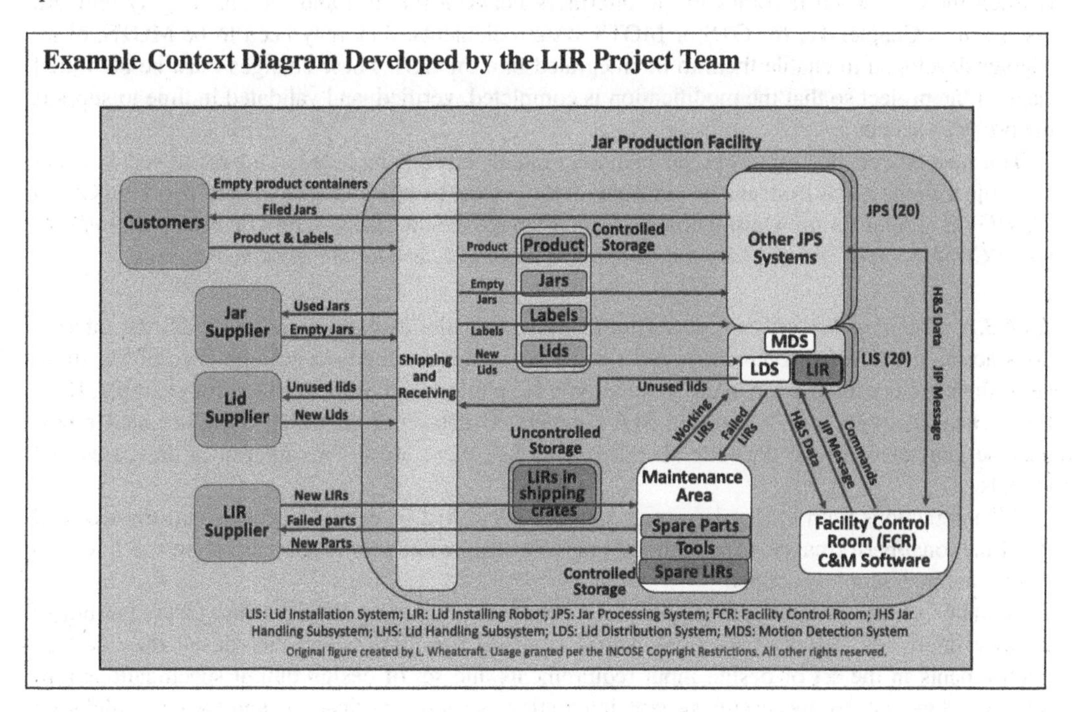

4.3.6.3.1 Enabling Systems An important source of interfaces that must be identified, defined, and managed includes interactions with enabling systems. Enabling systems are systems external to the SOI needed to "enable" certain lifecycle activities or enable the SOI to operate.

Enabling systems include support systems and services across the SOI lifecycle, for example, development, production, integration, system verification, system validation, deployment, training, storage, transportation, operations, maintenance, and disposal. They could consist of laboratories, test equipment, simulators, test fixtures, power supplies, clean rooms, transportation, storage, and integration facilities, to name a few.

During operations, the facility's heating, ventilation, air conditioning (HVAC), space, lighting, power, video monitoring, and secure access are examples of enabling systems.

Other system elements internal to a system could also be enabling systems for a given SOI. If an organization is developing an imaging system for a diagnostic instrument, the structure, power, cooling, sampling handling mechanism, and data collection system are all enabling systems for the imaging system. If someone is assigned to develop the diagnostic analysis software application, then the imaging system, microprocessors, operator display and keyboard, and data communications systems are enabling systems for the software application. *In today's software-intensive systems, arguably the hardware system elements could be considered enabling systems for the software.*

Project responsibility extends to the acquisition of services from organizations responsible for the relevant enabling systems in each lifecycle phase. Since enabling systems are often supplied by other parts of the organization or external organizations, their use must be included in the project's budget and schedule. Acquiring services from the organization responsible for an enabling system can be a large budget and schedule driver. In some cases, the enabling systems may need to be modified – this is often the case when it comes to the interfaces between the SOI and the enabling system. As discussed in Chapter 12, for OTS or MOTS system elements, they may need to be MOTS, or an adapter developed to enable them to be integrated into the SOI. These changes must be identified early in the project so that the modification is completed, verified, and validated in time to support the project's schedule.

The stakeholders that represent the enabling systems must be included in the list of stakeholders the project elicits needs from and become part of the project team's Interface Control Working Group (ICWG) or similar forum whose purpose it is to manage the interfaces across the lifecycle. *Refer to the INCOSE SE HB, Section 3.2.4 for more information concerning interface management.*

4.3.6.3.2 Defining Interfaces Once the interface boundaries have been identified, the interactions across the interface boundaries must be defined. In a document-centric practice of SE, these definitions are commonly recorded in some type of interface definition artifact, for example, ICD, IDD, Data Dictionary, or within the SOI's integrated dataset from which the associated report may be generated. In the data-centric practice of SE, these are often captured in databases and models.

All interactions with external systems must be identified in the set of needs and transformed into functional/performance or operational interface requirements for each interaction and traced to where that interaction is defined.

A major source of anomalies found during system integration involves interfaces. Failing to identify interfaces, define interactions across interface boundaries, and include all the interface requirements in the set of design input requirements and set of design output specifications represents a key risk to successful system integration, system verification, and system validation activities. *Refer to Section 6.2.3 for a more detailed discussion on interface definition and defining interface requirements.*

Because of the critical nature of interactions across interface boundaries, it is extremely important that the project define lifecycle concepts for how it will make sure the system will work safely and securely with the external systems with which it must interact in the intended operational environment and is protected from outside threats across those interfaces.

For new systems being developed concurrently with the SOI, the definitions concerning the interactions evolve as the design evolves. At the beginning of the project, the project team will need to define what is crossing the interface boundary (inputs and outputs) and the characteristics of what is crossing the boundary. Later, the design team will determine the specific media to be used, the communications protocols involved for each interaction, and the physical characteristics of each system at the interface boundaries.

When this information has not yet been defined, needs may have TBDs as placeholders until the information has been defined. As this information is known, the ICDs (or similar definition documents) will need to be updated.

The evolutionary nature of defining the interactions across interface boundaries must be addressed in the SOW, PO, or SA for systems that are outsourced to a supplier – failure to do so often results in costly contract changes and issues during integration, which can add significantly to development risk of the project.

See also Section 14.2.8 concerning interface management across the system lifecycle.

4.3.6.4 Higher-Level Lifecycle CNR A major driver for the project team, as well as the SOI to be developed, is the lifecycle CNR that were defined at the previous level and allocated to the SOI. During elicitation, the stakeholders representing these higher-level lifecycle CNR will make them known.

When there is a customer/supplier relationship, there will be higher-level lifecycle CNR developed by the developing organization as well as those developed by the customer organization that are allocated to the project team and shown earlier in Figure 2.8.

Both the customer and supplier sets of lifecycle CNR must be assessed for consistency and applicability to the SOI under development.

Some of these lifecycle CNR could be on the project team developing the SOI (for example, policies, processes, and work instructions); lifecycle CNR concerning the SOI; and lifecycle CNR dealing with system verification and system validation.

As part of the stakeholder elicitation activities, stakeholders could identify needs for the SOI in the form of requirement statements, "The SOI shall ...." These stakeholders could be internal to the developing organization, an external customer, or regulatory agencies. From a user perspective, some needs may be communicated in different forms – user stories, use cases, or operational scenarios for example.

For SOIs that are lower in the system architecture, there will be requirements allocated from higher levels of the architecture. Some of these allocations include budgets for performance and quality. The lower-level SOI's will derive child requirements from their set of needs and trace them to their allocated parents at the previous level. *Refer to Section 6.4 for a more detailed discussion concerning levels, allocation, and budgeting.*

Even though these higher-level requirements may be written as "shall" statements, they should only be considered as inputs to the *Lifecycle Concepts and Needs Definition* activities for that lower-level SOI. Along with all the other stakeholder needs obtained during elicitation, these inputs will have to be assessed in terms of conflicts, ambiguity, correctness, completeness, consistency, and feasibility during the lifecycle concepts and analysis and maturation activities.

Lifecycle concepts concerning the SOI at the previous level will be elaborated upon when defining the lifecycle concepts for the SOI. The SOI lifecycle concepts must be consistent with the context and intent of the higher-level lifecycle concepts. No matter the level at which the SOI exists, traceability will need to be established to show compliance with the lifecycle CNR that exist at the previous level.

4.3.6.5 Budget and Schedule As with the old question: "*Which came first, the chicken or the egg?*"; budgets and schedules are problematic. It is common for the customer (internal or external) to specify the budget and schedule at the beginning of the project without sufficient knowledge as to whether their specific needs can be met (are feasible given the stated budget) or, worse yet, without knowing their specific set of needs. This is often the case for both outsourced systems and internally developed consumer products. Even though this approach is the most common, it is also a high-risk approach and a source of many project failures.

System integration, system verification, and system validation activities are key cost and schedule drivers. These are a function of the number of needs and resulting design input requirements, as well as the defined system verification and validation *Success Criteria*, *Method*, and *Strategy* as discussed in Chapter 10. When the budget and schedule are specified before the customer or operational stakeholders have knowledge of how many design input requirements the SOI will be verified against and needs the SOI will be validated against, there is significant development risk to the project.

Another cost driver, discussed previously, is the technological maturity of the key technologies needed to meet the needs and requirements. Specifying a fixed budget and schedule too early, can constrain the solution space, limiting the technologies that can be considered.

Maturing a technology takes time and money. A common complaint made by developers is when management sets a launch date for a product before feasibility has been determined. When the budget and schedule are determined prior to establishing feasibility, the development risk is high.

In other cases, like in some government managed projects, the initial budget and schedule are notional (ballpark estimates), and the final budget and schedule are set within the time frame of a key decision point, usually the preliminary design review (PDR). At PDR, the project will have a more mature knowledge of both the number of requirements the system will be verified against, the number of needs it will be validated against, and the maturity of key technologies. With this knowledge, theoretically, the project should be able to estimate remaining technology maturation, development costs, system integration, verification, and validation costs more accurately.

In either case, budget and schedule will be key measures concerning feasibility during lifecycle concepts analysis and maturation.

4.3.6.6 Operational Environment System validation helps to ensure that the designed, built, and verified realized system meets its intended use in its operational environment when operated by the intended users and does not enable unintended users to negatively impact the intended use of the system as defined by the integrated set of needs.

As part of defining the lifecycle CNR, the intended use, conditions for use, personas/user classes of the intended users, operational environment, operating procedures, Instructions for Use (IFU), and the HMI must clearly be understood and defined.

For example, when developing a semi-truck, when defining the operational environment, the conditions for use must be made clear. Is the intended purpose to operate the truck in artic conditions (sub-zero temperatures and snow covered or icy roads) or in tropical conditions (high temperatures, high humidity, and dirt or muddy roads)? How does the various operational environments drive the design in term of performance, use, and maintainability? What features are needed to aid the driver and maintenance personnel in each of these conditions? What skills are needed for the drivers and maintenance personnel? What drivers and maintenance personnel anthropometrics are assumed so the truck can be used and maintained as intended for each of these environments?

Operational environment involves not only the other parts with which the SOI must interact across the interface boundaries but also the natural environment (temperature, humidity, pollutants) as well as induced environments (temperature, vibrations, mechanical loads, electrical and electronic emissions, acoustics).

Often overlooked, induced environments also include outputs of the system to the natural environment, which could include pollutants and waste. Pollutants include light, thermal, chemical, greenhouse gases, electromagnetic, hazardous materials, biological waste, or radiation. Requirements concerning pollutants and waste are usually addressed when defining what the SOI is allowed to output into the natural environment as constraints. These include constraints across all lifecycle stages, including manufacturing and disposal. Often, these constraints are included in a standard or regulation. If a system includes hazardous materials, this could result in the system being considered hazardous waste when disposed of. For a spacecraft landing on a planet, there are planetary protection requirements concerning the contamination of the natural environment by the spacecraft itself or during operations.

The operational environment may impose constraints that may increase complexity and constrain the types of components that can be used. For example, additional supervisory systems may need to be added to constrain a simple system to operate within a set of state boundaries. An extreme radiation environment will result in the need for the use of hardened components as well as additional shielding, resulting in increased mass. Extreme operating pressures will result in more robust physical designs, adding mass to the SOI.

The operational environment also has a human element as well. The human element includes social, cultural, ethical, and political considerations, as well as the abilities and knowledge of the intended users/operators. Because of that, the project team must address human factors that will drive the design. A major reason for defining personas and user classes is so that the intended users are well understood and defined, helping to ensure the system can be designed to be used and operated by the people represented by the personas/user classes. Induced environments and their impacts on users and operators must also be taken into consideration to ensure safe and comfortable working conditions.

To address human factors, it is common for one of the stakeholders to be a human factors expert trained and knowledgeable in HSE and HSI. They will be a key stakeholder during elicitation as well as key participants in the maturation of the lifecycle concepts and design activities to realize those concepts. They will also be involved in the system verification and validation activities associated with any needs and design input requirements associated with human factors.

An especially important part of the operational environment are the threats that could result in unintended users negatively impacting the intended use of the system. This includes security issues that are increasingly of concern in modern software-intensive systems. These threats represent significant operational risk to the SOI.

From a needs and design input requirements definition perspective, the operational environment defines the "under what conditions" part that should be included within needs and requirements expressions.

Example Drivers and Constraints Documented by the LIR Project Team

Drivers and constraints for the LIR Project include:

Regulations. OSHA, EPA, FDA, others?

Standards. Quality/Workmanship, Human Systems Integration/Engineering, others?

Enabling systems. Interfaces with existing systems, number of jars to be processed per shift per JPS, rate of processing jars, space, power, HVAC, communications.

Project schedule, budget, and resources.

Higher-level Needs and Requirements allocated from the LIS to the LIR.

Technology. Are there existing (COTS) LIRs that can meet our needs? If not, what is the TRL of needed technologies? What is the increased cost and time to modify existing COTS or develop new robots? Can we change our needs such that there are COTS LIRs that can meet our needs?

Customer and other stakeholder needs.
MGOs, measures.
Human/machine interfaces concerning the interaction with the LIR during operations, transportation and maintenance.

4.3.7 Identify, Assess, and Handle Risk

During the elicitation activities, the project team should be recording issues and risks within the SOI's integrated dataset. To help ensure completeness, it is helpful for the project team to focus on issues and risk as an additional activity. In doing so, issues and risks not communicated during elicitation can be identified and accessed.

Risks are anything that could prevent the delivery of a winning SOI, where "winning" can be defined as an SOI that delivers what is needed, within budget and schedule, with the needed quality. Once delivered, risk is anything that could impact the intended use of the SOI in its intended environment, by its intended users, or anything that would allow unintended users to prevent the intended use of the SOI or allow unintended users to use the SOI in an unintended manner, for example, hack into an aircraft's control system and use the aircraft as a weapon.

Stakeholders should be asked specifically about any issues and risks they think could prevent the SOI from being developed and delivered within budget, schedule, or risk during operations. Failing to address risk will result in an incomplete set of needs and resulting design input requirements, resulting in an SOI that will fail system validation.

4.3.7.1 Types of Risk Types of risk to be addressed are described briefly in the following sections.

4.3.7.1.1 Management Risk There is project risk that includes risk concerning budget, schedule, resources, and upper-level management support throughout the system life. A key risk to projects is a failure to include sufficient budget, schedule, and resources needed to support all lifecycle stage activities. Failure to do so puts the project at risk of cost overruns, schedule delays, and unavailability of people, facilities, and required resources.

To help mitigate these types of risk, it is important for the project to include sufficient margins and reserves for cost and schedule to address the *unknown unknowns*. To help maintain upper-level management support, the project team should have a project champion at higher levels of management that can communicate the value of the project and ROI to other upper-level management stakeholders throughout the lifecycle.

4.3.7.1.2 Development Risk There are development risks concerning problems that can occur due to a failure to follow PM and SE lifecycle process activities discussed in this Manual, resulting in work products and artifacts that are not complete, correct, consistent, and feasible.

Development risk also concerns the feasibility of performance and quality based on the maturity of critical technologies. If the TRL associated with the technologies needed to meet critical or high priority requirements is low, those requirements are high-risk and must be managed closely (as discussed in Section 4.3.6.2).

Performing TRAs [35], establishing the TRLs, doing AD^2 [36] assessments, and developing TMPs are key tools to identify and manage technology maturation risk. Basing project success on low TRL technologies can result in lifecycle CNR that may not be feasible. Consequently, the SOI may fail system validation.

Defective integrated sets of needs and the resulting design input requirements are also development risk factors [37]. To help reduce development risk, the project must avoid technical debt and focus on establishing consistency, correctness, completeness, and feasibility early in the development lifecycle.

The lifecycle concepts analysis and maturation activities will inform (feedback) the identification of additional development risk based on the knowledge gained during the lifecycle concept analysis and maturation activities.

A lack of traceability across the lifecycle is a major risk in that a failure to do so results in inconsistencies in work products and a lack of capability to do effective change assessment.

Assessing development risks helps to ensure that need statements and resulting requirements have the GtWR characteristic *C6—Feasible*, and integrated sets of needs and resulting requirements have the GtWR characteristic, *C12—Feasible*.

4.3.7.1.3 Production Risk A common problem is for the project team to fail to consider the needs for production. To address this problem, representatives from the production organization must be included as part of the project team from the beginning of the project.

The role of production is to transform the design output specifications into the physical SOI. There is production risk concerning the technologies needed to produce the SOI, facilities, and equipment. While it is lower risk when the SOI can be produced using the existing production capabilities, there is higher risk when new technologies are needed, and modifications need to be made to the existing production capabilities in terms of cost, schedule, and resources.

In addition, there is risk concerning problems that may occur during the actual building/coding of the system. Scalability is often an issue; building and testing one copy of a system in a laboratory is a lot different than when manufacturing is turning out hundreds or thousands of copies that will be operated in the intended operational environment, by the intended users.

Addressing quality and production verification presents unique issues that may generate derived needs for test or inspection points for components, subassemblies, assemblies, and subsystems that are only used during production, system verification, and system validation. These test points are used by quality personnel during production verification to help ensure the system was "built to spec" (meets the design output specifications) and the expected yield is achieved, where "yield" is defined as the percent of copies of an SOI that pass production verification.

The data from the test points may also be used as part of design testing and maturation, during design verification and validation, and maintenance activities during operations.

Refer to Chapter 9 for a detailed discussion on Production Verification.

4.3.7.1.4 System Integration, System Verification, and System Validation Risk A major risk faced by many programs and projects is a failure to address or underestimate what it will take (resources, cost, and schedule) to successfully complete system integration, system verification, and system validation activities with the required level of confidence.

The authors have seen many cases where a project will allocate only 20–30% of their budget and schedule to these activities. Experience has shown that these projects are most likely going to fail.

Assuming an organization follows good SE practices, such as those defined in this Manual, experience has shown that as much as 50% or more of the project's development budget and schedule should be allocated to system integration, system verification, and system validation activities. This percentage could be higher or lower depending on the complexity of the project and whether it is a brown-field or a green-field project. If the organization fails to follow SE best practices, and if the quality of the integrated sets of needs, sets of design input requirements, and sets of design output specifications is poor, this percentage could be even greater (primarily due to rework).

Another major risk during system integration, system verification, and system validation involves interfaces. Given that a key part of system integration involves interface boundaries and interactions across those boundaries, special attention must be paid to identifying all interface boundaries (external and internal), defining the interactions across the interface boundaries, and ensuring the design input requirements and resulting design output specifications include interface requirements addressing all interactions across the SOI interface boundaries.

As advocated in this Manual, integration, verification, and validation should be practiced continuously across the lifecycle. With the increased use of language-based models and the ability to run simulations of the modeled system, projects have the capability to do early system verification and validation activities before the system is built or coded. As a result, issues are identified and corrected earlier in the lifecycle, significantly reducing the risk of issues found during system integration, system verification, and system validation.

4.3.7.1.5 Compliance Risk As discussed previously, failing to show compliance with standards and regulations is a major source of risk for a project.

For example, the FDA publishes findings concerning their audits. Unfortunately, it is common for organizations to fail to receive approval to market their medical devices due to non-compliance. In 2017, *The Most Common FDA Audit Findings From 2017* [38] was reported as follows: "*Last year FDA conducted 17,487 audits worldwide, and issued 5045 FDA 483s (letter of non-compliance with FDA regulatory requirements), meaning that about 30% of the audits led to non-compliance in more than one regulated area. There were 498 Warning Letters issued during the same time, showing that 1 in 10 audits found repeat violations that required more serious enforcement action by FDA. Standard Operating Procedures (SOPs) and documentation deficiencies were the dominant findings in FDA audits for almost all areas of compliance. Poor Corrective and Preventive Action (CAPA) practices and inadequate complaint handling were found at more than half of the non-compliant medical device manufacturing facilities. Overall, the medical device manufacturing facilities had almost double the number of 483s issued compared to drug manufacturing sites, while biologics had the least number of 483s. More than half of the non-complaint clinical trial sites had Principal Investigators who did not follow the written protocol, followed by poor patient records and source documents*".

To minimize this risk, organizations must research and understand what is necessary for acceptance, including specific sets and types of data required by the *Approval Authority* and regulatory agencies to use as objective evidence for acceptance.

If an organization is developing a space system that will include humans, they need to address specific questions, such as: What are the licensing requirements for commercial space vehicles? US Department of Defense (DoD) controlled space systems? US National Aerospace and Aeronautics Administration (NASA) controlled space systems? What requirements will NASA place on a space system to be human rated? What extra testing and reviews will be required before humans are allowed to fly in, live in, work in, and interact with the space system? What US Department of Energy (DoE) requirements will the project have to meet if the space system includes radioactive materials?

As for medical devices, there are standards and regulations for both the process and the project, as well as standards and regulations for the space system itself. Requirements in these standards and regulations relevant to the project and system must be complied with before the system will be approved for use.

Some systems may require outside experts for certification. Does the system require an Underwriters' Laboratories (UL) certification? Or if the SOI includes a pressure vessel that requires an American Society of Mechanical Engineers (ASME) Class 1 Certification?

To avoid compliance issues, even if not explicitly stated by the customers or other stakeholders, the project team must be aware of and research all the possible relevant standards and regulations in the domain of the SOI. Compliance is expected to proceed to the operational phase, and failure to address relevant standards and regulations can result in failure during system verification and validation and system rejection by the customers or regulating agencies, *even if the customer did not explicitly identify the relevant standards and regulations to which the project must comply.*

4.3.7.1.6 Operational Risk There are operational risks involved in the performance of the system for its intended use by the intended users/operators in its operational environment. A failure to address operational risk during elicitation and lifecycle concept and need definition activities can result in failures during system validation. When doing system validation, it is important to focus on the intended use in the operational environment by the intended users/operators as defined in use cases, user stories, operational concepts, and site visits.

Off-nominal performance, inputs, outputs, failures, and unexpected changes in the operational environment represent significant risks during operations. Is the system agile, robust, or resilient enough to cope with unexpected changes to inputs or the environment? Is the system robust enough to still function when anomalies occur?

From a performance and safety perspective, sufficient performance margins and safety margins need to be included in the design input requirements as well as the realized design. Failure to do so could result in the system failing system validation, qualification, certification, acceptance for use, or failure during operations.

For example, if designing a spacecraft to land on the surface of the moon or another planet, off-nominal cases must be considered, and sufficient fuel margins must be included to address these off-nominal cases.

Of particular concern during operations are the threats that could result in unintended users negatively impacting the intended use of the system or using the system in an unintended way, resulting in loss. This includes security issues that are increasingly of concern in today's software-intensive, online systems. In today's world, cybersecurity threats are becoming more common. To mitigate these risks, the project team must address misuse cases as well as loss scenarios, define concepts for addressing these issues, define needed capabilities within the integrated set of needs, and transform the needs into design input requirements that will result in the SOI having these capabilities.

These are issues for which the stakeholders will have expectations – even if not explicitly stated. The challenge for the project team is to address these risks during lifecycle concept and needs definition activities.

Refer to Section 6.4 for a more detailed discussion margins and reserves to mitigate risk.

4.3.7.1.7 Interface Risk The project must not only identify and document interface boundaries but also assess each interaction across those boundaries in terms of stability, documentation, threats, and risk. Serious problems can, and all too often do, arise at the interface boundaries. These problems represent risk to the project during both development and operations.

The SOI is particularly vulnerable when interacting with external systems over which they may have little or no control. Thus, the SOI is vulnerable to undesirable events at or across the interface boundaries, so identifying interface boundaries and interactions across those boundaries is key to exposing potential risk. As discussed early, in the modern connected world, cyber security is a major concern that must be addressed for each cyber-related interface.

For each external interface boundary, the following questions should be asked:

- Which external systems already exist?
- Are any external systems being developed concurrently with the SOI?
- Who are the stakeholders for the external systems, and have they been involved in the elicitation activities?
- Who "owns" the interface control definition for different aspects of the system? Is there a well-defined process for stakeholder involvement in the development and maintenance of the interface definitions?
- Have all the specific interactions (inputs and outputs) between the SOI and each external system been defined?
- From a threat perspective, has risk been assessed concerning bad things that could happen across this interface in either direction? Specifically, assess misuse cases and loss scenarios that could be enabled if not addressed for each interface boundary.
- For existing external systems, are they compliant with the latest security standards and regulations? If not, they could be a vulnerability by allowing threat actors to access your system via the external system.
- Likewise, are external systems being developed concurrently with your SOI, compliant with the latest security standards and regulations?
- Is an existing external system likely to change how it interacts with the SOI across the interface boundary during the development or after the SOI is in use? How will the project team know if it changes? How will the proposed changes impact the SOI?
- For existing systems, is the documentation that defines the interface boundary and interactions across that boundary available and current? If not, how will this information be obtained?
- For new systems being developed concurrently, what is the process to be followed to document and agree on the specific interactions. Who is responsible for recording the interactions and getting approval? Who will have configuration control over those definitions? What is the schedule for doing so?
- For software systems, what standard or API applies to the interactions? Are they compliant with the latest security standards and regulations?

The resulting information needs to be assessed concerning each interaction between the SOI and external systems for each lifecycle stage. Failure to do so adds risk to the project.

Example Risks Identified by the LIR Project Team

Acronyms used in the following scenarios: LIR, Lid Installing Robot; FCR C&M, Facility Control Room Command & Monitoring; JHS, Jar Handling Subsystem; LHS, Lid Handling Subsystem; LDS, Lid Distribution System; ICD, Interface Control Document.

The following is a list of potential operational risks addressed by the LIR Project Team, grouped by phase.

Start-up

Lids have not been supplied to the LIR LDS prior to shift start.

Wrong size of lids supplied to the LIR LDS for the jars being presented.

Insufficient number of lids supplied to the LIR LDS.

LIR is not able to obtain lids from the LDS.

LIR fails to power on when commanded to do so by the FCR.

LIR fails self-test.

LIR is not able to report health, status, or messages to the FCR.

LIR fails to enter "Ready to Install Lids" state.

Off-Nominal Operations

The LIR fails to position the LIR LHS to obtain a lid from the LIR LDS.

The LIR fails to position the LIR JHS to be ready to grasp a Jar.

The LIR fails to receive from the FCR C&M software a "Jar In Place" message.

The jar on the conveyor belt is not positioned per the ICD.

Either the LIR or some other JPS system had a failure during the shift.

The LIR LHS fails to obtain a lid.

The LIR LHS fails to install a lid on a jar.

The LHS installs a lid at an improper torque.

The JHS fails to grasp the jar correctly.

The JHS grasps the jar with insufficient force to keep the jar from rotating during lid installation.

The JHS grasps the jar with too much force, breaking or deforming the jar.

The JHS moves or knocks down the jar.

The Fire Suppression System within the LIS is activated.

The LIR power source fails during operations.

The operator in the control room falls asleep or is not present when a failure occurs.

Someone enters the room where the LIR is operating.

Problem with the FCR C&M software.

The LIR receives a command from an unintended source.

Communication with the FCR C&M software is lost.

Shutdown

The LIR fails to automatically shut down at end-of-shift.

The LIR fails to return a lid to the LIR LDS it may have been holding at shutdown.

The LIR fails to notify the FCR C&M software of its shutdown status.

The LIR internal clock is not synchronized with the facility clock.

4.3.7.2 Risk Assessment and Handling The project team must do a risk assessment of each of the classes of risk discussed above. For each class of risk, the identified risks need to be recorded within the SOI's integrated dataset and handled (accepted, monitored, researched, or mitigated) during the system lifecycle concepts analysis and maturation activities.

Because risk management, risk assessment, and risk treatment/handling are critical crosscutting activities that occur concurrently with the *Lifecycle Concepts and Needs Definition* activities as well as across all other lifecycle activities, a brief discussion is warranted. *Refer also to the INCOSE SE HB Section 2.3.4.4 Risk Management Process.*

For highly regulated systems like medical devices, risk management is a major design control process that is key to a medical device being approved for use. This process is defined in ISO 14971:2019 *Application of Risk Management for Medical Devices*, which emphasizes the need to practice risk management across the system lifecycle. Of particular concern is risk associated with operations once a medical device is approved for use, including the prevention of unintended use of the medical device, unintended users preventing intended usage of the device or using it in unintended ways. While the focus is on medical devices, the guidance within this standard can and should be applied to all systems and products.

4.3.7.2.1 Risk Assessment *Risk assessment* involves the evaluation, analysis, and estimation or quantification (likelihood versus consequence) of each identified risk. Key questions that should be addressed as part of risk analysis. These include:

- What is the likelihood of this risk occurring?
- What are the consequences if this risk does occur?
- How soon do we need to act on this risk?
- How does this risk compare with other similar risks?
- Do we know enough to quantify the risk and determine the urgency to act?

Combining the likelihood with the consequence results in a quantification score for each risk. Assessing the timeframe for action helps establish a degree of urgency for handling the risk and effective communication of the risk and how the risk is to be handled.

4.3.7.2.2 Risk Treatment/Handling Risk treatment/handling includes planning, tracking, and controlling a risk across the lifecycle. The first thing that should be considered is the project's response to a risk, such as the decision to accept, research, monitor, or mitigate it. Credible rationale and criteria used to make the decision for a given response must be clearly defined.

Due to a variety of reasons, such as low likelihood, marginal or negligible consequence, politics, culture, or cost, the project may choose to *accept the risk*, i.e., do nothing. In other cases, the risk is not well understood, and urgency has not been established, so the project may want to do more *research and analysis* and collect additional data before deciding on a response. Or the project may choose to *accept yet monitor* a given risk. In this case, they recognize there is a risk, understand the risk, but there is no urgency in a response. Because they do recognize the risk, they will track, survey, or monitor trends and behavior of risk indicators over time. Based on this information, they reserve the right to change their response to the risk.

4.3.7.2.3 Risk Control/Mitigation For a risk with higher likelihood and more severe consequences where a degree of urgency has been established, the project will choose to control or *mitigate the risk*. Key considerations for mitigation include how a risk associated with a hazard or associated threat could be eliminated, the likelihood reduced, and/or the severity of the impacts/consequences reduced.

The result is a set of mitigation actions that will be determined, evaluated, approved, and implemented. Implementation includes developing plans and policies, assigning responsibility for managing implementation of those plans and policies, and expectations for those who must implement and follow those plans and policies.

When the risk mitigation involves the SOI under development, how the project team plans to mitigate the risk must be included in the lifecycle concepts analysis and maturation activities and reflected in the resulting integrated set of needs and set of design input requirements. It is also critical that complete traceability is established for all artifacts associated with mitigating the risk: the risk, lifecycle concepts to mitigate the risk, resulting need statements, design input requirements, the design, design output specifications, system verification artifacts, and system validation artifacts.

This traceability is needed not only for tracking but also as part of managing changes throughout the development lifecycle. Needs and requirements that focus on risk mitigation will normally have both the priority and critically attributes defined as well as the attribute that indicates rationale based on risk mitigation. Increasingly, regulatory agencies are requiring developers to show traceability across the lifecycle for all risks identified [39]. *Refer to Chapter 15 for a more detailed discussion on the use of attributes.*

When risk mitigation involves the user/operators, the risk mitigation activities are often addressed in both the training and certification of the users/operators as well as the IFU, operating procedures, checklists, and maintenance documentation.

4.3.7.2.4 Use of Margins and Reserves to Mitigate Risk There are several different uses of margins and reserves.

- *Development margin* or *technical margin* is defined as the difference between the estimated budgeted value and the actual value at the end of development when the system is delivered. Margins allow for both expected and unexpected change as the design matures over the system development lifecycle. Development margins for resources like mass, power, and time or margins for performance like, accuracy, precision, or rate are defined at the system level and allocated to the parts of the system architecture.
- *Operational margin* is defined as the difference between what is required during operations and what is provided. Operational margin provides additional capability to address unexpected changes, anomalies, security issues, or errors in defining the expected operational environment that may occur during operations. (*The need to address operational margins and issues is discussed in several different subject areas including agile systems and resilient systems. The reader is encouraged to explore each of these areas in more depth depending on the needs of the project and customer.*)
- *Management reserve* is defined as the portion of the available quantity held back or kept "in reserve" by management or the quantity owner during development and not made available through allocation. Reserves allow management to deal with the unexpected events such as out of-scope demands, unplanned changes, uncertainties as to what is feasible, and other uncertainties.

Margins and reserves are common approaches to mitigating development, operations, and management risk [40]. When the ability to meet performance expectations is based on a new or developing technology (low TRL), there is a risk that the needed performance will not be able to be achieved. Rather than base needs and design input requirements on best estimates or theoretical projections of performance of unproven technologies, include sufficient margins for budget, schedule, needs, requirements, and the design to address these types of risks. Failure to do so could result in the system failing system verification, system validation, qualification, certification, and acceptance for use. *Refer to Section 6.4 for a more detailed discussion on the use of margins and reserves to manage risk.*

For example, if the theoretical or overly optimistic conversion of sunlight to electricity efficiency is zz%, the needs and design input requirements should be written for a lower value that is more likely to be achieved given the maturity of the needed technologies at that time.

The size of the margins and reserves is based on the risk associated with projects. Brown-field systems are normally lower risk, so the size of the margins and reserves can be lower. Green-filed systems are at higher risk, especially those whose critical technologies are at low TRLs, so the size of the margins and reserves will need to be higher.

The early establishment of adequate margins and reserves and the effective management of them throughout the project's lifecycle play a critical role in the ability of the project to deliver a winning system.

From the beginning, the project team is strongly encouraged to define adequate margins and reserves – especially for programs with significant complexity and high risk. Failing to define these margins places the project at great risk of cost overruns and schedule slips. When defining the values within the design input requirements, it is critical that these values take into consideration the margins and reserves defined and managed by the project.

Projects that fail to define and manage margins and reserves for all allocated quantities are doomed to fail from the beginning.

Refer to Sections 6.2.1.3.6 and 6.4.6 concerning the use of margins when defining values within requirement statements and when allocating/budgeting requirements to lower levels of the architecture.

Note: Risk identification, assessment, and handling are continuous activities that occur over the life of the SOI. Known risks are periodically re-assessed in terms of likelihood, consequences, and urgency. If a project is doing risk management properly, project team members should be able to list the top ten risks being mitigated from memory. If not, risk mitigation may not be happening within the project.

4.4 CAPTURE PRELIMINARY INTEGRATED SET OF LIFECYCLE CONCEPTS

The results of the proceeding activities are integrated into a preliminary set of lifecycle concepts and supporting data. The set of lifecycle concepts can include concepts for acquisition, development, design, verification, validation, operations, deployment, support, and retirement concepts. These preliminary sets of lifecycle concepts and supporting data and information are inputs into the lifecycle concepts analysis and maturation activities. These are preliminary in the sense that the detailed analysis concerning completeness, consistency, correctness, and feasibility, has not yet been completed. The specific lifecycle concepts defined depend on the organization, its product lines, processes, and culture.

As shown in Table 4.4, the elicitation effort will produce a preliminary integrated set of lifecycle concepts, which will drive the lifecycle concepts analysis and maturation activities discussed in Section 4.5. Each column of the table represents an SOI lifecycle stage (Lx). Each row represents an operational scenario, concept, or use case for a specific stakeholder (Sx) (or group of stakeholders) for the lifecycle stages in which that stakeholder has a stake or involvement.

The task of the project team is to combine the various operational scenarios, concepts, and use cases into an integrated set of operational scenarios, concepts, or use cases for each lifecycle stage. This task can be challenging in that each stakeholder or stakeholder group may have a different perspective based on their role, unique needs, experience, and political environments both internal and external to the organization.

TABLE 4.4 Preliminary Integrated Set of Lifecycle Concepts.

Stakeholder (S)/ Lifecycle (L)	L1	L2	L3	L4	L5
S1	X		X		X
S2		X	X		
S3	X		X		X
S4		X		X	X
S5			X	X	X
Combined	XXX	XXX	XXX	XXX	XXX

The project team will need to validate stakeholder assumptions, as well as make tradeoffs to prioritize, deconflict, and ensure the integrated set of stakeholder needs are truly "required" and necessary for acceptance rather than simply preferences or desirability. There will be conflicts and inconsistencies that may not yet have been resolved as part of the elicitation activities. These issues must either be resolved or at least identified for future work during the lifecycle concepts analysis and maturation activities.

An analysis of the various sets of stakeholder needs for a given lifecycle stage, along with the MGOs and measures, allows the project team to identify the capabilities, features, functions, performance, quality, and compliance needed and expected by the stakeholders, as well as interactions (interfaces) between the SOI and external systems. This information will be used to help analyze and mature the preliminary set of lifecycle concepts.

The knowledge gained from this information will be used by those on the project team who participate in developing functional architectural and analytical/behavioral models. This modeling effort will be a major activity of the lifecycle concepts analysis and maturation activities discussed further in Section 4.5.

A major part of recording the preliminary lifecycle concepts is for the project team to identify processes, process activities, enabling systems, enabling organizations, facilities, and resources needed as part of the SOI development effort across all lifecycle stages.

Example Elicitation Outcome Scenarios During LIR Operations

Acronyms used in the following scenarios: LIS, Lid Installation System; LIR, Lid Installing Robot; JPS, Jar Processing System; FCR, Facility Control Room; JHS, Jar Handling Subsystem; LHS, Lid Handling Subsystem; LDS, Lid Distribution System; MDS, Motion Detection System; ICD, Interface Control Document.

After talking with all the various stakeholders, the LIR Project Team documented the following nominal operations scenarios. These are just a portion of the different scenarios developed as part of documenting the results of elicitation. Other scenarios concerning operations in the FCR, off-nominal operations, operations environment, safety, security, shipping, storage, uncrating and assembly, routine maintenance, movement within the facility, fault detection and recovery, lid logistics, jar logistics, recycling, human factors, and compliance were also defined. Each is a source of needs. Each perspective is needed for completeness.

Start-up: Prior to shift start, a technician removes unused lids from LIS and provides a shift's worth of lids (960) of the proper size to the LIS. Once the technician has left the LIS room, the LIR is powered up remotely by the operator in the FCR. Upon power up, the LIR completes a self-test and reports its health and status to the FCR as defined in the FCR ICD. The LIR positions itself to be ready to obtain a lid from the LDS and grasp the jar when commanded to do so.

Lid Installation: The JPS presents jars (1 every 30 seconds, starting at 8 pm and continuing to 4 pm, 355 days per year) on a conveyor belt located on the other side of the LIS/conveyor belt system access opening. The JPS pauses each jar for 10 seconds for lid installation and takes up to 20 seconds for the next jar to be in place.

At shift start, the JPS positions a jar in the LIS/conveyor belt system access opening per the JPS ICD and sends the FCR a message that the jar is in place within 1 second. The FCR notifies the LIR within 1second of when the first and subsequent jars are in place for lid installation. The LIR will grasp the Jar to keep it from rotating and install the lid during the jar's pause with a torque as recommended in Table 3. Once the lid is installed, the LIR obtains another lid from the LDS and positions itself to be ready to grasp the next jar and install a lid.

Concurrent LIR Functions During Operations

- *Commands*: During lid installation operations, the LIR will always be listening for valid commands from the FCR. When a command is received, the LIR will acknowledge receipt of the command, execute the command, and report the execution of the command.
- *Health & status*: During operations, the LIR will monitor system health and report any changes in status to the FCR C&M software. If a critical error is detected, the LIR will notify the FCR C&M software and perform the end-of-shift shutdown sequence. *(A critical error is an error that could result in damage to the LIR, conveyor belt system, a person, the LIS room, or render the LIR incapable of lid installation.)*
- *Change of state*: The LIR will notify the FCR C&M software any time its state changes (Powered up, Self-test complete, Ready to install lids, executing end-of-shift sequence, position, executing power down sequence). Changes of state can occur as a result of an internal software action or in response to a command from the FCR C&M software.
- *Controlled entry to the LIS*: There is an interlock built into the LIS room door that will notify the FCR C&M software when the door is open, resulting in the FCR C&M software sending the LIR the "execute end-of-shift shutdown sequence" command as well as all other JPS systems. In addition, the LIS is monitoring for motion within the LIS space that is not occupied by the LIR. If, for some reason, the door interlock doesn't work and motion is detected within the LIS room; the LIS will notify the FCR of the detected motion, which will trigger the FCR C&M software to send and "execute end-of-shift shutdown" command to all JPS systems, including the LIR.

LIR end-of-shift shutdown: The JPS stops jar processing operations at 4 pm. At this time, the LIR stops lid installation automatically based on its internal clock. The LIR shutdown sequence includes automatically ceasing lid installation, returning a lid to the LDS if it is holding one, positioning itself for start-up for the next shift, notifying the FCR C&M software of its status and state, and powering itself down (sleep state, ready to accept a power up command).

The following sections include several topics and activities that should be considered when defining the preliminary set of lifecycle concepts.

4.4.1 Perspectives

Lifecycle concepts can and are developed from several perspectives. This information is often recorded in various plans. It is important that the project team address each of these perspectives rather than just focusing on a single perspective. Each of the perspectives is equally important, as each is a source of needs and the resulting design input requirements. Assuming all the relevant stakeholders are identified and involved in the elicitation activities, each of these perspectives should be addressed. To help ensure consistency, correctness, and completeness, all three perspectives must be considered as part of the lifecycle concepts analysis and maturation activities.

4.4.1.1 Business/Project Management Perspective The *business/project management perspective* addresses how the project will conduct PM and SE activities to develop the SOI, with a focus on development of concepts, needs, and requirements. Of particular importance are the project's concepts for acquisition, security, safety, resilience, sustainability, development, manufacturing, system integration, system verification, system validation, maintenance, and retirement. Acquisition concepts include the make/buy decisions (what will be done in-house and what will be contracted out to a supplier). The project team will need to decide who will be responsible for system integration, system verification, and system validation. This information is needed to plan the project, define a Work Breakdown Structure (WBS) and Product Breakdown Structure (PBS), develop schedules and budgets, and issue contracts. *See also Chapter 13, Supplier Developed SOIs.*

The business/project management perspective is commonly documented in a PMP, SEMP, and other supporting plans. If parts of the development of the SOI are contracted out to a supplier, the customer will reflect this information in a SOW, SA, or PO in accordance with their acquisition process. The business *management perspective* is historically recorded in an enterprise-level ConOps or similar form.

Operational-level lifecycle CNR is documented at the operations layer, as shown in Figure 2.6. Their focus often concerns branding, market share, business-level standards, regulations, costs, and schedules. For example, considering how a product relates to the competition, cost per unit to manufacture, and other products developed by the enterprise, could yield the following need statement: *The operational-level stakeholders need the SOI to have a production cost of less than $xx and a yield of xx%.*

Applicable operational-level lifecycle CNR are allocated to the SOI and the project team. The project team therefore needs to work with the operational-level stakeholders to understand their lifecycle CNR and how each will be addressed.

4.4.1.2 Operational Perspective The *operational perspective* addresses how the stakeholders view the SOI in terms of how they will interact (interface) with the SOI during the various SOI lifecycle activities. The operational perspective also addresses how the SOI will interact within its operational environment during operations as well as with the external systems that make up the macro system of which it is a part. The operation perspective is historically recorded in an operational-level ConOps or similar form.

4.4.1.3 System Perspective The *system perspective* addresses the SOI's expected features, capabilities, functionality, performance, quality, safety, security, resilience, sustainability, interfaces, compliance, architecture, and physical characteristics. This is a technical view of the SOI, whose lifecycle concepts are historically recorded in an operations concept or similar form. Once the SOI lifecycle concepts have matured, they are then implemented via an integrated set of needs and transformed into a set of design input requirements.

Part of this transformation involves a change in perspectives where the integrated set of needs represents more of an operational, stakeholder perspective of the SOI when viewed as a black box, and the set of design inputs transformed from those needs represents a systems perspective of what the system must do to meet those needs.

Lifecycle CNR for the system, subsystems, and system elements is recorded at the appropriate layers shown in Figure 2.6. The focus is on the SOI and how the SOI will support the operational-level lifecycle CNR recorded in the operational-level ConOps.

Each of the above perspectives is equally important, as each is a source of needs and the resulting design input requirements. Assuming all the relevant stakeholders were identified and involved in the elicitation activities, each of these perspectives will have been addressed. To help ensure consistency, correctness, and completeness, all three perspectives must be considered as part of the lifecycle concepts analysis and maturation activities.

Gathering this information helps ensure the project team has the knowledge to define a preliminary set of lifecycle concepts on which they will perform the necessary analysis to mature the lifecycle concepts to a level of maturity such that a complete, consistent, correct, and feasible integrated set of needs can be defined.

4.4.2 Concept of Operations versus Operations Concept

Historically, many projects document the preliminary lifecycle concepts in a Concept of Operations (ConOps) or Operations Concept (OpsCon) type of format as discussed above. Unfortunately, the terms ConOps and OpsCon are often used interchangeability, even though each of their perspectives is different; this can result in confusion.

To avoid this confusion, one approach is to avoid these terms and develop a set of lifecycle concepts that represent each of the above perspectives and document them in a Systems Concept (SysCon) type of document or electronic equivalent. When doing this, it is advisable to separate each perspective into separate parts of the SysCon.

Standards concerning the development of ConOps and OpsCon documents include IEEE Std 1362–1998 "IEEE Guide for Information Technology—System Definition—Concept of Operations (ConOps) Document", ISO/IEC/IEEE 29148 "*Systems and software engineering. Lifecycle processes. Requirements engineering*", and ANSI/AIAA G-043A-2012 "*Guide to the Preparation of Operational Concept Documents*". NASA's Procedural Requirements document, NPR 7123.1B, "*NASA Systems Engineering Processes and Requirements*" includes a discussion and definitions of ConOps and OpsCon are closely aligned with G-043A.

The INCOSE SE HB provides specific definitions of ConOps versus OpsCon based on ISO/IEC/IEEE 29148.

Note: These standards were developed from a document-centric perspective rather than a data-centric perspective. From a data-centric perspective, the information in the ConOps, OpsCon, or SysCon should be recorded within the SOI's integrated dataset in a form that allows traceability to specific data and information items. Once feasible lifecycle concepts have been defined, traceability between the lifecycle concepts and the resulting integrated set of needs can be established.

4.4.3 Obtain Stakeholder Agreement

At this stage of *Lifecycle Concepts and Needs Definition*, the preliminary lifecycle concepts and supporting information representing each of the above perspectives must be presented to the stakeholders in a form suitable for review and feedback.

After the elicitation activities and the integration of the results, it is common to revisit the original problem statement, higher-level lifecycle CNR and make modifications as needed based on the latest information and knowledge gained up to this point. At the system level or lower levels, after the lifecycle concepts analysis and maturation activities are defined, higher-level lifecycle CNR may need to be revised not only from a feasibility standpoint but also from the standpoint of missing or incorrect information.

Defining a set of lifecycle concepts requires the integration of several disparate stakeholder views, which may not be harmonious. It is critical that the project team has confirmation from the stakeholders that they understand and agree with the problem statement, MGOs, measures, higher-level lifecycle concepts, needs, requirements, drivers and constraints, and risk as currently recorded within the SOI's integrated dataset. This is especially true concerning how the project team dealt with, and resolved inconsistencies, disagreements, ambiguity, and feasibility issues during stakeholder elicitation and resulting integration activities.

To get this confirmation and agreement, the project team must provide this information to the higher-level organization stakeholders and SOI-level stakeholders for their review and comment before proceeding with the lifecycle concepts analysis and maturation activities defined in Section 4.5.

This activity is critical to ensuring the resulting individual needs have the GtWR characteristics *C1—Necessary, C3—Unambiguous, C4—Complete*, and *C8—Correct* and sets of needs have the GtWR characteristics *C10—Complete, C11—Consistent, C13—Comprehensible*, and *C14 Able to be Validated.*

Note: In the past, using a document-based approach, it was common for organizations to baseline the set of lifecycle concepts in a OpsCon/ConOps type document as part of defining the scope of the project and then develop a set of design input requirements directly from the preliminary lifecycle concepts, as shown by the red lines in Figure 4.2*b.*

Lifecycle concepts analysis and maturation (as defined in Section 4.5) were historically performed under the heading "requirements analysis" to define the set of design input requirements. In cases where the customer provided a set of design input requirements to a supplier, "requirements analysis" was a supplier activity to ensure they understood the customer-supplied system requirements.

This approach is common for SOIs that will be outsourced to a supplier, where the supplier is given the set of design input requirements as part of the contract without also supplying them with the underlying analysis from which the requirements were derived.

Bypassing lifecycle concepts analysis and maturation and going directly to requirements analysis often results in the set of needs missing from the SOI's integrated dataset, and thus system validation would be loosely defined if at all. Additionally, feasibility of the preliminary lifecycle concepts, and resulting design input requirements, may not have been established, resulting in increased technical debt and risk for the project.

4.5 LIFECYCLE CONCEPTS ANALYSIS AND MATURATION

The activities performed up to this point resulted in a preliminary set of lifecycle concepts and inputs to those concepts shown in Figure 4.4. These preliminary lifecycle concepts have provided a broad description of system behavior, which is a starting point for lifecycle concepts analysis and maturation activities.

As a result of lifecycle concepts analysis and maturation activities discussed in this section, functional architectural and analytical/behavioral models are developed or expanded, and a preliminary physical architecture is defined. Based on the resulting information, the preliminary set of lifecycle concepts is transformed into a mature set of lifecycle concepts that are consistent, correct, complete, and feasible. These concepts are then transformed via needs analysis into an integrated set of needs.

FIGURE 4.4 Inputs to the Lifecycle Concepts Analysis and Maturation Activities.

Nomenclature notes: The use of "logical" can be confusing: does it mean a "logical model" or "logical architecture"? To avoid ambiguity, at this stage of development and level of abstraction, this Manual uses "analytical/behavioral model" when referring to language-based models that contain logic (XOR, AND, XOR, IF, WHEN) to show the flow of information, triggers for functions, and assess behavior common to language-based model development. When referring to diagrams showing functions, and basic inputs/outputs for those functions, "functional model" is used. When referring to architecture views of the SOI, "functional architecture," "functional architecture model," or "physical architecture" are used. Physical architecture refers to the physical subsystems and system elements included within the physical integrated system architecture, including electrical, hardware, chemical, mechanical, firmware, and software elements. It is the physical subsystems and system elements that will be procured, built, coded, integrated, verified, validated, and delivered.

The following topics should be considered during lifecycle concepts analysis and maturation.

4.5.1 Feasibility

Addressing feasibility during lifecycle concepts analysis and maturation activities is critical to being able to ensure the individual need statements and integrated set of needs are feasible. A set of feasible lifecycle concepts is key to having an integrated set of needs that are feasible.

A key expectation at the gate review to baseline the integrated set of needs and resulting set of design input requirements is that *there is at least one feasible design concept and physical architecture defined from which the integrated set of needs were derived.* Failing to establish feasibility before defining an integrated set of needs and transforming them into a set of design input requirements for an SOI results in a significant amount of technical debt.

Feasibility takes into consideration key drivers and constraints, including cost, schedule, technical, legal, ethical, and safety, with acceptable risk for this lifecycle stage. An SOI may not be feasible [41] if, to meet the MGOs and measures:

- It needs to break the laws of physics to achieve the stated intent (cannot be done).
- Its existence or use violates laws or regulations in an applicable jurisdiction.
- It leads to excessive program risk because of technical immaturity or inadequate margin with respect to program cost and schedule as a function of lifecycle phase.
- Its existence or use will violate social norms in terms of harm to humanity or the environment.
- Its use represents an unacceptable safety risk to the users, operators, or public at large.
- It is not sustainable, over its life.

Because technology is related to subsystems and system elements that make up the physical architecture of a system (rather than the functional architecture or analytical/behavioral model of the integrated system), the project team members responsible for design will need to be involved in the TRA. Based on this assessment, the project team will assign TRLs to parts of the system physical architecture associated with a critical technology needed to achieve a specific capability and associated performance. To do this, the functional architecture model will need to be mapped to the physical architecture prior to defining the integrated set of needs. This mapping must be done as part of the lifecycle concepts analysis and maturation activities. Without this knowledge, the feasibility of the resulting needs will be uncertain, adding technical risk to the project.

To establish feasibility, the project team must do the analysis and activities needed that will result in *defining a feasible set of lifecycle concepts* and *a preliminary physical architecture* that will meet the MGOs and measures representing the higher-level stakeholder needs within the defined drivers and constraints with acceptable risk. Maturing the lifecycle concepts and defining a preliminary physical architecture helps ensure the project team has addressed completeness, correctness, consistency, and feasibility prior to baselining the integrated set of needs.

4.5.2 Design as Part of the Lifecycle Concepts Analysis and Maturation Activities

It is a myth that design activities do not begin until after the integrated set of needs and set of design input requirements are baselined. Given that feasibility is an attribute of the physical architecture, the project team must define and mature a design concept and preliminary physical architecture concurrently with the maturation of the lifecycle concepts and associated artifacts.

Lessons learned indicate that preliminary design activities, including the development of prototypes and models, during lifecycle concepts analysis and maturation activities can identify critical issues that are difficult to determine until a physical design concept and architecture have been defined. This preliminary physical architecture must be of sufficient maturity to identify critical technologies and assess their maturity.

Note: Refer to the INCOSE SE HB Section 3.2.2 for more information concerning prototypes.

Using the knowledge gained from prototypes, models, and trade studies, a feasible design concept and preliminary physical architecture can be matured to allow a more complete understanding of both the programmatic and SE implications, development challenges, costs, schedule, and risk. With this knowledge, the project team can more accurately evaluate development, verification, validation, operations, sustaining engineering, and disposal considerations. Addressing these early in the SOI lifecycle will help mitigate the risk of expensive and time-consuming rework and accumulating excessive amounts of technical debt.

The preliminary design work at this stage enables validation of the preliminary integrated set of needs and set of design input requirements, providing analysis to determine whether something is missing or not needed. This validation is a key part of assessing completeness, correctness, consistency, and feasibility of the preliminary physical architecture and lifecycle concepts.

From a PM perspective, the development of a preliminary physical architecture enables the project team to define a PBS, which, in turn, results in more credible cost and schedule estimates and the ability to track and manage development or acquisition of the subsystems and system elements reflected within the PBS.

4.5.3 Use of Diagrams and Models for Analysis

A key analysis approach used to mature the lifecycle concepts is the use of diagrams and models. Models and diagrams are visualizations of the underlying data and information model of the SOI and the SE process activity artifacts. As such, they represent the underlying analysis from which the integrated set of needs is derived. Further elaboration of these models and diagrams is then used for the analysis needed to transform the integrated set of needs into the set of design input requirements. As the models mature, functions are decomposed, lower-level architectures are defined, and subfunctions, performance, quality, and physical attributes are allocated to the lower-level system elements.

These models can be refined as the physical architecture is defined and used to help define lifecycle CNR for the subsystems and system elements that are included within the SOI physical architecture. In doing so, the models are key to establishing traceability and supporting allocation and budgeting. These models are also key in identifying dependencies and interactions (interfaces) between parts of the physical architecture as well as interactions between the SOI, external systems, and the operational environment.

Analysis using models and diagrams enables the project team to demonstrate that the integrated set of needs, set of design input requirements, and the set of design output specifications are necessary, sufficient, and feasible [41] prior to baselining.

Without this underlying analysis, actual validation of the needs will likely only happen late in the development lifecycle during system integration, system verification, and system validation. As a result, the delivered SOI will often fail to meet its intended use when operated in the actual operational environment by the intended users (that is, it will fail system validation).

From a holistic, integrated view (as discussed in Chapter 2), the project team must develop an integrated system model from the beginning of the project. This integrated model can then be used during early system verification and validation activities as well as during design verification and design validation by using descriptive models and simulations to uncover issues before the SOI is built or coded (discussed further in Chapter 8).

The project team can use their models to clearly identify the functions, associated inputs, their outputs, as well as the dependencies and interactions between subsystems and system elements within the SOI, and between the SOI and the environment/external systems.

A key tenet of systems thinking is that the behavior of a system is a function of the interactions between the parts that make up the system as well as interactions between the system and external systems. A major advantage of using models is that the analytical/behavioral models can be transformed into mathematical models that can be used to assess the behavior and performance of the integrated system.

Example: Having a simulation capability enables the project team to perform design verification and design validation early to discover design issues within the analytical/behavioral model rather than waiting to discover these issues to appear during system integration, system verification, and system validation of the physical SOI.

The use of models and diagrams helps establish that individual needs and requirements have the GtWR characteristics *C1—Necessary* and *C8—Correct*; and the sets of needs and requirements have the GtWR characteristics *C10—Complete* and *C11—Consistent.*

4.5.3.1 Types of Diagrams and Models It is risky to attempt to address all characteristics in a single model because the model will either not address specific elements with enough definition and accuracy, and/or a "complete" model will become too complicated to generate, view, and comprehend, as well as to define and execute simulations [36, 42].

There are a variety of diagrams, models [43], and simulations that can be used to help with the lifecycle concepts analysis and maturation activities based on the needs of the project team and the type of analysis being performed. These are often domain-specific depending on the type of system being developed, the degree of data-centricity practiced by the project team, the specific analysis being performed, as well as the capabilities of the project team, their toolset, and culture.

As stated in the INCOSE SE HB, when using models and simulations, it is important that they go through a verification process to ensure they are formed correctly, as well as a validation process to ensure they correctly represent the entity they are modeling.

Some examples of diagrams are shown below (note that this is not an exhaustive list):

- Functional Flow Block Diagram (FFBD)
- Activity Flow Diagram (AFD)
- System Architecture Diagram (SAD)
- Use Case Diagram
- Context Diagram (external interface diagram)
- Activity Diagram
- Workflow Diagram
- Swim-lane Diagram
- Sequence Diagram
- States/Mode Diagram
- Data Flow Diagram (DFD)
- Entity Relationship Diagram (ERD)
- N^2 Diagram
- IPO (input/process/output) Diagram
- SIPOC (source, input, process, output, customer) Diagram

Examples of several types of models include (this is not an exhaustive list):

- Physical Model
- Structural Model
- Descriptive Model
- Behavior/Functional Model
- Temporal Model
- Mass Model

- Cost Model
- Probabilistic Model
- Parametric Model
- Layout Model
- Network Model
- Visualization
- Simulation
- Mathematical Model
- Prototype

A detailed discussion of each is beyond the scope of this Manual; however, each is well known and described on the Internet and in the literature.

4.5.3.2 Considerations When Developing Diagrams and Models A key feature when forming the project toolset is the capability of the toolset to support diagramming and modeling. The choice of a toolset will depend on which of the methods and types of diagrams and models the project team plans to use as part of their analysis and maturation activities. *Refer to Chapter 16 for a more detailed discussion on the features a project toolset should have.*

4.5.3.2.1 Perspectives As discussed earlier, diagrams and models can be developed from the *business/project management perspective* (activities), *customer/user operations perspective* (activities, interactions with the SOI), and *SOI perspective* (features, capabilities, functions, performance, and interactions). While the form of the diagrams and models is similar, the actor is different.

- If the actor is an organization (project or supplier), then diagrams of the activities can be developed, focusing on the activities used to develop the SOI as well as the inputs and outputs of each activity.
- If the actor is a user/operator, the diagram would focus on the activities performed by the user/operator in the process of using or interacting with the SOI.
- If the actor is the SOI, the diagram would focus on the functions performed by the SOI, inputs and outputs for the functions, sources of the inputs, and customers of the outputs.

Each perspective is necessary to mature the set of lifecycle concepts and to ensure a complete integrated set of needs and design input requirements. The business/*project management perspective* is needed to clearly understand the activities, work products, deliverables, costs, and schedules. The *user/operator perspective* is needed to clearly understand and define the human/system interactions, interfaces, use cases, and operational scenarios. The *SOI perspective* is needed to clearly understand the capabilities, functionality, performance, behavior, and interactions both within the SOI and with external systems.

To avoid confusion, it is important to make clear which perspective the diagram/model is communicating, the type of analysis the model is supporting, the need and requirements for the model, and expected outcomes from the use of the model.

4.5.3.2.2 Supplier Developed SOIs and Modeling A key consideration in outsourcing development to a supplier is the role that diagrams and modeling will have in the supplier development activities. In this context, the customer project team must consider the integration and data sharing aspects of the diagramming and modeling tools used by the both the customer and suppliers. If the customer is responsible for the integrated model of the SOI, either provisions will need to be made for the supplier to work within this modeling environment or use modeling tools that enable their models to be integrated within the customer's integrated model.

Another consideration is that when the supplier is given a set of design input requirements for a SOI as part of a contract, whether the customer should also provide access to the supplier the models that contain the underlying analysis the customer used to mature the lifecycle concepts, transform those into an integrated set of needs, and transform those needs into the set of design input requirements that were provided to the supplier. Doing so would be beneficial to both the customer and supplier by reducing the time the supplier spends on requirement analysis and architecting prior to design. If the supplier is not provided with this information, they would have to spend time to redo the analysis, adding time and cost to the contract.

4.5.4 Examples of Functional/Activity Analysis Using Diagrams and Models

Below are generic examples concerning the use of basic diagrams and models for analysis. While these diagrams were produced using a standard office application, in practice, the project team would develop these diagrams within the project toolset, enabling the underlying data to be managed within a sharable integrated data and information model. In this context, each of the diagrams or models would be a visualization of that data and information model. Which visualization is selected would depend on the analyst's purpose and need.

4.5.4.1 Fundamental System Model At a fundamental level, there are 1) external inputs, 2) functions, or activities to apply those inputs, 3) functions to transform those inputs into outputs, 4) the outputs, and 5) the functions to export those outputs to some destination. Each input has a source, and each output has a destination. The source or destination could be another function, a user/operator, or another system, subsystem, or system element. The functions and associated inputs and outputs exist in an external operational environment that contains other systems.

For the SOI, types of inputs and outputs consist of information (data, commands, messages, and measurements), energy (electrical power, heat), or matter (solids, gases, and fluids). For an organization (project team or supplier), the inputs and outputs consist of work products, artifacts, and deliverables. For operations, the inputs, and outputs involve the user/operator interactions across the HMIs.

This basic model shown in Figure 4.5 is "The Receive Inputs," "Transform Inputs to Outputs," and "Export Outputs." These functions or activities are decomposed as needed to capture the specific

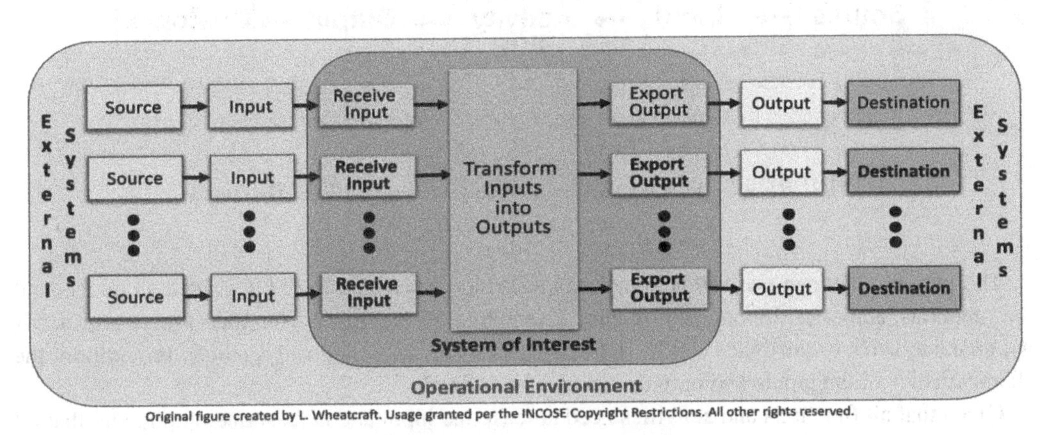

FIGURE 4.5 Fundamental System Model.

functionality, activities, and performance needed for the SOI to accomplish its intended purpose/use in its operational environment.

Given the fundamental system model, it is common to start analysis activities by reviewing the preliminary set of lifecycle CNR that resulted from the stakeholder elicitation activities. During this review, the functions are identified. A function is a task, action, or activity that must be performed by some entity to achieve a desired outcome. One method is to identify the verbs used to describe use cases, user stories, or operational scenarios and determine which apply to the project or user/operator (activities) or to the SOI (functions).

The result is a list of functions to be implemented by the SOI and activities to be implemented by the users/operators, or the project team. Each of these verbs will result in need statements in the integrated set of needs which will be transformed into SOI functional/performance, or operational requirements that will be included within the set of design input requirements.

For the list of functions and activities, a common diagram used for analysis is referred to as a SIPOC diagram, as shown in Figure 4.6. Where SIPOC stands for **S**ource-**I**nput-**P**rocess (function/activity)-**O**utput-**C**ustomer. For each function or activity, its inputs and outputs are identified. For each input, the source is listed, and for each output, the customer or destination of the output is listed. In the function/activity blocks, the functions/activities are stated as verb/object pairs. *Note: the use of "customer" here is because the definition of customer includes "entity that has requested or needs the output."*

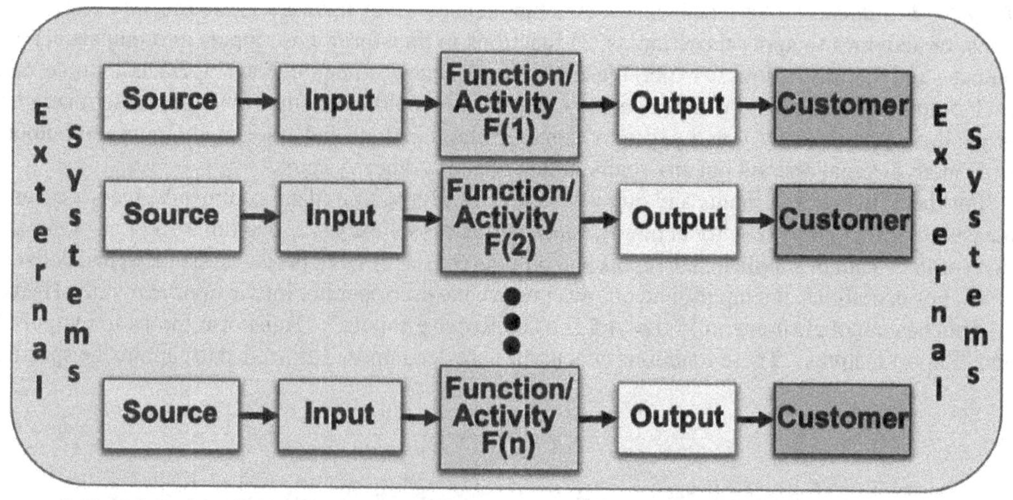

Original figure created by L. Wheatcraft. Usage granted per the INCOSE Copyright Restrictions. All other rights reserved.

FIGURE 4.6 Functional/Activity Analysis using a SIPOC Diagram.

The project team can use SIPOC diagrams to better understand the SOI as well as understand the functions, activities, inputs, and outputs. Expanding on this model, for each function or activity, characteristics or attributes can be defined (such as performance and quality). In addition, the characteristics of the inputs and outputs can also be defined.

Given that all functions and activities need at least one input and at least one output, and that all inputs need a source and all outputs need a destination or customer, part of the functional/activity analysis is focused on "completeness," i.e., discovering what is missing and correcting these deficiencies.

Example List of Functions Associated with LIR Start-up

Acronyms used in the following scenarios: LIR, Lid Installing Robot; C&C, Command and Control; FCR, Facility Control Room; C&M, Command & Monitoring.

Using the operational scenarios that resulted from the integration of the elicitation activities, the following functions were identified concerning start-up at the beginning of the shift.

Initial conditions – LIR connected to power source and communication network, LIR in power down, sleep state, ready to receive a command. LHS positioned to obtain lid from LDS, JHS in position to grasp next jar.

(F1) Power up LIR when "Power up" command is received from FCR C&M software.

(F2) Send Power Status message to FCR C&M Software once power up is complete.

(F3) Perform self-test once power up is complete.

(F4) Send Health & Status (H&S) data to the FCR C&M Software upon completion of the self-test.

(F5) Enter "Ready for Lid Installation" State upon successful completion of self-test.

(F13) Report State Change to the FCR C&M Software.

Example SIPOC Diagrams Developed by the LIR Project Team

Using the functions identified for start-up, the following SIPOC diagrams were developed.

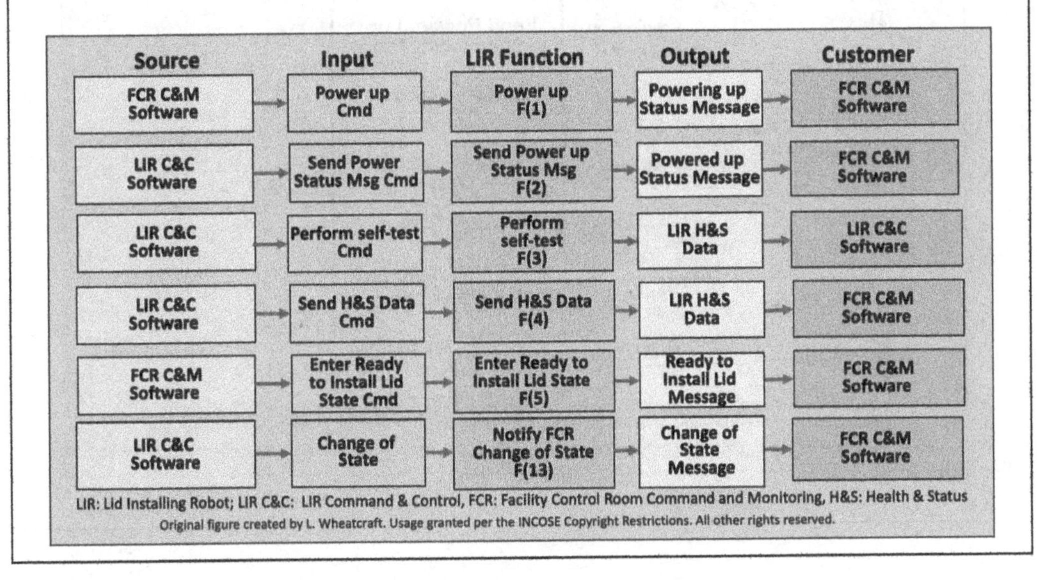

4.5.4.2 *Functional/Activity Flow Block Diagram* With the knowledge gained from developing the SIPOC diagrams and resulting analysis, the functions or activities can be combined to form a FFBD, or an AFD, as shown in Figure 4.7.

Because the source of an input could be another function or activity, and the customer or destination for each output could be another function or activity, developing a FFBD is like "connecting the dots." Many graphical modeling tools allow this to be done by drag and drop, connecting inputs to their source and outputs to their destination.

The FFBD and AFD are representations of SOI functions or user/organizational activities, where the blocks represent the inputs, outputs, and functions/activities, and the lines illustrate how they are connected. The focus is on the flow of information, energy, and matter, not how the function or activity is performed.

The FFBD can be used to show functional flows internal to the SOI, user/operator interactions with the SOI, interactions of the SOI with external systems, as well as manufacturing processes and workflows.

Because the focus of the AFD is on activities performed by people, AFDs can be used to develop operating procedures or IFU. Again, each perspective is needed, but it is best to use separate diagrams for activities performed by the organization and functions performed by the SOI.

Example LIR FFBD Developed by the LIR Project Team

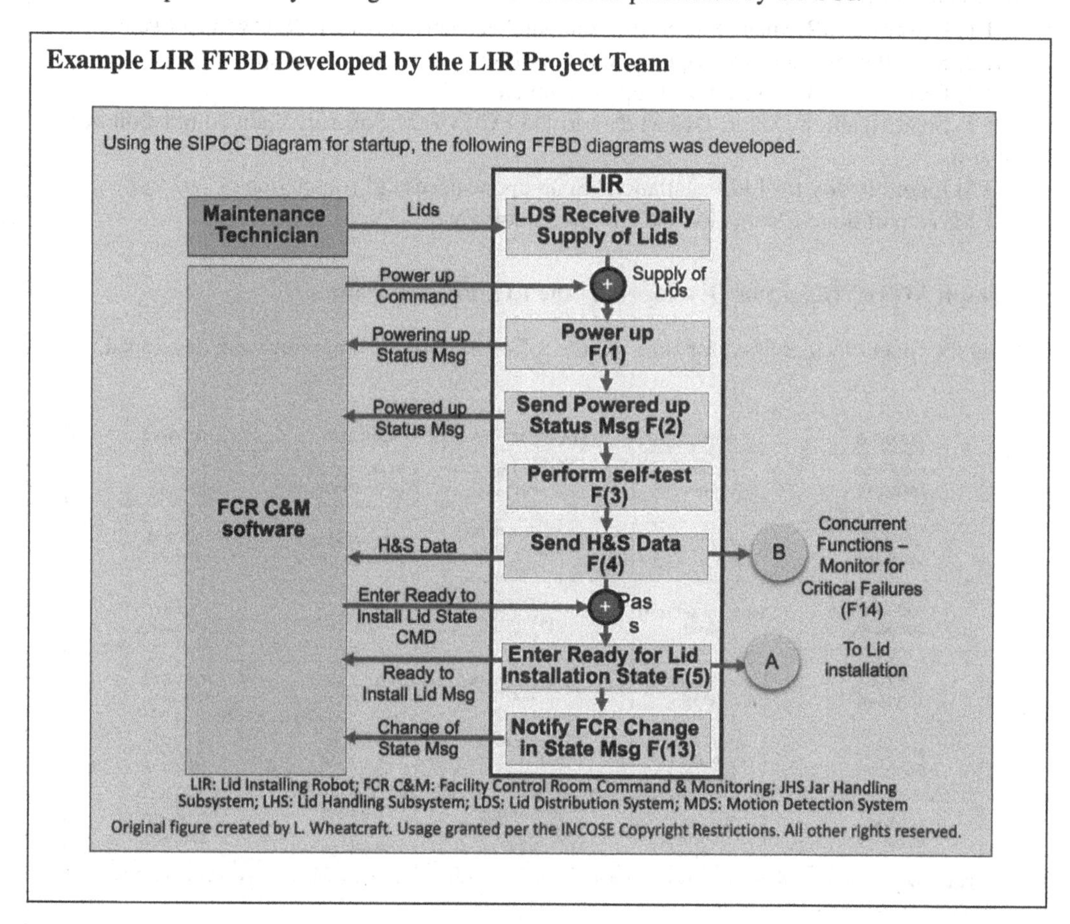

4.5.4.3 System Architecture Diagram A function or activity may be accomplished by one or more subsystems or system elements. The objective of creating a SAD is to provide the foundation for defining the physical architecture through the allocation of system-level functions first to functional architectural entities and then to physical hardware/software, facilities, and personnel.

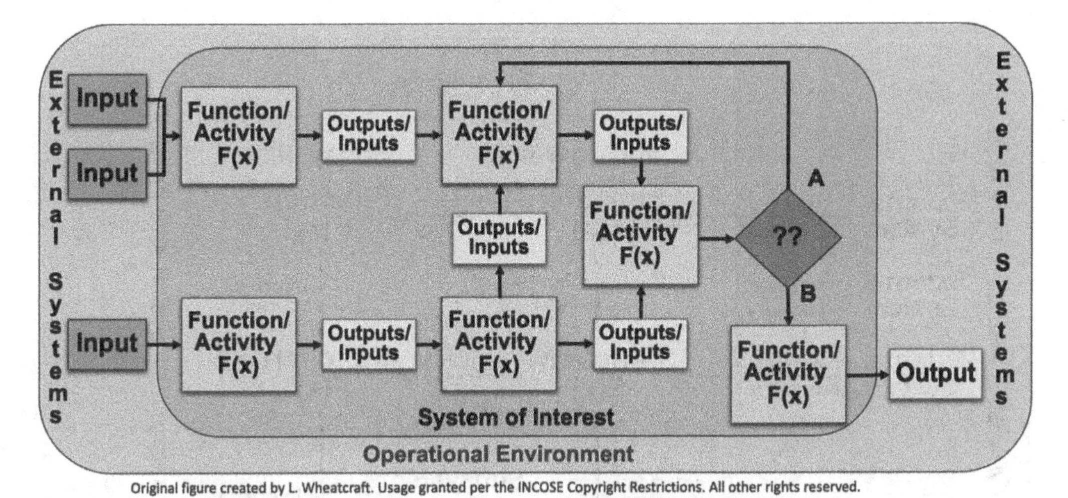

FIGURE 4.7 Generic Functional Flow Block Diagram or Activity Flow Diagram.

In this context, the functions shown in Figure 4.7 are allocated to subsystems. Based on these allocations, the system-level functions are decomposed into subfunctions, and additional derived functions are defined, if appropriate. These can then be mapped to the subsystems, as shown in Figure 4.8.

The input/output blocks shown in Figure 4.7 represent interactions across an interface boundary. There will be interactions between external systems and the system, as well as internal interactions between the functions within the subsystems and system elements across an interface boundary. Each interaction represents either an input or an output, which must be clearly defined. These definitions will be included within an ICD or similar type of repository and referenced within the applicable interface requirements.

Similar functions are grouped together into functional system elements referred to as "logical elements." The functional architectural model can be transformed into more detailed analytical/behavioral models to help show and understand internal and external behaviors and interactions.

Once the analytical/behavioral model has been reviewed for correctness and completeness, the functions, performance, interactions, and quality can be allocated to the physical architecture. The allocation of these functions will eventually be realized by design and communicated to the builders/coders via the design output specifications.

For greenfield SOIs, the physical architecture will be based on the project team's experience in developing similar systems, top-down. For brownfield SOIs, the physical architecture may already be defined, in which case the functional architecture and analytical/behavioral models would be created bottom-up (if they do not exist) and would be used for analysis of the modified system.

4.5.5 Levels of Detail and Abstraction

An important consideration when developing functional architectural and analytical/behavioral models and the resulting physical architecture is the concept of levels. As discussed in Chapter 2, levels

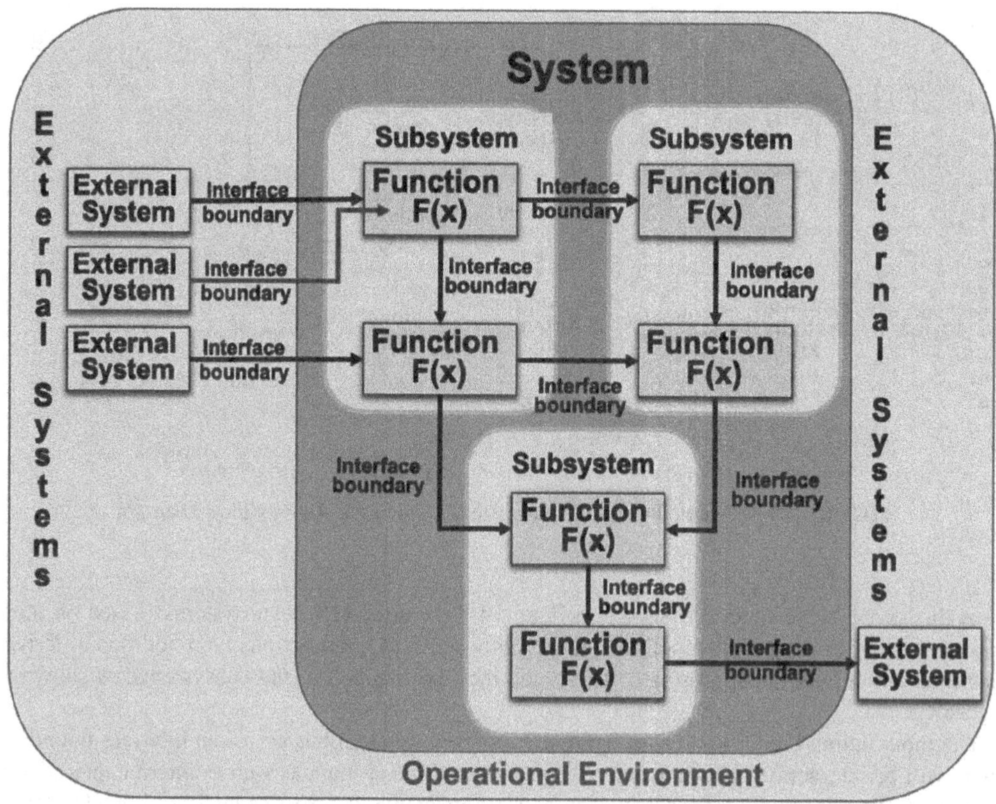

FIGURE 4.8 System Architecture Diagram.

can refer to levels of architecture, levels of detail (high or low), levels of decomposition, or levels of abstraction.

At this lifecycle stage, the focus is on levels of abstraction and decomposition. Initially, high-level functions are defined, for example, "Process Inputs." Then that function is decomposed into subfunctions, that together result in the parent function being realized, for example "Receive inputs." "Store Raw Inputs," "Transform Inputs," "Store Transformed Inputs," "Display Transformed Inputs," and "Export Transformed Inputs".

When addressing architecture, these levels of abstraction will be the basis for defining the levels of the architecture – first functional, then physical. From a design input perspective, the challenge is determining the level of decomposition and detail most appropriate for what is needed for a given SOI lifecycle stage. It is important to understand where the "line" shown in Figure 4.9 is between design inputs (what) and the design outputs (how). The goal of decomposition at this stage of development is to decompose functions such that a function can be allocated to a single subsystem or system element.

As a rule of thumb, to obtain the right level of abstraction, decomposition should continue down to provide a level of detail that will allow the project team to validate the functional architectural and the logical/behavior models demonstrate that the lifecycle concepts will result in the MGOs, measures, and higher-level needs and requirements to be met within the drivers and constraints with acceptable risk. And then STOP – lower levels of detail are addressed in later lifecycle stages via iteration and recursion.

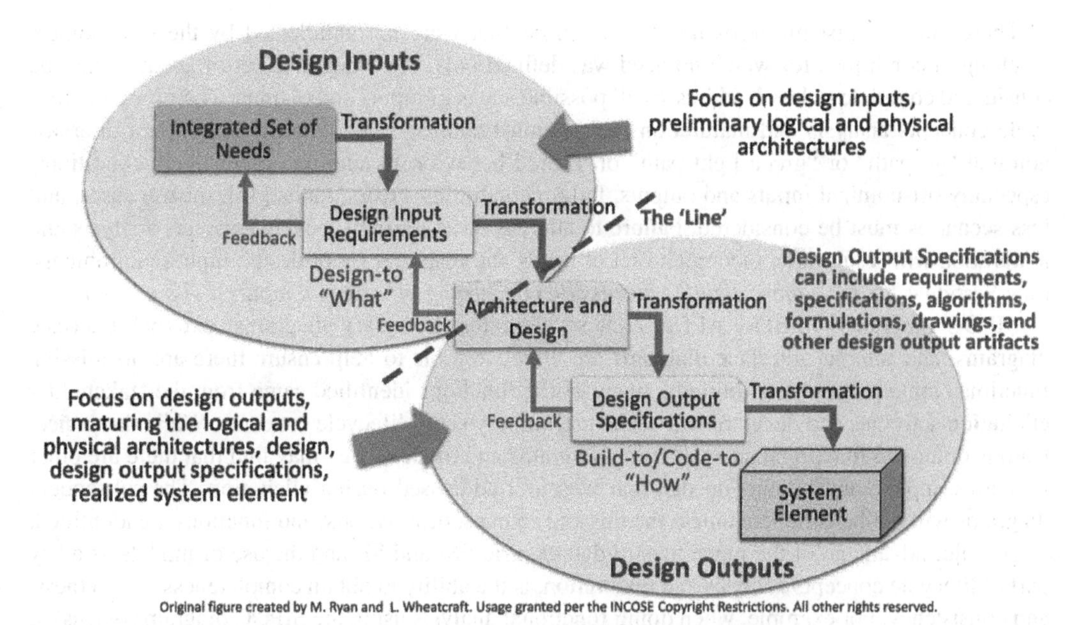

FIGURE 4.9 The Line Between Design Inputs and Design Outputs.

In the beginning of the modeling effort, focus should be on the development of functional architectural models of externally observable functions, performance, quality, and interactions with external systems (the "black box" view). It is what is externally observable (what) that represents the integrated set of needs that will be transformed into the set of design input requirements. The higher-level (of abstraction) models will be a major source of needs and the resulting design input requirements.

If it is not externally observable, it is the responsibility of the project team members performing physical architecture definition and design definition to concurrently define the analytical/behavioral models and resulting physical architecture, to address (how). These logical/behavioral interactions within the SOI enable the architects and designers to dive down to a more detailed logical understanding of the flow of information and interactions within the SOI (the "white box" view).

From a holistic perspective, for whichever level the project team is addressing, they should have a working knowledge of the entities at the level above and the level below, as well as any system, subsystems, or system elements with which they interact at the current level (internal or external).

It is important that, when defining the integrated set of needs, the project team returns to the stakeholder external view perspective. A common error when defining the integrated set of needs and resulting design input requirements for an SOI is including needs and requirements dealing with internal aspects of the SOI, resulting in requirements "level" issues where the requirements are communicated at the wrong level.

This activity helps establish the GtWR characteristic C2—Appropriate.

4.5.6 Completeness of the Lifecycle Concepts

If the lifecycle concepts are not complete, the integrated sets of needs and resulting design input requirements will not be complete, resulting in a realized SOI that fails system validation. As stated in the GtWR, well-formed integrated sets of needs and their resulting set of design input requirements have the characteristic *C10—Complete.*

There may be missing needs due to conditions that were not anticipated by the stakeholders resulting in conditions for which no need was defined [44]. The integrated set of needs cannot be considered complete unless it addresses all possible states of inputs and outputs. Therefore, the lifecycle concepts analysis and maturation process must address more than the nominal or alternate nominal "go path" or "green light path" of desired behavior. In addition, other likely conditions, especially off-nominal inputs and outputs, failure conditions, errors, faults [45], misuse cases, and loss scenarios must be considered. Failure to address these during lifecycle concepts analysis and maturation will result in an incomplete set of needs and resulting set of design input requirements. (*See Section* 4.5.8 *for a more detailed discussion concerning off-nominal inputs and outputs.*)

SIPOC diagrams, FFBDs, AFDs, context diagrams, boundary diagrams, external interface diagrams, and internal interface diagrams are effective tools to help ensure there are no missing functions, inputs, or outputs. Initially, many of the functions identified came from the stakeholder elicitation activities and development of the preliminary set of lifecycle concepts. While the project team is doing the functional analysis using diagrams and models, they may find that there are other functions, inputs, and outputs needed that were not addressed during elicitation. The subsequent diagrams will not be complete unless the missing sources, destinations, and functions are identified.

A major advantage of the move toward data-centric PM and SE and the use of models as a key part of lifecycle concepts analysis and maturation, is the ability to obtain completeness, correctness, and consistency. For example, when doing functional analysis using the SIPOC diagrams discussed earlier, every input must have a source, and every output must have a customer (destination). It is often the case when developing the SIPOC diagrams that inputs are missing, the source of inputs is missing, and there are outputs that have no destination (customer). All these cases must be addressed before completeness can be achieved.

Another characteristic of well-formed needs and requirements is that each is characteristic *C1—Necessary*. In the case above, where outputs have no destination (customer), it may be that the output is not necessary, and therefore the function that generated it is not necessary. If there is no customer who has a need for the output, why does it exist along with the function that produced that output? If this is the case, then the function should be removed.

Once these issues have been addressed, each input must be analyzed for possible behaviors as well as combinations of inputs, considering the range of conditions that may influence these behaviors. Likewise, each output must be analyzed for possible behaviors, considering the range of conditions that may influence these behaviors. While it may be impractical to address all possible conditions of inputs or outputs as well as combinations of conditions, the project team should consider which combinations of input conditions, output conditions, alternate nominal, off-nominal, and misuse conditions should be addressed. This will help ensure that an acceptable response to each of these conditions or combination of conditions is reflected in the integrated set of needs and resulting set of design input requirements.

A key tool to aid in this analysis is the ability to use models to run simulations, enabling the project team to assess the correctness and completeness of the modeled system. Simulations also aid in identifying combinations of inputs, outputs, and operating conditions that can lead to failures, as well as approaches to address those failures or error conditions. Another tool is the descriptive model, which enable queries to show incomplete data (such as missing inputs), which enables the SE to address those issues.

An additional aspect to assess is the concept of cascading failures. As systems become more complex and software-intensive, there may be cases where one abnormal input may not impact the immediate function that it supplies but may result in an output that could impact another function, resulting in that function's output to be problematic. For example, a sensor failure could result in an undesirable system response that causes another system (or human system) to respond inappropriately, resulting in a failure of the integrated system. The result of this analysis can lead to additional needs and resulting design input requirements.

Other factors affecting completeness of the lifecycle concepts (and resulting integrated set of needs) include ensuring that:

- There is a concept defined for each lifecycle stage, including nominal, alternate nominal, off-nominal operations, misuse cases, and loss scenarios.
- Relevant stakeholders from each lifecycle stage participated in the elicitation activities.
- For all lifecycle stages, interface boundaries and interactions across those boundaries between the SOI and external systems during each lifecycle stage have been addressed. Of critical importance are interactions across interface boundaries that could prevent the intended use of the SOI by the intended users or allow an unintended user to use the SOI in an unintended way.
- For each function, required performance has been defined.
- Stakeholder needs for quality (-ilities) have been addressed.
- All applicable standards and regulations (or analogs) have been identified, and concepts for compliance have been defined.

4.5.7 Completeness, Correctness, Consistency, and Feasibility Considerations for SOIs Being Developed to Customer-Supplied System Requirements

An important consideration is when a set of requirements is defined by the customer and included in the contract with a supplier. In this case, it is critical that the supplier assess the completeness, correctness, consistency, and feasibility of the set of requirements. (*Note: Ideally this should have been done as part of the proposal process, however, that may not always be the case.*)

For suppliers developing a product to a customer provided set of system requirements, it is tempting to look at that set of requirements as the only set of requirements. In reality, there are other stakeholder needs that must also be considered to ensure the SOI is able to be developed, integrated into the macro system it is a part, and enable system validation. In addition, the customer set of system requirements may not have the characteristics of well-formed requirements. For example:

- The production team needs the product to be manufacturable.
- The test team needs the product to be testable (able to be verified to meet the requirements and validated to meet the set of needs).
- The users need the product to be easy and safe to interface with from a human factors' perspective.
- The supplier may need the product to conform to their company's strategic development effort aligned with technology maturation.
- The product may need to address standards and regulatory compliance that were not considered by the customer.
- The customer and public need the product to be safe and secure from a cyber perspective, even if they fail to adequately address misuse cases and loss scenarios.
- There is the potential that the customer may have missed a use case related to some compatibility with existing infrastructure within the operational environment.
- There is the potential that the customer may have missed an external system or enabling system with which the SOI must interact.
- There is the potential that the customer may not have completely assessed the feasibility of the individual requirements or set of requirements.

- There is the potential that the customer may not have completely communicated their expectations for quality, safety, security, resilience, or sustainability.

Any of these considerations could lead to additional derived needs that must be captured in some fashion to define a set of design input requirements that is complete, correct, consistent, and feasible.

Key point: If the suppliers only follow the customer-supplied requirements included in the contract, they are likely to generate an SOI that may not work as needed when integrated into the macro system or reduce the capability to validate the integrated system in its operational environment when used by the intended users.

Rather than accepting the customer's requirements as-is, the supplier should consider the customer's supplied set of system requirements as inputs to the lifecycle concepts analysis and maturation activities discussed in this section. During this assessment, gaps could be identified that the customer failed to address when defining, analyzing, and maturing the lifecycle concepts, defining the integrated set of needs, and transforming them into their set of system requirements.

Failing to address these gaps will result in an incomplete set of needs and resulting design input requirements. Resolution of those gaps can be addressed during the lifecycle concepts analysis and maturation activities using data and information gathered from the system-level stakeholders during elicitation activities.

In some cases, the resulting supplier developed design input requirements may compete with or be inconsistent with the customer-supplied requirements. When this is the case, the differences will have to be reconciled to determine whether the customer requirements take priority or the supplier's derived requirements from their lifecycle concepts and needs analysis, possibly requiring the negotiation of waivers and contract changes.

By only providing an SOI that addresses the customer-supplied system requirements that were supplied in the contract, the SOI may pass system verification (it meets the customer-supplied system requirements) but fail system validation (fails to meet the customer expectations even if not clearly communicated within the customer-supplied system requirements).

4.5.8 Risk Assessment

As part of establishing completeness (as well as robustness and resilience), the project team must do an assessment of the operational and interface risks discussed previously. The integrated set of needs and resulting set of design input requirements will not be complete until operational and interface risk assessments are complete and all identified risks are managed in some way.

A common tool to use in assessing operational risk is an FMEA. For systems with user interaction, in addition to a system FMEA to discover risks internal to the system or product, the FMEA may focus on the user's interaction during operations, identifying each activity and what hazards and threats may exist as part of the user/operator interaction. This type of FMEA is often referred to as a "Use" or "User" FMEA (UFMEA).

Operational risk can also result from an unexpected change in the operational environment. There could be cases when the natural environment (temperature, humidity, particulates, air flow) could exceed what is nominal, or the induced environment (temperature, vibrations, acoustics, EMI, EMC) is not well-defined or may exceed what is defined as nominal.

For example, an engine is produced for a general-use truck to be marketed around the world. When manufactured, nominal conditions of use were defined, and the truck was designed assuming these nominal conditions plus a reasonable tolerance in case these conditions were exceeded for some duration of time. In addition, requirements on the quality of the cooling liquid in the radiator were specified. These trucks were extremely popular and were sold and used in a variety of settings. In some cases, failures were reported within the warranty period. Root cause analysis revealed that some trucks were used in mountainous, jungle settings with dirt roads and elevated temperatures and humidity. Often, the loads carried by the trucks exceeded the recommended load limits, especially when going up steep hills. As a result, the trucks overheated and radiators lost cooling liquid. Often, the only replacement liquid readily available was muddy water from a stream along the road; this resulted in cooling system failures.

A FMEA should be done on every external and internal input and output, to identify possible off-nominal or misuse cases [36, 42]. Failure to do so will result in an incomplete set of needs and associated set of design input requirements as well as a system that is not resilient and will not meet stakeholder's expectations even if possible, off-nominal or misuse cases were not explicitly stated.

The project can use the various diagrams discussed to assess what can go wrong for each input, output, and interface. Each off-nominal or misuse case represents a risk that must be mitigated depending on its likelihood and impact/consequence. This type of FMEA is often referred to as an Interface FMEA (IFMEA). For example, during functional analysis using the SIPOC diagram shown in Figure 4.6, the project team should do an IFMEA for each input and output. As part of the IFMEA, they can define nominal, alternate nominal, and off-nominal inputs and outputs, each of which can be assessed for both external and internal interactions as shown in Figure 4.10 by simulating each within the model. A simulation capability improves the effectiveness of the FMEA from a table-based to a simulation-based analysis.

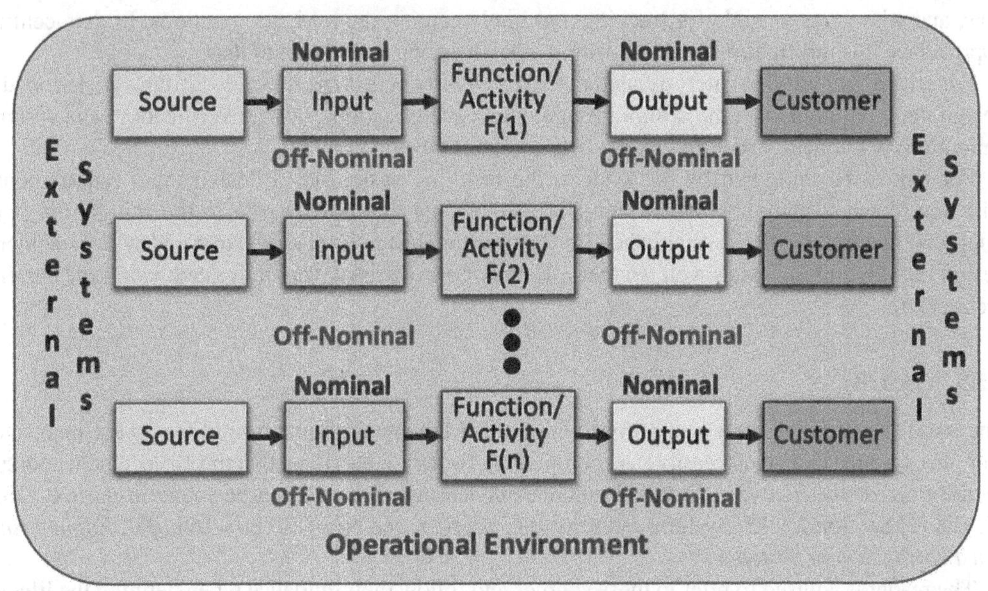

FIGURE 4.10 Input/output Risk Assessment.

For each off-nominal input or output, the possible causes can be assessed. In addition, the project can assess possible misuse cases. The FMEA assessments will identify each hazard and associated threat, the likelihood, the consequence, and urgency. The urgency takes into consideration priorities, criticalities, security, and safety of key activities and functions, as well as losses that may occur. Based on this information, the project team can determine how they will handle the risk.

For the risks to be mitigated, the mitigation action could involve people (training/skills), processes, procedures, IFU, and/or the system or product itself.

- The focus of mitigation actions for people or processes is on preventing or minimizing errors and hazards associated with user/operator interactions with the system that may result in harm to the user/operator or other stakeholders during operations.
- If the risk will be mitigated by the system or product, part of the lifecycle concepts analysis and maturation will be concepts for how best to mitigate the risk via design. The focus is on faults, failures, or misuse that may prevent the system from fulfilling its critical functions and performance. The project team must decide on the form of mitigation: will the approach be to completely remove the hazard and associated threats, reduce the likelihood, and/or reduce the consequences – or a combination?

The concept of mitigation and resulting needed capabilities will be communicated as part of the integrated set of needs, which will be transformed into design input requirements that will result in the system having these capabilities. To help manage the needs and requirements associated with risk mitigation, a risk mitigation attribute should be defined for each need and resulting design input requirement that is involved in the mitigation of the risk (as discussed further in Chapter 15).

Some projects communicate the results of their FMEA assessments in reports, often in the form of a spreadsheet that includes a list of hazards, their risk, how the risk will be managed, if mitigated, how, and who is responsible for tracking and implementing the mitigation actions. In data-centric approaches, this information is stored within a database via the project toolset.

For critical, security, or safety-related risks, traceability across the lifecycle artifacts (risks, needs, design input requirements, the design, design output specifications, system verification, and system validation) is frequently required by the *Approval Authority*.

For risk to be mitigated by the system, the resulting system-level design input requirements will flow down to the subsystems and system elements that have a role in the mitigation. It is important that traceability is established for the resulting lower-level subsystem and system element design input requirements as well, so that it is clear how the risks were mitigated within the system architecture.

4.5.9 Security

The word "security" is used many times within this Manual. With increasingly software-intensive, complex systems that are accessible via computer networks, it is critical that the project team address security across the lifecycle. Detailed guidance concerning security can be found in the INCOSE SE HB [1] Section 3.1.12, Systems Security Engineering, and NIST SP 800-160 [46], *Engineering Trustworthy Secure Systems*.

The reader is advised to refer to these sources and follow their guidance when defining the lifecycle concepts, identifying capabilities needed, and reflecting these within the integrated set of needs and resulting design input requirements.

4.5.10 Human Systems Integration and Human Systems Engineering

HSI and HSE are important considerations when defining the lifecycle concepts, an integrated set of needs, and a resulting set of design input requirements for the SOI. In addition to defining detailed scenarios for the user/operator interaction with the system, there are human factors that must be taken into consideration, including vision, hearing, anthropometrics, education, training, skills, and certifications. These should have been addressed when defining the personas/user classes discussed previously.

System validation determines whether the realized SOI can be used as intended within the operational environment by the intended users. That is one reason for defining personas/user classes so that the intended users are understood and defined, and the system is designed to be used/operated by these the people represented by these personas/user classes. "Used as intended" implies the use or operation is defined by operating procedures, or IFU. Thus, the conditions for use must be defined to include not only the realized system but also the personas/user classes, operating procedures, IFUs, and the HMI. This is one of the reasons why the lifecycle concepts must be developed from different perspectives, which include the user/operator interactions with the system as well as the failure analysis performed to identify potential hazards, threats, and risks associated with the use and operations.

For example, labels need to be able to be read by the user/operator, the equipment may need to be operable for a person wearing gloves and other personal protection equipment (PPE), or a biological sample prepared in a specific manner before being inserted into a diagnostic instrument.

Security and safety-associated risks are heavily dependent on the people interacting with the SOI, and the defined processes (procedures and IFUs) developed to address those risks. The people must be trustworthy, have the knowledge and skills to properly interact with the system, and be trained in the processes to do so. Effective, secure, and safe operation of the system may depend on the user/operator following a precise sequence of events and actions. Security and safety risks identified during the UFMEAs and IFMEAs may be mitigated through people (training) and/or processes (procedures or IFUs).

Refer to the INCOSE SE HB [1] *Section 3.1.4 and the INCOSE HSI Primer Volume 1* [47] *for additional information concerning HSI.*

4.5.11 Systems of Systems (SoS)

A SoS is formed when several systems are combined to perform some capability none of the systems that are part of the SoS can do by themselves. If the SOI is being developed such that it may be part of a SoS, then key considerations to be addressed when defining lifecycle concepts, an integrated set of needs, and resulting design input requirements for your SOI include:

- SoS requirements allocated to the SOI.
- Interactions between the SOI and other systems within the SoS and associated risks.
- Interactions between the users and operators of the SOI with users and operators of the other systems within the SoS.
- Security implications associated with those interactions (misuse and loss).
- Alternate and off-nominal interactions.
- Sustainability and resilience considerations.

Refer to the INCOSE SE HB [1] *Section* 4.3.6 *for additional information concerning SoS.*

4.5.12 Standards and Regulations Compliance

For all relevant standards and regulations identified as drivers and constraints, the project team must define feasible lifecycle concepts for how they will comply with the applicable requirements within the relevant standards and regulations, as well as how they will provide objective evidence of compliance.

To do this, the project team must do an analysis of each relevant standard and regulation and assess which apply:

- to the project's processes and methods,
- to the SOI, either as design inputs or design outputs.
- to system verification and validation that will result in objective evidence of compliance, and
- to qualification, certification, and acceptance.

A key consideration when assessing standards and regulations is the feasibility of compliance within project constraints as well as identifying the resources, cost, and time required to provide objective evidence of compliance.

Refer to Sections 4.3.6, 4.3.7, 4.6.2 and Chapter 6 for additional information concerning standards and regulations compliance.

4.5.13 Iteration to Mature Lifecycle Concepts

SE is a knowledge-based practice that is iterative and recursive by nature. With each iteration, valuable knowledge is gained. Based on this knowledge, the next iteration can be performed, increasing the resolution and maturity of the models, lifecycle concepts, physical architecture, and associated work products and artifacts. These iterations can be thought of as a series of cycles, zeroing in on lifecycle concepts, needs, and an initial physical architecture as shown in Figure 4.11.

Lifecycle concepts analysis and maturation starts with the preliminary set of lifecycle concepts and inputs to those concepts, performing functional/performance analyses, developing analytical/behavior models, and defining alternate physical architectural concepts resulting in trade space. The project team can then assess the candidate physical architectures in terms of cost, schedule, technology, feasibility, risk, value, and ROI. Trade studies are conducted to zero in on at least one

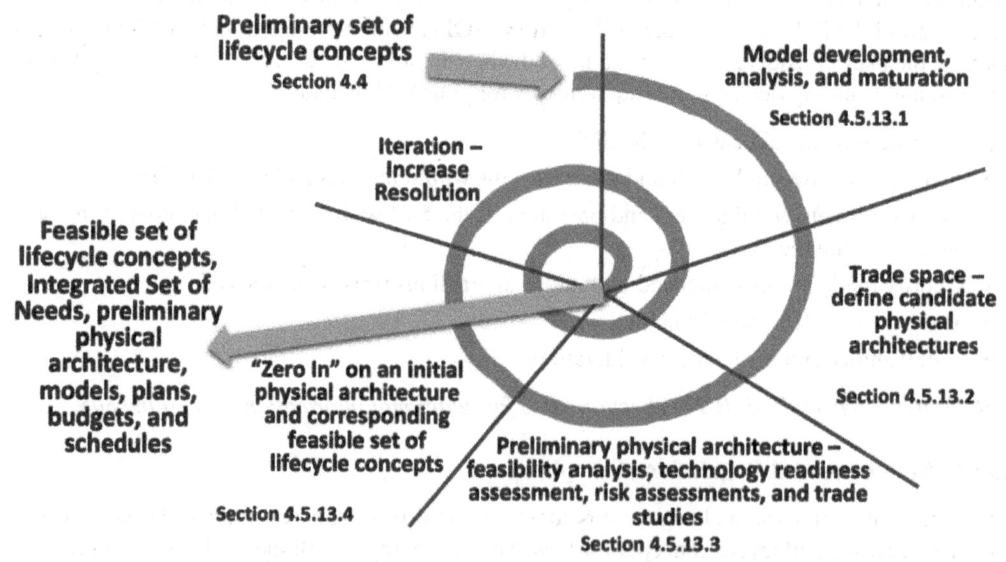

FIGURE 4.11 Zeroing in on a Set of Feasible Lifecycle Concepts, Needs, and Architecture.

feasible physical architecture concept. These activities result in a smaller set of candidate physical architectural concepts to consider, reducing the trade space.

Concurrently, project and SE management team members are developing a WBS, PBS, preliminary budgets and schedules, draft management plans, acquisition plans, and maturing the lifecycle concepts for development, procurement, design, system integration, system verification, system validation, transportation, installation, maintenance, and retirement. With the knowledge gained during each iteration, these concepts will be fine-tuned.

The resolution is increased for the next iteration. For each iteration, the definition of the functional architectural model, analytical/behavioral models, preliminary physical architecture, and design concept are better understood and refined. These refinements may require updates to the MGOs, measures, operational-level and system-level needs and requirements, and preliminary OpsCon/ConOps/SysOps based on the feasibility assessments completed during each spiral. These updates will be coordinated with the stakeholders and approved.

Additional iterations are repeated until the project team has zeroed-in on at least one set of feasible lifecycle concepts and a preliminary physical architecture they feel is mature enough, with acceptable risk for this lifecycle stage to proceed further in the lifecycle.

Concurrently, project team members are maturing plans, WBS, PBS, budgets, schedules, and related work products. With this maturity, the project team can define an integrated set of needs and finalize the PM and SE work products and artifacts as appropriate to this stage of the lifecycle. Together, this information represents the scope of the project.

As discussed in Chapter 8, for design verification and validation a similar cyclic process is used for maturing and finalizing the physical architecture, design concept, and design output specifications.

The following sections provide a more detailed description of the lifecycle concepts analysis and maturation activities that occur during each iteration shown in Figure 4.11.

4.5.13.1 Model Development, Analysis, and Maturation Based on the preliminary set of system lifecycle concepts and the inputs shown in Figure 4.4, functional analysis is continued as needed, and the project team defines or refines the functional architectural and analytical/behavioral models for the SOI.

During the first spiral, the models will be developed – if not already started as part of defining the preliminary set of lifecycle concepts. During subsequent iterations, the models will be updated and refined based on the knowledge gained during the previous iteration. Initially, the level of detail should be limited to a stakeholder view of the integrated system – what is observable externally (functionality, performance, quality, physical attributes, and interfaces). It is common to start with a focus on capabilities, features, functionality, performance, and interactions with external systems. Then a functional architecture can be defined to aid in the transformation into a preliminary physical architecture.

It is important that the models address the external interactions (connectivity, interaction, and flow of information, energy, and matter) between the SOI and the macro system of which it is a part. The models need to reflect the interactions and interfaces across all lifecycle stages, not just during operations. There may be interfaces needed during testing, system verification, system validation, production, transportation, installation, maintenance, or disposal that are not used during operations. Omitting consideration of a lifecycle stage will most likely result in missing needs and requirements.

It is seldom the case that a single analytical model can be constructed, though this is desirable. Rather, the project team will need to establish a set of integrated models to demonstrate *why* and to *what degree* certain characteristics and capabilities are necessary for satisfying the operational-level and system-level needs and requirements within the defined drivers, constraints, and risks. Such considerations as optimization, availability, reliability, resilience, sustainability, and effectiveness (achieving the desired outcomes) may need an integrated set of models that relate the desired outcomes to characteristics of the SOI and the project (for example, budget, schedule, and risk) [48].

As part of developing and maturing the models, the project team will identify and resolve consistency, correctness, and completeness issues that may exist (a major benefit of using diagrams and models for analysis). Part of this analysis will involve identifying the source of each input and the "destination/customer" for each output.

A key consideration at this stage is understanding the "partials," that is, how ROI or value changes as a function of some KPP (for example, mass, data volume, performance, quality attribute, communication bandwidth, computing power, operational environment, or technologies used). How will the ability of the SOI to meet the needs be affected as a function of changes in cost, schedule, technology, or another key parameter? For example, how would the cost, schedule, and proposed technologies be affected by a change in any of the needs?

For complex systems where parameters across the system architecture have a dependency (directly proportional or inversely proportional), it is difficult to track and understand the partials "manually." With the analytical/behavioral models, the project team can evaluate changes to assess the impact of a change on meeting the needs and resulting design input requirements. This knowledge will be of great benefit in choosing or optimizing the physical architecture within the defined drivers and constraints.

As a result of this analysis, schedule, resource, and cost estimates are developed or revised. Inputs shown in Figure 4.5 are addressed and revised based on the knowledge gained during these activities. The lifecycle concepts are revised as needed based on the information gained during this analysis. Plans are developed or revised. These revisions may require updates to the operational-level and system-level needs and requirements. These changes must be coordinated with the stakeholders and approved.

Concurrently with the model development, analysis, and maturation, the project team should be identifying a preliminary integrated set of needs that will result in meeting the higher-level needs contained within the MGOs within the identified drivers and constraints with an acceptable level of risk for this lifecycle stage.

The goal is a demonstrably complete, correct, consistent, and feasible integrated set of needs that include functions and observable outputs from the functional analysis with needed performance within the conditions of operation (states, modes, and environments) and triggering events. The use of models and diagrams for analysis helps establish that individual needs and requirements have the GtWR characteristics *C1—Necessary* and *C8—Correct*; and the sets of needs and requirements have the GtWR characteristics *C10—Complete* and *C11—Consistent*.

4.5.13.2 Trade Space – Define Candidate Physical Architectures Based on the knowledge gained from the functional architecture and analytical/behavioral model development, analysis, and maturation activities, the project team will define one or more candidate physical architectures and design concepts for implementing those architectures. Part of defining the physical architectures is to map the functions and interfaces contained in the functional architecture model to the candidate physical architectures.

In a greenfield development environment, when a new SOI is being developed, there are often multiple candidate architectures and corresponding design concepts to be considered. At this stage of development, each of these needs to be defined at a level that will allow feasibility to be accessed.

In a brownfield development environment, when a legacy or heritage system exists, the physical architecture and design concept will already be defined. The key issue is the impact of changes to the inputs shown in Figure 4.4 and their impact(s) on the system elements that are contained within the physical architecture. At a minimum, any retained (or legacy) system elements must have their interfaces satisfied. A key consideration for the project team is whether they want to continue to evolve the existing system or start over with a "blank piece of paper" greenfield development. This approach allows for more innovation, especially given new technologies, the move to software-intensive systems, new or changed stakeholder needs, and changes to the operational environment.

In either case, the project team will identify which system elements may already exist either within the enterprise or from external suppliers as OTS, which may exist but would need to be modified (MOTS), and which they think may need to be developed. From a cost and schedule perspective, reuse of existing OTS parts within the enterprise normally takes priority over MOTS or OTS from suppliers. Developing new parts from scratch is a higher risk, especially if a technology is needed that is beyond current state-of-the art. *Refer to Chapter 12 for a more detailed discussion on using OTS and MOTS system elements.*

Based on the choices, one or more candidate physical architectures and design concepts are defined, and alternate system elements are identified. These candidates will be assessed based on their feasibility to meet the MGOs, measures, and other key system parameters within the drivers and constraints with acceptable risk.

4.5.13.2.1 Considerations When Defining the Trade Space There are several important considerations when identifying and evaluating candidate physical architectures, design concepts, and alternate system elements:

- *Supply chain.* In the modern world, the supply chain is a key issue. Supply chain considerations must include resilience, sustainability, availability, cost, and risk. Factors that can influence these considerations include whether the system element is produced domestically or non-domestically and whether there is a single or multiple suppliers, and their trustworthiness, availability, and sustainability. If non-domestic, politics, tariffs, and trade restrictions must be considered as part of the project's risk assessment. Social considerations must also be considered in context of human rights and the environment. For system elements dependent on rare-earth materials, the availability of those materials must be considered, whether used directly by the organization or their suppliers.
- *Use of existing/heritage system elements.* A common mistake is assuming the TRL of an existing or heritage system element is higher than it really is. TRLs are a function of a specific use in a specific operational environment. The environment includes both the operational and induced environments within the system of which the system element is a part. If an existing/heritage system element was at TRL 9 (common use) for a specific use and operational environment in another system, it is a best practice for the project team to assume an initial TRL 5 or 6, until proven otherwise, for their specific use and operational environment. This can be an issue when evaluating OTS system elements. Was the OTS element designed for the exact same use in the exact same operational environment? If not, the TRL for the OTS system element will have to be assessed for specific use and specific operational environment for the system under development. *Refer to Chapter 12 for a more detailed discussion on the use of OTS system elements.*
- *Production capability.* Another key issue is the production capability of the organization, or applicable external organizations if the production is being outsourced. Being able to use an existing production capability is much more cost effective than choosing a design that requires the use of new technologies for production. This approach will result in the need for new or modified facilities, equipment, skills, processes, and procedures, which will need to be treated

as separate projects and go through their own system development. This additional development effort will have a significant impact on cost and schedule. Closely related to TRLs for critical technologies used in the SOI, there are Manufacturing Readiness Levels (MRLs) used to manage the risk associated with production of a new system or product. An example is pharmaceuticals or vaccines that were of high quality when developed in low quantities in a laboratory setting, but scaling factors resulted in quality issues when large quantities were being produced in a factory setting.

- *Availability and cost.* For those system elements whose development, manufacturing, or coding will be outsourced to an external organization, the project team needs to research what is available and from which suppliers. Government projects will often issue a Request for Information (RFI) to find out this information. The project team members responsible for procurement and contracting will issue the RFI and the resulting RFP. Feasibility considerations include the cost and schedule of the suppliers to deliver the system element consistent with the project's budget and schedule.

4.5.13.3 Physical Architecture Selection and Analysis It is the subsystems and system elements within the system physical architecture that will be built, integrated, verified, validated, and delivered to a customer or provided to consumers.

The project team will perform trade studies between the candidate physical architectures, design concepts, system elements, support systems, and enabling systems with elaboration to better understand the relationship between the parts, cost, schedule, risk, and value.

Value improvement is achieved by assessing tradeoffs of the candidate architectures. Analytical models should be developed to address an optimized solution that adds the most value. Additionally, these assessments may be used to hybridize innovative solutions, resulting in value creation. Having a single candidate means there is no decision necessary, and the solution is only guaranteed to be satisfactory but not necessarily optimal.

For each candidate, a TRA is performed to determine the TRL of the candidate subsystems and system elements that are proposed to be part of the physical architecture, a common approach is the process outlined in the US GAO TRA Guide [35]. A TRA report is generated for each candidate physical architecture, subsystems, and system elements that are part of it. In addition to the TRLs, the MRL of the production facility should also be assessed as it relates to the candidate physical architectures and design concepts.

The most promising, candidate physical architectures, design concepts, and system elements are identified that will best implement the SOI lifecycle concepts selected to achieve the integrated set of needs and desired ROI with risk appropriate for this lifecycle stage. These physical architectures and design concepts are defined to the level of detail appropriate for this lifecycle stage. Lower levels of detail are performed in later lifecycle stages via iteration and recursion. More in-depth analysis and trade studies are performed for each candidate physical architecture. De-scoping and contingency options are defined in case critical technologies cannot be matured as needed based on the established priorities, criticality assessments, cost, and schedule.

The interface analysis and risk assessment activities discussed earlier should be completed or updated for each iteration, candidate architecture, and associated design concept being considered.

An analysis of enabling systems required during the different lifecycle stages, including design, production, system verification, system validation, and operations, is performed. Of concern is the availability of the enabling systems at the times needed to support project schedule and whether any of the enabling systems will need to be upgraded or modified to meet the needs of the project. This analysis is important for the project to develop their budget and schedule, as well as define and evaluate the interfaces and mature their acquisition plans.

Interactions between the system and stakeholders (operators, maintenance, and update personnel) across the system lifecycles need to be assessed, and failure analysis, such as a UFMEA, needs to be completed for each candidate design concept. From a stakeholder perspective, the focus is on roles, functions, performance, and quality – how they will interact with the system to accomplish the mission. These considerations are part of the HSI and HSE activities.

As an aid to the analysis of the candidate physical architecture and design concepts, the project team may also use rapid prototyping. With advances in 3D printing and additive manufacturing, study/design teams can develop prototypes quickly to be used as part of the architectural and design concept trade activities. Similarly, for software, prototypes of user interfaces (displays or screens) can be developed.

When prototypes are developed, various configurations can be made available to the actual users. These prototypes can be functional, non-functional, or form, fit, and function. A prototype is useful to help the project team better understand stakeholder needs, which will drive the selection of the lifecycle and design concepts. Prototypes are developed based on the currently known needs and requirements.

By using prototypes, the users can get an "actual feel" of the SOI in the representative operational environments. Also referred to as "discovery learning," user interactions with the models and prototypes can enable both the user and study/design team to better understand the needs and requirements for the desired SOI. Models and rapid prototyping allow for user feedback much earlier in the development lifecycle concerning issues, such as missing functions, substandard performance, safety, security, quality, and non-intuitive user interfaces. This is an interactive process that allows stakeholders to understand, modify, and eventually approve a model of the system that meets their needs.

The project team will use all this data in a trade study to down select to the most promising physical architectures and design concepts based on criteria such as feasibility, value, and ROI. Schedule, resource, and cost estimates are assessed and refined during each iteration.

4.5.13.4 Zeroing in on a Feasible Architecture and Design The outcomes of the system lifecycle concepts analysis and maturation activities include functional architecture definition and analytical/behavioral models of the SOI, as well as at least one preliminary feasible physical architecture and design concept. The information from these activities will be fed back to the inputs shown in Figure 4.4 which may need to be adjusted based on the knowledge gained during these activities, requiring coordination with and approval by the stakeholders.

From the use cases, operational scenarios, and user interactions with SOI models or prototypes, a preliminary integrated set of needs can be developed addressing core functions and associated performance that will result in the stakeholder real-world expectations being met as communicated within the MGOs and measures. Because feasibility has been addressed, the integrated set of needs and resulting set of design input requirements should also be feasible with acceptable risk for this lifecycle stage.

Based on the results of the TRA and AD^2 assessments for the selected physical architecture and design concept, a TMP will be developed to mature the critical technologies needed in accordance with the processes outlined in the GAO TRA Guide. For US Government projects, the GAO [35] recommends that the critical technologies should be at least TRL 3 to be considered during the concept stage and be capable of being matured to at least TRL 6 by the time of the PDR, and TRL 7 by the time of the CDR. As stated earlier, technology maturity relates to development and operational risk. Failing to follow these guidelines puts the project at risk of failure.

Based on the MRL assessment of the selected physical architecture and design concept, a Production Maturation Plan (PMP) will be developed for the production organization to develop the capability needed to produce the SOI. The implementation of this plan will be a separate project, with its own PM and SE activities that will result in a production capability that has successfully passed its own system integration, system verification, and system validation program in time to support the production of the SOI.

In addition to TRLs and MRLs, it is best practice to include development costs, schedule reserves, and operational margins to help mitigate risk from unknown unknowns.

With the knowledge gained to this point, activities completed, and the maturity of the resulting artifacts, the project team needs to determine whether or not they are ready to proceed with the activities associated with completing the definition of the mature set of lifecycle concepts and integrated set of needs or go back and do another iteration, repeating the activities to further refine and mature the lifecycle concepts and the preliminary physical architecture as shown in Figure 4.11. This activity helps ensure individual needs and requirements have the GtWR characteristic *C6—Feasible* and sets of needs and requirements have the GtWR characteristic *C12—Feasible*.

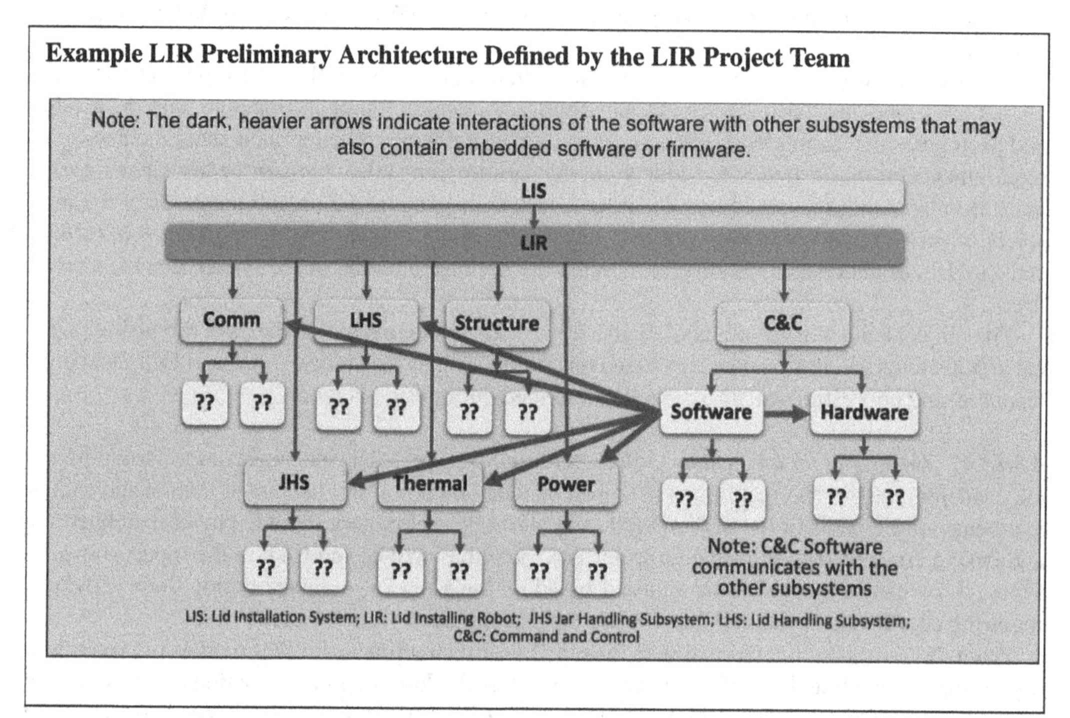
Example LIR Preliminary Architecture Defined by the LIR Project Team
Note: The dark, heavier arrows indicate interactions of the software with other subsystems that may also contain embedded software or firmware.
LIS
LIR
Comm
LHS
Structure
C&C
??
Software
Hardware
JHS
Thermal
Power
Note: C&C Software communicates with the other subsystems
LIS: Lid Installation System; LIR: Lid Installing Robot; JHS Jar Handling Subsystem; LHS: Lid Handling Subsystem; C&C: Command and Control

4.5.14 Activities Concurrent with Lifecycle Concepts Analysis and Maturation

Concurrently with the lifecycle concepts analysis and maturation activities, project and SE management must define their implementation approaches and concepts, including PM activities and plans, SE activities and plans, contracting, procurement, production, integration, system verification and validation strategies, transition to operations, budget, and schedule.

The project team must assess and define strategies and plans concerning relationships, dependencies, and partnering with other organizations. This is extremely important if the SOI is part of a larger SoS. The project team must also document key risks and risk mitigation plans. From an acquisition perspective, strategies must be defined for when decisions are to be made to: buy (from an external source), build/code (make internally), reuse/modify (existing systems), or buy/try/decide.

PM and SE work products and artifacts that should be in various stages of development during these lifecycle concepts analysis and maturation activities include the following list of examples (note that many of these artifacts will be required in some form of maturity at a gate review such as MCR or SR where the lifecycle concepts and resulting integrated set of needs are baselined).

Examples of completed activities and artifacts include:

- Stakeholders were identified and included in the Stakeholder Register.
- A clear and concise problem statement.
- MGOs and measures were defined and recorded.
- Stakeholder needs and requirements elicitation is complete and recorded.
- Use cases, user stories, and operational scenarios were captured.
- Drivers and constraints were identified and recorded.
- Risks were identified and a Risk Management Plan (RMP) was developed.
- External interface and context diagrams were created.
- Functional architecture and analytical/behavioral models.
- Proof-of-concept prototypes.
- Trade studies were completed.
- At least one preliminary physical architecture was identified.
- A Preliminary PBS.
- A Preliminary PMP.
- A Preliminary budget and schedule.
- A Preliminary WBS.
- A Preliminary SEMP.
- A Preliminary Document Tree.
- Technology assessment reports.
- Technology maturation plans are in work.
- A Production maturation plan is inwork.
- An Information management plan in work.
- Acquisition, procurement, and contracting plans are in work.
- Master Integration, Verification, and Validation (MIVV) and System Integration. Verification and Validation (SIVV) Plans are in work.
- Project ontology and database schema definition are in work.
- Draft integrated set of needs for the SOI.
- System validation planning attributes (*Method*, *Strategy*, *Success Criteria*, and *Responsible Organization*) have been defined for each need statement.
- Draft set of design input requirements is in work.
- Draft system verification planning attributes (*Method*, *Strategy*, *Success Criteria*, and *Responsible Organization*) are in work for each requirement statement.
- Trace records for needs and design input requirements and their sources are in progress.

Practicing PM and SE from a data-centric approach, the data and information contained in these work products and artifacts will be represented within the SOI's integrated dataset, as discussed previously. *Refer to Chapter 14 concerning actions and products concerning needs, requirements, verification, and validation management that must also be addressed.*

4.6 DEFINE AND RECORD THE INTEGRATED SET OF NEEDS

4.6.1 People and Process Needs

The real system is more than just the product or SOI being developed. The ultimate purpose of SE is to provide a *capability* in response to a problem, threat, or opportunity. This capability is fulfilled using a triad of people, processes, and the product to be developed, as shown in Figure 4.12. The "real system" is a combination of the three interacting to provide the needed capability.

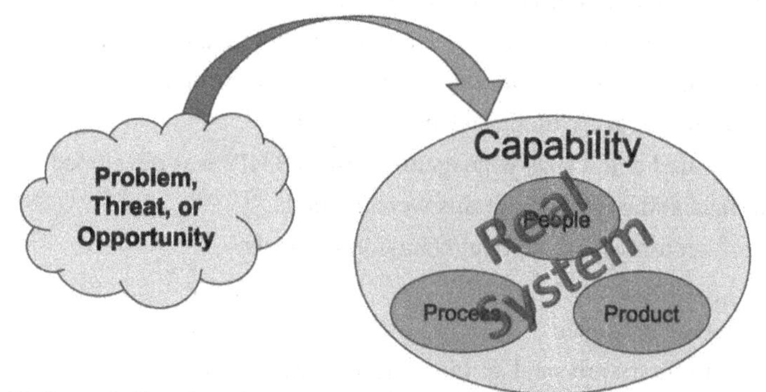

FIGURE 4.12 Provided Capability Includes People, Process, and Products.

Outputs of the lifecycle concepts analysis and maturation activities include concepts for PM, SE (developing the SOI), as well as concepts for development, procurement, production, transportation, storage, operations, sustaining engineering, and retirement. From these concepts, the organization can identify people (organizational) needs and process/procedure needs. In this context the "people needs" represent what stakeholders at the organization's strategic and operational levels need from the people and organizations responsible for performing the activities associated with each of the above concepts. People needs will be the basis for the PMP, SEMP, and other plans that contain requirements for the organizations (people) and processes to which the needs apply.

For system development and production that will be contracted out to suppliers, specific concepts, activities, processes, and deliverables should have been identified as part of the lifecycle concepts analysis and maturation activities discussed previously. The activity SIPOC analysis and AFDs help identify what activities and deliverables the project needs from the suppliers. These needs can then be transformed into specific requirements for the suppliers and communicated via a SOW, SA, or PO.

A common issue and risk for projects is developing a SOW, SA, or PO without doing the analysis needed to clearly define the acquisition process, identify what is expected from the suppliers, and define the customer/supplier roles, responsibilities, and interactions. This analysis helps identify and define the specific PM and SE processes expected to be followed by the suppliers, activities performed, required deliverables, expected performance, and what is necessary for acceptance. *See also, Chapter 13, Supplier Developed SOIs. Refer also to the INCOSE SE HB Section 2.3.2, Agreement Processes, Section 2.3.2.1, Acquisition Process, and Section 2.3.2.2, Supply Process.*

From a holistic practice of SE, the lifecycle concepts and need definition activities discussed in this section result in needs associated with people, processes, and the product to be developed. As discussed in Chapter 2, it is important to document and manage these sets of needs and requirements separately. While important, further discussions concerning the people and process needs are beyond the scope of this Manual. In the subsequent sections, the focus is on the product (SOI) needs and requirements.

4.6.2 Needs and Requirements from Standards and Regulations

Many standards and regulations contain at least three types of requirements: Process, System Verification/System Validation, and Product. The project team will need to determine where they will document needs associated with invoking standards and regulations based on type and applicability. It is important that the three types of requirements are documented separately according to the entity they apply:

1. *Process.* Requirements dealing with activities, processes, deliverables, workmanship. These types of requirements should be included within the project's PMP, SEMP, design controls, MIVV Plan, SIVV Plan, system certification plan or system qualification plan, quality management plan, RMP, and SOWs, POs, or SAs.
2. *System verification and system validation.* Requirements dealing with obtaining objective evidence that the needs or requirements have been met, the form of documentation, and the form and content of the *Approval Package* to be provided to the *Approval Authority.* These include requirements concerning qualification, certification, and acceptance of the product. These types of requirements should be included within verification or validation procedure requirements. *See Chapter 10 for a more detailed discussion on these topics.*
3. *Product.* Requirements dealing with the characteristics, capabilities, and performance of the product. Needs that invoke these types of requirements should be invoked within the integrated set of needs and transformed into well-formed design input requirements.

Based on the lifecycle concepts associated with the product requirements within the standards and regulations that will be addressed by the SOI, the applicable standards and regulations will be invoked within the integrated set of needs and then transformed into well-formed design input requirements that meet the intent of the requirements within the standards or regulations.

This approach of invoking standards and regulations within the integrated set of needs is a departure from how this has been done in the past. In the past, it has been a customary practice to call out entire standards or regulations or sections of a standard or regulation within a requirement statement:

"The <SOI> shall comply with all applicable ISO standards."

"The <SOI> shall comply with all requirements within FDA regulation xyz."

"The <SOI> shall meet all requirements within OSHA regulation xyz, Section a.b.c.d."

However, there are major issues with this approach. In the first example, who determines which requirements are applicable? Because the first two requirements are written in passive voice, to which entity do they really apply (SOI or project?) In the second and third examples, how certain is it that all requirements in the standard or regulation are applicable, and if applicable, to which entity are they applicable?

While well-intended, each of these statements is too vague to verify or validate the system against because each is ambiguous as to which specific requirements within the standards and regulations are applicable. Using this approach is like kicking the can down the road and letting someone else determine which requirements are applicable or not.

From a cost and schedule perspective, it is dangerous to call out a complete or even a section standard or regulation when only a portion of the requirements apply. Many standards and regulations included hundreds of requirements. All requirements invoked within the set of design input requirements will have to be implemented in the design and the system verified to meet those requirements.

This is a real issue if contracting to a supplier who is directed to implement all requirements in the set. To address this issue, the project team must read each standard and regulation applicable to the compliance expectations for their type of SOI and identify specifically which requirements apply to their project, and which apply to the system elements that may be developed by other organizations/suppliers.

Using the approach of invoking the applicable standards and regulations within the integrated set of needs, the stakeholder's expectations are clearly communicated and baselined as part of the scope of the project. *Refer to Section 4.3.6.1 for additional information concerning Standards and Regulations.*

4.6.3 Define the Integrated Set of Needs

Defining and agreeing on a set of feasible lifecycle concepts enables the project team to define an integrated set of needs based on those concepts and the information used to define those concepts.

As shown in Figure 4.13, the project team derives an integrated set of needs that reflect the set of feasible system lifecycle concepts, MGOs, measures, operational-level and system-level needs and requirements, drivers and constraints, and risk mitigation.

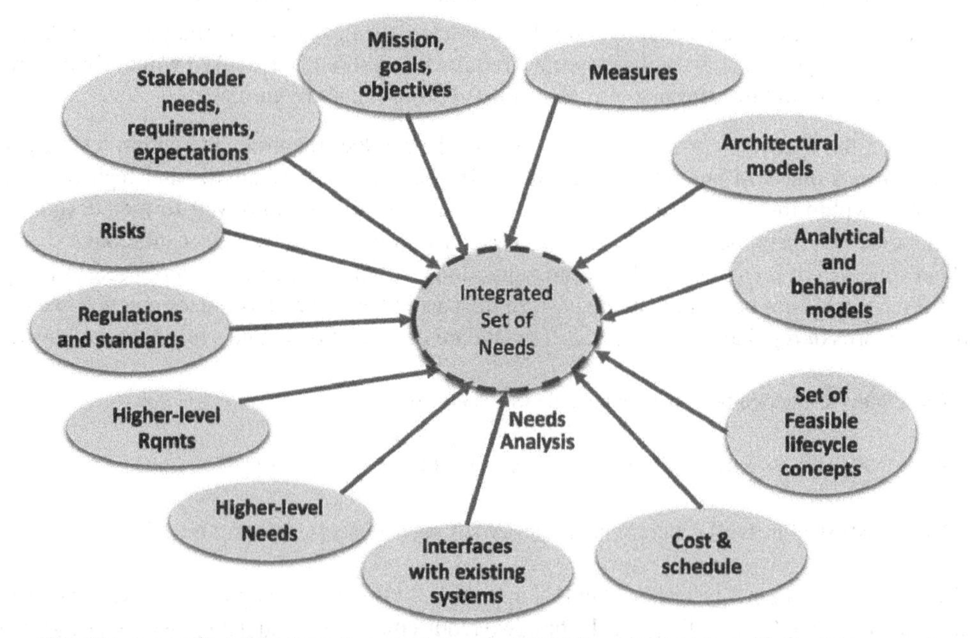

FIGURE 4.13 Sources of the Integrated Set of Needs.

This integrated set of needs represents the scope of the project and is what will be transformed into the set of design input requirements. In addition, it is this integrated set of needs against which the design input requirements, the design, the design output specifications, and the realized SOI will be validated.

4.6.3.1 The Form of Need Statements *Needs are Not Requirements* As discussed in Chapter 2 and in the GtWR, the need statements are written in a structured, natural language from the perspective of the what the stakeholders need the SOI to do, while the design input requirements transformed from the needs are written from the perspective of what the SOI must do to meet the need(s) from which they were transformed.

To help distinguish needs from the requirements, the need statements do not include the word "shall." Using "shall" in need statements results in confusion as to whether the statement is a need, or a design input requirement transformed from a need. One approach to avoiding this confusion is to use the format: "The stakeholders need the system to … … … ." Or, for a goal, "The stakeholders would like the system to … … ".

These statements are in contrast to the form of the SOI's technical design input requirements, "The SOI "shall" … " or for a goal, "The SOI "should" … … "

Using these distinct formats helps make it clear what a need statement is and what a design input requirement is. The GtWR defines the structure and characteristics of well-formed need statements and sets of needs, as well as rules that help to write need statements that have those characteristics.

4.6.3.2 Including Goals Within the Integrated Set of Needs As discussed earlier, in the context of MGOs, the goals and objectives are higher levels of abstraction of needs. It is common in the English language, some words like "goals," are used with different meanings, adding ambiguity and confusion. Goals communicated as "would like" or "should" statements are not requirements, are not binding, and are not subject to design and system verification or validation, so why include them?

Historically, goals have been included in sets of requirements in various forms, communicating the customer or stakeholder desire for some feature or capability that, while not critical, adds some value, as long as it is low risk and does not impact critical functions or drive cost and impact schedule. For example, Marketing personnel may want a non-critical feature added to help separate the product from the competition.

By developing an integrated set of needs as advocated in this Manual, the communication of goals should be addressed and resolved when defining the needs and baselining the needs as part of scope definition. With this approach, as part of the transformation of the needs into requirements, for each of the "The stakeholders would like the system to … … " goals, those doing the transformation and resulting engineering analysis would decide which goals would be addressed in the set of design input requirements and expressed as requirement "shall" statements. These requirements would be low priority, not critical, low risk, and trace back to the goal within the integrated set of needs.

4.6.3.3 Use of Attributes to Define and Manage Needs As defined in Chapter 2, a need expression includes a need statement plus a set of attributes. Attributes that can aid in definition and management of needs include those shown below. *Refer to Chapter 15, for a more detailed discussion on attributes, definitions of each, and guidance concerning the use of attributes for need statements.*

A1: *Rationale*. Reason for the need's existence, intent of the need.

A3: *Trace to Source*. Where the need originated: stakeholder, user story, scenario, use case, constraint, risk, lifecycle concept, analysis, or model as examples.

A6: *System Validation Success Criteria*.

A7: *System Validation Strategy*.

A8: *System Validation Method*.

A9: *System Validation Responsible Organization*.

A12: *Condition of Use*: Operational conditions of use expected in which the need applies and will be validated.

A26: *Stability*: stable, likely to change, and incomplete.

A28: *Need Verification Status*: true/false, yes/no, or not started, in work, complete, and approved.

A29: *Need Validation Status*: true/false, yes/no, not started, in work, complete, and approved.

A30: *Status of the Need* (in terms of maturity): draft, in development, ready for review, in review, and approved. (A28 is a prerequisite for A29 and A30).

A31: *Status of implementation*: One or more design input requirements have been defined that will result in the intent of the need being met.

A34: *Priority*. Relative importance of the need.

A35: *Criticality*. Achievement of the need is critical to the SOI being able to meet its intended use in the operational environment.

A36: *Risk* (of implementation): One or more risk factors (cost, schedule, and technology) associated with being able to achieve the need.

A37: *Risk* (mitigation). The need is linked to a risk the project has decided to mitigate. Often related to safety, security, and quality.

A38: *Key Driving Need* (KDN). Implementing the need could have a significant impact on cost and/or schedule.

It is a best practice to define the attributes for need statements when the need statements are formulated. For example:

A1: *Rationale*. If rationale cannot be defined when the need statement is being formulated, the existence of the need statement is questionable.

A3: *Trace to source*. If there is no source to trace the need statement to, the existence of the need is questionable.

A6: *System validation success criteria*. If the Success Criteria for successful validation that the verified and built system has met the need cannot be defined, the need statement wording will have to be changed.

4.6.3.4 Organizing the Integrated Set of Needs The integrated set of needs represents the agreed-to outcomes of the needs and requirements elicitation activities, definition of drivers and constraints, and lifecycle concepts analysis and maturation activities. These outcomes include functionality (what the stakeholders need the system to do), expected performance and quality ("how well" characteristics), the conditions of action, including triggering events, system states, and operational environment ("under what operating conditions"), as well as compliance (with standards and regulations).

To help with the development and organization of the design input requirements that will be transformed from the integrated set of needs, it is useful to organize the integrated set of needs into the following categories: function/performance, fit, form, quality, and compliance.

- *Function/Performance*. The primary functions and associated performance that the SOI needs to perform in terms of its intended use. The functions address the capabilities and features the stakeholders expect the SOI to have; performance addresses how well, how many, and how fast attributes of the function. For example, for a medical diagnostic device, needs concerning what is to be measured, accuracy, and precision would be included in this category. Many of the primary functions involve interactions (interfaces) between the SOI and systems external to the SOI. All critical and high priority needs would be included in this category.
- *Fit (operational)*. Needs dealing with functions that deal with a secondary or enabling capabilities, functions, and interactions between the SOI and external systems needed for the system to accomplish its primary functions. This includes functions concerning the ability of the system to interface with, interact with, connect to, operate within, and become an integral part

of the macro system it is a part. Fit includes human system interactions and interfaces as well as both the induced and natural environments (conditions of operations, transportation, storage, and maintenance). For example, needs associated with safety, security, power, cooling, transportation and handling, storage, maintenance, and disposal.

- *Form (physical characteristics).* The shape, size, dimensions, mass, weight, and other observable parameters and characteristics that uniquely distinguish a system. For software, form could address programming language, lines of code, and memory requirements.
- *Quality.* Fitness for use. For example, various "-ilities" such as reliability, testability, operability, availability, maintainability, operability, supportability, manufacturability, and interoperability.
- *Compliance.* Conformance with design and construction standards and regulations.

A key advantage of using these categories is to help ensure the integrated set of needs has the GtWR characteristics *C9 – Conforming*, and *C10 – Complete*. Each of the above categories represents a unique perspective and source of needs. Failing to consider each perspective will result in missing needs and corresponding design input requirements.

One method of recording which category a need is being grouped is the use of attribute A40-*Type/category*. Using this attribute will help the project team to export the needs into a needs document organized by the category.

- The function/performance and fit needs come from both stakeholder elicitation, risk assessment and mitigation, as well as the diagrams and models developed during lifecycle concepts analysis and maturation activities. While performance is included in the analytical/behavioral models, the technical ability to achieve a given performance is a function of the physical architecture (for example, physics, chemistry, biology, or thermodynamics).
- Form and quality needs are system attributes that are more associated with the physical system than a functional architecture or analytical/behavioral models of the physical system.
- Compliance needs are identified from both the stakeholder elicitation activities and their identification as part of defining the drivers and constraints.

Note: The above suggested approach to organizing needs and requirements using these categories, is just one of many approaches. No matter the approach used, the main consideration is to ensure completeness of the integrated set of needs and resulting set of design input requirements.

Example Integrated Set of Needs Defined by the LIR Project Team

The following represent an initial integrated set of needs for the LIR. Not shown are the attributes that must be defined and combined with the need statements to form need expressions. Each need statement has a unique identifier to enable traceability.

Acronyms used: LIS, Lid Installation System; LIR, Lid Installing Robot; FCR, Facility Control Room; LDS, Lid Distribution System; JPS, Jar Processing System. (Phrases and terms with the form Xx_Yy_Zz are defined in the project glossary or data dictionary.)

Functional/Performance

N001: The stakeholders need the LIR to obtain Lids from the LDS.

N002: The stakeholders need the LIR to install Lids of various sizes on Jars of various sizes and materials.

N003: The stakeholders need the LIR to install Lids on Jars with the proper torque depending on lid size and jar material.

N004: The stakeholders need the LIR to install Lids on Jars per the JPS timing.

N005: The stakeholders need the LIR to install a Lid on a Jar when notified a Jar is in place.

N006: The stakeholders need the LIR to maintain the Jar position on the conveyor belt.

N007: The stakeholders need the LIR to receive and execute commands received from the FCR.

N008: The stakeholders need the LIR to report the execution of all commands received from the FCR.

N009: The stakeholders need the LIR to perform a self-test during power up and report the results to the FCR.

N010: The stakeholders need the LIR to install at least 99% of the Lids on the Jars per Shift.

N011: The stakeholders need the LIR to monitor and Health_and_Status during operations.

N012: The stakeholders need the LIR to report Health_and_Status data to the FCR periodically.

N013: The stakeholders need the LIR to report any Critical_System_Failure that occurs during operations to the FCR.

N014: The stakeholders need the LIR to shutdown automatically whenever a Critical_System_Failure is detected.

N015: The stakeholders need the LIR to report any change in System_State to the FCR.

N016: The stakeholders need the LIR to shutdown automatically at the end-of-shift.

Fit (Operational)

N017: The stakeholders need the LIR to operate using Facility_Power currently available within the LISs.

N018: The stakeholders need the LIR to operate in the existing LIS environment.

N019: The stakeholders need the LIR to operate in the presence of electrical emissions existing in the existing LIS rooms.

N020: The stakeholders need the LIR to operate after being shipped in shipping crates in typical shipping environments.

N021: The stakeholders need the LIR to operate after being stored in shipping crates in an un-conditioned storage location environment.

N022: The stakeholders need the LIR to continue to operate after being exposed to water from the Fire Suppression System.

Form (Physical Characteristics)

N023: The stakeholders need the LIR to limit operations to within the LIS_Operations_Area.

N024: The stakeholders need the LIR to fit through LIS door without disassembly.

N025: The stakeholders need the LIR to allow a single maintenance technician to move an assembled LIR within the production facility.

N026: The stakeholders need the LIR to install Lids on the Jars located on the JPS conveyor belt through the opening in the wall between the LIS and the conveyor belt.

Quality (-ilities)

Security

N027: The stakeholders need the LIR to accept only Valid_Commands from the FCR.

Safety

N028: The stakeholders need the LIR to cease operations whenever commanded by the FCR C&M software.

N029: The stakeholders need the LIR to shutdown whenever commanded by the FCR C&M software.

N030: The stakeholders need the LIR Cease_Operations whenever motion is detected within the LIS.

N031: The stakeholders need the LIR to limit noise during operations such that it is not a danger to human hearing.

Maintainability

N032: The stakeholders need the LIR to have a MTTR not to exceed 15 minutes.

N033: The stakeholders need the LIR to identify and display failures to the LRU level.

N034: The stakeholders need the LIR to allow the use of common tools for repair that are currently in the maintenance area or can be procured from common commercial tool vendors.

N035: The stakeholders need the LIR to have parts, components, and assemblies that can be interchanged with the same parts, components, and assemblies in other LIRs.

Reliability

N036: The stakeholders need the LIR to have an operational lifetime of at least 5 years.

N037: The stakeholders need the LIR to have a MTTF of no more than 3 months, assuming 8 hours/day of powered operations.

Availability

N038: The stakeholders need the LIR to install Lids during the shifts of operation.

Transportability

N039: The stakeholders need the LIR to be able to be disassembled for shipment to accommodate crated the size and weight limits for shipping.

N040: The stakeholders need the LIR to survive after being dropped 4 feet onto a concrete surface when in shipping crates.

Compliance (Standards/Regulations)

N041: The stakeholders need the LIR to comply with [TBD] EPA requirements concerning the use and handling of hazardous materials.

N042: The stakeholders need the LIR to comply with [TBD] OSHA requirements concerning human/machine interfaces.

N043: The stakeholders need the LIR to comply with [TBD] OSHA requirements concerning the safety of humans working with robots.

N044: The stakeholders need the LIR to comply with [TBD] FDA requirements concerning the handling and prevention of contamination of food.

N045: The stakeholders need the LIR to meet the requirements of MIL-STD-461E for radiated emissions.

N046: The stakeholders need the LIR to use power connectors for obtaining facility power that meet UL standards.

N047: The stakeholders need the LIR to use lubricants, fluids, and fuels that meet UL 546xx non-flammability standard.

N048: The stakeholders need the LIR to be labeled in English.

N049: The stakeholders need the LIR to use only parts whose physical characteristics are measured using English Units of measure.

4.6.4 Record the Integrated Set of Needs

The integrated set of needs must be recorded in a form and media suitable for review and feedback from the stakeholders, as well as in a form that allows traceability across the lifecycle. Traceability is critical in support of need verification and validation as discussed in Chapter 5, as well as change assessment and management, as discussed in Chapter 14.

The project team must obtain confirmation from the stakeholders that the project team understands and is addressing their expectations, needs, requirements, MGOs, measures, drivers and constraints, and risk as communicated by the integrated set of needs. Confirmation with the stakeholders is critical because the integrated set of needs represents the agreed-to scope of the project and what is necessary for acceptance, against which the built or coded SOI will be validated.

Needs can be recorded and managed via either a document-centric or data-centric approach; however, based on the discussion in Chapter 3, it is highly recommended that the organization use a data-centric approach and document the integrated set of needs within the project's toolset.

The chosen method should allow:

- traceability to the sources of the needs;
- traceability between each need and the design input requirements transformed from the need;
- the inclusion of attributes for the need statements resulting in need expressions (Section 4.6.3.3); and
- traceability to system validation artifacts (Section 4.7).

4.6.4.1 Managing Unknowns During the definition of the integrated set of needs, there may be unknowns, resulting in the project team having to make assumptions regarding performance and functional criteria to allow the subsequent lifecycle activities to proceed. This often happens when the project team skips the lifecycle concepts analysis and maturation activities discussed earlier prior to defining the set of needs.

In other cases, further analysis or research is still required that would interrupt the overall workflow to develop and baseline a complete integrated set of needs, for example, the maturation of critical technologies. While this work will continue, it is critical to capture the unknowns and resulting ongoing work to ensure the associated activities to address the unknowns are funded, tracked, and managed. *Refer to Section 14.2.4 for a more detailed discussion on managing unknowns.*

This activity helps ensure individual needs have the GtWR characteristics *C3 – Unambiguous*, *C4 – Complete*; *C6 – Feasible*, *C7 – Verifiable*, and *C8 – Correct*.

4.6.4.2 Appropriate to Level One of the characteristics of well-formed needs and sets of needs stated in the GtWR is the characteristic *C2 – Appropriate.* The specific intent and amount of detail communicated within a need or requirement statement must be appropriate to the level (the level of abstraction and level of architecture) of the entity to which it refers. Section 4.5.5 discussed levels of detail, abstraction, and architecture, which will help determine the appropriate level to record need expressions.

Needs tend to be written at a higher level of abstraction than the requirements that are transformed from them.

For example, it is acceptable for a need statement to state,

"The stakeholders need the system to meet the requirements contained in government safety Regulation xyz, Section 1.2.3."

This clearly communicates the stakeholder expectations concerning this safety regulation. However, this level of abstraction is too high for what would be communicated within a well-formed design input requirement statement. The design input requirements would be more specific as to exactly which requirements in those standards or regulations are applicable, and then the project team would do the analysis to determine what the SOI must do to meet the intent of those requirements. The result of this analysis would be design input requirements that, when implemented by design and the realized system, would meet the intent of the applicable regulation requirements invoked by the need statement.

Another example of a medical diagnostic system:

- System need statement:
  "The stakeholders need the diagnostic system to [measure or detect] [something] with an accuracy as good as or better than similar devices in the market."
  This is an appropriate level of abstraction for a need statement, clearly stating the expectation the stakeholders have concerning accuracy, however, this would not be a good design input requirement.
- Design input requirement that was transformed from the need statement:
  "The diagnostic system shall [measure or detect] [something] with an accuracy of [xxxxx]."
  The developers have explored various concepts for meeting the need for accuracy, examined candidate technologies, assessed their TRL, and decided the value [xxxxx] is feasible with acceptable risk for this lifecycle stage. As stated, this is an acceptable design input requirement. The developers may also define other design input requirements dealing with precision, false positives, false negatives, time to determine a result, or allowable degradation over the life of the system. Each of these requirements will be traced to the need statement from which it was derived.

Example: Need Versus Requirement for the LIR System

LIR Need statement: N010: "The stakeholders need the LIR to send status messages to the FCR."

This is an appropriate level of abstraction for a need statement, clearly stating the expectation the stakeholders have concerning the sending of messages to the FCR; however, this would not be a good design input requirement in that this implies an interaction between the LIR and the FCR, and thus the resulting requirement would be written as an interface requirement referring to an ICD.

LIR Design input requirement transformed from the need statement: "The LIR shall send Status_Messages to the FCR C&M software using the message format defined in the FCR C&M Software ICD xyz, paragraph 1.2.3."

In this example, the term Status_Message is formatted to indicate it is a term defined within the project glossary or data dictionary, and, as an interface requirement, it refers to the ICD in which the required format is defined. There would be at least one other interface requirement needed to address the media in which the message is being communicated. Note that this is a generic requirement concerning the format of messages sent. There would also be other requirements dealing with specific types of messages being sent that would include conditional statements concerning what would trigger the sending of the message.

4.6.4.3 Completeness of the Integrated Set of Needs As stated in the GtWR, well-formed sets of needs have the characteristic *C10 – Complete*. Section 4.5 discussed approaches to ensuring the set of lifecycle concepts and the resulting integrated set of needs are complete. Completing the activities associated with stakeholder needs elicitation, identification of drivers, constraints, and risk, and defining, analyzing, and maturing a complete set of lifecycle concepts are key to having a complete integrated set of needs.

From a completeness perspective, it is helpful for the project team to address the following questions before baselining the integrated set of needs:

- Does the problem or opportunity statement to which the lifecycle concepts and resulting integrated set of needs addresses communicate the "right" problem or opportunity?
- Does the integrated set of needs adequately address the real-world stakeholder expectations concerning the defined problem or opportunity?
- Have all relevant stakeholder viewpoints been included?
- Have all conflicting needs and requirements been resolved?
- Have all product lifecycle stages been considered?
- Have all needs and requirements been prioritized?
- Have all critical functionality, performance, quality, and compliance needs been identified?
- Have all ambiguous and incorrect statements from the stakeholders been resolved?
- Have the corresponding performance measures been defined for each function?
- For each of the MGOs, are there corresponding needs for each objective?
- Is there a need for each defined measure?
- Are there needs defined for each applicable standard and regulation?
- Have all internal dependences/relationships been identified?
- Are the system lifecycle concepts from which the needs are transformed feasible (cost, schedule, technology, legal, regulatory, ethical)?
- Does the integrated set of needs address risk that the project team has decided to mitigate by design and traceability established?
- Have the stakeholder expectations of the SOI in respect to the competition been addressed?
- Have the stakeholder expectations concerning security, safety, resilience, and sustainability been addressed?
- Have all dependences/interactions/interfaces with external and enabling systems been addressed?
- Have all assumptions been recorded and validated?
- Have the key terms used within the integrated set of needs been defined and definitions agreed to?

- Have all the agreed-to attributes that are part of each need expression been defined, including those associated with system validation?

While these questions should have been addressed during the elicitation, lifecycle concepts analysis and maturation activities, it is a best practice to revisit these questions before the integrated set of needs is submitted for baselining. If the answer is "no" to any of these questions, the integrated set of needs is not complete, which will result in an incomplete set of design input requirements as well as risk to the project that the system will fail system validation.

4.6.4.4 Needs Feasibility and Risk The Needs Feasibility/Risk Bucket shown in Figure 4.14 is one approach the project team can use to address feasibility as well as manage risk when defining the integrated set of needs.

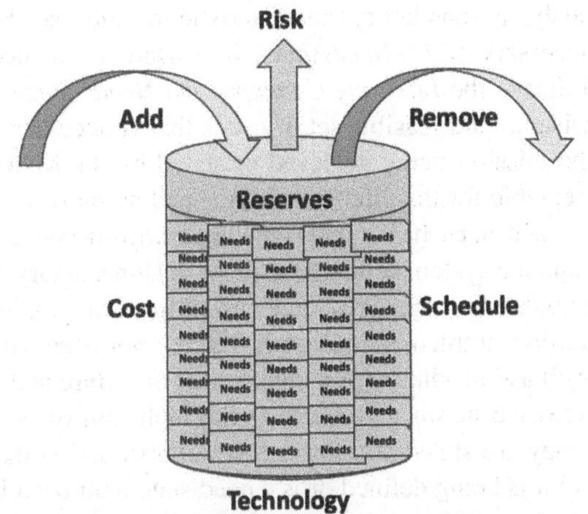

FIGURE 4.14 Needs Feasibility/Risk Bucket.

As stated in the GtWR, well-formed need statements have the characteristic *C6—Feasible*, and integrated sets of needs have the characteristic, *C12—Feasible*. Section 4.5 discussed approaches to ensuring the set of lifecycle concepts is feasible. A set of feasible lifecycle concepts is key to having an integrated set of needs that is feasible.

If it is not feasible with acceptable risk, the need should not be included in the set. Doing so can negatively impact cost and schedule and increase the risk that a need will not be met (fail system validation).

The assessment of feasibility is not binary. A feasibility analysis of each need that is a candidate for inclusion in the integrated set of needs must be made in terms of the cost, schedule, technology, and risk *before* adding the specific need to the integrated set of needs.

The following sections describe the Needs Feasibility/Risk Bucket process in further detail.

4.6.4.4.1 Feasibility of the Set of Needs Given the cost and schedule constraints, only so many needs will fit in the bucket without undue risk to the project. While individual needs may be feasible, implementing the integrated set of needs within budget and schedule constraints may not be.

As part of baselining the integrated set of needs, an overall risk assessment is made of the set of needs, and risk mitigation plans are defined based on the level of risk. It is common for the risk mitigation plans to include a de-scope plan as well as a management reserve to help mitigate the development risk for the unknown unknowns that frequently occur during a development project, especially for a project that depends on the maturity of critical technologies that have a low TRL for this point in the lifecycle.

4.6.4.4.2 Controlling the Size of the Bucket A major challenge is limiting the number of needs that are put in the "needs bucket." It is common to see cases where projects define too many needs for a product or cases of scope creep where needs are added without considering the feasibility, value, or risk of doing so. Rather than focusing on the MGOs, the project tries to develop a system that does something for everyone. The result is often a system that is "a jack of all trades, but a master of none," resulting in a high-risk set of needs that may not be feasible.

As discussed previously, a mandatory characteristic of any need expression is that it is needed—that is, it is necessary *(C1—Necessary)*. If a need is not necessary, why is it in the integrated set? The goal during the *Lifecycle Concepts and Needs Definition* activities is to define a complete, correct, consistent, and feasible set of needs that is necessary and sufficient and, when realized, will result in the mission being achieved as stated by the MGOs, within the drivers and constraints, with risk acceptable for this lifecycle stage – and no more.

By its very existence, each need has a cost associated with it because it must be maintained, managed, implemented, and the system validated to meet it. Unnecessary needs result in work being performed that is not necessary and takes resources away from the implementation of those needs that are necessary. In addition, implementing needs that are not necessary may result in degraded system performance as well as introduce a potential source of failure and conflict.

One reason for an excessive number of needs is the inclusion of needs that are really design input requirements, i.e., they are stated at a lower-level of abstraction that is not appropriate for a need statement at the level it is being defined or is a need statement for a lower-level element being defined for its parent system. In either case, the result is a need statement that is not consistent with the GtWR characteristic of well-formed need statements, *C2—Appropriate*. The need statement may be valid and necessary, but it is being communicated at the wrong level or in the wrong form.

Gold plating is another major cause of unnecessary needs being included in the set. Gold plating is the act of adding features to a system that are not necessary to achieve the mission.

How can gold plating be identified? If a need statement cannot be traced to one of the sources, it could mean that the need statement is not necessary, and thus, is gold plating. The project team needs to be sensitive to gold plating, as this is a major source of scope creep, can impact the cost and schedule of the project, adding to the risk of project failure. *Refer also to Section 14.2.5.1 for more details concerning controlling needs and requirements creep.*

To help determine if a need is necessary, requiring that the proposed need can be traced to a source (*Attribute A3*) as well as requiring that all needs have rationale (*Attribute A1*) will help prevent unnecessary needs from being included in the set.

Another method to help combat uncontrolled growth of needs or including needs that are not necessary, is for the project team to adopt a zero-based approach to defining the integrated set of needs and maintain this approach throughout project execution. A zero-based approach means that all needs are thoroughly evaluated by the project not only for feasibility but also for applicability and value regarding meeting the agreed-to MGOs, measures, drivers, constraints, risk mitigation, and lifecycle concepts and thereby "earn" their way into the set of needs. The use of the concepts of criticality and priority will aid in this evaluation. If the requirement is not critical or high priority, why add the requirement to the set?

In addition to the feasibility evaluation, ask:

- *Why is the need statement necessary?* What role does it play in the realization of the MGOs, measures, drivers and constraints, risk mitigation, and lifecycle concepts for the SOI? If there is no rationale, or a weak rationale for its existence, consider not adding it to the set or removing it from the set.
- *What value does the need statement add as compared to the other need statements?* If the value is low or questionable, consider not adding the need to the set or removing it from the set.
- *What would happen if this need was not included in the set – would the developing organization address this concern anyway?* If nothing would happen or the developing organization would have to address this concern anyway to meet other needs, then why should it be included in the set?
- *Is there a trace from the need to the MGOs, measures, drivers and constraints, risk mitigation, and lifecycle concepts?* If it cannot be traced to a source, why should it be included in the set?
- *Is the need appropriate to the level of architecture at which it is being stated?* The need may be valid, but not appropriate for the system, subsystem, or system element at that level. If the need is not being stated at the appropriate level, move it to the appropriate level subsystem or system element to which it applies.
- *Is the need statement really a design input requirement, i.e., states what the SOI must do to meet a need rather than what the stakeholders need the SOI to do?* If so, determine what the real need is and state it as such. Asking "why" often will result in the identification of the real need that is appropriate to the level at which it is being stated.

Adopting a zero-based approach to defining the integrated set of needs, assessing feasibility, and asking the above questions will help avoid gold plating and avoid defining needs that are not needed.

4.6.4.4.3 Managing Change, Feasibility, and Risk The Needs Risk/Feasibility Bucket is also useful for managing change. Once the bucket is full and the integrated set of needs is baselined, change will happen. If any of the needs change, or a new need is proposed to be added, a feasibility/risk analysis must be made as part of the change process, as well as addressing the above questions.

When a stakeholder wants to add a new need to the already full bucket, responses include:

- *Remove a lower priority, non-critical need* to make room in the bucket for the need being added without adding risk or eating into the management reserves. (This is a key reason for assigning priority and critically to each need statement.)
- *Reject the addition of the new need.* However, politically, this may not be advisable. A better, passive-aggressive response would be "Yes, but … … " and explain the possible consequences of implementing the added need. If management still wants the need added, then accept the addition and risk of doing so.
- *Accept the addition of the new need.* Often, management will want to accept the new need without removing any of the existing needs. In this case, the management reserves could be impacted, as well as the risk to the project potentially increased.

Note: Any proposed change to a need will involve an assessment of the sources from which the need was transformed as well as any dependent peer needs. For a more mature project, the project team will also need to look down the trace chain to assess the potential impacts on the design input requirements that were transformed from that need.

Refer to Section 14.2.5 for a more detailed discussion concerning managing change and assessing the impacts of a change.

4.7 PLAN FOR SYSTEM VALIDATION

The successful outcome of a project is a validated system. System validation is obtaining data that can be used as objective evidence that the verified and validated SOI satisfies the integrated set of needs, MGOs, and measures that, together, define what is necessary for acceptance as discussed at the beginning of this section.

The GtWR includes the characteristic, *C14—Able to be Validated*, for a well-formed set of needs. Able to be validated means that the project team or customer will be able to validate that the realized SOI has met the intent of each need within the integrated set of needs.

A best practice to ensure the needs can be validated, is to plan for how the project will validate that the system will meet the integrated set of the needs during design validation and system validation activities, as discussed in Chapters 8, 10, and 11.

The following attributes should be defined for each need statement when it is formed (along with the other attributes discussed earlier).

A6: System Validation Success Criteria.

A7: System Validation Strategy.

A8: System Validation Method.

A9: System Validation Responsible Organization.

Refer to Chapter 10 for a detailed discussion concerning system verification and validation Success Criteria, Strategy, and Methods. Also, refer to Chapter 15 for a detailed discussion concerning the use of attributes.

Defining the system validation *Success Criteria, Strategy*, and *Method* at this lifecycle stage is important, in that it will help ensure the need statements are worded properly and clearly state what the intent is in terms of stakeholder real-world expectations. It is the intent that must be shown to have been met as part of both design validation and system validation activities.

Additionally, addressing system validation early in the project during *Lifecycle Concepts and Needs Definition* activities will provide important project requirements in terms of facilities, test equipment, environments, resources, and enabling systems that will be needed to support the system validation activities. This information is used by project managers in their budget and schedule planning.

For cases where there is a customer/supplier contractual arrangement, it must be made clear during *Lifecycle Concepts and Needs Definition* activities which organization(s) is responsible for the planning and conduct of the system validation activities and recording the results: the customer, the supplier, or a combination of both, as discussed in Chapter 13.

In customer/supplier development projects, there may be a formal OT&E where users and other stakeholders use the SOI as they intend in the intended operational environment as defined in the use cases and operational scenarios. In commercial in-house development projects, the organization will conduct internal validation activities as well as involve beta users to exercise the SOI in its operational environment. For highly regulated products, such as medical devices, system validation requirements and resulting artifacts to be supplied to the *Approval Authority* are defined within the federal regulations.

4.8 BASELINE AND MANAGE LIFECYCLE CONCEPTS AND NEEDS DEFINITION OUTPUTS

The outputs of the *Lifecycle Concepts and Needs Definition activities,* as shown in Figure 4.2 are listed below. This list is not inclusive, as it does not include all the PM work products listed in Section 4.5 that are developed concurrently with the SE artifacts. Some of the many outputs of *Lifecycle Concepts and Needs Definition* activities for a SOI include:

- Draft Integrated Set of Needs.
- Risks, Risk Management Plan.
- Set of feasible lifecycle concepts
- Functional architectural and analytical/behavior models.
- Preliminary physical architecture.
- Technology Readiness Assessment and Technology Maturation Plan.
- Design and system validation planning artifacts.
- Draft PM and SE management plans, WBS, PBS, budget, and schedules.
- For each need:
  - Traceability to each source.
  - Traceability to system validation planning artifacts.

Prior to baselining, the project team must verify and validate the integrated set of needs as discussed in Chapter 5. The need statements are verified against the need statements quality standards defined in the GtWR (or similar organizational needs quality definition document) and validated against the MGOs, measures, operational-level and system-level needs and requirements drivers and constraints, risk, and lifecycle concepts from which they were derived.

Once the integrated set of needs has been verified and validated, they will be baselined, along with other PM and SE work products and artifacts, at a gate review such as a SR, System or Mission Concept Review (SCR or MCR) in accordance with the activities discussed in Section 14.2.1.

Applications within the project toolset should allow tracking of the status of the *Lifecycle Concepts and Needs Definition* activities using a dashboard that communicates key metrics related to these activities. These metrics can be obtained using the need attributes discussed earlier. Making this information accessible will enable stakeholders to have the same view of the maturity of the integrated set of needs definition activities. This information also represents a single authoritative source of truth (ASoT) for the project.

A major activity in managing needs definition activities is managing change. Establishing traceability and keeping that information current is key to being able to manage impacts of changes throughout the system development lifecycle. The details of needs management are included in Chapter 14, *Needs, Requirements, Verification, and Validation Management.*

This activity is critical to ensuring the resulting individual needs have the GtWR characteristics *C1—Necessary*, *C3—Unambiguous*, *C4—Complete*, C6—Feasible, and *C8—Correct* and sets of needs have the GtWR characteristics *C10—Complete, C11—Consistent, C12—Feasible, C13—Comprehensible, C14—Able to be Validated, and C15—Correct.*

5

NEEDS VERIFICATION AND VALIDATION

This chapter provides a detailed discussion on the planning, activities, and documentation associated with needs verification and validation.

In this chapter, "needs verification" and "needs validation" refer to verification and validation of the need expressions themselves, rather than validation that the System of Interest (SOI) meets the needs as discussed in Chapters 2, 10, and 11.

Needs verification and validation assess the quality of the individual need expressions and integrated set of needs to ensure they have the characteristics of well-formed need statements and sets of needs as defined in the Guide for Writing Requirements (GtWR) (needs verification) and correctly communicate the intent of the sources from which they were derived as shown in Figure 5.1 (needs validation).

Needs verification and validation should be conducted both continuously as the individual need statements are defined and discretely as part of their baseline activities. Using a data-centric approach to needs definition, rather than waiting to do needs verification and need validation on the completed integrated set of needs, individual needs can be verified and validated during needs definition activities.

The status of needs verification and validation can be tracked using the attributes associated with need verification status (A28) and need validation status (A29). These attributes allow the project team to track the progress of the needs verification and validation activities. Once verification and validation of each individual need expression in the set is complete and issues have been resolved, the integrated set of needs will be ready to be baselined as was discussed previously in Chapter 4 and managed for change as discussed in Chapter 14.

While needs verification and validation are discussed separately in this Chapter, the activities will often be performed concurrently. If a Natural Language Processing (NLP)/Artificial Intelligence (AI) application is used for some of the needs verification activities, that analysis should be completed prior to needs validation, enabling the reviewer to focus on content and intent rather than on structure.

INCOSE Needs and Requirements Manual: Needs, Requirements, Verification, Validation Across the Lifecycle, First Edition. Louis S. Wheatcraft, Michael J. Ryan, and Tami Edner Katz.

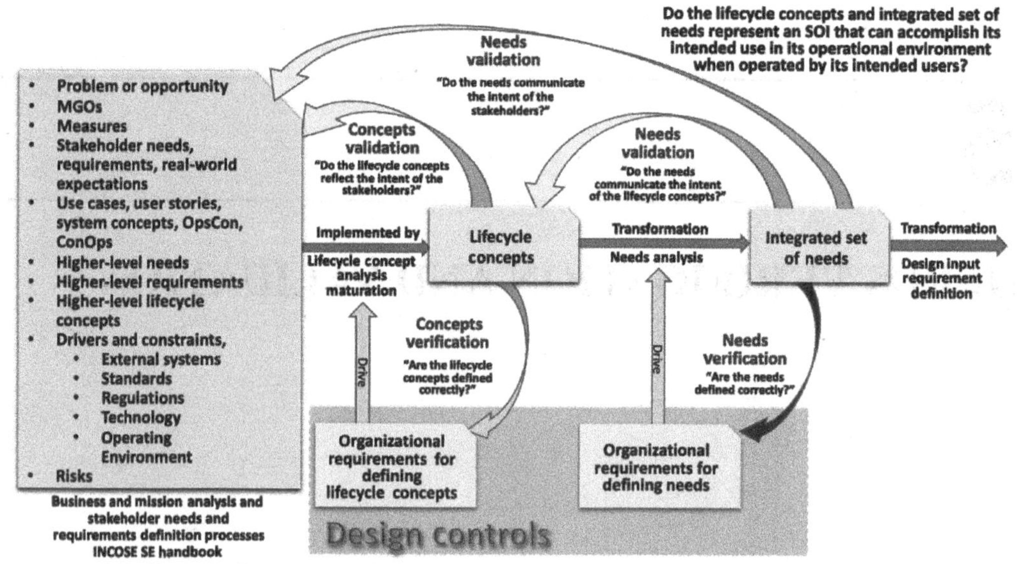

FIGURE 5.1 Needs Verification and Needs Validation Overview.

Needs verification and validation may seem to be one-time activities, but they are not. While the integrated set of needs will be baselined, it does not mean they will not change. During the design input requirement definition activities, it is common to discover issues with the needs, requiring them to be updated when the set of design input requirements is baselined.

During architecture and design analysis, maturation, and definition of the design output specifications, it is also common to discover issues with the needs (and resulting set of design input requirements), requiring them to be updated when the physical architecture, design concept, and design output specifications are baselined. These updated needs will be the focus of system validation activities. In either case, any changes to the integrated set of needs must go through the needs verification, need validation, and Configuration Management (CM) activities prior to approval, as discussed in Sections 14.2.3 and 14.2.5.

Even though the focus of this section is needs verification and validation, it is important that the project team adapts a systems-thinking view and considers the relationships, interactions, and dependencies associated with needs and other Systems Engineering (SE) artifacts and Project Management (PM) work products, especially the higher-level lifecycle Concepts, Needs, and Requirements (CNR). As a result of the needs verification and validation activities, any changes made must be reflected in these other artifacts and work products.

5.1 NEEDS VERIFICATION

A summary of the needs verification activities is shown in Figure 5.2.

During the needs verification activities, the project team will review the need expressions and the set of needs to verify they have the characteristics and adhere to the rules that result in those characteristics as defined by the GtWR or similar organizational guide. In terms of needs verification, the GtWR represents the organizational requirements for writing well-formed need

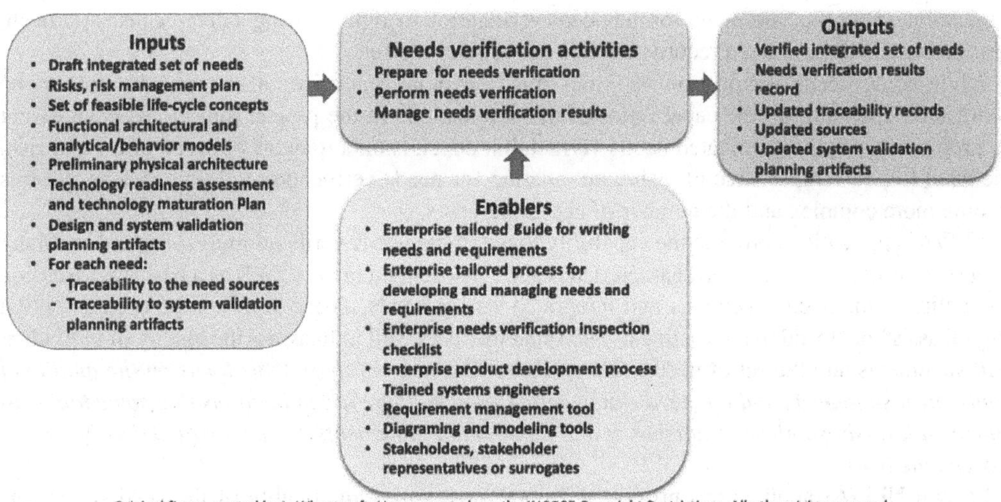
Inputs
• Draft integrated set of needs
• Risks, risk management plan
• Set of feasible life-cycle concepts
• Functional architectural and analytical/behavior models
• Preliminary physical architecture
• Technology readiness assessment and technology maturation Plan
• Design and system validation planning artifacts
• For each need:
- Traceability to the need sources
- Traceability to system validation planning artifacts
Needs verification activities
• Prepare for needs verification
• Perform needs verification
• Manage needs verification results
Outputs
• Verified integrated set of needs
• Needs verification results record
• Updated traceability records
• Updated sources
• Updated system validation planning artifacts
Enablers
• Enterprise tailored guide for writing needs and requirements
• Enterprise tailored process for developing and managing needs and requirements
• Enterprise needs verification inspection checklist
• Enterprise product development process
• Trained systems engineers
• Requirement management tool
• Diagraming and modeling tools
• Stakeholders, stakeholder representatives or surrogates
Original figure created by L. Wheatcraft. Usage granted per the INCOSE Copyright Restrictions. All other rights reserved.

FIGURE 5.2 Needs Verification IPO Diagram.

statements and sets of needs. The needs verification activities will also include inspection of traceability records from the sources from which the needs were transformed to ensure traceability with those sources.

5.1.1 Prepare for Needs Verification

As shown in Figure 5.2, there are several enablers to successful needs verification. These include an enterprise-tailored GtWR and a process for the development and management of sets of needs and sets of design input requirements. There should also be an enterprise product development process, including verification and validation across the system lifecycle, and systems engineers trained in and knowledgeable of what is in the guide, as well as how to perform the needs verification and validation activities.

Of particular importance are the needs and requirements management and modeling/diagramming applications within the project's toolset used for the underlying analysis to produce the input artifacts from which the integrated set of needs was transformed and recorded.

Using a data-centric approach to SE, traceability can be established within the project toolset where the traceability matrices are created as reports generated by the toolset. With a well-defined traceability strategy and a well-defined needs attribute strategy, needs verification can, to some extent, be automated using these tools, enabling advanced analysis involving traceability of needs to the input artifacts shown in Figure 5.2. As a minimum, these tools can identify missing traceability or attributes via reports. For example, a report could be generated to list all needs that do not trace to a source, or report on a source that does not have an implementing lifecycle concept and resulting need.

Note: While applications within the project toolset can produce trace reports, these applications cannot yet determine the accuracy and correctness of the traces between entities. This assessment must be done manually.

Assuming the processes are in place and applications are in use, the artifacts listed as inputs in Figure 5.2 should have been produced and matured during *Lifecycle Concepts and Needs Definition* to the point where they are ready for the needs verification activities. Using a data-centric approach, these artifacts will have been recorded within the project toolset.

While some needs verification activities must be done manually, others may be able to be automated depending on the capabilities of the applications in the project toolset. To what extent the project can rely on automated needs verification depends on a tradeoff between effort and risk. The need for, and importance of, using automation for needs verification will increase as systems become more complex and the number of needs increases.

NLP/AI applications provide the capability to automate needs verification to some extent. Many of these tools use a subset of the characteristics and rules defined in the GtWR as a basis for assessing the quality of the need statements and integrated sets of needs. These tools can be used as both a "digital assistant" to aid in the writing of need statements as well as to assess the quality of individual need statements and the set of needs. *(Note: Currently, many of these tools focus on the quality of requirement statements and not need statements. However, there are indications that some tool vendors are addressing both need statements and requirement statements and understand the differences between the two).*

Several NLP/AI applications provide a "score" concerning the quality of the need statements based on criteria defined by the project team as well as identifying specific defects. The project team will determine how this score will be used in the definition and management of the needs and need verification activities.

Because these applications only assess requirements against a subset of the characteristics and rules defined within the GtWR, the project team must understand the application limitations and not assume that the set of needs is well-formed, having all the characteristics defined within the GtWR, based solely on the quality score from the NLP/AI applications. For example, current NLP/AI tools are not capable of assessing the necessity, correctness, nor feasibility of individual needs nor completeness or feasibility of the integrated set of needs.

A key preparation activity for needs verification is the creation of a *Needs Verification Inspection Checklist* if one does not already exist. If the organization has a generic checklist, they can tailor the checklist to the SOI being developed and the project's processes. This checklist serves as a standard to measure the needs against and will help guide all the needs definition and verification activities. Having addressed each of the areas and questions within the checklist will help lead to the successful completion of the *Lifecycle Concepts and Needs Definition* activities and needs verification activities. *An example of a Needs Verification Inspection Checklist is included in the GtNR Appendix D.*

5.1.2 Perform Needs Verification

There are several activities that should be performed to ensure the individual need expressions and set of needs are well-formed and provide objective evidence they have been verified.

Guided by the *Needs Verification Inspection Checklist*, the project team should:

- Manually verify individual need expressions and set of needs have the characteristics defined in the GtWR [5] or a similar guide. This could be done by using the *Needs Verification Inspection Checklist* as a guide to inspect each need expression by individuals or as part of a tabletop or peer review. This task is difficult to do manually, especially for large sets of needs. given the number of characteristics and the number of rules discussed in the GtWR, as well as the underlying analysis and activities discussed in Chapter 4 that must be considered to help determine that need statements have those characteristics. *This activity helps establish that each need has the GtWR characteristic C9—Conforming.*
- Verify that the set of needs contains individual needs that are unique, do not conflict with or overlap with other needs in the set and that the units and measurement systems they use

are homogeneous. *This activity helps establish that each need has the GtWR characteristics C9—Conforming and C11—Consistent.*

- Verify that the language used within the set of needs is consistent (i.e., the same words are used throughout the set to mean the same thing) and that all terms used within the requirement statements are consistent with the architectural model, project glossary, and data dictionary. *This activity helps establish that each need has the GtWR characteristics C9—Conforming and C11—Consistent.*
- If included in the project toolset, use an NLP/AI application that provides the capability to partially automate the verification of the need statements in terms of how well they adhere to the rules for writing well-formed needs and sets of needs. For need statements with defects, members of the project team will have to examine the defective need statements and fix the defects identified by the application. *This activity helps establish that each need has the GtWR characteristic C9—Conforming.*

- Verify that individual need expressions have the set of attributes defined and agreed to by the project team. The project toolset should be able to produce a report concerning whether any of the attributes are "null," i.e., no values have been defined for a given attribute. While a report can tell if an attribute has been defined, it cannot assess the quality or accuracy of the information in the attribute—that assessment will have to be done manually. For example, does the text in the rationale statement include the information expected to be in the rationale? Does the rationale state why the need statement is necessary? Is the source of any numbers explained? *This activity helps establish that* each *need has the GtWR characteristic C4—Complete.*
- Verify that individual need expressions have the set of system validation attributes defined and agreed to by the project team. The project toolset should be able to produce a report concerning whether any of the system validation attributes are null, i.e., no values have been defined. While a report can tell if a system validation attribute has been defined, it cannot assess the quality or accuracy of the information in the attribute—that assessment will have to be done manually. For example, does the text in the validation attribute that defines the validation success criteria include the information expected? *This activity helps establish that each need has the GtWR characteristics C9—Conforming and the set of needs has the GtWR characteristic C14—Able to be Validated.*
- Use the project toolset to generate reports to confirm traceability of each need to one or more input artifacts (sources). Each need must be traced to at least one source from which it was derived. In a document-based approach, trace matrices are often developed and managed manually requiring a lot of time and effort to keep them current and accurate. *This activity helps establish that each need has the GtWR characteristic C1—Necessary.*
- Use the project toolset to generate reports to confirm that each source shown in Figure 4.11 has at least one derived need that addresses that source. *Bidirectional traceability—if the tool allows a trace from a need to its source, it should also include the capability to trace each source to its implementing lifecycle concept and associated need statement(s). This activity helps establish that the set of needs has the GtWR characteristic GtWR C10—Complete.*
- Confirm that the project has done risk assessments and, for each risk that will be mitigated by the SOI, the project has established traceability between the risk and the lifecycle concepts that define a concept for mitigation of that risk and traceability to the need that addresses that mitigation concept. *This activity helps establish that the set of needs has the GtWR characteristic C10—Complete.*
- Referring to the SOI external interface diagrams, context diagrams, boundary diagrams, or functional models, verify that there are needs that address each of the interfaces and interactions

across the interface and that each need that addresses an interface traces back to the source that identified that interface. *This activity helps establish that the set of needs has the GtWR characteristic C10—Complete.*

5.2 NEEDS VALIDATION

The GtWR includes the characteristic, *C14—Able to be Validated*, for a well-formed set of needs. "Able to be validated" means that the set of needs is formed such that the project team will be able to obtain objective evidence during system validation that the integrated set of needs will lead to the achievement of the MGOs, measures, operational-level and system-level needs and requirements, risk mitigation, and lifecycle concepts within the constraints (such as cost, schedule, technical, legal, and regulatory compliance) with acceptable risk.

The needs validation activities validate that each need and the set of needs clearly communicate the intent of the sources from which they were transformed, in a language understandable by all project team members, customers, and other key stakeholders.

The needs validation activities also determine whether the integrated set of needs will result in an SOI that does what it was intended to do in its operational environment when operated by its intended users and will be acceptable to the customers, users, regulators, *Approval Authority*, and other stakeholders.

A summary of the needs validation activities is shown in Figure 5.3.

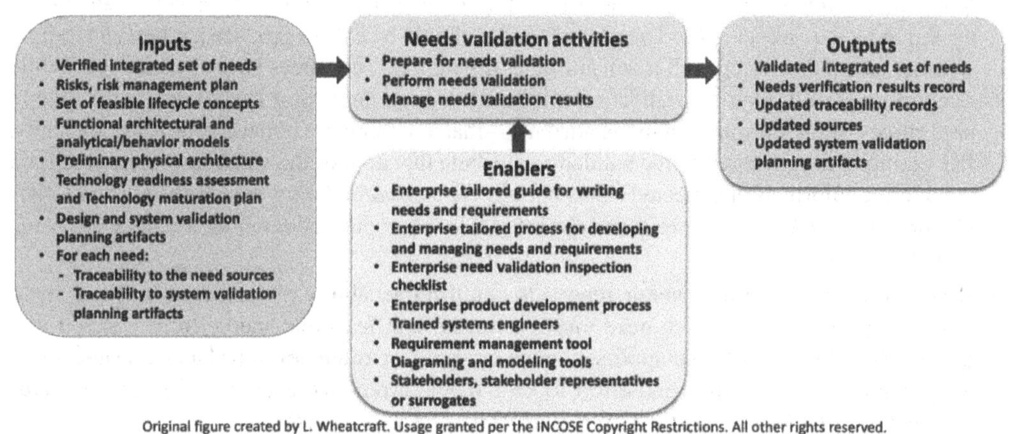

Inputs
• Verified integrated set of needs
• Risks, risk management plan
• Set of feasible lifecycle concepts
• Functional architectural and analytical/behavior models
• Preliminary physical architecture
• Technology readiness assessment and Technology maturation plan
• Design and system validation planning artifacts
• For each need:
- Traceability to the need sources
- Traceability to system validation planning artifacts
Needs validation activities
• Prepare for needs validation
• Perform needs validation
• Manage needs validation results
Outputs
• Validated integrated set of needs
• Needs verification results record
• Updated traceability records
• Updated sources
• Updated system validation planning artifacts
Enablers
• Enterprise tailored guide for writing needs and requirements
• Enterprise tailored process for developing and managing needs and requirements
• Enterprise need validation inspection checklist
• Enterprise product development process
• Trained systems engineers
• Requirement management tool
• Diagraming and modeling tools
• Stakeholders, stakeholder representatives or surrogates

FIGURE 5.3 Needs Validation IPO Diagram.

5.2.1 Prepare for Needs Validation

The first step in planning for needs validation is to ensure the enablers shown in Figure 5.3 are in place, and those project team members doing the need validation activities have access to the inputs.

Even though individual need statements and the set of needs may be well-formed, and traceability has been verified, the message they communicate may not be as intended. It is important that organizations do both needs verification and needs validation. Failing to do so adds the risk that the SOI being developed will fail to do what it was intended in its operational environment when operated by its intended users and not meet the validation *Success Criteria* that define what is necessary for acceptance.

Assuming the *Lifecycle Concepts and Needs Definition* activities were completed, and tools are in use, the artifacts listed as inputs in Figure 5.3 should have been produced and matured to the point where they are ready for the need validation activities. Using a data-centric approach, these artifacts will have been recorded within the project toolset.

The *Lifecycle Concepts and Needs Definition* activities should have resulted in analysis records, diagrams, and models that provide the underlying analysis and rationale for the transformation that resulted in each of the need expressions. Each need expression should include a trace to its source and rationale to help understand the source of the need and the intent of what the need statement is communicating.

For example, a stakeholder may have an expectation for how long it should take to do some task like replacing a tire. Assuming the expected time was stated as no longer than 3 minutes, what assumptions did the stakeholder make when stating that value? Is there sufficient analysis and is a feasible concept available from which the need statement was transformed? What is the variation in the conditions of use for replacing a tire considering different vehicles and tire types, operational environment, tools needed, location of the spare tire, location of the tools, availability, ease of use, and availability of the instructions for changing a tire using the supplied tools, as well as the capability of the person changing the tire? Is the organization targeting 100% satisfaction for a given audience or is there a tradeoff range? Were personas/user classes defined for the class of stakeholders that would be expected to change the tire in this amount of time? Were use cases defined for changing a tire that this need traces to? Was there a Use or User Failure Mode and Effects Analysis (UFMEA) done for each step of the use case?

A key preparation activity for needs validation is the creation of a *Needs Validation Inspection Checklist* if one does not already exist. If the organization has a generic checklist, then tailor the checklist to the SOI being developed and the project's processes. This checklist serves as a standard to measure the needs against and will help guide the needs definition and validation activities. Having addressed each of the areas and questions within the checklist will help lead to the successful completion of the *Lifecycle Concepts and Needs Definition* activities and needs validation activities. *An example of a Needs Validation Inspection Checklist is included in the GtNR Appendix D.*

5.2.2 Perform Needs Validation

There are several activities that should be performed to provide objective evidence that the individual need expressions and integrated sets of needs are validated to accurately communicate the intent of the sources from which they were transformed.

Note: Unlike needs verification, needs validation confirms that the intent of the source is effectively communicated and cannot be done currently without the project team doing the analysis manually—none of the NLP/AI applications currently available have the capability to do this type of analysis.

Guided by the *Needs Validation Inspection Checklist*, the project team should:

- Use the trace matrices to perform an analysis to validate that each need expression clearly communicates the intent of those source(s) from which it was derived. *This activity helps establish that each need has the GtWR characteristic C8—Correct.*
- Assess whether the needs are necessary and sufficient to meet the intent of the sources they are traced to and derived from. (*Note: There will be cases where a need or group of needs satisfies more than one source.*) *This activity helps establish that the set of needs has the GtWR characteristic C10—Complete.*

- For each need that references a capability or performance value dependent on a critical technology, confirm the risk (of implementation) attribute has been defined indicating this dependency. Also, confirm a Technology Readiness Level (TRL) has been assigned to the technology and there is a Technology Maturity Plan (TMP) to mature this technology. *This activity helps establish that each need has the GtWR characteristic C6—Feasible.*
- Confirm that the project documented their assessment that the set of needs is feasible in terms of cost, schedule, and technology maturation and has included key product development activities in their cost and schedule estimates, including the use of enabling systems, lifecycle concepts analysis and maturation, design input requirement definition, design verification and design validation, system integration, system verification, validation, and procurement. *This activity helps establish that the set of needs has the GtWR characteristic C12—Feasible.*
- Confirm with the stakeholders associated with each source that the message being communicated by each need statement is correct and acceptable to ensure the set of needs are the right needs, i.e., do they accurately represent the agreed-to sources from which they were transformed? Do the needs correctly and completely capture what the stakeholders need the system to do in the context of its intended use in its operational environment, when operated by its intended users, in terms of form, fit, function, quality, and compliance? There is no substitute for the stakeholders, validating that individual needs and the set of needs represent what is necessary for acceptance. *This activity helps establish that each need has the GtWR characteristics C1—Necessary, C3—Unambiguous, and C8—Correct and the set of needs has the GtWR characteristics C10—Complete, C13—Comprehensible, and C14—Able to be validated.*
- For each need expression, ensure system validation attributes have been defined and included in the need expression addressing the validation *Success Criteria, Strategy, Method,* and *Organization* responsible for system validation. This information must be defined and documented before the integrated set of needs is baselined. Doing so ensures that what is necessary for acceptance has been defined and agreed to as well as that the project has allocated the necessary budget, time, and resources for completing system validation activities. This also helps ensure each need is worded such that the design input requirements, design, and the system can be validated to meet the need. *This activity helps establish that the set of needs has the GtWR characteristic C14—Able to be validated.*
- For each need that addresses an interaction with an external system, confirm the specific interactions have been defined and documented in some type of interface definition type document or record, for e.g., an Interface Control Document (ICD), Interface Definition Document (IDD), data dictionary, or there is a plan in place to define these interactions. *This activity helps establish that the needs associated with interfaces have the GtWR characteristic C14—Able to be validated.*

5.3 MANAGE NEEDS VERIFICATION AND VALIDATION RESULTS

Needs verification and validation records are created by documenting the results and outputs of the needs verification and needs validation activities. The outputs include the updated input SE artifacts and PM work products shown in Figures 5.2 and 5.3, Outputs.

For each need expression, update the attributes listed in Section 5.4. Applications within the project toolset should allow tracking of the status of the needs verification and validation activities using a dashboard that communicates key metrics related to the results of these activities using these attributes. Making this information accessible will enable all project team members and key stakeholders to have a shared view of the maturity and status of the needs verification and need validation activities.

Defects found in the set of needs, as well as any defects found in any of the input SE artifacts or PM work products, must be addressed and corrected before they are baselined.

Changes to any of the system validation information contained in the system validation attributes should be updated in the system validation planning artifacts discussed in Chapter 10.

If there is insufficient data to complete needs verification and validation activities, a lack of a set of feasible lifecycle concepts from which the needs were transformed, or a lack of traceability records within the project toolset, the issue should be addressed as a risk that could cause subsequent requirement, design, and system validation to be deemed unsuccessful.

5.4 USE OF ATTRIBUTES TO MANAGE NEEDS VERIFICATION AND VALIDATION

Attributes that can aid in needs verification and validation management include:

A1: *Rationale*. Intent of the need, reason for the need's existence.

A3: *Trace to Source*. Where the need originated—stakeholder, concept, MGO, measures, user story, scenario, use case, constraint, risk, lifecycle concept, architectural model, or diagram as examples.

A26: *Stability*: stable, likely to change, and incomplete.

A28: *Need Verification Status*. Binary—true/ false, yes/no, or incremental—not started, in work, complete, and approved. (A28 is a prerequisite for A29 and A30).

A29: *Need Validation Status*. Binary—true/ false, yes/no, or incremental—not started, in work, complete, and approved.

A30: *Status of the Need* (in terms of maturity). Draft, in development, ready for review, in review, and approved.

A36: *Risk (of implementation)*. Indicating the degree of risk of the need or requirement not being met. This could be a result of the instability (A26) of the requirement or due to feasibility in terms of the TRL of the technology needed for a requirement to be jet, cost, or schedule.

Refer to Chapter 15, for a more detailed discussion on attributes and definitions of each.

6

DESIGN INPUT REQUIREMENTS DEFINITION

The focus of the *Design Input Requirements Definition* activities is on transforming the baselined integrated set of needs into a well-formed set of design input requirements expressed as "shall" statements. The resulting set of design input requirements are inputs to the *Architecture Definition Process*, flowing the requirements down (allocating) from one level of the architecture to the next, and implementing a design solution per the *Design Definition Process*.

This chapter is an elaboration of the concepts in the *System Requirements Definition Process* described in ISO/IEC/IEEE 15288 and INCOSE SE HB. However, unlike those sources that imply that the stakeholder requirements are directly transformed into the system-level design input requirements for the System of Interest (SOI), this Manual treats the stakeholder requirements as inputs into *Lifecycle Concepts and Needs Definition*. The result is an integrated set of needs for the SOI that represents the project scope against which the resulting set of design input requirements, design, and realized SOI will be validated.

It is this integrated set of needs that is transformed into the set of system requirements for the SOI, referred to in this Manual as design input requirements, which address *what the SOI must do* to satisfy the needs from which they were transformed.

Given that the SE technical lifecycle processes are applied iteratively and recursively as the project team moves down the physical architecture, what is described in this section can be applied to the development of an SOI (system, subsystem, and system element) set of design input requirements—no matter the architectural level at which the SOI exists.

It is important to understand the I-NRDM approach defined in Chapter 3 and the *Lifecycle Concepts and Needs Definition* activities discussed in Chapter 4. The result of completing these activities is not only an integrated set of needs but also the *underlying analysis and associated artifacts* from which they were formed. As part of these activities, the project team will have been concurrently defining a preliminary set of design input requirements.

The *Design Input Requirements Definition* activities discussed in this chapter build upon this work, resulting in a mature set of design input requirements that are baselined and allocated/budgeted to the subsystems and system elements at the next level of the architecture, stopping when the SOI is

INCOSE Needs and Requirements Manual: Needs, Requirements, Verification, Validation Across the Lifecycle, First Edition. Louis S. Wheatcraft, Michael J. Ryan, and Tami Edner Katz.

a system element that needs no further decomposition before being realized by the *Design Definition Process*.

Using a concurrent approach enables issues that occur while defining the preliminary set of design input requirements to be addressed in a less formal, agile manner earlier in the lifecycle during *Lifecycle Concepts and Needs Definition* activities. The resulting lifecycle concepts, models, and integrated set of needs will address these issues prior to being baselined, avoiding technical debt associated with a more serial document-centric "waterfall" approach.

The reader is cautioned to be aware of the risks of using an existing set of requirements that were developed for a similar system or were developed by a customer who did not perform the precursor activities as advocated in this Manual. In either case, without the *Lifecycle Concepts and Needs Definition* activities, the real stakeholder intent and expectations of the stakeholders may not be well understood nor completely reflected within these sets of requirements, resulting in an SOI that may be verified to meet those requirements, yet fail system validation.

Organizations that do not use the approach advocated in this Manual have an increased risk of accumulating a considerable amount of technical debt and will still have to do the activities discussed for lifecycle concepts analysis and maturation and needs definition as part of the *Design Input Requirements Definition* activities. An up-front investment in performing these activities is essential in preventing the technical debt and potential rework associated with developing a design and resulting system that does not conform to the lifecycle concepts or meet stakeholder needs.

A summary of the *Design Input Requirements Definition* activities is shown in Figure 6.1.

6.1 PREPARE FOR DESIGN INPUT REQUIREMENTS DEFINITION

Preparing for *Design Input Requirements Definition* activities consists of gathering or obtaining access to the required input artifacts, as shown in Figure 6.1.

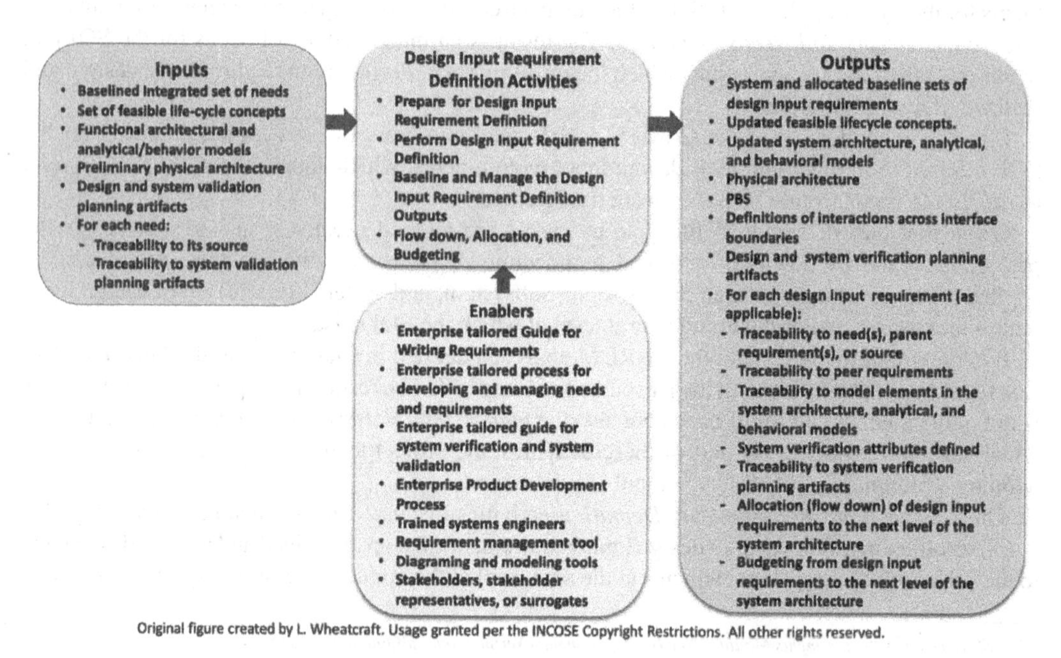
Inputs
• Baselined integrated set of needs
• Set of feasible life-cycle concepts
• Functional architectural and analytical/behavior models
• Preliminary physical architecture
• Design and system validation planning artifacts
• For each need:
- Traceability to its source
Traceability to system validation planning artifacts
Design Input Requirement Definition Activities
• Prepare for Design Input Requirement Definition
• Perform Design Input Requirement Definition
• Baseline and Manage the Design Input Requirement Definition Outputs
• Flow down, Allocation, and Budgeting
Enablers
• Enterprise tailored Guide for Writing Requirements
• Enterprise tailored process for developing and managing needs and requirements
• Enterprise tailored guide for system verification and system validation
• Enterprise Product Development Process
• Trained systems engineers
• Requirement management tool
• Diagraming and modeling tools
• Stakeholders, stakeholder representatives, or surrogates
Outputs
• System and allocated baseline sets of design input requirements
• Updated feasible lifecycle concepts.
• Updated system architecture, analytical, and behavioral models
• Physical architecture
• PBS
• Definitions of interactions across interface boundaries
• Design and system verification planning artifacts
• For each design input requirement (as applicable):
- Traceability to need(s), parent requirement(s), or source
- Traceability to peer requirements
- Traceability to model elements in the system architecture, analytical, and behavioral models
- System verification attributes defined
- Traceability to system verification planning artifacts
- Allocation (flow down) of design input requirements to the next level of the system architecture
- Budgeting from design input requirements to the next level of the system architecture
Original figure created by L. Wheatcraft. Usage granted per the INCOSE Copyright Restrictions. All other rights reserved.

FIGURE 6.1 Design Input Requirements Definition IPO Diagram.

As shown in Figure 6.1, there are several enablers to successful definition of the set of design input requirements. These include an organizationally tailored Guide for Writing Requirements (GtWR) as well as tailored processes for the definition and management of well-formed sets of needs and sets of design input requirements. There should also be an organizational process for product development, procedures, work instructions, and experienced systems engineers trained in and knowledgeable in how to perform the design input requirements definition activities concurrently with the *Architectural Definition Process*.

Of particular importance are the Requirement Management Tools (RMTs) and modeling/diagramming applications within the project's toolset used to develop and record the SOI lifecycle concepts, the resulting integrated set of needs, and traceability among all artifacts across the lifecycle. This section assumes the project will be moving toward a data-centric approach to SE and recording and managing the set of design input requirements and establishing traceability using applications within the project's toolset that support the data-centric approach discussed in Chapter 3.

A key preparation activity is the creation of a Design Input Requirements Verification and Requirements Validation Inspection Checklists if they do not already exist. If the organization has a generic checklist, they can tailor the checklist to the SOI being developed and the project's specific processes. This checklist serves as a standard to guide the design input requirements definition activities as well as provides a standard to measure the quality of the design input requirement statements and sets of requirements (activities to perform requirement verification and requirement validation are discussed in Chapter 7). Addressing the activities and questions within the checklist will lead to successful completion of the *Design Input Requirements Definition* activities. *An example Design Input Requirements Verification Inspection Checklist is contained in the GtNR, Appendix D.*

6.2 PERFORM DESIGN INPUT REQUIREMENTS DEFINITION

Design Input Requirements Definition involves the activities shown in Figure 6.2. Each activity results in data and information needed to define the design input requirement expressions.

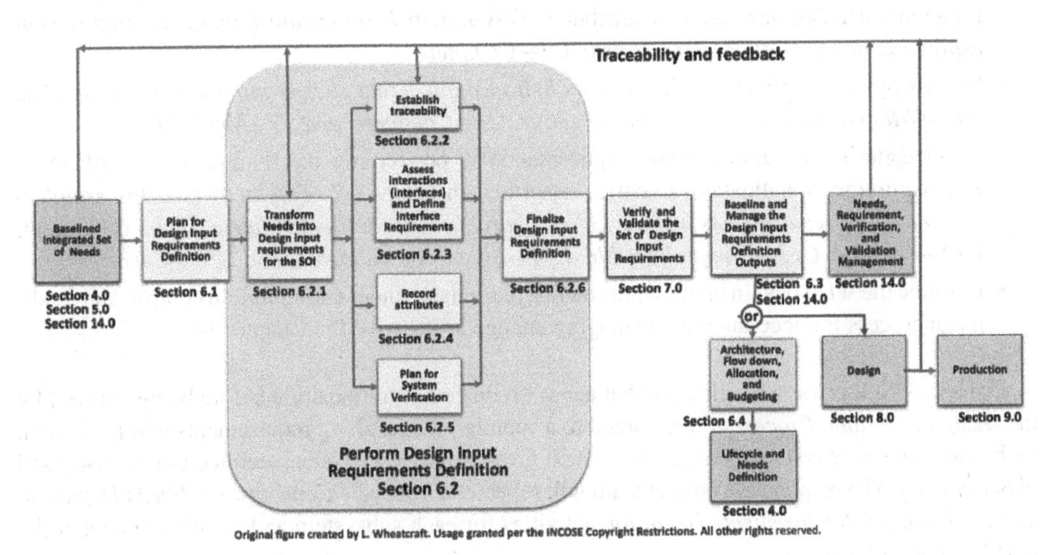

FIGURE 6.2 Design Input Requirement Definition Activities.

Key activities include:

- Transform the integrated set of needs into a set of design input requirements.
  - Ensure there is at least one requirement that implements the intent of the need from which it was transformed and that the implementing requirement(s) is/are *sufficient* to satisfy the need [41]. *This activity helps establish that each requirement has the GtWR characteristics C1—Necessary and C8—Correct.*
  - Ensure each requirement is stated at the appropriate level of the architecture. *This activity helps establish that each requirement has the* GtWR *characteristic C2—Appropriate.*
  - Ensure each requirement is necessary and not gold plating. The design input requirements should specify the requirements necessary for the minimum viable product that satisfies customer needs. *This activity helps establish that each requirement has the* GtWR *characteristic C1—Necessary.*
  - Ensure each requirement and resulting set of requirements have the characteristics of well-formed requirements as defined in the GtWR. *This activity helps establish that each requirement has the* GtWR *characteristic C9—Conforming.*
  - Ensure each functional requirement is defined in terms of its performance characteristics. *This activity helps establish that the set of design input requirements has the* GtWR *characteristic C10—Complete.*
- Establish traceability:
  - Between each requirement and the need from which it was transformed, parent requirement, or source. *This activity helps establish that each requirement has the GtWR characteristic C1—Necessary.*
  - Between dependent peer requirements. *This activity helps establish that each requirement has the GtWR characteristics C1—Necessary, C8—Correct, and C11—Consistent.*
- Assess interactions (interfaces) and define interface requirements. *This activity helps establish that the set of design input requirements has the GtWR characteristics C10—Complete and C11—Consistent.*
- Record attributes as part of each requirement expression. Refer to Section 6.2.4 and Chapter 15 for a more detailed discussion on attributes. *This activity helps establish that each requirement expression has the GtWR characteristic C4—Complete.*
- Plan for system verification. *This activity helps establish that each requirement expression has the GtWR characteristics C3—Unambiguous, C4—Complete, and C7—Verifiable.*
- Finalize the set of design input requirements by performing the requirements verification and requirements validation activities described in Chapter 7. *This activity helps establish that the set of requirements has the GtWR characteristics C10—Complete, C11—Consistent, C12—Feasible, C13—Comprehensible, C14—Able to Be Validated, and C15—Correct.*
- Baseline the set of design input requirements, place them under configuration control, and establish a process for accessing and managing change as described in Chapter 14.

Unless the SOI is a system element that needs no further decomposition before being realized by the *Design Definition Process* or outsourced to a supplier, the resulting requirements will flow down to the subsystems or system elements at the next level of the SOI physical architecture, as discussed in Section 6.4. The responsible project team will repeat the *Lifecycle Concepts and Needs Definition* and *Design Input Requirements Definition* activities for each subsystem and system element at the next level of architecture.

6.2.1 Transforming Needs into Design Input Requirements

The *Design Input Requirements Definition* activities begin with the transformation of the integrated set of needs into a set of design input requirements appropriate for the level of the SOI that communicate "what" the system must do to meet the needs, avoiding requirements stating "how" to achieve the design realization of the physical SOI. *This activity helps establish that each requirement has the GtWR characteristics C1—Necessary, C2—Appropriate, C4—Complete, and the sets of design input requirements have the GtWR characteristic C10 Complete.*

The resulting design input requirements are recorded, agreed-to, baselined, and placed under CM in accordance with the *Needs and Design Input Requirements Management* activities discussed in Chapter 14. For systems that are outsourced or procured, the requirements on the organization supplying the service or developing the SOI are recorded in a Statement of Work (SOW), Supplier Agreement (SA), and Purchase Agreement (PO) separate from the design input requirements.

An approach that is helpful is to use the needs-to-requirements transformation matrix shown in Table 6.1.

Using the needs-to-requirements transformation matrix, Column A consists of need statements written from the stakeholder's perspective and Column B reflects design input requirements communicating what the system must do such that the intent of the needs in Column A will be met. Each need statement in Column A will be transformed into one or more well-formed design input requirement statement(s) that are written at a level of abstraction appropriate for the level of architecture the SOI exists.

If a need in Column A invokes a standard or regulation, multiple requirements will be addressed, as discussed later in the Section "Compliance." The resulting requirements are the project team's response to the requirements in the standard or regulation whose implementation will result in compliance with the applicable requirements in the standard or regulation invoked by the need.

TABLE 6.1 Needs-to-Requirements Transformation Matrix.

Column A Needs	Column B Design Input Requirements	Column C External Interface
Functional/performance (function)		
Need 1	Rqmt 1 (could be more than one)	Interface (if applicable)
Need 2	Rqmt 2 (could be more than one)	Interface (if applicable)
Operational (fit)		
Need 3	Rqmt 3 (could be more than one)	Interface (if applicable)
Need 4	Rqmt 4 (could be more than one)	Interface (if applicable)
Physical Characteristics (form)		
Need 5	Rqmt 5 (could be more than one)	...
Need 6	Rqmt 6 (could be more than one)	...
Quality (-ilities)		
Need 7	Rqmt 7 (could be more than one)	...
Need 8	Rqmt 8 (could be more than one)	...
Compliance (standards/regulations)		
Need 9	Rqmt 9 (could be more than one)	...
Need 10	Rqmt 10 (could be more than one)	...

Column C is used to record instances where the SOI needs to interact with another system to implement the requirement stated in Column B. Entries in Column C will be used to validate the context diagrams and models developed during the *lifecycle concepts analysis and maturation activities* discussed in Chapter 4. If Column C indicates the requirement in Column B is an interface requirement, then the Column B interface requirement will need to address the specific interaction across an interface boundary between the SOI and the external system and include a reference (pointer) to where that interaction is defined, such as an ICD, IDD or data dictionary, as discussed in Section 6.2.3.

Example Transformation of LIR Needs into Design Input Requirements

The following represents an initial transformation of several of the LIR needs into well-formed design input requirements. Each of the resulting requirements will be traced to the parent need. Attributes for each requirement will also need to be defined.

Acronyms used: LIS: Lid Installation System; LIR: Lid Installing Robot; FCR: Facility Control Room; LDS: Lid Distribution System; JPS: Jar Processing System.

N001: The stakeholders need the LIR to obtain Lids from the LDS.

R01: The LIR shall obtain Lids from the LDS as described in the LIR/LDS ICD xyz Section 2.4.3.

Note: The resulting requirement is written on the LIR in active voice. Because the LIR must interact with another system, this is an interface requirement and to be complete, the requirement must include a reference to where the interaction is defined—in this case an ICD.

N002: The stakeholders need the LIR to install lids of various sizes on jars of various sizes and materials.

R02: The LIR shall install on the Jars, Lids having the characteristics defined in Lid Specification LID1234.

R03: The LIR shall install Lids on Jars having the characteristics defined in Jar Specification JAR1235.

Note: The resulting requirements are written on the LIR in active voice. The focus of R02 is on the specific lids to be installed on the jars, The focus of R03 is on the specific jars the lids are to be installed. To be complete, the requirements must refer to where the characteristics of the lids or jars are defined—in this case, the lid or jar specification that describes the material characteristics and sizes as well as the location each is to be grabbed during installation.

N003: The stakeholders need the LIR to install lids on jars with the proper torque depending on lid size and jar material.

R04: The LIR shall install Lids on the Jars with a torque within the torque ranges specified in Table 6 for the Jar size and material.

Note: The actual torque values consistent with the jar size and material must be clearly defined else the requirement is ambiguous (C3), not verifiable (C7), and not complete (C4). One way of communicating the various values is within a table. Given the table contains multiple parameters defining the required torque based on jar characteristics, some would say that to address the characteristic C5—Singular and rule R18—single thought, there should be multiple requirements each addressing a single parameter. In doing this, others have a valid concern associated with the characteristic C13—Comprehensible. No matter the approach taken, the system must be verified for each of the parameters. It is up to the organization to decide which approach they choose to take—and then be consistent in the application of this approach.

N004: The stakeholders need the LIR to install lids on jars per the JPS timing.

R05: The LIR shall install Lids on Jars per the timing defined in JPS ICD xyz, Section 2.3.5.

Note: Because the LIR is interacting with an existing system, the timing is defined in that system's ICD. Rather than pulling out the value from the ICD, it is better to refer to the ICD that defines the timing, otherwise the specific number would be documented in two places which could get out of synchronization (the value in the ICD is changed, but that change is not reflected in the requirement).

Defining requirements is not just an exercise in writing but is also an exercise in engineering. Every requirement represents an engineering decision as to what the SOI must do or a quality the SOI must have to meet the needs from which they were transformed. The engineering determination as to what the system must do to meet a need is a result of detailed requirements analysis (ideally using models and simulations) as well as composing the requirement statements such that they are well-formed (having the characteristics and following the rules defined in the GtWR).

The resulting set of design input requirements represents the analysis and agreed to transformation from the baselined integrated set of needs. Areas addressed within the resulting set of requirements include functionality (what the system must do), expected performance and quality ("how well" characteristics), the conditions of action, including triggering events, system states, interactions with other systems (interfaces) and operational environment (under what operating conditions), compliance (with standards and regulations), and physical characteristics of the SOI.

6.2.1.1 Organizing the Sets of Design Input Requirements Organizations need to define a specific approach to organize and manage their design input requirements that are best suited to their processes, toolset, culture, and product line—especially concerning how they will manage requirements within design and construction standards to which they must comply. This approach should be addressed within their organizational needs and requirements development and management process documentation, work instructions, and the project's System Engineering Management Plan (SEMP).

It is useful to include a template or outline showing the specific method of organization expected. This template/outline serves as a checklist to help ensure all the various categories of design input requirements have been addressed. A key advantage of using a template/checklist is to help ensure the set of design input requirements has the GtWR characteristic *C10—Complete.*

When organizing the sets of design input requirements, it is useful to have more granularity than just "functional" and "non-functional" requirement categories. A major issue with this grouping is that "performance" is considered "non-functional" by some people, resulting in functional requirements being written without performance included within the requirement expression or set of requirements. However, stating a functional requirement without some performance expectation is ambiguous, and it may not be possible to verify the system meets that requirement, i.e., the requirement does not have the *GtWR* characteristics *C4—Complete, C7—Verifiable, and C8—Correct. See Rule R1 in the GtWR for a more detailed discussion on the structure and form of requirement statements.*

To help with the development and organization of the design input requirements, it is useful to organize them using the same categories as was discussed in Section 4.6.2.3, and is shown in Table 6.1, for grouping the integrated set of needs and resulting design input requirements: function/performance, operational (fit), physical characteristics (form), quality, and compliance. Each of the categories represents a distinct perspective and source of the design input requirements. Failing to consider each perspective could result in missing requirements.

One method of recording which category a requirement is being grouped into is the use of attribute *A40—Type/category*. Using this attribute will enable the export of the requirements into a requirement document organized by the category a requirement is in.

For each of these requirement categories, the project team should define a standard form or template by which the requirement statement is written. This allows consistency in how each type is communicated as well as helping ensure the requirement has the GtWR characteristic *C9—Conforming*. For more information about templates and patterns, refer to the GtWR Rule R1 and Appendix C.

Note: The above-suggested approach to organizing requirements using these categories is just one of many approaches. No matter the approach used, the main consideration is to ensure completeness of the resulting set of design input requirements.

6.2.1.2 Considerations for Each Category of Requirement

6.2.1.2.1 Functional/Performance Functional/performance requirements focus on the primary functions associated with the intended use of the SOI. This would include functional/performance requirements dealing with a core function or capability of the SOI. These requirements would include those that were transformed from a need for functions that have the attribute "Critical" defined. Secondary or enabling functions can be included within the Operational (fit) category.

Functional requirements and their associated performance characteristics are unique and are the heart of what actions and capabilities the stakeholders need the SOI to provide. Each function will have one or more functional/performance requirements that address performance characteristics (such as *how well*, *how many*, or *how fast*) a function needs to perform.

In many cases, there are multiple performance characteristics for a function. In a model, the function would appear as a single entity with multiple performance characteristics defined separately along with any conditional statements, triggers, or constraints. Each performance characteristic would be linked to its own verification attributes, as shown in Figure 6.3. These attributes include the verification *Success Criteria, Method, Strategy*, and *Responsible Organization.* In some cases, for a given performance characteristic, there may be more than one target performance criterion defined as part of the *Success Criteria* that must be met as part of the verification activities.

For cases where a function has multiple performance characteristics, some would say that to address the characteristic *C5—Singular* and rule *R18—Single Thought*, there should be multiple requirements each addressing a single performance characteristic and target performance criterion, resulting in a "family" of functional/performance requirements, each stating the functional name, expected performance value, and condition or trigger associated with the execution of that function.

"The [SOI] shall [perform function A] [with a performance of xxxx] [when/if some trigger or condition]."
"The [SOI] shall [perform function A] [with a performance of yyyy] [when/if some trigger or condition]."
"The [SOI] shall [perform function A] [with a performance of zzzz] [when/if some trigger or condition]."

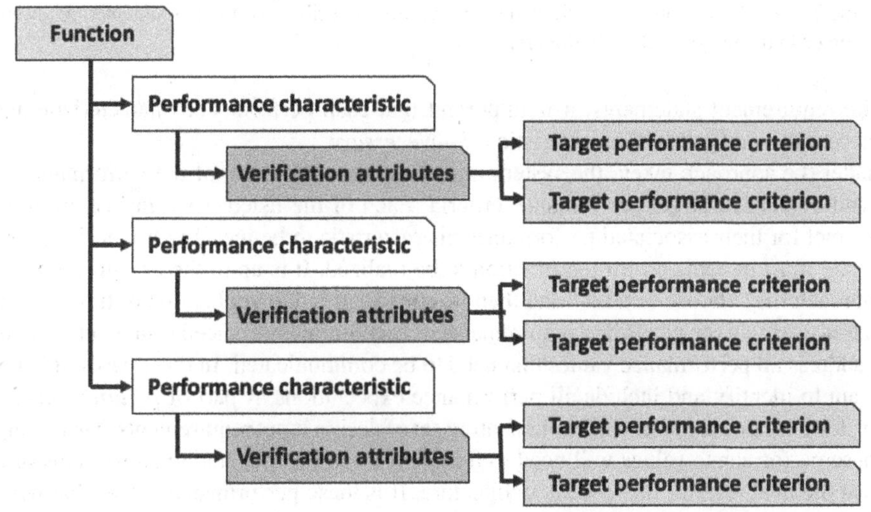

FIGURE 6.3 Function with Multiple Performance Characteristics.

The "functional/performance requirement family" will have similar attributes and system verification activities and system verification procedures that will address the family of requirements. *Refer to Section* 6.2.4 *and Chapter 15 for more information on attributes.*

An advantage to using this approach is that it is easier to allocate the individual one-thought requirements, trace a resulting child requirement to its allocated parent, and plan verification activities.

However, others have a valid concern associated with the characteristic *C13—Comprehensible.* Their point of view is that having multiple requirements for a single function can be confusing, making the set of requirements less comprehensible. In addition, having individual requirements results in a larger number of requirements in the set.

An alternate approach to addressing cases where there are multiple performance characteristics for a function is to list the performance characteristics in a table, as attributes, or in another form that is referred to by the functional/performance requirement.

"The [SOI] shall [perform function A] with performance characteristics shown in Table yyyy [when/if some trigger or condition]."

Given that the table would contain multiple parameters defining the required performance characteristics, each must be the focus of its own verification activity. Consequently, using this approach, allocation to the next level of the architecture and the planning for verification will be more complex.

When there is more than one target performance criterion for a performance characteristic, the target performance criteria could be communicated in a *target performance criteria attribute* that contains each of the target performance criteria for a given performance characteristic. Using this approach, the verification activity associated with that performance characteristic would address each of the target performance criteria, as shown in Figure 6.3, as part of meeting the *Success Criteria* defined for that performance characteristic.

"The [SOI] shall [perform function A] [with a performance of zzzz] that meets the Target_Performance_Criteria [when/if some trigger or condition]."

Where the Target_Performance_Criteria is defined within an attribute of the performance requirement expression or in the project's data dictionary.

As with requirement statements, it is important that each performance characteristic and each performance criterion have the characteristic *C1—Necessary*.

No matter the approach taken, the system must be verified for each of the performance characteristics and associated target performance criteria. Each of the listed performance criteria would have to be met for their associated performance characteristic to be met. In turn, each performance characteristic must be realized for the function to be realized. It is up to the organization to decide which approach they choose to take—and then be consistent in the application of this approach.

When transforming the needs into requirements, it is common that a need statement for a function does not address all performance values that need to be communicated. In these cases, it is up to the project team to identify and include all performance expectations as part of *requirements analysis* when transforming the need into the implementing set of design input requirements. This is important in that these performance values will need to flow down and be budgeted to subsystems or system elements at the next level of the system architecture. It is these performance values that will be the focus of the design and system verification activities.

Functions stated in a need statement may be at the proper level of abstraction for a need statement, but that level of abstraction may not be appropriate for a design input requirement.

For example, "The stakeholders need the SOI to monitor the physical environment at the operator's workstation." Looking at the rationale for the need statement, the stakeholder's intent is to provide a "comfortable" environment for the operator to work in. In this case, the project team will need to transform this need statement into a set of design input requirements that meet this intent. Given that a comfortable environment is a function of temperature, humidity, and airflow, the system will need to address all three.

In the example mentioned above, a problem with the word "monitor" is identifying a characteristic that is observable and therefore "verifiable." For "monitor" to be a valid function or action verb in a requirement, it must be clearly defined in terms such that the requirement is verifiable. One way of doing this could be to decompose the function "monitor" into a set of subfunctions: "measure" or "receive," "store," "display," and "control" for each of the environmental factors that are to be monitored. For each of these subfunctions, specific performance values will be defined concerning how frequently the data are sampled and displayed, how much history is expected to be maintained, how the data will be displayed and in what form, the frequency of updating the displayed data, and specific range within which the measured parameters are to be kept that will result in a "comfortable" environment. Some of these subfunctions likely involve an interaction with another system and will therefore be communicated as an interface requirement.

When defining the performance values, it is important that the project team involve the stakeholders to ensure these values will meet their needs. In the example mentioned above, the function "control" would need to be further decomposed such that each parameter is compared to the range it needs to be within and issuing commands to the subsystems or system element responsible for controlling each parameter to increase or decrease the value until it is within the proper range. The requirements resulting from the decomposition could be communicated at the level of the SOI or at a lower level of architecture.

As can be seen in this example, through a detailed analysis, the result will be multiple requirements to meet the intent of the single need statement.

It is important to understand that functional requirements include interface requirements. Each function has inputs and outputs, as discussed in Section 4.5.3. Because of this, functional/performance requirements include interface requirements when the function involves an interaction with an external system. In these instances, the performance part of the requirement relates to a specific interaction across the interface boundary as defined in some sort of interface definition type document such as an ICD, IDD, or data dictionary.

In the example mentioned above, the functions "receive," "display," and "control" involve interactions with external systems across an interface boundary and thus must include interface requirements with a form of:

> "The [SOI] shall [interact in some way] [with external system xyz] [in accordance with the definition documented in ICD xxxxx] [when/if some trigger or condition]."
> (Depending on the intent, a performance value may have to also be included in the requirement statement. For example, display the received parameter within some time value.)

Refer to Section 6.2.3 *for a more detailed discussion concerning defining interactions across interface boundaries and defining "interface requirements" that address those interactions.*

Some design input requirements can be classified into more than one requirement category. For example, a functional requirement may address an operational consideration, as discussed next. The important point is that the requirement has been defined; how the project team classifies it is up to the organizational guidelines, and the project team applies the guidance consistently for placement of that type of requirement.

6.2.1.2.2 Operational (fit) Functions that deal with a secondary concern during operations that are needed for the system to accomplish its primary functions. This would include functional/performance requirements dealing with enabling functions (functions that enable the primary functions to be performed). For example, a function that involves installation or maintenance, but does not have anything to do with the primary purpose of the system, would be documented as a Operational (fit) requirement. Functions involving power, cooling, heating, or other resources to enable the primary functions could also be included in this category.

Operational (fit) requirements deal with the ability of the system to operate within its operational environment by its intended users. Operational (fit) includes human system interactions as well as interactions with other external entities, and thus will be interface requirements enabling the SOI to become an integral part of the macrosystem it is a part.

Operational (fit) requirements are not limited to just operations and the operational environment, they can also relate to other lifecycle stages such as transportation, storage, facilities, personnel interactions, maintenance, and logistics.

The sources of Operational (fit) requirements should be able to be traced back through the set of needs to information that resulted from both stakeholder elicitation activities as well as the external interface, context, and diagrams and models developed during lifecycle concepts analysis and maturation activities.

Each of the areas mentioned below should have been addressed during the lifecycle concepts analysis and maturation activities and reflected within the resulting integrated set of needs. Many of the areas mentioned below involve interactions (interfaces) between the SOI and external systems. *Refer to Section* 6.2.3 *for a more detailed discussion concerning defining interactions across interface boundaries and defining "interface requirements" that address those interactions.*

Areas to consider include:

- *Transportation, handling, and installation.* Given that the SOI may need to be moved, transported, and installed, what this will involve? Are specific features needed to allow the SOI to be moved around within a facility or transported between facilities? Often the size and weight of an SOI are predicated on the mode of transportation. Are there any special interfaces needed for handling, transportation, or installation that are not needed during normal operations? Are special packaging or containers needed during transportation to protect the SOI from the environment associated with handling, transportation, or installation?

- *Facilities*. What are the facility requirements that enable the system to operate and perform as needed? Examples include space, power, lighting, and environmental control. Facilities need to be considered for every lifecycle phase including manufacturing, testing, and storage as well as operations. Each may have unique handling, lifting, support/test equipment, power, space, and operational environment requirements.
- *Training and personnel*. What are the training and personnel considerations that reflect on how the product is designed and built? The military does this well ... it is not useful for a three-year enlistee to have to undergo two years of training to operate or maintain a system. This area also involves determination of anthropometric constraints based on an analysis of the typical users (personas as discussed in Chapter 4), for example, vision, grip, reach, mass, language of labels and displays, training, expertise, or certifications. Requirements dealing with training and personnel are challenging to write but are needed to reduce operational costs. Because of the importance and difficulty of addressing these types of requirements, there are specialty engineering subject matter experts who focus on HSE and HSI-type requirements. These types of requirements should focus on what SOI characteristics need to be communicated as requirements ("The SOI shall....") such that the personnel needs are addressed, and not requirements on users, operators, or maintainers ("The users shall ... ").
- *Maintenance*. How will the system be maintained and updated? What access is needed, what information needs to be generated by the system to help with diagnostics and troubleshooting, and what type of tools or test equipment will be used? Are there any unique interfaces that may be needed to support maintenance, i.e., a test port to connect diagnostic equipment not used during normal operations?
- *Environments*. Address operational, non-operational, transportation, and storage environments. Address both induced environmental conditions as well as natural environment. Also, include requirements related to what the SOI is allowed to introduce into its operational environment, such as vibration, noise, heat, or EMI/EMC. During movement, transportation, or installation, the SOI may be subjected to environments (loads, temperature, humidity, particulates, or electromagnetic emissions) that are quite different from those to which it will be exposed during normal operations or storage. Does the SOI need to operate during movement or transportation? Does the SOI need to meet all its requirements (survive) after being subjected to the transportation environment?
- *Logistics*. To meet its requirements over its operational life, what supplies and consumables are needed over the operational lifetime for operations and maintenance? In what quantities?
- *Security*. Consult with security SMEs early in the project to understand which security-related requirements should be included. For many software-intensive systems, security is a major concern, especially at the interfaces, to stop non-intended users from preventing the system from being used as intended or using the system in an unintended way. There are many standards and regulations dealing with system security. *Note: Security requirements may involve critical functionality of the SOI or deal with the way the SOI interacts with external entities across interface boundaries. As such, they could be recorded within the function/performance or compliance categories.*

- *Safety*. Consult with safety SMEs early in the project to better understand which safety-related requirements should be included. There are many standards and regulations dealing with safety. Safety can address both safety with respect to the users and operators as well as the operational environment and systems in which the SOI interacts. Many of the HSI and HSE requirements will be related to safety. *Note: Safety requirements may involve critical functionality of the SOI or deal with the way the SOI interacts with external entities across interface boundaries. As such, they could be recorded within the function/performance or compliance categories.*

The areas mentioned above need to be defined and included in the project's checklist/template for design input requirements to ensure that all these areas are addressed as applicable to the SOI being developed. Each of these areas often results in unique requirements and involves interactions across interface boundaries between the SOI and external systems. Missing or incorrect "Fit/Operational" requirements often cause problems, not only during operations but also with system integration, system verification, and system validation, resulting in budget overruns and schedule slips. Not addressing this class of requirements often results in an inability if the SOI to be able to meet the primary functional/performance requirements.

6.2.1.2.3 Physical Characteristics (Form) Requirements that address the shape, size, dimensions, mass (weight), and other observable characteristics that uniquely define the SOI. For software, "form" requirements may address the number of lines of code, memory requirements, or programming language as examples.

Physical characteristics requirements address SOI characteristics associated with the physical system that are not always included within functional, architectural, or analytical/behavioral models. Requirements addressing physical characteristics usually are constraints that are driven by the macrosystem the SOI is a part of. Physical characteristics requirements may originate directly from the stakeholder elicitation activities or from the lifecycle concepts analysis and maturation activities as well as the architectural definition activities.

The project team needs to ensure that each of the physical characteristic requirements is needed and that there is a good rationale for including them in the set. It is easy to cross the line into implementation; stay at the "what" level, not "how." *Do not over-constrain the SOI.* It is especially important that traceability is established to the source of all physical characteristics related requirements to ensure they are indeed "necessary."

Physical characteristics requirements will also be allocated/budgeted to the next lower level of the architecture for implementation. *Refer to Section* 6.4 *for a detailed discussion on moving between levels, allocation, and budgeting.*

6.2.1.2.4 Quality Quality requirements address "fitness for use" and are often referred to as the "-ilities" (such as reliability, testability, availability, maintainability, operability, supportability, manufacturability, and interoperability, to name a few).

Some considerations when writing quality requirements include:

- Quality requirements associated with the physical architecture are not always identified from the functional architecture or analytical/behavioral models.
- In most cases, quality requirements tend to apply to the system as a whole and will need to flow down via allocation and budgeting to lower-level subsystems and system elements.
- Quality requirements tend to be difficult to define and implement. Because of this, they are often defined, managed, and implemented under the heading "specialty engineering" by SMEs with the proper training and experience.

- Quality requirements are difficult to verify that the system meets them, often involving statistical analysis best performed by specialty engineers with the proper knowledge, training, and experience.
- Quality requirements can be cost and schedule drivers and classified as "Key Driving Requirements." *Refer to Chapter 15 concerning needs and requirements attributes.*
- System verification and validation against quality requirements presents a challenge that must be addressed early in the project in that some may involve extensive testing over a period of time to gain the confidence needed for acceptance. For example, the system's ability to meet lifetime requirements will be verified by a combination of test and analysis to determine if the lifetime requirements can be met. *Refer to Chapter 10 for a detailed discussion on defining verification success criteria, verification strategy, and verification methods.*
- Key expectations of the stakeholders include quality; however, these expectations are often not stated explicitly, if at all. Because of this, it is up to the project team to collaborate with the stakeholders during the elicitation activities to ensure the stakeholder expectations for quality are explicitly stated and addressed during lifecycle concepts analysis and maturation activities and communicated within the integrated set of needs.
- There are many "-ilities" that can be defined, below are some examples. (*The reader can search the internet for quality or non-functional requirements to get a listing of the most frequently used "-ilities" along with definitions and example requirements.*) *Note that not all "-ilities" apply to all subsystems, system elements, or systems.*
  - o *Maintainability.* Address the ability of a system to be maintained (to keep in a desired state). Maintainability can be expressed in terms of maintenance times, maintenance frequency factors, Mean Time to Repair (MTTR), maintenance labor hours, maintenance cost, and tools needed to repair. Specific attributes of software that relate to ease of maintenance of the software itself, mode for troubleshooting, diagnostics, and the ability to update the software/firmware as examples.
  - o *Operability.* Address the expectations of everyday operations (such as automatic start or the number of actions needed from an operator/user to perform a particular task).
  - o *Availability.* What are the expectations for the percentage of time the product is available for use during the defined time for operations? 24 × 7, xx% of time during peak periods? Availability is dependent on maintainability, reliability, and periodic maintenance procedures.
  - o *Supportability.* What features and characteristics need to be defined to keep the SOI operational?
  - o *Manufacturability.* Address requirements/constraints for producing the SOI (such as need to retool, design for assembly, or design for manufacture).
  - o *Reliability.* Establish the ability needed to achieve lifetime, operational lifetime, storage lifetime, allowable failure rate, Mean Time to Failure (MTTF), Mean Time Between Failures (MTBF), etc.
  - o *Interoperability.* Address the ability of the SOI to interact with other elements (such as software and hardware on different machines from different vendors to share parts and data). The ability of parts from one system to be used "as is" in another similar system, operational environment, and purpose.
  - o *Resiliency.* Address the ability of the SOI to withstand or recover quickly from off-nominal conditions or events. What features or capabilities should the SOI have to make it more resilient?
  - o *Sustainability.* What is the ability to share, reuse, repair, and recycle parts and materials within the SOI as much as possible, expanding the life of products, minimizing waste and pollution, and creating a closed-loop system? Sustainability includes considerations

concerning the environmental and social aspects as the key elements in product design to reduce the harmful impacts of the product throughout its lifecycle. *Refer to the INCOSE SE HB concerning product sustainability and disposal.*

Missing or incorrect "quality" requirements are often the cause of failed system validation, product recalls, warranty work, and negative reviews on social media. Each organization needs to define the types of quality requirements they will include within the set of design input requirements based on the domain and expectations of their customers, needs of other stakeholders, and ensure they are included in the checklist/template to ensure that all are addressed when developing the design input requirements. This determination should be done within the strategic and operational levels of the organization and reflected within the business requirements and stakeholder needs that are allocated to the SOI.

Modern RMTs allow an organization to develop and maintain a library of -ility requirements that apply to their domain and product line. This allows projects to pull applicable requirements from this library, rather than each project from "reinventing the wheel" for these types of requirements. This approach also helps ensure there will be no missing quality requirements helping to achieve the characteristic of a set of design input requirements *C10—Complete.*

6.2.1.2.5 Compliance Compliance requirements include the applicable design and construction standards and regulations the project team must address with the design, build/code, operation, and disposal of the SOI. For many systems, relevant standards and regulations can represent a substantial portion of the design input requirements. Compliance with standards and regulations are drivers and constraints that come directly from the stakeholder elicitation activities, as discussed in Chapter 4. All relevant standards and regulations should have been identified and invoked within the integrated set of needs. *Refer to Sections 4.3.6.1, 4.3.7.1, 4.5.11, and 4.6.2 for additional information concerning compliance with standards and regulations.*

When these needs are transformed into design input system requirements, the project team responsible for the transformation will do an engineering analysis to determine which specific requirements apply to the SOI under development (versus project or process requirements). They will also determine whether the requirements are design inputs for the SOI, requirements on the design team and resulting design output specifications, or requirements on production. For those that apply to the SOI, they will determine specifically what the SOI must do to meet the intent of the applicable requirements within the standard or regulation.

Based on that analysis, the project team will then derive individual well-formed design input requirements that, when implemented by the design, will result in the intent of the parent requirements within the standard or regulation to be met. Using a data-centric approach, each of these requirements will be traced to the parent requirement within the standard or regulation; this is important as this traceability will be used to demonstrate compliance.

While this approach could result in a larger number of requirements, each would be well-formed making it clear what is expected to be addressed by the SOI design. Because these well-formed requirements will include the verification attributes, the project team will be able to better plan for verification resulting in more accurate budgets and schedules.

Example Transformation of an LIR Need That Invokes Requirements Within an FDA Standard into Well-Formed Design Input Requirements

N044: The stakeholders need the LIR to comply with requirements in [TBD] FDA Standard xyz concerning the handling and prevention of contamination of food.

The LIR project team read the applicable FDA standard concerning handling of food and prevention of contamination of food. In the regulation, there were requirements concerning the properties of the hardware coming into contact with the jars containing food, the ability to sanitize those parts, and requirements concerning the use of lubricants. Based on their analysis, they derived the following design input requirements that they feel meet the intent of the regulation. They also traced these requirements to the need statement that invoked the regulation as well as to the applicable requirement within the regulation.

R063: The LIR shall have Parts that come in contact with the Lids and Jars that are non-porous, meeting the requirements defined in FDA Standard xyz, Section 4.3.2, Part a.

R064: The LIR shall have Parts that come in contact with the Lids and Jars that can meet their design input requirements after being sanitized using sanitizing liquids or materials approved in FDA Standard xyz, Section 4.3.3, Part b.

R065: The LIR shall meet the LIR design input requirements when lubricants that meet non-contamination lubricant FDA Standard xyz, Section 4.3.4, Parts a and b are used for lubrication of the parts that come in contact with the lids and jars.

The project team should collaborate with the *Approval Authority* to define the system verification and validation attributes for each requirement such that verification that the SOI meets those requirements would provide objective evidence that the intent of the source requirements within the standard or regulation was met and that the SOI is in compliance with the source requirements.

Note: In many cases, the resulting design input requirements may be placed in one of the other categories. For example, a safety or security standard invoked within the set of needs. To meet the intent of the requirements within the standard, the derived requirements may involve certain capabilities, functionality, and performance. In these cases, the requirements would be placed in the most appropriate category consistent with the project's categorization of requirements.

Below are some general considerations, best practices, and guidance to use for defining requirements dealing with standards and regulations during the transformation activities:

- In most cases, the need statements will call out standards and regulations by document number. As part of the transformation activities, the project team will have to determine the specific version of each standard and regulation and its date.
- Requirements within standards and regulations are written "generically" at a level of abstraction that is applicable to a class of products, but not necessarily the specific products being developed by the organization or a specific project. As a result, the requirement statements often contain wording that seems ambiguous or not appropriate (in terms of system verification) for the level the project is recording the requirements. Because of this, the project team should not copy and paste individual requirements from applicable standards and regulations into their requirement set verbatim, they should apply the only applicable requirements and transform them into well-formed and appropriate requirements for the SOI using the guidance from the GtWR.
- The requirements in some standards and regulations do not have the characteristics of well-formed requirements as defined in the GtWR. As a result, the requirement statements often contain wording that seems ambiguous or not appropriate (in terms of system verification) for the level the project is recording the requirements. Again, the project team should not

copy and paste individual requirements from applicable standards and regulations into their requirement set verbatim, they should transform them into well-formed and appropriate requirements for the SOI using the guidance from the GtWR.

- There are cases where standards contain specific design implementation that addresses a particular concern, without communicating what the actual concern is or the reason for that specific implementation. In these cases, the design input requirement is a design constraint that is an exception to the "avoid design implementation" within the design input requirements.
  - Alternately, requirements that are design solutions can be allocated directly to the design team as design controls rather than being included within the set of design input requirements. Ideally, the "why" for these design implementations would be communicated with rationale statements for these types of requirements.
  - An important concern for these types of requirements is that the real issue (design input) they are addressing is not stated. As a result, the intent may not be addressed. It may be that that solution may have been valid at the time the standard or regulation was written, but now in the present, there are other more effective solutions that meet the intent. To allow more innovation, the project team must address this issue and work with the customer and regulatory agencies to address this concern (and perhaps negotiating a tailoring to the standard by showing they can address the concern with a different solution).
- In some cases, requirements in one standard or regulation may be inconsistent or contradict a requirement in another similar standard or regulation. The project team will need to work with their compliance office or the *Approval Authority* to determine which one is applicable or takes precedence.

6.2.1.3 Guidelines When Formulating Design Input Requirements As the integrated set of needs are transformed into design-input requirements, there are recommended approaches to ensure the resulting requirements are well-formed, understandable, able to be implemented, and provide clear criteria to enable system verification. Below are several guidelines to consider when defining design input requirements.

6.2.1.3.1 Design Input Requirements Versus Needs To help distinguish design input requirements from needs, design input requirement statements include the word "shall" and need statements do not. "The stakeholders need the system to … " or for a goal "The stakeholders would like the system to … ". versus design input requirement, "The SOI "shall" … " or for a goal, "The SOI "should" … "

Using these distinct formats helps make a clear distinction between design input requirements and needs. The GtWR defines the structure and characteristics of well-formed design input requirement statements and sets of design input requirements, as well as rules that help to write design input requirement statements that have those characteristics.

6.2.1.3.2 Including Goals Within the Requirements Set As discussed earlier, when defining sets of needs and requirements, the concept of including goals can be confusing. Because goals by their nature are not requirements, are not binding, and are not subject to design and system verification, why include them?

Historically, goals have been included in sets of requirements in various forms communicating the customer or stakeholder desire for some feature or capability that, while not critical, adds some value, is low risk, and does not impact critical functions nor drive cost and impact schedule.

By developing an integrated set of needs as advocated in this Manual, the communication of goals should be addressed and resolved when defining the needs and agreed to when baselining the needs as part of scope definition. With this approach, as part of the transformation of the needs into requirements, for each of the goals "The stakeholders would like the system to …," those doing the transformation and resulting engineering analysis would decide which goals will be addressed in the set of design input requirements and expressed as requirement "shall" statements. These requirements would be, by definition, low priority, not critical, and low risk and would trace back to the goal within the integrated set of needs they were transformed from.

However, a case can still be made for including the concept of goals within the set of design input requirements, in various forms. Some stakeholders, for example, marketing, may have a valid reason for including goals (should statements) within the set of requirements, as long as addressing the goal does not increase the cost of developing the product, nor increase time to market. A customer may also have a valid reason, especially when their knowledge of what is possible is limited. In this case, they may define requirement in the form or a threshold value and a goal value: "The system shall perform [some function] with a minimum performance of xxxx with a goal performance of yyyy." In this case, bidders would bid either xxxx, yyyy, or some value in between based on their capability to deliver that level of performance; whatever they bid becomes the binding requirement against which the system will be verified.

In either case, the intent must be clearly addressed in the rationale. In addition, where there is a customer/supplier relationship, how the supplier is expected to respond to the goals must be addressed in the contract.

6.2.1.3.3 Tolerances *Tolerance* refers to a permissible deviation from a specified value. Tolerances are usually used when specifying the physical characteristics or performance of the SOI. There are several considerations when defining tolerances for values within a requirement statement.

First, if the tolerance is too "tight" the cost to design and manufacture a system that meets that tolerance will be more expensive, as will the cost to verify the system meets that tolerance. It is important to consider design, manufacturing, and system verification when stating tolerances in a requirement statement.

Second, if the tolerance is too "loose" the possible variability during manufacturing may result in an issue with "tolerance stack up." This is when the combination of tolerances may result in an SOI that does not work. For example, there may be a performance requirement for the overall SOI that has been allocated to several subsystems and system elements at the next level of the architecture. At the SOI level, the performance requirement has a specified tolerance. Each of the child requirements at the next level will also have to have a tolerance specified, and each child requirement may be further allocated to lower-level subsystems and system elements, each deriving its own requirements with a tolerance. If these parts and components within the system elements are designed and manufactured independently, there is a high likelihood that when integrated together, the combination of tolerances could result in the parent system requirement failing to perform within its specified tolerance. For mechanical parts, the result may be interfaces that do not align properly.

Tolerances are a major consideration when defining the design output specifications. Verifying the system mechanical parts are manufactured within these tolerances is addressed during production verification (as discussed in Chapter 9).

6.2.1.3.4 Accuracy and Precision For requirements that deal with taking a measure, the concepts of accuracy and precision (or imprecision) must be considered as a way to express expected performance. While both are similar to tolerance, the concepts of accuracy and precision are different.

Accuracy is a measure of how close an average of measures is to a baseline value expressed as a percent. *Precision* is a function of the distribution of measurements taken, i.e., closeness of agreement between independent, repeated measures obtained from the same sample under specific conditions. Using a target as an example, accuracy is the average of the distance each arrow or bullet hits in relation to the bullseye. Precision is represented by the standard deviation (in units of the test) or coefficient of variation (in units of percent). Again, using the target as an example, precision deals with how "tight" the cluster of arrows or bullets are—are the arrows or bullets all over the target versus they are in a tight grouping? *Imprecision* (random error or random variation) represents a lack of repeatability or reproducibility of the same result; represented by the standard deviation (in units of the test) or coefficient of variation (in units of percent).

When addressing accuracy and precision, there are several important considerations:

- If the accuracy or precision values are too "tight" or "narrow," the cost to design and manufacture a system that meets those performance values will be more expensive, as will the cost to verify the system meets those values. It is important to plan ahead for design, manufacturing, and system verification when stating requirements that deal with accuracy and precision.
- If the expectations for accuracy and precision apply over the life of the system.
- If both accuracy and precision should be stated versus accuracy alone. When stating both, should both be included within a single requirement statement or separately? In reality, both accuracy and precision are quality characteristics of the performance of a function dealing with measurement.

In many cases, precision is as important as accuracy. When looking ahead to verification, both will be addressed during the same verification activity and, as such, it would be acceptable to include both in the same requirement statement as the acceptance of the measurement requirement would be based on both characteristics to be within the stated values.

6.2.1.3.5 System Lifetime and Expected Performance An issue that must be addressed when defining the design input requirements is whether the performance or quality values in a requirement statement apply to expected performance at the beginning of life (BOL), end of life (EOL), or both. Degradation is a certainty in most physical systems.

For the medical instrument example discussed previously, the customers would expect the accuracy and precision requirements to apply to both BOL and EOL. In some cases, a specific amount of degradation may be allowed in the system performance; if so, there should be a requirement stating the amount of allowed degradation over the required operational life of the system. This could result in a system whose performance is better than necessary for BOL to meet the EOL requirement. In other cases, it may be best to address performance for both BOL and EOL.

If there is no degradation requirement and the expectation for performance is EOL, this must be stated clearly so the project team can make allowances during design for natural degradation within the system components so the verification *Success Criteria* can be met and the system will pass system verification and validation when the system is produced. When there are requirements for EOL performance, do these requirements assume preventative maintenance that will result in the EOL requirements to be met? If so, this must be clearly communicated to the design team and reflected in the preventive maintenance plans.

There is also what is referred to as "design" lifetime versus "operational" lifetime. Consumer product warranties are often based on the design life. There are many cases when the actual operational lifetime is greater than the design life (for example, a rover on the surface of Mars or cars that are driven over 200,000 miles); there are also examples for some consumer products where the operational life is remarkably close to the design life (*that is, where a consumer product seems to fail just after the warranty period ends*).

Another consideration is how lifetime performance requirements are allocated or budgeted to parts when multiple parts of the system play a role in overall system performance. If one part does not meet its budgeted value, can the other parts make up the difference, such that the integrated system can be verified to meet its performance requirements across the entire design lifetime?

There can exist issues when lifetime requirements are stated incorrectly (or incompletely). It is common to have a design input requirement for lifetime that states: "The [system] shall have a lifetime of at least 5 years," which raises the following questions:

- If there are 50,000 copies of the system, what does this requirement imply?
- After 5 years, does this mean all 50,000 copies of the system are expected to meet all the design input requirements or some percentage of the copies?
- Does lifetime include time spent during shipping, storage, and operations, or only operations?
- How does routine preventative maintenance impact the lifetime requirement and expected performance?
- Is "years" an appropriate unit of measure? If the system is only operated 8 hours/day, 300 days/year over the 5-year period, then "hours" of operation may be a more appropriate unit of measure for stating operational lifetime.

Lifetime for a space vehicle must be treated differently than for a terrestrial system that can be maintained. For a long-duration space exploration mission, there will be both beginning of mission (BOM) and end of mission (EOM) requirements for key system performance (such as a radio isotope-driven power supply).

If a customer extends the system lifetime need and requirement after the product is released, this is really a change in needs and requirements that could result in a need to requalify the system for the extended lifetime, especially for an EOL/EOM requirement.

6.2.1.3.6 Use of Margins to Address Uncertainties A major problem when defining and managing design input requirements is the failure of systems engineers to appreciate the concept of managing resource margins [40].

During the lifecycle concepts analysis and maturation, definition of the needs, and transforming those needs into requirements, the project team defines values based on the most current information and analysis to compute a "best estimate" (or theoretical projections) of the performance of those values. A key issue is a failure to understand that these initial computations are only approximations based on assumptions that may or may not be correct and may change as the system design matures.

When defining performance values within a requirement, including margins is especially important for systems whose performance is based on new, immature technologies where there is additional risk in being able to meet the stakeholders' needs for critical capabilities. In addition, margins are useful for managing production and operational risks when the knowledge and understanding of a resource to be consumed or supplied is limited.

While feasibility and risk management were mentioned several times in Chapter 4 concerning lifecycle concepts analysis and maturation activities, there will still be uncertainties when defining specific performance values within requirement statements. Margins are a means to help manage these uncertainties across the lifecycle.

Experienced systems engineers have learned that it is common that projected resource usage, like mass and resource consumption (for example, electrical power or fuel), tends to be underestimated and that *the actual amount needed tends to grow as the project moves across the lifecycle*. On the other hand, the ability of a system to produce (supply) resources (such as power, fuel, water, food, or breathable air) is often overestimated, and the production capability and performance of the system *tend to be lower than promised as critical technologies and the architecture and design mature across the lifecycle*. This under- and over-estimation can lead to issues concerning meeting resource supply, meeting the defined needs, and could lead to cost and schedule challenges.

Knowledge concerning utilization and production is directly proportional to the maturity of the technologies being used. The lower the TRL, the larger the uncertainty of successfully integrating the desired technological implementation or system capability. As the design matures, parts are procured and integrated together; the knowledge of the actual mass, consumption, and production numbers will mature.

When writing requirements concerning physical characteristics (like mass or volume) and performance requirements concerning resource consumption/use, a margin should be included with the estimated base value to account for possible growth as the project matures and to account for uncertainties during development. Thus, the performance value in the requirement would be formed by adding these margins to the base value as shown:

Resource demand requirement value = base value + (growth margin + development margin).

The form of the resulting requirement would be:

> The [SOI name] shall limit [resource] [usage/consumption] to less than or equal to [Resource Demand Requirement value] during [applicable operating conditions].

For requirements containing performance values concerning the ability of a system to supply a resource, in addition to a ceiling value or best estimate, a margin for possible shortfalls as well as a development margin to account for uncertainties during development should be included. In this case, the performance value in the requirement would be formed by subtracting these margins from the ceiling value as shown:

Resource production requirement value 1 = Ceiling Value – (Shortfall Margin + Development Margin)

The form of the resulting requirement would be:

> The [SOI name] shall produce a quantity of [resource] of at least [Resource Demand Requirement value (1)] per [time value] during [applicable operating conditions].

This value is computed from the perspective of a system responsible for producing or supplying a given resource.

From the perspective of a customer of that system, they would form the requirement from the perspective of what the customer needs the system to supply. In this case, in addition to the minimum value needed, an operational margin for unexpected use should also be included:

Resource production requirement value 2 = Resource Production Minimum + Operational Margin

The form of the resulting requirement would be similar to the example mentioned above, but the value is computed from a different perspective.

The [SOI name] shall produce a quantity of [resource] of at least [Resource Demand Requirement value (2)] per [time value] during [applicable operating conditions].

Resource production requirement "value 1" is formed from a *capability* perspective, while resource production requirement "value 2" is formed from a *need* perspective ("what can be provided" versus "what is needed"). Both perspectives are valid and must be considered. Hopefully, the "gap" between the two values is positive and value (1) is greater than or equal to value (2) such that the system responsible for supplying the resource is able to meet or exceed the needs of the customer. If not, the customer's need is not being met, which could result in cost overruns and schedule slips. In cases where there is a customer/supplier relationship, there is also the risk of expensive contract changes.

During operations, consumption (or utilization) needs and production efficiencies must consider not only nominal operations but also alternate and off-nominal operations and degradation of performance over time. During production, consumption/utilization needs and production capability are based on assumptions, and these assumptions may not be valid due to a lack of actual operational experience for the specific project or the reliance on systems using technologies with lower levels of technical maturity (low TRLs); for example, this is the first time the system is actually used for this specific purpose in this specific operational environment. There are also the unknown unknowns that are likely to occur.

For example, the project team is rolling out an updated version of a software application. This application turns out to be immensely popular. The number of people wanting to acquire the application overwhelms the servers and network capabilities, and there are either long wait times, slow download speeds, or the servers crash.

As a related example, a new online government system in put online that crashes because of the unexpected number of concurrent users that are trying to set up new accounts and log in, or the external systems this system interfaces with were not designed for that peak volume. The application was designed based on an estimate of the average number of users on any given day. The initial peak of the number of users (demand) at the beginning of the project or surges in demand or periodic decreases in supply were not considered, or at best, underestimated, thus the ability to supply the services during peak periods was exceeded, causing the system to fail.

Operational Capability provides additional resources during surge or peak consumption periods, as well as enabling the system to succeed even when the unexpected happens. Operational capability also provides additional resources when operational conditions or the environment are different from those assumed when the system resource requirements were defined.

Including an operational margin is necessary to have an operational capability that enables the system to meet its intended use even when the unexpected happens during operations, providing additional resources when operational conditions or the operational environment are different from those assumed when the system resource requirements were generated. Operational Capability is directly related to the concept of resilience: "*The ability of the SOI to withstand or recover quickly from off-nominal conditions or events*" [1].

Including margins when defining performance values within requirement statements and effective management of them throughout the project's lifecycle play a critical role in the success of a project. Failing to do so places a project at risk of cost overruns and schedule slips.

Note 1: When including margins, it is important that what they are and how they were used and computed are included within the rationale for each requirement.

Note 2: The size of the margin to be included is a complex issue beyond the scope of this Manual. However, as a rule of thumb used by some organizations, for an average project, a margin of ~30% is often used. For more complex and riskier projects, a greater margin should be used; for less complex and less riskier projects, a smaller margin can be used. Within an organization, lessons learned and best practices for developing similar SOIs should provide guidance as to the size of margins that are best for a given project.

Note 3: As the project progresses across the lifecycle, the degree of uncertainty becomes less. As a result, at key milestone reviews, the size of the margins can be decreased (released).

6.2.1.3.7 Appropriate to Level One of the characteristics of well-formed requirements and sets of requirements stated in the GtWR is the characteristic *C2—Appropriate.* As stated in the INCOSE GtWR, the specific intent and amount of detail of a need or design input requirement statement are appropriate to the level (of abstraction) of the entity to which it refers as well as appropriate to the level of architecture it exists. *Refer to Sections 2.3.2, 4.5.6, and 6.4 for a more detailed discussion on "levels."*

An example of a medical diagnostic system:

- *System need statement.* "The stakeholders need the [xyz] diagnostic system to [measure or detect] [something] with an accuracy better than similar devices currently in the market."

  This is an appropriate level of abstraction for a need statement, clearly stating the stakeholder's expectation concerning accuracy; however, this would not be a good design input requirement.
- *Requirement transformed from the need statement.* "The [xyz] diagnostic system shall [measure or detect] [something] with an accuracy of at least [xxxxx] [unit of measure]."

  The project team has determined the current state-of-art accuracy in the market, explored various concepts for meeting the stakeholder's need for accuracy, examined candidate technologies, assessed their TRL, and decided that the value of at least [xxxxx] is feasible with acceptable risk for this lifecycle stage.

  As stated, this is expressed at an appropriate level of abstraction for a design input requirement. The developers will also define other related design input performance and quality requirements such as precision, total allowable error, false positives, false negatives, and time to determine a result. Each of the requirements will be traced to the need statement from which it was derived.

Requirements must also be appropriate to the level of the architecture they are defined, and the level of the system will be verified to meet that requirement. For example, system-level requirements are written as "The [System name] shall ...." The integrated system will be verified to meet these requirements. As discussed in detail in Section 6.4, these system-level requirements are allocated/budgeted to subsystems and system elements at lower levels of the architecture for implementation. The subsystems and system elements at the next level will elaborate the allocated requirement depending on the role the subsystem or system element has in the allocated parent requirement's implementation. Requirements for a subsystem are written as "The [subsystem name] shall ..." and requirements for a system element are written as "The [system element name] shall ...."

For example, there may be an SOI level requirement that states, "The SOI shall operate at a voltage of 28 VDC ± 4 VDC." This requirement would be allocated to the lower-level subsystems and system elements that use electrical power. One of these would be the Power Subsystem. The Power Subsystem would go through the lifecycle concepts and needs definition activities discussed in Chapter 4, and the resulting set of Power Subsystem needs would be transformed into a set of

Power Subsystem design input requirements of the form "The Power Subsystem shall ... " These requirements are unique to the Power Subsystem. As such, during system integration, the Power Subsystem would be verified to meet all its "Power Subsystem shall ... " requirements. It would not be appropriate to communicate any unique Power Subsystem requirements at the system level, nor would it be appropriate to communicate the system-level power requirement at the Power Subsystem level.

A common "appropriate to level" issue is stating design implementation details (design outputs: "how," build-to/code-to) within the set of design input requirements. Design input requirements are meant to be a "why/what" design-to level of abstraction. The detailed design requirements belong in design outputs: design output specifications, drawings, algorithms, or diagrams. Stating design output level of abstraction requirements as design inputs results in several problems:

- **First is the issue of innovation**. By stating the "how," it is like saying: "I do not care if there is a better way, do it my way." In most cases, it is best to state the "why/what" and leave it up to the designers to determine the most effective "how." Of course, there are exceptions, in that there may be a good reason for stating implementation. In these cases, the implementation requirement is treated as a constraint. To avoid confusion, the rationale for the implementation requirement (constraint) must be clearly stated.
- **Another problem with requirements that state the "how" concerns allocation** (reference Section 6.4). When design details are shown as design inputs, there is often no definition of the "why and what" concerning the intent the detailed design requirement is addressing. In some cases, the reason "why/what" for the implementation reveals that more than one subsystem or system element has a role in meeting the intent. The parent (why/what) needs to be captured at the system level and then allocated to the subsystems and system elements that have a role in meeting that requirement. Failing to do so could result in missing requirements for the subsystems or system elements. *This issue is the main reason for this Manual to use the phrase "design input requirements" rather than just "requirements" or "technical requirements."*

As shown in Figures 6.4 and 6.5, a useful construct is "the line," which separates design inputs from design outputs. In Figure 6.4, the line is shown between the design inputs and the *Design Definition Process* where the "what/design to" requirements appropriate as design inputs are communicated "above the line" and "how/build-to/code-to" requirements are communicated "below the line" in the design output specifications.

In Figure 6.5, the line is applied to a system architecture consisting of subsystems and system elements. Requirements for the system flow down to subsystems and system elements at the next level of the system architecture for further decomposition and elaboration. For system elements that need no further decomposition to be realized by the *Acquisition Process* or *Design Definition Process* (buy, make, or reuse), "the line" shows the boundary between design inputs and design outputs.

The concept of "the line" is useful during design input requirement definition activities to help ensure the requirements are appropriate to level. During the definition of the design input requirements, when someone proposes a requirement that is a level of abstraction associated with design definition (design outputs), project team members can say: "While that is a good design output requirement, it belongs below the line in the design output specifications, not above the line in the set of design input requirements."

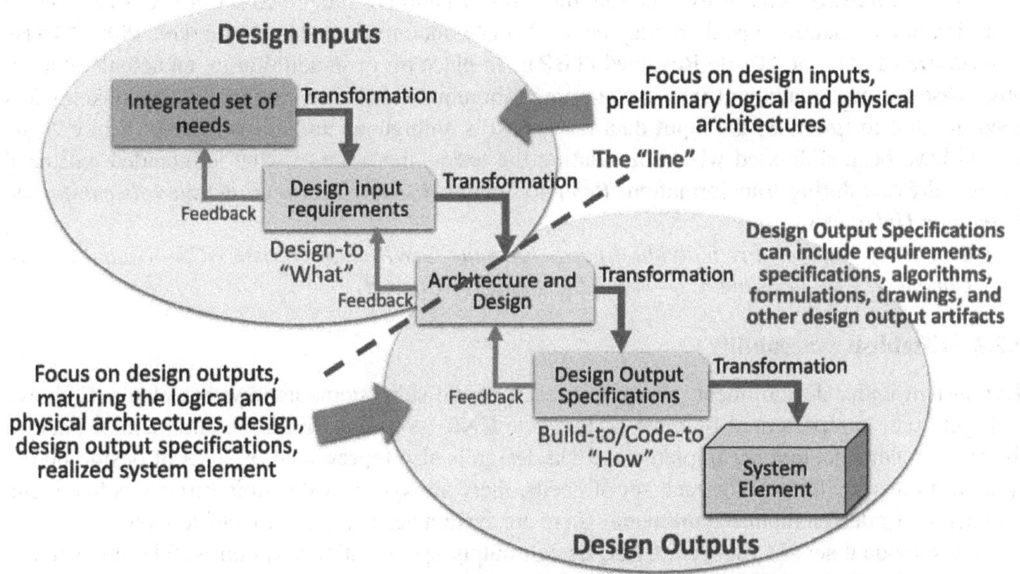
Design inputs
Integrated set of needs
Transformation
Focus on design inputs, preliminary logical and physical architectures
The "line"
Design input requirements
Feedback
Transformation
Design-to "What"
Design Output Specifications can include requirements, specifications, algorithms, formulations, drawings, and other design output artifacts
Architecture and Design
Transformation
Feedback
Focus on design outputs, maturing the logical and physical architectures, design, design output specifications, realized system element
Design Output Specifications
Feedback
Transformation
Build-to/Code-to "How"
System Element
Design Outputs
Original figure created by M. Ryan and L. Wheatcraft. Usage granted per the INCOSE Copyright Restrictions. All other rights reserved.

FIGURE 6.4 The "Line" Between Design Inputs and Design Outputs.

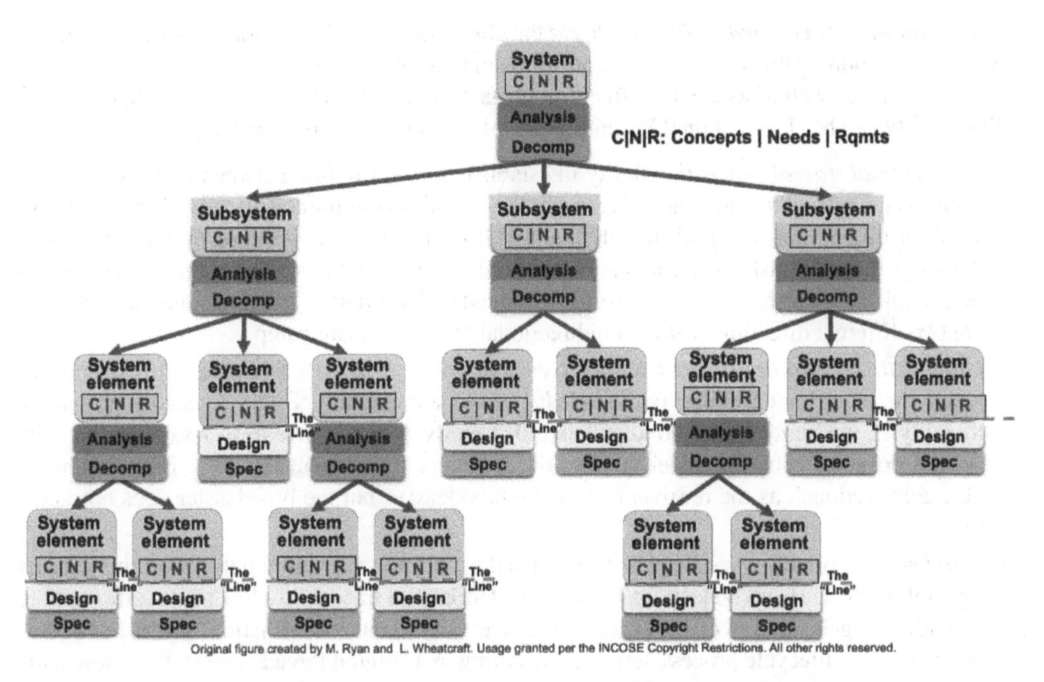
System
C|N|R
Analysis
Decomp
C|N|R: Concepts | Needs | Rqmts
Subsystem
System element
Design
Spec
The "Line"
Original figure created by M. Ryan and L. Wheatcraft. Usage granted per the INCOSE Copyright Restrictions. All other rights reserved.

FIGURE 6.5 The "Line" Within a System Architecture.

6.2.1.3.8 Managing Unknowns During the transformation of integrated set of needs into the set of design input requirements, there may be needs that contain values that are marked with a "To Be Determined (TBD)" or "To Be Resolved (TBR)," in place of, or in addition to, an actual value. In other cases, a need statement may contain an ambiguous value. For example, "The stakeholders need the SOI to [*process*] the input data fast." Fast is ambiguous and not verifiable. While "fast" should have been addressed when formulating the needs, the issue of what is intended will need to be addressed during transformation. *Refer to Section 14.2.4 for a more detailed discussion on Managing Unknowns.*

This activity helps ensure individual needs have the GtWR characteristics C3—Unambiguous, C4—Complete; C6—Feasible, C7—Verifiable, and C8—Correct.

6.2.2 Establish Traceability

The system under development, along with each of the subsystems and system elements shown in Figure 6.5, is represented by a set of lifecycle CNR (concepts, needs, requirements). Each of the system elements that are implemented via design is also represented by a set of design output specifications. In addition, for each set of needs, there are system validation artifacts defined, and for each set of design input requirements, there are system verification artifacts defined.

The individual sets of lifecycle CNR, design output specifications, system validation artifacts, and system verification artifacts do not exist in isolation, rather they represent a multi-level, "spider web" of relationships that represents a data and information model of the integrated system (refer to Chapter 3). These relationships are documented via traceability connections (i.e., links) that allow the relationships to be traced between the entities that are linked both vertically across levels and horizontally across the lifecycle. *This is the fundamental concept of traceability.*

6.2.2.1 Traceability Defined *Traceability* is the ability to establish an association or relationship between two or more entities and to track entities from their origin to the activities and deliverables that satisfy them, as well as assess the effects on artifacts across the lifecycle when change occurs.

Traceability can be "bidirectional," "unidirectional," "horizontal," or "vertical."

- *Bidirectional traceability* is the ability to establish a two-way link between entities such that each has knowledge of the other. This capability enables practitioners to move forward, backward, or up and down digital threads that result from establishing bidirectional traceability. Although the relationship is created in both directions, it is not the same relationship. One direction should be indicated as "traced from" or "traced to." Similarly, "satisfies" and "satisfied by" are two different directions within a bidirectional "satisfy" relationship.
- *Unidirectional traceability* is the ability to establish a one-way trace from one entity to another, where the receiving entity has no knowledge of the source entity. Entity A establishes a trace to entity B, but entity B has no knowledge that entity A has a trace to it. Examples include: A GPS does not know the receivers, but the receivers know about the GPS. Broadcasting is also unidirectional, as the receivers know the broadcaster, but the broadcaster does not know its receivers.
- *Horizontal traceability* involves the forward and backward traceability between entities across the SOI lifecycle (from concept to retirement). Horizontal traceability links data, information, and artifacts generated in one lifecycle process activity to data, information, and artifacts generated in other lifecycle process activities, resulting in a "digital thread" connecting these data, information, and artifacts across the lifecycle.
- *Vertical traceability* is most often referred to in the context of levels of organization and architectural levels of the system or product under development. Each level has lifecycle CNR

defined for each entity at that level. Organizational-level business requirements drive the development of the operational-level lifecycle CNR. As the operational lifecycle CNR are defined, bidirectional traceability is established with the higher-level business requirements.

Various levels exist from a hierarchical architecture view of an SOI. Higher-level needs and requirements are allocated to the SOI, and bidirectional vertical traceability is established as these requirements are defined. The SOI system-level requirements are allocated to the lower-level subsystems. Again, bidirectional, horizontal, and vertical traceability are established, and the process repeats for each system element within the hierarchy. *Refer to* Section 6.4 *for more information concerning vertical traceability between levels.*

Traceability helps establish that the set of design input requirements has the GtWR characteristics *C1—Necessary*, *C10—Complete*, and *C11—Consistent*.

From a requirements perspective, traceability enables the following relationships to be established:

- *Parent/child*—shows connection of a higher-level allocated requirement to one or more child requirements that, when realized, will result in the intent of the parent to be met.
- *Requirement to a source*—shows connection to where requirement content was derived (such as a constraint, standard, regulation, MGOs, measures, concept, risk, a model, analysis, or need).
- *Flow down of a requirement to a lower-level entity (allocation)*—allows a link from a higher-level requirement to a lower-level subsystem or system element the parent requirement is allocated or budgeted to. *Note: This is not a requirement-to-requirement link, but a requirement to a lower-level subsystem or system element within the SOI architecture, for which child requirements will be defined and then traced back to the allocated parent requirement.*
- *Peer to peer*—establishes relationships among requirements at the same level, either within the same set or requirements contained in different subsystem or system element sets of requirements. The relationship could be general or grouped by topic (such as connecting requirements of a particular theme, connecting interface requirements, requirements sharing a common budgeted quantity, and connecting performance requirements that relate to a common function).
- *Interface requirement to an interface definition*—provides traceability between an interface requirement and the agreed-to definition concerning the interaction across an interface boundary between the SOI and another entity.
- *Dependency of a requirement on another*—shows a relationship of one requirement to one or more requirements in which there is a dependency—for example, items that must be completed together to be satisfied (can be other requirements, or representation of specific conditions) or a relationship to another requirement such that a change in one will result in the need to change the other. The other requirement could be in a separate set of requirements for a different system, subsystem, or system element.
- *Design input requirement to implementing design artifacts*—shows a relationship between a design input requirement and its design implementation and resulting design output specifications.
- *Requirement to system verification artifacts*—displays relationship of requirements to verification artifacts or activities that will provide evidence of requirement satisfaction.
- *Textual requirement in an RMT to the equivalent requirement within a model*—a trace between a requirement and an entity within a model. For example, a functional requirement would be linked to a function within a functional model; a performance characteristic of a function would be linked to that function within the model.

This is not an exhaustive list but shows the kind of traceability that can be established within a set of requirements, requirements in other sets of requirements, and other artifacts within the SOI's integrated dataset (for example, needs, models, diagrams, design artifacts, system verification, and system validation planning artifacts).

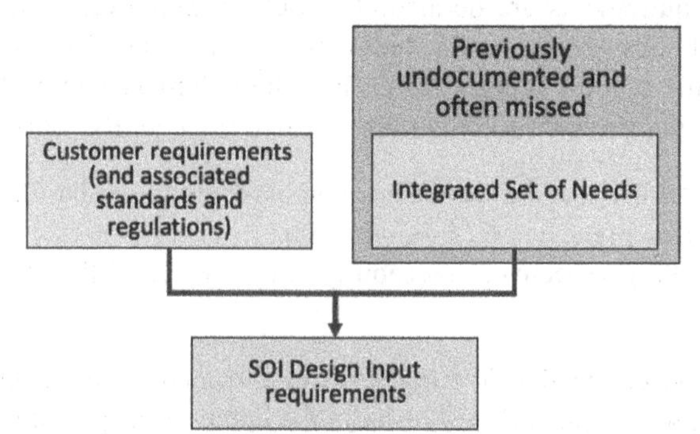

FIGURE 6.6 SOI Requirements "Trace to Source" Includes Parent Requirements and Needs.

Note: While all requirements must trace to a source, not all requirements must trace to an allocated parent. Allocated parent requirements are not the only source of requirements. As shown in Figure 2.9 and the example requirement flow in Figure 6.6, *design input requirements are a result of a transformation from one or more needs in the integrated set of needs (which are transformed from the set of lifecycle concepts as well as multiple other sources as shown in Figure 4.13).*

While all the allocated requirements will have child requirements that trace back to the parent, other requirements are derived based on the *Lifecycle Concept and Needs Definition* activities and, as such, may not trace to an allocated parent requirement. Developing design input requirements solely based on allocated parent requirements will result in an incomplete set of requirements.

Traceability can be supported by use of attributes (such as *A2—Trace to Parent, A3—Trace to Source, A5—Allocation/budgeting, A32—Trace to Interface Definition, A33—Trace to Peer Requirements*, and *A37—Risk [mitigation]*) as defined in Chapter 15.

6.2.2.2 Using Traceability to Show Compliance with Standards and Regulations Many standards and regulations require traceability to be established across the lifecycle of the product/system; examples include ARP4754, "Guidelines for Development of Civil Aircraft and Systems"; ISO13485, "Medical devices—Quality management systems—Requirements for regulatory purposes"; ISO26262, "Road vehicles—functional safety"; and USC Title 21 Part 820, "Quality System Regulation" for medical devices; traceability references in these are outlined below.

- ARP4754A Section 5.3.1.1 requires requirements dealing with safety to be "uniquely identified and traceable" to "ensure visibility of the safety requirements at the software and electronic hardware design level."
- ISO13485 Section 7.3.2 requires organizations to document "methods to ensure traceability of design and development outputs to design and development inputs."

- ISO26262 Section 6.4.3.2 requires "Safety requirements shall be traceable with a reference being made to:
  a) each source of a safety requirement at the next upper hierarchical level;
  b) each derived safety requirement at the next lower hierarchical level, or to its realization in the design; and
  c) the verification specification in accordance with 9.4.2."
- USC Title 21 Part 820 requires developing organizations of medical devices to develop and maintain a Design History File (DHF) that "shall contain or reference the records necessary to demonstrate that the design was developed in accordance with the approved design plan and the requirements of this part." Traceability is a critical part of the DHF.

Failure to show required traceability can lead to a product not being approved for its intended use in the marketplace. For example, when developing medical devices, the US Food and Drug Administration (FDA) requires developers to perform a Use Failure Mode and Effects Analysis (UFMEA) to identify hazards to users of the device. These hazards represent a risk that must be mitigated for the FDA to approve the device for its intended use. For those hazards that will be mitigated via design, the FDA requires traceability from the UFMEA, to needs, to design input requirements, to design, to design output specifications, and finally to the design and system verification artifacts that provide objective evidence that the realized system meets the requirements, as well as to system validation artifacts that provide objective evidence that the realized system meets the needs associated with the device (when used for its intended use in its intended operating environment as specified by its instructions for use). Without this traceability clearly communicated, the FDA will not approve the device for use.

6.2.2.3 Managing Traceability Within a Project's Toolset For complex systems consisting of multiple subsystems and system elements, establishing, maintaining, and managing all the traces effectively can be difficult if done manually using standard office applications. A major reason RMTs were developed was so that traces could be established and managed within a relational database. With the move toward a data-centric approach to SE and increased use of models, applications have been developed to allow not only traceability within and between sets of requirements but also traceability between requirements and all other SE artifacts across all lifecycle stages. Some RMTs enable the user to define a set of traceability rules within the tool; the RMT then can help the user to not only establish traceability per the rules but also alert the user when a rule is not being followed.

Not all projects require the same amount of traceability data, so the specific deliverables that are traceable for the project are determined during the project's needs and requirements management planning. Generally, the more complex or regulated a system, the greater the need for traceability. A project in a heavily regulated industry, or one with numerous components, interfaces, risks, and stakeholders, requires more detailed traceability than a project without those characteristics.

Given the importance of traceability to project success, many SE tools enable practitioners to define a traceability relationship meta-model. This model is established at the beginning of a project, defining the artifacts to be managed within the project's toolset and the traceability and relationships between those artifacts. Many modern tools enable the project to define rules to ensure the model is followed and flag discrepancies when the rules are not followed. An example model is shown in Figure 6.7.

Within the SOI's integrated dataset, the established traceability can be thought of as a "Traceability Record." *Note: As used in this Manual, the traceability record is not a specific thing but is a*

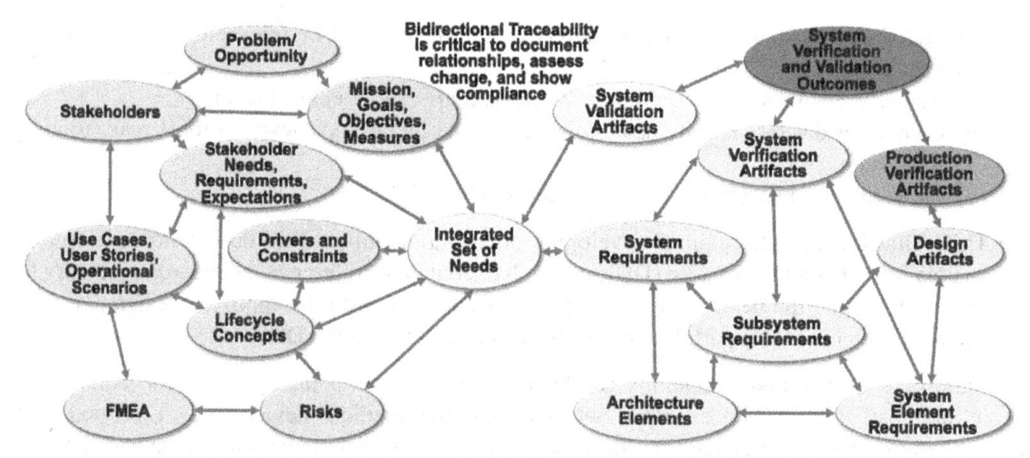

FIGURE 6.7 Traceability Relationship Meta-model.

representation of the traceability links established within the SOI's integrated dataset. See also the discussion on the use of traceability matrices below.

This is a key concept when adopting a data-centric approach to SE and the Information-based Requirement Definition and Management (discussed in Chapter 3). Multiple tools can build links between requirements, needs, and other artifacts, and direction and type of links can be established to describe unique relationships between requirements and other artifacts represented within the SOI's integrated dataset; it is this information that is part of the Traceability Record. *Refer to Chapter 16 concerning recommended features for a project toolset.*

6.2.2.4 Parent/Child Traceability It is a common convention to have child (lower-level requirements) link to parent requirements (higher-level requirements), as these are often created after the parent requirements are generated. An example of a parent/child trace that can be established from linking is shown in the Lid Installation Robot (LIR) example shown in Figure 6.8.

Starting at the bottom, the LIR Power Supply system element requirements are traced to their allocated parent LIR Power Subsystem requirement, which is traced to its allocated parent LIR system-level design input requirement, which is traced to the LIR need from which it was transformed.

These parent/child relationships can be established and maintained within the RMT as a set of links from the lower-level requirements to the higher-level requirements. In a data-centric practice of SE, the need can be traced to its source. Once these links are established, bidirectional traceability is also established.

Traceability is established concurrently with the transformation of needs into design input requirements and continues as lower levels of subsystem and system element needs and requirements are defined. It is important that when a need or design input requirement is written, the associated trace attributes and linkages are captured within the RMT traceability record at that time. Going back later and trying to establish traceability after the fact can be difficult and time-consuming.

Conventionally, it is expected that a parent requirement, need, or source will have one or more subordinate requirements tracing to it. A potential question arises in whether it is appropriate for a single requirement to trace to multiple parent requirements, needs, or sources. Assessing each link independently will help determine the validity of those traces. If a child requirement is necessary to satisfy multiple parent requirements independently, the traces are valid and should continue to be tracked within the RMT.

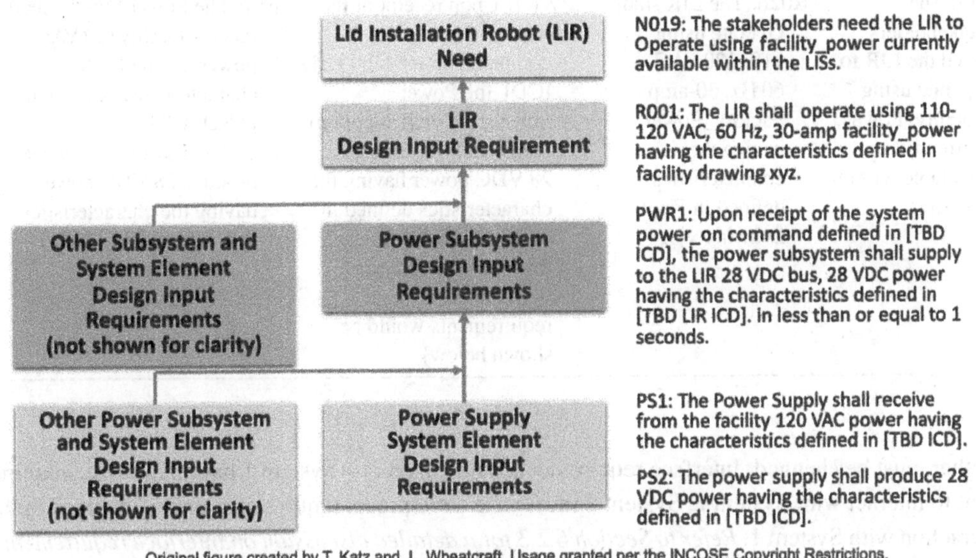

FIGURE 6.8 Example Requirement Parent/Child Trace.

6.2.2.5 Using Trace Matrices Being able to visualize the parent/child relationships that result from traceability can be a challenge. A "Trace Matrix" is a common table visualization of the traceability record showing the above-mentioned relationships in tabular form. The display of a trace matrix supports the analysis needed to show that each need has been transformed into one or more design input requirements, that each requirement is needed by linking it to the need from which it was transformed, a parent, or source, as well as showing the requirement's implementation at the next level of the system architecture by displaying its child requirements for each subsystem or system element it was allocated to. *As discussed earlier and shown in Figure* 6.6, *a subsystem requirement may not always trace to an allocated parent but to a subsystem-level need or source.*

An example Trace Matrix is shown in Table 6.2.

The trace matrix provides a visualization of the traceability record maintained within the project's data and information model of the SOI and is often an artifact that can be generated automatically. The trace matrix is useful to help assess and manage changes, as well as assessing consistency and completeness of the requirement set and artifacts the requirements are linked to. Rather than a trace matrix, some RMTs have the capability to show the linkages between requirements in a diagram or tree structure.

6.2.2.6 Establishing Traceability Among Dependent Peer Requirements Requirements are often related to other requirements at the same level. In fact, in today's increasingly complex, software-intensive systems, dependency between requirements for a system, subsystem, or system element, as well as between subsystems and system elements, has increased exponentially.

There are cases where a requirement cannot be satisfied in a solution without the other requirements being satisfied. There may be dependent requirements such that a change in one could mean

TABLE 6.2 Example LIR Trace Matrix.

System Need	System Requirement	Subsystem Requirement	System Element Requirement
N017: The stakeholders need the LIR to operate using Facility_Power currently available within the LISs.	R026: The LIR shall Operate using 110–120 VAC, 60 Hz, 30-amp Facility_Power having the characteristics defined in Facility Drawing xyz.	PWR1: Upon receipt of the System Power_On command defined in [TBD ICD], the Power Subsystem shall supply to the LIR 28 VDC bus, 28 VDC power having the characteristics defined in [TBD LIR ICD] in less than or equal to 1 sec [Additional subsystem requirements would be shown below]	PS1: The Power Supply shall receive facility120 VAC power having the characteristics defined in [TBD ICD]. PS2: The Power Supply shall produce 28 VDC power having the characteristics defined in [TBD] ICD.

another must be changed. Interface requirements come in pairs—System 1 has an interface requirement to interact with System 2; System 2 must have a reciprocal requirement that is involved in the interaction with System 1. *Refer to Section* 6.2.3 *for a detailed discussion on interface requirements.*

In other cases, a resource may have been allocated (budgeted) to multiple subsystems or system elements at the next level of the architecture. The resulting requirements form a dependent set, such that any changes in a budgeted parameter for one subsystem or system element could require changes in another budgeted parameter for another subsystem or system element to be consistent with the parent budgeted parameter. In these cases, all the budgeted parameters must be managed as dependent variables. *Note: This is a major issue in a document-based practice of SE where these variables are frequently treated as independent variables. Refer to Section* 6.4 *for a detailed discussion on allocation and budgeting.*

Dependent requirements are often contained within different sets of design input requirements across different elements within the system architecture. To properly capture and manage these dependencies, all subsystems and system elements within the system architecture need to be identified, defined, and managed within the SOI's data and information model of the integrated system under development. This is especially important when different internal and external organizational units are developing parts of the architecture. *Note: Again, this is a major issue in a document-based practice of SE where dependent requirements may exist in separate documents often developed at different times and managed by different organizational units.*

Dependency analysis is a technique that is used to discover dependent relationships between requirements and between requirements and other work products. Once analyzed, the set of requirements is recorded by linking dependent requirements together. As part of this, for each dependent requirement, it is important to indicate within its rationale the nature of the dependency. In addition to traceability matrices, some RMTs allow the dependencies to be viewed visually either in tabular form or in a graphic form, for example, a traceability tree.

Dependency analysis can be used to help ensure alignment and consistency between needs, design input requirements, design output specifications and the project's plans, PM work products, and other SE artifacts generated across all system development lifecycle phases.

Dependency analysis can also be used to help ensure bidirectional traceability and consistency between each level lifecycle CNR, architecture elements, design output artifacts, system verification artifacts, and system validation artifacts.

The management of dependencies is another major reason for using a data-centric approach to SE, and the advantage of using RMTs that support traceability across the lifecycle and language-based modeling tools. Being able to manage dependencies is also critical in managing change and assessing the impacts of change throughout all work products and artifacts generated across the system lifecycle.

Dependency analysis helps establish that the set of design input requirements has the GtWR characteristic *C11—Consistent.*

6.2.2.7 Guidelines for Recording and Managing Traceability There are several useful guidelines for recording and managing traceability:

- There is a range of tools available for establishing and managing traceability, from simple spreadsheets and tables to high-end applications that control the requirements and provide full traceability across the lifecycle. Unless there is a small set of requirements, it is highly recommended that the project team uses an RMT or other application that allows the establishment and management of traceability relationships between requirements and their parents, sources, and needs from which they were transformed. The tools used to establish traceability and create and manage traceability records should be identified at the beginning of the project, as traceability can quickly become complex; switching tools mid-project could present major challenges.
- When models are used as an analysis tool to identify needs and design input requirements, the tools in the project toolset need to enable traceability between requirements and the models from which they were derived, even if the data are defined and maintained within different software applications. This is key to being able to maintain consistency and correctness of the needs and requirements no matter which tool is used to view them. This enables the ability to change a need or requirement in one tool and have that change reflected in the other tool, so there is an ASoT. *Refer to Chapter 16 concerning features a project toolset should have.*
- Design input requirements are transformed from the integrated set of needs, the trace between the requirement and its related need should be established when the requirements are initially defined and recorded within the RMT.
- Child requirements created in response to an allocated parent requirement should establish a trace to its parent as it is initially defined and recorded within the RMT. *Refer to Section* 6.4 *for a detailed discussion concerning allocation.*
- All requirements must trace to a need, a source, or a parent. Requirements that do not have this trace, are referred to as "orphan requirements." *Note: in some literature, the concept of "self-derived requirements" is discussed. The justification is that the engineers know the requirement is needed even if it cannot be traced to a parent. This is a bad practice as it represents an independent assumption on the part of the engineer(s). Even if the requirement does not trace to an allocated parent, there should be a source or need from which it was transformed.* There should be no orphans.
- It is a best practice to define the system verification attributes when a requirement is defined. Traceability to the system verification attributes should be established when the requirement is initially defined and recorded within the RMT (note that the verification data may exist in other applications in the digital ecosystem).

- It is also a best practice to trace requirements to the test cases within a verification plan or procedure that executes the system verification activity for that requirement.
- When a child requirement traces to multiple parents, needs, or sources, each trace must be assessed for its validity.
- The trace of all requirements to a parent, source, or need should be verified, independently, if possible, to ensure that the requirements trace is correct during requirement verification (discussed in Chapter 7).

6.2.2.8 Use of Traceability to Manage Requirements The traceability information within the traceability records allows for the monitoring and controlling of needs and design input requirements, including approving and baselining the integrated set of needs and set of design input requirements, managing changes, monitoring needs and requirements status, and communicating results.

As discussed in Chapter 7, traceability is a major tool used during the verification and validation of the sets of needs and sets of design input requirements prior to baseline. Once the baseline is established, the project team members responsible for the needs and requirements definition and management can use traceability to help ensure changes are addressed throughout the lifecycle as defined in the project's change management process (discussed in Chapter 14).

Traceability, along with allocation, can be used to uncover possible issues within a requirement set, which can be a source of change (as discussed in Section 14.2.7).

6.2.2.9 Requirement Relationships Within Models As discussed in Chapters 3 and 4, the use of language-based models is highly encouraged when performing lifecycle concepts analysis and maturation activities and definition of the integrated set of needs. When this approach is used, traceability between the needs and model elements representing those needs must be established as each need statement is defined.

During the transformation of the needs into the design input requirements, the project team elaborates these same models in the identification of the requirements, their allocation, and the definition of child requirements concurrent with the architecture definition as the team moves down to lower levels of the architecture. As this is done, traceability relationships between requirements are established within the tool. As these models mature, the project team can use the models to help validate the requirements [48, 49].

Many of the benefits of using traceability relationships are additionally enhanced when requirements are developed using Model-Based Systems Engineering (MBSE) techniques, such as with SysML [29] or other language-based modeling tools.

When using requirement content in MBSE tools, there are additional ways requirements can show relationships to other requirements or other elements within the model using requirements diagrams. For example, in SysML applications, requirement diagrams are used to display textual requirements, the relationships between requirements, and the relationships between requirements and other model elements. Types of relationships [50] within SysML include:

- *Containment*—showing how a model element is contained within a larger package, such as showing how requirements are contained within a specific theme package. For requirements, this could be used to indicate a requirement is contained within a given set for an entity within the system architecture.
- *Trace*—showing that requirements have a dependency, changes to a connected requirement could result in the need to change a dependent requirement.

- *Derive requirement*—type of dependency showing a requirement was derived from the connected requirement. Derived requirements can correspond to child requirements at the next level of the architecture, indicating a parent/child relationship.
- *Refine*—type of dependency, connecting a requirement to a model element that provides more details, such as a use case, misuse case, loss scenario, user story, or operational concept.
- *Satisfy*—type of dependency, connecting a requirement to a model element that provides fulfillment of the requirement within the design, such as a design output specification.
- *Verify*—type of dependency, connecting a requirement to a model element that provides objective evidence of requirement satisfaction, such as a test case, verification *Activity*, verification *Procedure*, or other verification artifacts such as verification *Method*, *Success Criteria*, or *Strategy*.

Current SysML and other language-based modeling tools provide the ability to make these relationships and display them graphically or in tabular form, such as a traceability matrix. These tools also allow allocations to be shown. This is most often done by showing allocation from structural or behavior elements from one model element to another. Allocations are commonly shown graphically through various types of requirements diagrams, showing the elements associated with the requirements through a "satisfy" relationship.

Note 1: Most requirement relationships in SysML are based on the UML dependency. The arrow points from the dependent model element (client) to the independent model element (supplier). Hence in SysML, the arrow's direction is opposite to that typically used for requirements flows, where the higher-level requirement points to the lower-level requirement [50]. *In some RMTs, the arrow points from the child to the parent. This is because the set of requirements that includes the parent may be under configuration control, preventing real-time changes. In SysML this is often referred to as a "derived from" relationship. Ideally, the parent/child link should be bi-directional.*

Note 2: Currently, SysML does not include an entity "needs" and thus does not support the development of "need diagrams" as it does for "requirement diagrams." However, as part of the meta-model, a "need" stereotype can be created based on the requirement element within SysML. Also, a "validate" stereotype can be created based on the verify relationship. The flexibility of SysML makes it easier to model both needs and requirements. Stereotypes such as these are part of the metamodel. This ensures that everyone in the project team working within the model does so in accordance with the metamodel.

6.2.3 Defining Interactions and Recording Interface Requirements

Because of the criticality of properly addressing interactions across interface boundaries and managing these interactions, this section is included to help the reader better understand the concepts of interface boundaries, identifying and defining interactions across those boundaries, and defining resultant design input requirements associated with each interaction. *Refer to Section 14.2.8 for a discussion on managing interfaces.*

A key characteristic of today's increasingly complex, software-intensive systems is the number of internal interactions within a system, between its subsystems, and between each of the system elements, as well as between the system and external systems (shown in Figure 6.9). The increased number of interface boundaries and interactions across those boundaries relates directly to the complexity of a system, increasing the complexity of integration of the system elements and subsystems as well as assessing both positive and negative emerging behaviors of the integrated system as a result of those interactions.

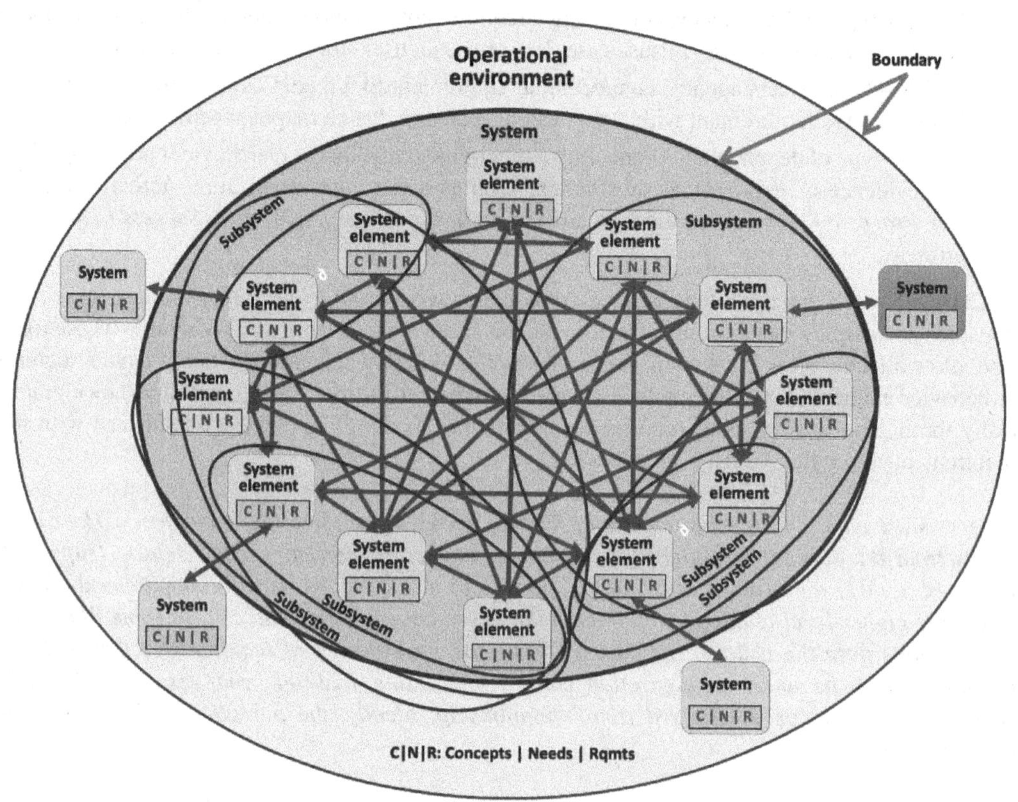

FIGURE 6.9 Complexity is a Function of the Number of Interactions Among System Elements.

Another key characteristic of today's software-intensive systems is the form of the interactions. In the past, when many of the systems were mostly mechanical, electrical, or fluid, the interfaces were more visible (for example, connectors and pin assignments, wires, pipes, cables, pressure lines, mechanical parts, or bolts) and could be easily shown on a drawing. In software-intensive systems, there can be multiple computer modules, each with embedded software that communicates (data, commands, and messages) across one or more communication buses. In modern automobiles, it is common to have more than 150 of these computer modules fed by multiple sensors and controlling multiple actuators.

Given that the behavior of a system is a function of the interaction of its parts, as well as interactions with the external systems and environment of which it is a part, it is critical that the project team identifies and defines each of the interactions between all subsystems and system elements within the system, as well as interactions with external systems. Failing to do so will result in costly and time-consuming rework during system integration, system verification, and system validation.

Chapter 4 discussed the identification of interactions (interfaces) between systems and assessing and mitigating risk associated with those interactions; the integrated set of needs includes each interaction at an appropriate level of abstraction. During the transformation from the needs to design input requirements, a lower level of abstraction is needed when defining each of the interface requirements.

While the set of needs identifies the need for an SOI to interact with external systems across an interface boundary, the resulting design input requirements must address each specific interaction (power, mechanical, data, messages, commands, etc.) along with a pointer to where those interactions are defined (for example, an ICD, IDD, or similar type document, location in a database, or data dictionary) for each interaction.

6.2.3.1 What an Interface "Is" *"An interface is a boundary where, or across which, two or more systems interact,"* as shown in Figure 6.10.

The systems could be engineered systems, human systems, or natural systems. The interaction could be:

- *Direct.* Actual connection exists between two systems.
- *Indirect.* No direct connection between systems, but there is some design feature of a system that can affect a design feature of another system (for example, induced environments or competition for a common resource).

At a highlevel of abstraction, the interaction could involve the exchange of energy, matter, or information.

An interface requirement is a functional requirement that involves a defined interaction of a system across an interface boundary with another system or entity. The interaction is communicated as a verb/object pair. An interface requirement includes a reference that defines the interaction.

In these definitions, the keywords are "boundary," "interact," and "define." From a requirements standpoint, any time the wording of a requirement indicates or implies an interaction with another system, there is an interface boundary involved across which the two systems interact. "Interact" involves a function verb/object pair. The function verb indicates some action concerning receiving

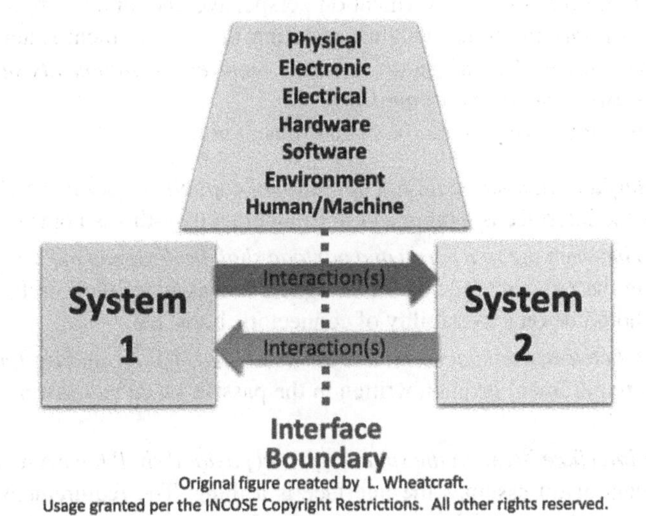

Original figure created by L. Wheatcraft.

FIGURE 6.10 An Interface is a Boundary, Not a Thing.

an input or supplying an output; the object is the actual "thing" that is crossing the interface boundary. The "performance" aspect is included in a definition of the interaction as discussed below in Section 6.2.3.4. *If there is an interaction across an interface boundary, the requirement dealing with the specific interaction is classified as an "interface requirement."*

Physical interactions can include mechanical, electrical, electronic, power, gases, fluids, and software/hardware. There are functional interactions that include information transfer (data, commands, and messages), human/machine, maintenance, and installation.

Induced environments are environments that are the result of the operation of a system; natural environments are the result of natural conditions. Induced environments include thermal, loads/vibrations, acoustic, electromagnetic, and radiation. As such, induced environments involve interaction across an interface boundary and are usually addressed along with other direct and indirect interactions when defining the interactions and writing interface requirements.

Requirements concerning natural environments, as well as induced environments from existing external systems, are usually covered when defining the operational environment as constraints. Interactions between the SOI and the natural environment are not normally shown in the external interface or boundary diagrams, where the focus is on interactions with other engineered or human systems.

Often overlooked, induced environments also include outputs of the system to the natural environment, which could include pollutants and waste. Pollutants include light, thermal, chemical, electromagnetic, hazardous materials, and radiation. Requirements concerning pollutants and waste are also covered when defining what the system is allowed to output to the natural environment, which are commonly couched as constraints contained in standards and regulations.

6.2.3.2 What an Interface "Is Not" It is important to understand what an interface is not. A general rule is that the word "interface" should not be used in a requirement statement either as a noun or as a verb. As a noun, it implies the interface is a thing, which it is not—it is a boundary across which two systems interact. As a verb, it is ambiguous, in that often there are multiple interactions between systems across a single interface boundary.

It is a best practice to focus on individual interactions when writing interface requirements. This is important from both a system verification perspective and an allocation perspective. Also, multiple interactions imply multiple thoughts, resulting in a requirement statement that does not have the GtWR characteristic *C5—Singular: needs or requirement statements should state a single capability, characteristic, constraint, or quality factor.*

Examples of how **not** to write an interface requirement are:

- *The* ***digital interface*** *shall maintain full operational capability after two failures.* This requirement assumes the interface is a thing and has functionality—this is not true.
- *The* ***interfaces*** *between the spacecraft and payload shall be designed to [xxxx].* This is a requirement written in the passive voice on the designers and assumes the interfaces are things. The requirement should be on accessibility of connectors, bolts, etc.
- *The* ***interfaces*** *between the spacecraft and payload shall have standard labels, controls, and displays.* This requirement is again written in the passive voice and assumes the interface is a thing.
- *The* ***electrical interface*** *between the spacecraft and payload shall have a reliability of 0.99999.* This requirement again assumes the interface is a thing. The requirement is on each of the systems and applies to any hardware or software of each system involved in the interaction on each side of the interface boundary.

- *The SOI shall **interface** with [XXX].* This requirement is ambiguous because it does not focus on a specific interaction. There are often multiple interactions; to be singular, there should be a requirement for each interaction.

Unfortunately, it is common to see statements like these. The bottom line: There should be no requirements that say, "The *interface* shall ... " or "The [SOI] shall *interface* with ...."

Writing interface requirements is a three-step process:

Step 1: Identify the interface boundaries and interactions across those boundaries.

Step 2: Define the interactions across the interface boundaries.

Step 3: Write the interface requirements.

The following sections are an elaboration of these steps.

6.2.3.3 Step 1: Identify the Interface Boundaries

Note: Step 1 should have been completed during needs elicitation, drivers and constraints identification, risk assessment, lifecycle concepts analysis and maturation, architecture definition, and needs definition activities discussed previously in Chapter 4. As part of these activities, models and diagrams were developed that identified each interface boundary and each of the interactions across those boundaries.

Before the interactions can be defined and interface requirements are written, an analysis needs to be done of the SOI under development and the context in which it interacts with the macrosystem it is a part (external interfaces) and an analysis of the parts that make up the SOI and how they interact with each other (internal interfaces).

A common diagram used to show these interactions is an external interface diagram, sometimes referred to as an external interface, context, or boundary diagram (an example is shown in Figure 6.11).

Using diagrams and models for the lifecycle CNR activities (as discussed in Chapter 4), all the external systems that the SOI interacts with across its lifecycle should already be defined and assessed. For the Spacecraft shown in Figure 6.11, there are fourteen external systems shown. There are multiple interactions between the Spacecraft and each of the external systems. The arrows indicate the directionality (inputs/outputs) of each interaction.

In some cases, there may be one or more intermediate systems between the SOI and another external system it interacts with. In Figure 6.11, intermediate systems are shown for communications between the Spacecraft and Ground Control via a Ground Communications Network and the Communication Satellite. While the focus may be on interactions (exchange of data and commands) between the Spacecraft and Ground Control, the interactions with the intermediate systems must be addressed as well.

For example, addressing interfaces associated with commanding the Payload and retrieving data back to Earth, there would be interface boundary definitions and requirements for each interacting pair of systems for a total of five sets of interaction definitions and interface requirements:

1. Spacecraft Payload to Communications System (internal interface)
2. Spacecraft to Communications Satellite (external interface)
3. Communications Satellite to Ground Comm Network (external interface)
4. Ground Communications Network to Ground Control (external interface)
5. Spacecraft to Ground Control (external interface)

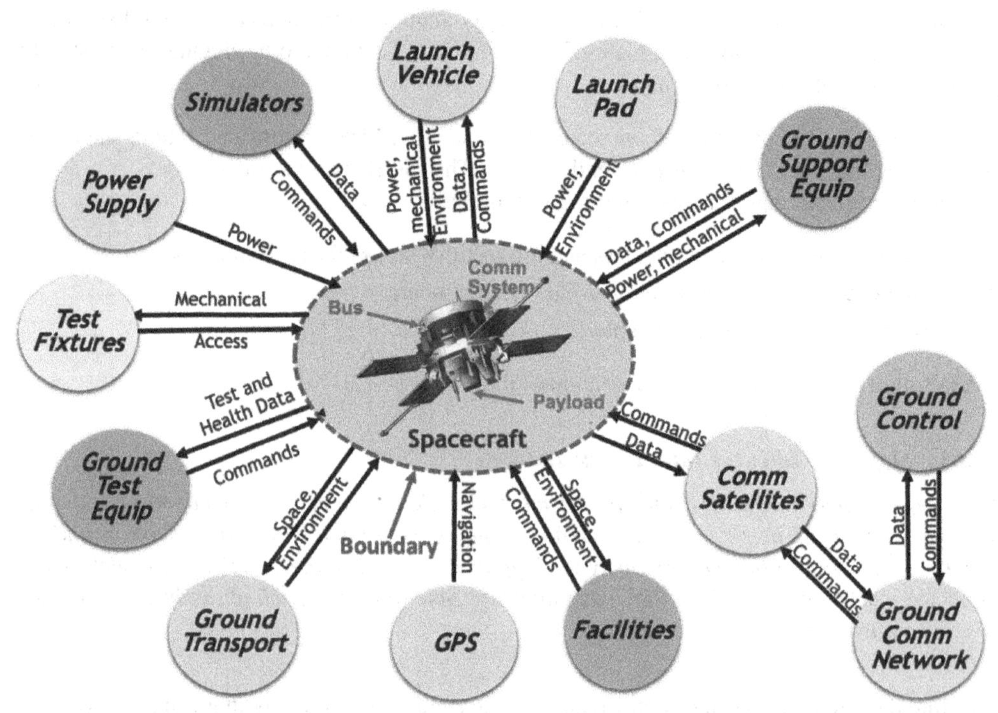

FIGURE 6.11 Example External Interface, Context, or Boundary Diagram.

For existing systems like the Communications Satellites and Ground Communications Network, which are included in several different SoS that support multiple Spacecraft and multiple Ground Control facilities, there would be an established interface definition type document (IDDs or ICDs) (see next section) that each Spacecraft and Ground Control facilities would comply with. Note that the third interface, Communications Satellite to Ground Communications Network, is an existing interface and is outside of the scope of the spacecraft development.

For each system external to the spacecraft, a one-on-one system-to-system interface diagram can be developed showing the specific interactions between the Spacecraft and each of the external systems. For the example shown in Figure 6.11, there would be at least eleven separate system-to-system Interface Diagrams like that shown in Figure 6.12.

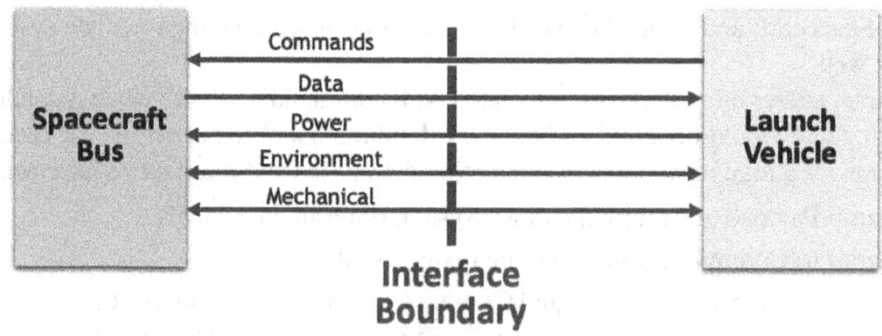

FIGURE 6.12 Example System-to-System Interface Diagram.

Within the system, there are also internal interfaces to address. As an example, consider the interactions between the Spacecraft subsystems Bus and the Payload. Each of the individual interactions must be defined and will result in one or more interface requirements allocated to the bus and the payload. Depending on the approach used by the project, rather than individual diagrams like that shown in Figure 6.12, the other SE tools and diagram types can be used to identify, define, and manage each of these interactions.

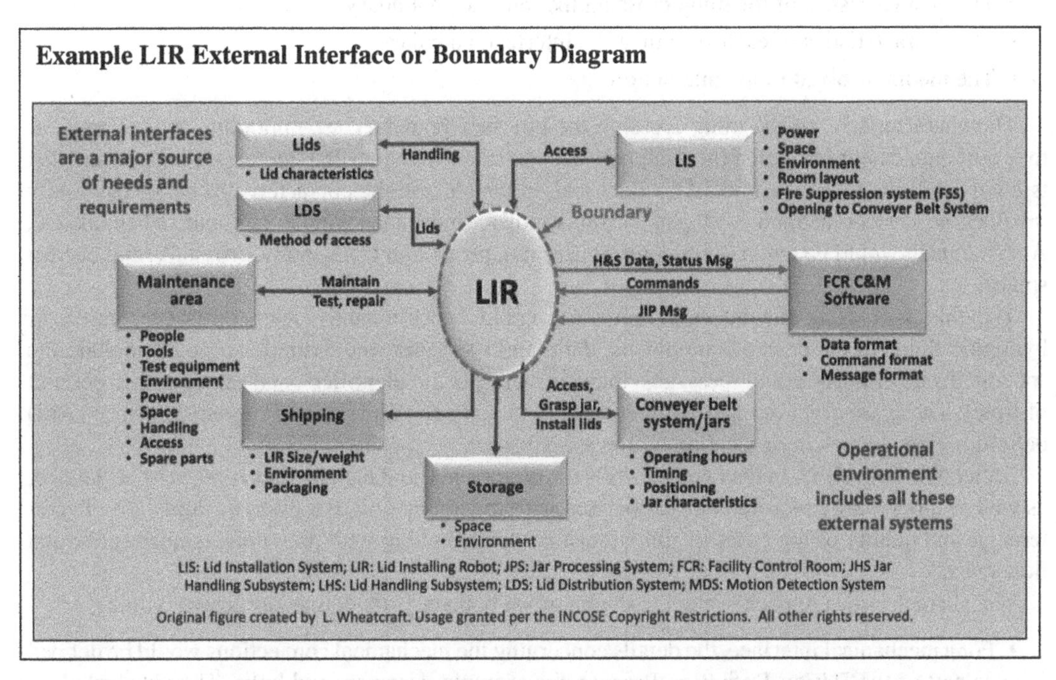

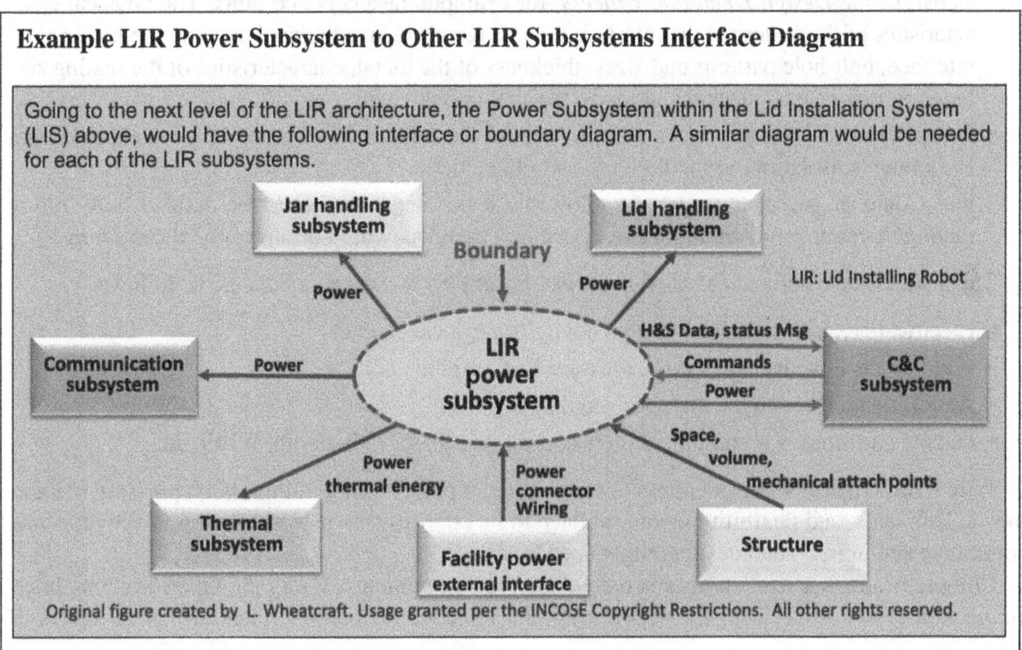

6.2.3.4 Step 2: Define the Interactions Across the Interface Boundary *An Interface Definition is a common agreement concerning a specific interaction across an interface boundary between the SOI and another system commonly recorded in an ICD, IDD, or similar type of interface definition type document or location in a database.*

Each interaction shown in the individual interface diagrams must be defined. To completely define an interaction at or across an interface boundary, three things must be addressed:

- The characteristics of the thing crossing the interface boundary.
- The characteristics of each system at the interface boundary.
- The media involved in the interaction.

The characteristics of the thing crossing the interface boundary are important design inputs as they communicate information the designers need to know to realize the interactions. The thing that is involved in the interaction could be electrical, in which case the characteristics of the electricity involved need to be defined: voltages, currents, noise, impedance, ripple, rise and fall times, etc. Also included would be any requirements concerning protection from shorts, crosstalk, and current spikes.

The things crossing the interface boundary could be commands, data, messages, gases, or hydraulic fluid. In the case of commands, data, and messages, the definitions would include the specific data, commands, and messages, their format, specific identifiers, and rate (number per unit of time). These are often defined within a data dictionary or Application Program Interface (API) definition type document or location within a database.

Protection from errors in the exchange of data, commands, and messages also needs to be defined, as well as any security requirements associated with the interaction. For gases or fluids, the characteristics and quality of the gases or fluids must be defined along with pressures, temperatures, and flow rates.

The characteristics of each system at the interface boundary are defined as design outputs.

- For a mechanical interface, the details concerning the mechanical connections would be defined as part of the *Design Definition Process*, for example, fasteners and bolts. The physical characteristics of the system at the interface may be recorded in a drawing showing the mechanical interface, bolt-hole patterns and sizes, thickness of the metal, characteristics of the mating surfaces, torque values, mechanical loads, thermal transfer, etc.
- For an electrical, data, or command interface; the type and part number of connectors, pin assignments, isolation, grounding, etc., would be defined.
- For a fluid or gas connection, the allowable leak rate, if any, must be defined both during mating/de-mating and while mated; actual part numbers would be shown on the drawings.

The media is also defined as a design output. Examples of various media are as follows:

- Electrical/electronic signals through a wire or physical contact
- Fluid or gas flow through a tube, pressure line, or pipe
- An RF signal through the air, space, and fiber optics
- Data via a common communication bus, electromagnetic, Bluetooth, Wi-Fi, etc.

Wire sizes, types of wire, shielding, wire covering, pipes, pressure ratings, burst pressure requirements, leak rates, and flexibility would all have to be addressed as part of the design activities and communicated in ICDs and design output specifications.

Communications across an interface boundary can be complex. Using the Open Systems Interconnection (OSI) model [51] for networking and telecommunications, there are seven layers of interactions (physical, data link, network, transport, session, presentation, and application) across interface boundaries that must be defined and addressed by the design.

In many cases, there is a standard that defines both the media and the characteristics of the systems at the interface boundary, for example, Ethernet, USB, networks (IEEE 802.xx), wiring, or pressure lines. These standards are important from safety, security, and interoperability perspectives. An operator does not want to be injured when disconnecting or connecting a high-pressure line; homeowners do not want unauthorized personnel able to access their home network; and someone buying a new printer expects it to work with their existing computers and network. Standards for interactions between systems address these issues.

Applicable standards associated with interactions across interface boundaries need to be invoked within the set of design input requirements; as such, these standards both constrain and drive the design. It is common to reference the standards in the interface definition type documents or database.

For developing systems, not all this information will be known until the design is complete. In the beginning, all that can be defined using design input requirements is the identification of what is involved in the interaction, its characteristics, and the role each system has depending on whether the interaction is an input or output.

The details of what the system looks like at the interface boundary and the media involved are design outputs and will be included as part of the design output specifications. Thus, the definition statements in an ICD, IDD, or other similar interface definition type document or database may start out with placeholders—TBDs/TBRs—for tables, drawings, or graphs. When the design-dependent information is known, the TBDs/TBRs can be replaced with the actual design information. Before the systems can be built or coded, there can be no TBDs/TBRs in the design output specifications nor in the document or location in which the definitions are recorded. In this sense, interface definitions evolve as the design evolves. *Refer to Section 14.2.4 for more details concerning managing TBXs.*

The evolutionary nature of interface definitions for developing systems must be addressed when contracting out the development of system elements to a supplier. The role of the supplier, their relationship with other suppliers their SOI interacts with, their role with the customer and/or integrating organization, and how they are to accommodate the evolution of the interface definitions must be addressed in the SOW, PO, or SA. Failing to do so often results in expensive contract changes as well as budget overruns and schedule slips.

The general format of an interface definition statement is as follows:

- [Thing being defined] is [are] [whatever the definition is]. or
- [Thing being defined] is as shown in [Drawing xxxxx] or [Figure yyyy].
- [Thing being defined] has the characteristics shown/defined in [Table or Graph zzzz].

It is important to understand the difference between interface definitions and the interface requirements that invoke those definitions. Notice the terms “shall,” ”must,” or “will” are not used for an interface definition because the definition is just stating an agreed-to fact. The design input requirement that will drive design and to which the design and system will be verified against will invoke the applicable definition for the specific interaction the requirement is addressing. Given that an interface requirement is a type of functional requirement, a reference to where the interaction is defined is the performance part of the requirement.

Examples of interface definition statements include:

- The DC voltage [supplied by System 1] has the characteristics shown in Table xyz or Figure 123.
- The mechanical attach points [between System 1 and System 2] are as shown in Drawing xyz.
- The fluid [supplied by System 1] has the characteristics defined in Table xyz (pressure, quality, flow rate, temperature).
- The leak rate at the connection [between System 1 and System 2] is less than [xxx units per unit of time].
- The commands [sent by System 1] are defined in Table xyz.
- The data stream [accepted from System 2] has the characteristics defined in ....
- The data parameters used within the SOI are defined in [Data Dictionary xxxxx].
- System 1 printer port complies with USB 3.0 standard.

These are statements of fact that need to be recorded in the data and information model, agreed to, baselined, and configuration managed in a form that can be referred to by an interface requirement.

6.2.3.4.1 Defining Interfaces with Existing Systems For existing systems, their design is complete, so their interactions with external systems will be defined in an existing ICD, IDD, or similar form in tables, figures, and drawings developed as outputs of the INCOSE SE HB *Architecture Definition Process* and *Design Definition Process*.

The interface definitions for the existing system define what some future developing system must do to interact with the existing system. As such, interface definitions for existing systems are constraints on the new developing system. Any developing system that wants to interact with the existing system must comply with the existing system's interface definitions for each interaction. The existing system's definitions do not contain "shall" statements because they are defining the as-built system's interface boundary, media, and characteristics of the things crossing the interface boundary—statements of fact.

In the example shown in Figure 6.13, when Existing System 1 was designed and built, the existence of New System 2 specifically was not known, but there was a requirement for Existing System 1 to interact in some way with future systems like New System 2.

In this case, there will be no specific Existing System 1 requirements stating, "*System 1 shall [interact in some way with System 2] as defined in [location the interaction definition is recorded].*" From Existing System 1's viewpoint, if any other future system wants to interact with it, that other system must do it in accordance with Existing System 1's interface definitions.

The owner of Existing System 1 will have recorded the definitions for interactions across the interface boundary in a configuration-controlled document or location for future systems. For example, New System 2 can be designed to interact with Existing System 1 in accordance with System 1's definitions for each interaction.

Developing New System 2 will include interface requirements in their set of design input requirements or design output specifications that refer to Existing System 1 interface definitions for each interaction that New System 2 will have with Existing System 1.

In the example shown in Figure 6.13, Existing System 1's definitions in its ICD for supplying power across the interface boundary could include a drawing that defines the wiring, connector, pin assignments, shielding, and grounding information for an external system to obtain power from Existing System 1 and a table that defines the power characteristics such as voltage, current, noise,

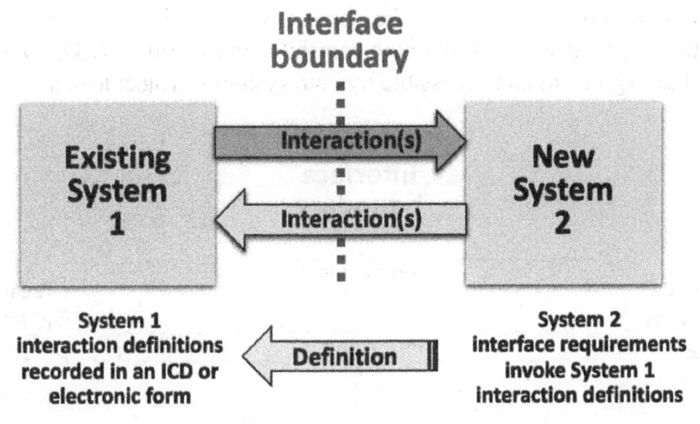

FIGURE 6.13 New System 2 Interactions with an Existing System 1.

and filtering. Because Existing System 1 is existing, these requirements are constraints on New System 2's design.

The New System 2 will define three separate design input requirements that are needed to completely address each specific interaction with the existing system.

System 2 shall operate on power obtained from System 1 having the characteristics defined in System 1 ICD 2345 Table 3.6. This requirement addresses the characteristics of the thing crossing the boundary.

System 2 shall obtain power from System 1 in accordance with the wiring defined in System 1 ICD 2345 Drawing 3–4. This requirement addresses the media.

System 2 shall obtain power from System 1 in accordance with the connections defined in System 1 ICD 2345 Drawing 3–4. This requirement addresses the boundary.

6.2.3.4.2 Defining Interactions for Two Systems Being Developed Concurrently In the case where both systems are being developed concurrently, defining the interactions gets more complicated. In this case, neither system currently exists, for example, subsystem or system elements within the system architecture or a new external system being developed concurrently with the SOI.

The definitions of the interactions between the two systems will evolve over time as the design matures. In the beginning, while addressing the problem space, the focus is on "what" not "how," resulting in design input requirements. Concerning the interaction between the two systems: What information must be defined so the design team can design each system so they can interact as required across the interface boundary? What do the designers need to know to do the design? What are the constraints?

As shown in Figure 6.14, an agreement is reached by the project teams for each system concerning the definitions for each of the interactions, resulting in applicable interface requirements included in each of the developing system's set of design input requirements that invoke those agreed-to definitions. Each system's interface requirements would invoke the same common, agreed-to, configuration-controlled interface definition.

In the example shown in Figure 6.14, New System 2 transmits a command to New System 3 when some trigger event occurs, and New System 3 must perform some action upon receipt of the command from New System 2. In this case, the characteristics of the command would be defined, recorded, agreed to, baselined, and configuration controlled in a common ICD, IDD, or similar document or in a database agreed to and accessible by both system's project teams.

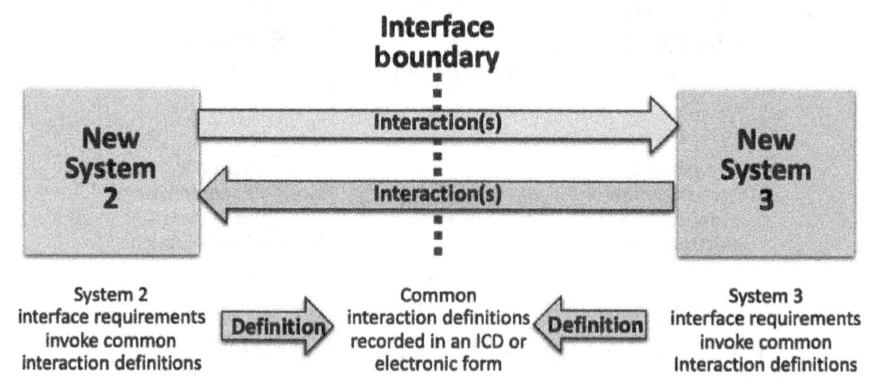

FIGURE 6.14 New System 2 and System 3 Interactions.

Common definition included in an ICD: "*The xyz command has the characteristics described in Table 2.2.*"

In this case, the resulting interface requirements would be:

When [trigger event], System 2 shall transmit to System 3 the [xyz command] having the characteristics defined in [ICD xyz, Section 4.5.6, Table 2.2] in less than or equal to 5 ms.

System 3 shall execute the [xyz command] having the characteristics defined in [ICD xyz, Section 4.5.6, Table 2.2] in less than or equal to 4 ms of receipt from System 2.

Once the design for both systems is complete, as in the case of an existing system, the definitions concerning the interactions, what the system's characteristics are at the interface boundary, and the media involved will be recorded in a configuration-controlled interface definition type document or electronically within a database in the form of figures, tables, and drawings, which will be referenced by both the design input requirements and design output specifications for both systems.

For systems that are being developed concurrently, the organizations responsible for these systems must mutually agree, record, baseline, and configuration control the definitions.

It is common to form an Interface Control Working Group (ICWG) to manage these definitions and the resulting agreements for the interactions across each interface boundary. Alternately, an organization may address the definition, agreement, and control of the interface boundaries within a SE and Integration Working Group (SE&I WG). Suppliers that are developing the system elements must be included as part of these working groups as defined in their SOW, PO, or SA.

From a configuration control standpoint, interface documentation is controlled by the common parent organization to both projects responsible for the systems under development. For larger complex systems, where various system elements development is contracted outside the organization, the integrating organization would control the interface definitions and include them with the set of design input requirements for each of the system elements.

As the design matures for each system element, the suppliers would provide detailed definitions for each interaction to the integrating organization as specified in the supplier SOW, PO, or SA.

Note: This can be a challenge when multiple suppliers' design activities are dependent on the design decisions concerning the interactions made by another supplier, especially if the supplier considers the information to be proprietary. Failing to address this issue adequately in the contract for each supplier can result in significant costs if the evolution in the design results in changes to one or more contracts.

6.2.3.4.3 Recording the Definitions Associated with Interactions Across Interface Boundaries The detailed information concerning interactions across the interface boundaries must be defined and recorded somewhere—no matter what the document is called or how the information is recorded.

When two systems that will interact with each other are being developed concurrently, it is critical that the interactions are defined, baselined, and configuration managed within common sets of definitions and not separately for each of the interacting systems. Doing so can have negative consequences if either of the separate definitions changes without the corresponding change being made in the other set of definitions. It is critical that there is an ASoT concerning the agreed-to definitions concerning how the two SOIs will interact across an interface boundary.

Take, for example, a case where two mechanical systems that will be bolted together were being concurrently developed by separate contractors. Rather than having a common interface definition type document and drawing to which each system was built to, each organization had their own, separate mechanical interface drawing as a deliverable as part of their contract. At the parent system's PDR, it was discovered that the number of bolt holes in the individual drawings for each system was different! If this issue was not identified until actual integration, the impact on the project would be severe.

It is common practice in a document-centric practice of SE to define and record interface definitions in some type of document whose purpose is to contain these definitions, for example, an ICD or IDD. However, documents that include these definitions can have a variety of names. Some organizations put the pre-design definition information in an ICD Part 1 and the post-design details in an ICD Part 2. Others may use different names for the document that contains the definitions: IDD, Internal IDD (IIDD), Interface Agreement Document (IAD), Interface Definition Agreement (IDA), API, or a Data Dictionary. The name does not really matter—what matters is that the interface definitions are recorded, baselined, and configuration managed.

Note: Some organizations develop an Interface Requirements Document (IRD). This is not recommended because the document title contains the word "requirements." As a result, some feel that the document must include requirements (shall statements). In some cases, the organization puts "shall" in the interface interaction definitions. This is problematic because interface definitions are not interface requirements, rather interface definitions are statements of fact invoked by interface requirements, as discussed in the next section.

In other cases, the organization pulls all interface requirements from the sets of design input requirements for the two systems that interact with each other and places all the interface requirements into a common IRD, which is managed and controlled by the parent organization of the two systems. If these requirements are removed from the sets of design input requirements, this can result in each system's set of design input requirements to be incomplete, which is problematic from a flow down, allocation, and design and system verification perspective, especially when most of the system's functional and operational requirements involve interactions with external systems. But if these requirements are not removed from the sets of design input requirements, then the requirement exists in multiple locations, with the risk of changes not being implemented in both locations, resulting in the interface requirements being inconsistent and contradictory. A better approach is to use a data-centric practice where all design input requirements are managed comprehensively, enabling views to specific sections (such as the interface requirements) when needing to address those topics.

The point of any "ICD" or "IDD" interface definition type document is to establish a common location or repository to record an agreement concerning information-defining interactions across an interface boundary used by two or more SOI's and their associated design input requirements and design output specifications to eliminate errors resulting from recording duplicate technical content in multiple locations with each location under configuration control by a different organization. In a data-centric practice of SE, the interface definitions are recorded electronically within a database associated with an RMT or other application used to manage the interface definitions, allowing the interface requirements to be traced to the definitions for the interactions as well as to any dependent interface requirements.

For an existing system, whatever the form, the interface boundary definitions reflect the existing system's "as built" configuration.

As the project team moves down the levels of the system architecture, interactions between the subsystems and system elements at each level of the architecture are identified, defined, and interface requirements written. In the end, all interactions will be implemented by the design for the system elements and communicated via the design output specifications for each system element. It is a best practice to record and manage the internal interface definitions separately from the external interface definitions. For the Spacecraft example shown in Figure 6.11, there would be a set of definitions for each external system boundary, as well as one or more internal sets of interface boundary definitions. For the internal boundary definitions, some organizations divide and manage the sets of definitions based on organization, technical discipline, mechanical, electrical, fluids, software, or human/machine as examples.

Special considerations need to be made when OTS, MOTS, or COTS system elements are to be used. Because they are "off-the-shelf"—they already exist. In this case, the project team will have to assess how the OTS, MOTS, or COTS system element will interact with other system elements and, based on this assessment, how it will be integrated into the SOI physical architecture. One option is to treat the OTS or COTS system element interface characteristics as constraints and design the other system elements to interact with the OTS or COTS system element based on its definitions as documented in its ICD or similar document. In other cases, the project team may be able to modify the OTS system element (MOTS) such that it can interact with other system elements within the SOI based on the project team's definitions. There may even be cases where the project team may build or procure an "adapter" to enable the OTS or COTS system element to be integrated into the SOI. *Refer to Chapter 12 for a more detailed discussion concerning OTS, MOTS, and COTS subsystems and system elements.*

6.2.3.5 Step 3: Write the Interface Requirements The phrase "interface requirement" refers to the specific form or template for a functional/performance requirement that deals with an interaction of a system across an interface boundary with another system.

As such, interface requirements should not be considered a separate type of requirement when organizing the set of design input requirements—doing so often leads to confusion and duplication of requirements. Interface requirements may be included with the functional/performance requirements or "Operational (fit)" requirements discussed previously. Safety, security, and HSI requirements often include interface requirements as well.

Once the interfaces have been identified and the interactions across the interface boundaries defined, the actual interface requirements can be written.

Following this three-step order is an ideal case. In the real world, often the interface requirements are written before the interactions have been defined, especially when two systems are being developed concurrently. In this case, the interface requirements will point to a TBD document or location in the database that will contain the definitions for each interaction across an interface boundary. When the interface is with an existing system, the definitions with the existing system should already be recorded and under configuration control. Again, these TBDs represent future work that must be completed before the system design output specifications can be completed.

6.2.3.5.1 Interface Requirement Definition and Format An *Interface Requirement* is a functional requirement that involves a defined interaction of a system across an interface boundary with another system or entity.

The format of the interface requirement is such that it includes a function/verb indicating directionality (input/output; supplier/receiver), the name of the object involved in the interaction, AND a reference (pointer) to the specific location where the definition of the specific interaction across the interface boundary is located. *A common error concerning interface requirements is not including the reference pointer to where the interaction is defined. Without this, the interface requirement is incomplete and unverifiable (the system will not be able to be verified to meet the requirement).*

All interface requirements have the same general form: *Note: If the interaction is in response to some trigger event, then that trigger event would be included in the requirement text.*

> *"[The System] shall [interact (function verb/object) with] [Another System] as defined in [location where the interaction is defined]."*
>
> *"[The System] shall [use/provide from/to] [Another System] [something] having the characteristics defined in [location where the something is defined]."*

As discussed earlier, the word "interface" is not included in an interface requirement, neither as a noun nor a verb.

Because each interface requirement has a "shall," the SOI will be verified that it was designed and built such that it can interact with the other system in accordance with the agreed-to definition for that interaction. Prior to system integration, each subsystem and system element on either side of an interface boundary must be verified to have met its interface requirements in accordance with the agreed-to definitions for the specific interactions. Having passed system verification, once integrated, the resulting parent system can be verified to meet its own set of design input requirements, including any of its own interface requirements.

As discussed earlier, for the case where the SOI has an interaction with an existing system, the interface requirement will include a pointer to the existing system's ICD or other location where the interaction is defined. This assumes that the existing system 1) has recorded the definition in some configuration-managed form and 2) that the definition is current (defines the current configuration of the existing system). This may not always be the case and represents an integration risk to the project.

6.2.3.5.2 Interface Requirement Examples Below are some examples concerning the existing Spacecraft Bus interactions with a new Payload being developed to be integrated into the Spacecraft. Interactions between the Spacecraft Bus and Payload are shown in Figure 6.15. In this example, the interactions between the Payload and the Spacecraft Bus across the interface boundary are defined in an existing Spacecraft Bus to Payload ICD 1234.

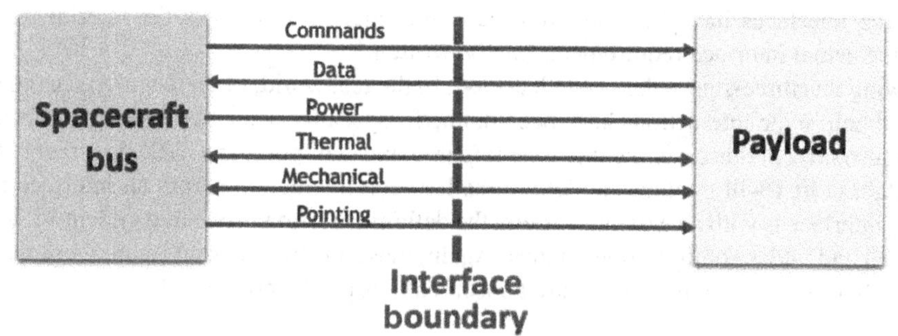

FIGURE 6.15 Interactions Between the Spacecraft Bus and the Payload.

- The Payload shall communicate with the Spacecraft Bus processor via a 1553 bus, as shown in Spacecraft Bus to Payload ICD 1234, Figure 6. *Specifies the media used for communicating payload sensor data to the Spacecraft.*
- The Payload shall send Sensor A data to the Spacecraft Bus having the characteristics defined in Spacecraft Bus to Payload ICD 1234, Table 5.4. *Specifies the format the payload needs to send the Sensor A data to the spacecraft over the 1553 bus.*
- The Payload shall receive from a [ground power supply] ground power having the characteristics described in Spacecraft Bus to Payload ICD 1234, Table 4.2. *Specifies the characteristics of power used during development and testing. The characteristics are the same as those provided by the Spacecraft Bus during operations.*
- The Payload shall receive from the Spacecraft Bus, 28-volt power having the characteristics described in Spacecraft Bus to Payload ICD 1234, Table 4.2. *Specifies the characteristics of the power the spacecraft is providing to the payload.*
- The Payload shall receive 28-volt power from the Spacecraft Bus in accordance with the connections defined in Spacecraft Bus to Payload ICD 1234, Drawing 2–5. *Specifies the characteristics of the physical connections the Payload needs to connect to the Spacecraft's 28-volt payload power bus.*
- The Payload shall receive from the Spacecraft Bus, pointing data that have characteristics defined in Spacecraft to Payload ICD 1234, Table 7.2. *Specifies the characteristics and format of the pointing data supplied by the Spacecraft Bus to the Payload.*
- The Payload shall mechanically attach to the Spacecraft Bus in accordance with the mechanical connections shown in Spacecraft to Payload ICD 1234, Drawing 5–5. *Specifies the mechanical connections, bolt holes, bolt sizes, and torque values to be used to mechanically attach the Payload to the Spacecraft Bus.*

If the Spacecraft Bus and Payload were being developed concurrently, each system would need to include in its set of design input requirements an interface requirement for each of its interactions with the other system. Because of this, interface requirements for these systems are developed in pairs. For the Spacecraft Bus/Payload example, for each of the "The Payload shall… " interface requirement there would be a corresponding "The Spacecraft Bus shall… " interface requirement.

The Spacecraft Bus shall [interact (function verb/object)] with the Payload [as defined in or having the characteristics shown in] [common location where the interaction is defined].

The Payload shall [interact (function verb/object)] with Spacecraft Bus [as defined in or having the characteristics shown in] [common location where the interaction is defined].

Prior to integration, during design and system verification, the Payload would be verified that it can interact with Spacecraft Bus in accordance with the interface requirements that point to the common interaction definitions. The Spacecraft Bus would also be verified that it can interact with the Payload in accordance with its interface requirements that point to the same common definitions. To enable these verifications prior to integration into the actual parent system, emulators or simulators are often developed.

Example LIR Interface Requirements

Acronyms used: LIS Lid Installation System; LIR: Lid Installing Robot; JPS: Jar Processing System; FCR: Facility Control Room; JHS: Jar Handling Subsystem; LHS: Lid Handling Subsystem; LDS: Lid Dispensing System; MMS: Motion Monitoring System. (Phrases and terms with the form Xx_Yy_Zz are defined in the project glossary or data dictionary.)

R001: The LIR shall obtain Lids from the LDS as described in the LIR/LDS ICD xyz, Section 123.

R006: The LIR shall install a Lid on a Jar within 10 sec of receiving from the FCR C&M software a Jar_in_Place message having the characteristics defined in the FCR ICD xyz, Section 123.

R007: The LIR shall report within 1 sec the completion of Lid_Installation to the FCR C&M Software as defined in the FCR ICD xyz, Section 123.

R008: The LIR shall install Lids on Jars positioned on the JPS CBS conveyor belt as defined in the JPS ICD xyz, Section 123.

R010: The LIR shall receive from the FCR C&M software Valid_Commands defined in the FCR ICD xyz, Section 123.

R012: The LIR shall perform the Power_Up_Sequence within 1 sec of receipt of a Power_Up_Command from the FCR C&M Software having the characteristics defined in the FCR ICD xyz, Section 2.4.5, Section 123.

R016: The LIR shall report LIR_Health_and_Status to the FCR C&M Software as defined in the FCR ICD xyz Section 123 within 1 sec of the completion of the Self_Test.

R026: The LIR shall Operate using 110-120 VAC, 60 Hz, 30-amp Facility_Power having the characteristics defined in Facility Drawing xyz.

R027: The LIR shall receive Facility_Power using an EIC 320-compatible electrical 3 prong connector defined in Facility Drawing xyz, Sheet 11.

R039: The LIR shall only accept Valid_ Commands from the FCR C&M Software that meet the Security Protocols defined in the FCR ICD xyz, Section 567.

6.2.3.5.3 Traceability and Interfaces If two system's interface requirements are in response to a common allocated parent, they will both have a trace to the common parent. For example, a system may have a requirement to perform some function. That function is decomposed into subfunctions, each of which is implemented within a different lower-level subsystem or system element. In this case, the implementing lower-level subsystems or system elements may have to interact with each other, thus there would be an internal interface boundary across which each would interact. The specific interactions would be defined, and each would define appropriate interface requirements invoking those common definitions.

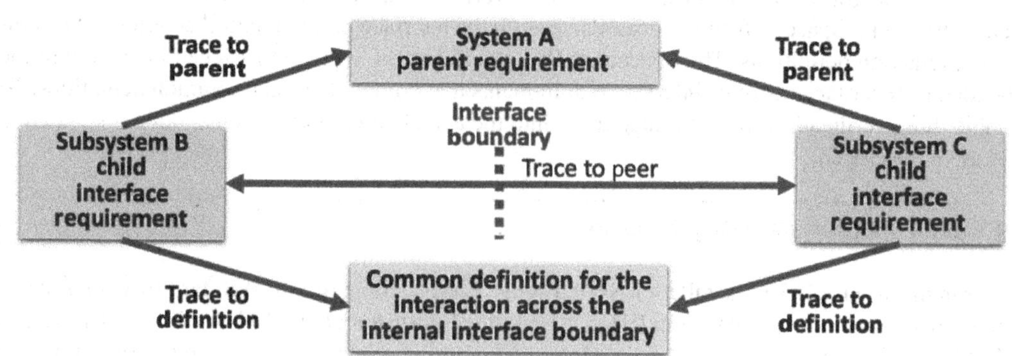

FIGURE 6.16 Interface Requirement Traceability.

As shown in Figure 6.16, each of the interfacing systems will develop its own half of the interface requirement pair, with each pointing to the common interaction definition, wherever that is recorded. From an RMT standpoint, each system's interface requirement will trace to its common parent, trace to the other system's corresponding interface requirement recording the dependency, and trace to the common interaction definition.

6.2.3.6 Using RMTs to Manage Interfaces Ideally, RMTs would be able to enforce the use of the template for interface requirements. In this case, when an internal lower-level project team generates an interface requirement, the RMT should have the following capabilities:

- Require a pointer to where the interaction is defined or indicate that the definition is missing.
- Notify the other internal subsystem or system element project team of the need for them to include the complement interface requirement for the pair in their set of design input requirements.
- Require the complement interface requirement to point to the same interaction definition.
- Include a flag to indicate the status of the pair of interface requirements and interaction definition, i.e., indicate if both requirements exist and point to a common interaction definition, if it exists. If it does not exist, the RMT could notify both project teams concerning the need to define the interaction.

These capabilities would help automate the definition and management of the internal interface requirements and help avoid many of the defects and issues discussed earlier.

This activity helps establish each interface requirement has the GtWR characteristics *C1—Necessary, C4—Complete, C8—Correct,* and *C7—Verifiable*. It also helps establish the set of design input requirements that has the GtWR characteristics *C10—Complete, C11—Consistent, and C14—Able to be Validated.*

6.2.3.7 Common Interface Requirement Defects and Issues As with any type of requirement, when defining interface requirements, there are several common defects and issues that need to be avoided. If left undetected and unresolved, these defects and issues can result in one or more of the following undesirable outcomes:

- Ambiguous requirements that the system cannot be verified to meet.
- A set of design input requirements that is incomplete.
- An SOI that cannot be successfully integrated into the macrosystem it is a part.
- An SOI that is rejected by the customer or that is not certified for release and used by the public.
- Costly budget overruns and schedule slips.

Common defects and issues concerning interface requirements include:

- Writing interface requirement statements that include the word "interface" either as a noun or verb (reference Section 6.2.3.2).
- Interface requirements not written in the form of an interface requirement. The most common defect is a failure to include a pointer within the requirement statement to where the interaction is defined and agreed to.
- Failing to identify all external systems in which the SOI interacts.
- Identifying an interface boundary with another system, but not addressing all interactions across that boundary.
- Failing to define, agree to, and configuration manage all interactions across an interface boundary.
- Failure to include all interactions (functional and physical) with external systems in the system model and assess overall system behavior as a function of the interactions with external systems.
- Failure to include all internal SOI interactions (functional and physical) within the system model and assess overall system behavior as a function of the interactions of the system elements that make up the SOI.
- Failure to verify a subsystem or system element meets all its interface requirements prior to integration into the macrosystem it is a part.
- Assuming that design verification of all functional interactions across system boundaries using a model of the system is an adequate substitute for verification of the actual physical system's interface requirements.
- Assuming that if a system is verified to meet all interface requirements, the system's behavior within the operational environment when operated by the intended users will meet the stakeholder needs.
- Failure to do a risk analysis for each interaction across all interface boundaries. What could go wrong? What happens if an input or output is not as defined and agreed to? For security-critical interactions, how could an unintended user interact across the system boundary such that the intended use of the system is compromised? How does the SOI address these issues?
- From an integrated system behavior, another issue is cascading failures. A non-nominal input results in a function to produce an off-nominal output, which is then an off-nominal input to another function, and so on. In today's software-intensive systems, cascading failures across multiple interface boundaries must be addressed.

The best practices advocated in this Manual are designed to help avoid these defects and issues. For today's increasingly complex, software-intensive systems, it is critical that the project team uses a data-centric approach to address all interactions across interface boundaries. From a completeness and correctness perspective, the project team must use diagrams and models to identify all functional and physical interface boundaries and interactions across those boundaries with both internal subsystems and system elements as well as with external systems—including the users and operators.

6.2.3.8 Interface Requirements Audit Due to the critical importance of correctly addressing all interactions between the SOI and the external systems, the common interface requirements defects and issues discussed above must be avoided or corrected before the set of design input requirements is baselined.

When developing subsystems and system elements within the system architecture, it is common to assign each to different groups based on discipline, organizational function, or procurement strategy. As a result, these subsystems or system elements are often developed in silos. From an interface perspective, each may develop an external interface diagram or boundary diagram for their specific SOI. Using this approach, there are often inconsistencies between diagrams.

> Take, for example, an audit for a system development effort that revealed that when integrating all the boundary diagrams for the subsystems and system elements that made up the system into an integrated system view of the subsystems and system elements, on average, there was a 40% mismatch.
>
> The boundary diagram for one system element showed an interaction with another system element, yet the other system element's boundary diagram did not show the same interaction with the first system element.
>
> This meant that for the integrated system, 40% of the internal interactions had not been defined, and the resulting interface requirements were missing. It is not surprising that once the results of this audit were made known, the Lead Systems Engineer made the statement, "No wonder we have so many integration problems!"

To help avoid inconsistencies and ensure there are no missing interface requirements or interactions that have not been defined, it is important that the project team conducts an "interface requirements audit" before baselining the set of design input requirements. Sadly, it is common to find numerous inconsistencies during such audits, confirming that the effort is time well spent. This audit can be one with either a data-centric approach or a document-centric approach, depending on how the project is capturing its interface content.

6.2.3.8.1 Data-Centric Interface Requirements Audit In a data-centric approach, the interface definitions are captured as part of a system model element. Many MBSE tools have these capabilities to support interface audits and generation of interface matrices.

Using MBSE approaches and tools, the SOI, its lower-level elements, and its connected external systems are captured as part of the system model. With these connections defined, interface definition tables can be automatically generated to confirm that all interfaces within the SOI and to external systems have been captured and defined within the system model.

Interface requirements can be captured in the system model application or an RMT and traced to the interface definition of the corresponding element to establish the trace-to-source. In cases where each side of the interface is a new subsystem or system element in development, an interface requirement is captured for each side of the interface.

At this point, querying the system model then enables the ability to capture whether: 1) all interfaces have a definition, and 2) all defined interfaces have corresponding interface requirements. For any missing or mismatched interfaces and interface requirements, the SE can make adjustments in the system model to ensure they are addressed.

6.2.3.8.2 Document-Centric Interface Requirements Audit The approach to conducting a document-centric interface requirements audit is relatively simple but time consuming compared to a data-centric approach using models. Create a spreadsheet with five columns and label them as "SOI Interface Requirements," "SOI Interaction Definitions," "External System Interaction Definitions," "External System," and "External Systems Interface Requirements," as shown in Table 6.3.

TABLE 6.3 Interface Requirements Audit Spreadsheet.

SOI Interface Requirements	SOI Interaction Definitions	External System Interaction Definitions	External System	External Systems Interface Requirements
—	—	—	—	—
—	—	—	—	—
—	—	—	—	—

Start with the external interface diagram or boundary diagram for the SOI for which the audit is being conducted. If done correctly, the diagram should show all external systems with an interface boundary across which or at which the SOI is interacting and an indication concerning what each of the interactions are. Given that there can be multiple interactions with a single system, each interaction needs its own interface requirement and corresponding agreed-to definition.

With this knowledge, review the set of design input requirements, copy all requirements where there is even a hint of an interaction between the SOI and any external system, and paste the requirement text in Column 1 "SOI Interface Requirements." If there is an interaction, there is an interface boundary. *Note: If there are multiple interactions across an interface boundary between the SOI and an external system but the interface requirement is not written properly as discussed earlier, the defects will need to be corrected to proceed with the audit.* There should be at least one interface requirement (as discussed earlier, there can be up to three requirements) for each interaction between the SOI and an external system across the interface boundary; each interaction must be defined and agreed to.

Next, assuming it is defined, copy the pointer in the interface requirement concerning the location of the interaction definition and paste the location of the definition in Column 2, "Interaction Definitions." If there is no pointer to the interaction definition, place "missing" or "not defined" in **BOLD** text in Column 2. If the definition refers to a "TBD" or "TBR," place that text in Column 2 in *ITALIC* text.

In Column 4, "External System" put the name of the external system that is involved in the interaction.

Next, fill out Column 5, "External System's Interface Requirements." This involves reviewing the set of design input requirements for the external system named in Column 4 and locating the "other half" of the interface requirement pair that "should" exist as a complement to the SOI interface requirement in Column 1. Ideally, the external system will also have an external interface diagram or boundary diagram that can be referred to indicate its external interactions with other systems—including the SOI for which the audit is being performed. If the complement interface requirement does not exist, flag this as "missing" in **BOLD** in column 5.

Next, assuming it is defined, copy the pointer in the external system's interface requirement concerning the location of the interaction definition and paste the location of the definition in Column 3

"External System Interaction Definitions." Again, if there is no pointer to the interaction definition, place "missing" or "not defined" in **BOLD** text in Column 2. If the definition refers to a "TBD" or "TBR," place that text in Column 2 in *ITALIC* text.

Note: If the SOI is interfacing with an existing system, there may not be an existing system interface requirement concerning the SOI; however, there should be a definition concerning the interaction(s) in that existing system's ICD or IDD.

Ideally, the external system complement interface requirement will point to the same agreed-to definition that the SOI's interface requirement points to. If the pointers to the interaction definitions in Columns 2 and 3 are different, flag the difference by turning the text of both definitions to **BOLD**, indicating the inconsistency.

Once all the SOI's interface requirements concerning the external system have been addressed in the spreadsheet, it is important to look at the interface diagram or boundary diagram and resulting interface requirements for the external system to determine whether the external system has indicated an interaction with the SOI, yet the SOI diagrams and interface requirements did not address that interaction. If these interactions are not addressed by the SOI, then fill out Columns 3, 4, and 5 for the external system. Then if the SOI's complement requirement does not exist, flag this case as "missing" in **BOLD** in column 1.

Repeat the above steps for every external system with which the SOI has an interaction. If there are any external systems the SOI has indicated an interaction with, but that external system has no indication of any kind of interaction with the SOI, type the name of the external system in Column 4 in **BOLD** text.

TABLE 6.4 Example Interface Requirements Audit.

SOI Interface Requirements	SOI Interaction Definitions	External System Interaction Definition	External System	External Systems Interface Requirements
SOI IR 1	Def 111	Def 111	ES1	ES1 IR 1
SOI IR 2	**Def 112**	**Def 123**	ES1	ES1 IR 2
SOI IR 3	Def 113		ES1	**MISSING**
SOI IR 4	***TBD***	***TBR***	ES1	ES1 IR 4
SOI IR 5	**MISSING**	Def 134	ES2	ES2 IR 1
MISSING		Def 135	ES2	ES2 IR 2
SOI IR 6	Def 114	Def 114	ES2	ES2 IR 3
SOI IR 7	Def 115		**ES2**	**MISSING**
SOI IR 8	**Def 116**	**Def 128**	ES3	ES3 IR 1
SOI IR 9	Def 117		**ES3**	**MISSING**
SOI IR 10	**MISSING**	Def 131	ES3	ES3 IR 1
SOI IR 11	***TBD***	***TBR***	ES3	ES3 IR 3
MISSING		Def 131	ES4	ES4 IR 1
SOI IR 12	Def 138	Def 138	ES4	ES4 IR 2
SOI IR 13	Def 119		**ES4**	**MISSING**
MISSING		Def 135	ES4	ES4 IR 3

When looking at the results of the initial audit, there will be a lot of **BOLD** and *ITALIC* distributed across the rows and columns. These indicate issues that need to be investigated and resolved. This is what makes the audit worthwhile. The project does not want to wait until the start of system integration, system verification, and system validation to find out these issues exist—the impact to cost and schedule would be significant.

Table 6.4 is an example of a finished audit spreadsheet with a few of the rows filled in. For an interface requirement audit for the spacecraft example shown in Figure 6.10, there will be multiple rows. Given there are 14 external systems and assuming an average of four interactions with each external system with three interface requirements for each interaction, there could easily be 168 rows, one for each interface interaction/requirement.

This example is for the systems external to the SOI. Referring to the internal interactions between system elements within an SOI, the number of interactions could be much larger. Because of this, it is critical for the project team to do an interface requirement audit for all internal system elements as well. Ideally, the project team would develop one or more internal interface diagrams to show all the internal interactions between system elements.

6.2.4 Use of Attributes to Develop and Manage Design Input Requirements

As defined in Chapter 2, a requirement expression includes a requirement statement and a set of attributes. Requirement attributes aid in the definition and management of the design input requirements: *Refer to Chapter 15 for a more detailed discussion on attributes, definitions of each, and guidance concerning the use of attributes.*

6.2.4.1 Inheritance When it comes to attributes, it is important to understand the concept of "inheritance." In the context of needs and design input requirements, when the needs expressions are formed, they include both the need statement and a set of attributes. When these needs are transformed into design input requirements, each design input requirement expression consists of a requirement statement and a set of attributes. In several cases, the attributes assigned to the design input requirement should be inherited from the need from which it was transformed. For example, owner, priority, criticality, risk (mitigation), and KDR. If a need is critical, the implementing design input requirements must also be critical. The concept of inheritance also applies to parent/child relationships where there may be some attributes of the parent that need to be passed down (inherited) by the child requirements.

The concept of inheritance can also be applied to cases where models were used to identify needs and design input requirements. Model elements can be assigned attributes as well. If a function within a model has an attribute of "critical," the corresponding needs and design input requirements linked to that function within the model should also inherit the "critical" attribute.

As in genetics, a design input requirement will inherit attributes from its source (need) and parent requirement. There may be cases where the attributes of the allocated parent requirement and the need for which the requirement was transformed are not the same. For example, a parent requirement may not have the attribute "critical" in the context of the SOI it represents higher in the architecture; however, the need for a subsystem or system element to which the requirement was allocated and from which the child design input requirement was transformed may have the attribute "critical" in the context of the element for which the need and requirement applies. In this case, the concept of "dominant" gene also applies to a "dominant" attribute. The "critical" attribute by its nature would be considered dominant (more important), thus the child requirement would inherit the more important "critical" attribute.

6.2.5 Plan for System Verification

The outcome of all successful projects is a verified and validated SOI that has been accepted by the customer or has been approved for use by the *Approval Authority*. System verification is obtaining the objective evidence needed to show that the SOI satisfies its set of design input requirements.

The GtWR includes the characteristic, *C7—Verifiable*, for well-formed design input requirement statements. "Verifiable" means each requirement statement is structured and worded such that its realization by the design and resulting SOI can be verified to have been met to the customer's or regulator's satisfaction at the level the requirement exists.

A best practice to ensure the design input requirements are "verifiable" is to plan for how the project will verify that the system will meet each requirement during system verification activities (discussed in Chapters 10 and 11).

Information that should be defined concerning system verification for each requirement statement includes system verification *Method, Strategy, Success Criteria,* and the *Organization Responsible* for the planning and execution of the system verification activities, and condition of use. This information can be defined within the system verification attributes that should be included within each design input requirement expression.

A6: System Verification *Success Criteria*

A7: System Verification *Strategy*

A8: System Verification *Method*

A9: System Verification *Responsible Organization*

A12: Condition of Use

Note: Refer to Chapter 10 for a detailed discussion concerning the system verification Method, Strategy, and Success Criteria and Chapter 15 for a discussion on the use of attributes.

Establishing this information when the requirement statements are defined is important, in that it will help ensure the requirement statements are worded properly and clearly state the intent; it is this intent that must be confirmed to have been met as part of both design and system verification activities.

Addressing system verification early will elicit important project requirements in terms of facilities, enabling systems, test equipment, environments, and resources that will be needed to support the system verification activities. This information is used by the project team in their budget and schedule planning. More detailed design verification and system verification planning will be defined concurrently during the architecture definition and design definition activities. *This activity helps establish that the needs associated with interfaces have the GtWR characteristic C7—Verifiable.*

For cases where there is a customer/supplier contractual arrangement, it must be made clear during design input requirement definition activities the roles and responsibilities for the planning and conduct of the system verification activities and recording the results: the customer, the supplier, or a combination of both as discussed in Chapter 13.

6.2.6 Finalize Design Input Requirements Definition

6.2.6.1 Record the Design Input Requirements The set of design input requirements must be recorded within the SOI's integrated dataset in a form and media suitable for review and feedback from the stakeholders, as well as support traceability across the lifecycle and the requirement verification and requirement validation activities discussed in Chapter 7.

Design input requirements can be recorded and managed via either a document-centric or data-centric approach; however, based on the discussion in Chapter 3, it is highly recommended the organization use a data-centric approach and document the set design input requirements within the project's toolset.

The chosen method should result in traceability records that show:

- Traceability to the needs from which they were transformed.
- Traceability to the parent requirements and other sources.
- Traceability to dependent peer requirements.
- Traceability between interface requirements and interface definitions.
- Traceability to system verification artifacts.
- Traceability to functional architectural and analytical/behavioral models.

When using a document-centric approach to communicating and managing requirements, it is useful to provide information about requirement relationships within the published document. Developing document publication methods within different requirement tools is an area that will need consideration when working with requirement management approaches that allow links and attributes to be exported along with the requirements themselves.

Examples of this are displaying information about the parent/child relationships, which can often be provided from a link or attribute within the RMT database in the form of a trace matrix, or about verification attributes that address how the SOI will be verified to meet the design input requirements in the form of a System Verification Matrix.

When a requirement document form is used as part of a contract, it is particularly important to ensure the correct amount of information is provided (or omitted), as the document will be used to inform another company's legally binding delivery of a product.

If the requirements are used entirely within the organization that is producing the SOI, there are many benefits to the project team working directly with the requirements within the RMT (a data-centric SE approach) and avoid incurring overhead on hardcopy documentation (a document-centric SE approach).

Once the set of design input requirements for the SOI has been recorded, the set can be baselined in accordance with the *Needs, Requirements, Verification, and Validation Management* activities discussed in Chapter 14. As part of the baseline activities, the set of design input requirements will go through the design input requirements verification and requirement validation activities discussed later in Chapter 7.

Below are some special considerations concerning completeness, consistency, feasibility, and risk that should be addressed when preparing the set of design input requirements for baselining.

6.2.6.2 Completeness, Correctness, and Consistency As stated in the GtWR, well-formed sets of design input requirements have the characteristic *C10—Complete*. Previously, Sections 4.5.6 and 4.6.4.3 discussed approaches to ensuring the set of lifecycle concepts and resulting integrated set of needs are complete. Well-formed requirements also have the GtWR characteristic *C8—Correct*, and sets of requirements have the GtWR characteristics *C11—Consistent* and *C15—Correct*. A complete, correct, and consistent set of lifecycle concepts is key to having a complete, correct, and consistent set of needs, which, in turn, is key to having a complete, correct, and consistent set of design input requirements. *If the project has not developed an integrated set of needs prior to defining the set of design input requirements as defined in this Manual, there is a greater risk that their set of design input requirements will not be complete, correct, or consistent.*

From a completeness, correctness, and consistency perspective, it is helpful for the project team to address the following questions when defining the set of design input requirements:

- Are there one or more design input requirements for each of the SOI's needs?
- Do all requirements trace to a parent or source? (No orphans.)
- Are there design input requirements defined for each of the areas: Function/performance, Operations (Fit), Physical Characteristics (Form), Quality (-ilities), and Compliance?
- Have all conflicting design input requirements been identified and resolved?
- Have all product lifecycle stages been considered?
- Have all design input requirements been prioritized and criticality established?
- Have all ambiguous terms been addressed?
- For each design input requirement that addresses a function, have all the corresponding performance requirements been defined?
- Has rationale been included in each design input requirement containing numbers to explain how the numbers were derived and a trace to the source of those numbers?
- Have all internal dependencies/relationships been identified and linked (traceability)?
- Are the functional/performance and quality requirements feasible (cost, schedule, technology, legal, regulatory, and ethical)?
- For all requirements dependent on a critical technology, has the TRL been defined for that technology and has a TMP been developed?
- For each risk whose mitigation involves the design of the SOI, have design input requirements been defined and traceability established?
- Are there interface requirements included in the set of design input requirements concerning all dependencies/interactions/interfaces with systems external to the SOI across all lifecycle stages? Have all the interactions been defined, agreed to, and baselined?
- Have all assumptions been recorded in the rationale attribute and validated?
- Are the terms used within the requirement statements consistent with the ontology defined for the project?
- Have all the attributes that have been agreed to by the project been defined within the requirement expressions, including rationale and the attributes associated with system verification?

While these questions should have been addressed during the lifecycle concepts analysis and maturation activities and during the definition of the integrated set of needs, it is a best practice to revisit these questions when formulating the set of design input requirements. If the answer is "no" to any of these questions, the set of design input requirements may not be complete, correct, or consistent, resulting in a risk that the system will fail system validation.

6.2.6.3 Requirements Feasibility and Risk As stated in the GtWR, well-formed requirement statements have the characteristic *C6—Feasible*, and sets of design input requirements have the characteristic *C12—Feasible*. Section 4.5 discusses approaches to ensuring the set of lifecycle concepts is feasible. A set of feasible lifecycle concepts is key to having an integrated set of needs that is feasible, which, in turn, is key to having a feasible set of design input requirements.

Similar to the Needs Feasibility/Risk Bucket discussed in Section 4.6.4.4, the Requirements Feasibility/Risk Bucket shown in Figure 6.17 is one approach the project team can use to address feasibility as well as to manage risk when defining a set of design input requirements.

Feasibility and risk assessment are key reasons for doing the SOI lifecycle concepts analysis and maturation activities prior to defining and baselining the integrated set of needs from which the

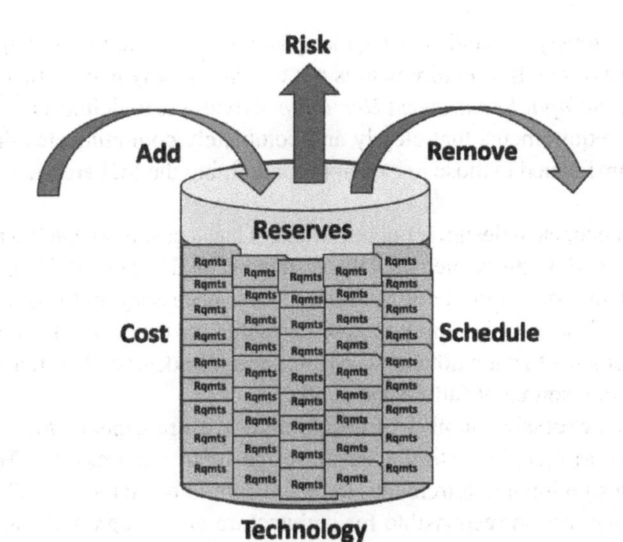

FIGURE 6.17 Requirements Feasibility/Risk Bucket.

design input requirements are transformed. If it is not feasible with acceptable risk, the requirement should not have been included in the set. Doing so can negatively impact cost and schedule and increase the risk that a requirement that cannot be met (fail system verification).

The assessment of feasibility is not binary. A feasibility analysis of each requirement that is a candidate for inclusion in the set of design input requirements must be made in terms of the cost, schedule and technology and risk *before* adding the specific requirement into the set of design input requirements.

6.2.6.3.1 Feasibility of the Set of Design Input Requirements Given cost and schedule constraints, only so many requirements will fit in the bucket without undue risk to the project. While individual requirements may be feasible, implementing the set of requirements within the budget and schedule constraints may not be.

As part of baselining the set of design input requirements, an overall risk assessment is made of the set of requirements, and risk mitigation plans are defined based on the level of risk. It is common for project risk mitigation plans to include a de-scope plan as well as a management reserve to help mitigate the risk for the unknown/unknowns that frequently occur during a development project, especially for a project that is dependent on the maturity of critical technologies that have a low TRL for this point in the lifecycle.

6.2.6.3.2 Controlling the Size of the Bucket A major challenge is limiting the number of requirements that are put in the "requirement bucket." It is common to see cases where projects over-specify their requirements for a product, or cases of scope creep where requirements are added without considering the feasibility, value, or risk of doing so. Rather than focusing on the MGOs, the project tries to develop a system that does something for everyone. The result is often a system that is "a jack of all trades, but a master of none," resulting in a high-risk set of needs that may not be feasible. *Take, as an example, the case where one government organization had defined over 4000 requirements for a single system element.*

As discussed previously, a mandatory characteristic of any design input requirement is that it is "needed" *(C1—Necessary)*. If a requirement is not necessary, why is it in the requirement set? The goal during the *Design Input Requirement Definition* activities is to define a necessary and sufficient set of design input requirements that clearly and completely communicates the set of needs from which they were transformed to those responsible for defining the SOI architecture and design—and no more.

By its very existence, each design input requirement has a cost associated with it because it must be maintained, managed, implemented, and the system verified to meet it. Unnecessary design input requirements result in work being performed that is not necessary and take resources away from the implementation of those design input requirements that are necessary. In addition, implementing design input requirements that are not necessary may result in degraded system performance as well as introduce a potential source of failure and conflict.

One reason for an excessive number of design input requirements is due to specifics of implementation (i.e., design) creeping into the set of design input requirements. When developing the "design-to" set of design input requirements, the focus should be on the "what" not "how." With this information, the organization responsible for architecture and design addresses the "how"—what hardware, mechanisms, and software will result in the set of design-to requirements being met. That is the major reason the set of "design-to" set of requirements are referred to as "design input requirements" in this Manual.

Another reason for an excessive number of design input requirements for a system, subsystem, or system element at a given level is the inclusion of lower-level requirements at too high a level, thus those requirements are not consistent with the GtWR characteristic *C2—Appropriate*. The requirements may be valid and needed, but they are being communicated at the wrong level.

Gold plating is another major cause of unnecessary design input requirements being included in the set. Gold plating is the act of adding requirements for features that are not necessary to address the problem or opportunity the SOI is being developed to address. This results in requirements that do not conform the GtWR characteristic *C1—Necessary*.

How can gold plating be identified? If a design input requirement cannot be traced to a need from which it was transformed, an allocated parent requirement, or a source, it could mean that the design input requirement is not necessary and, thus, is gold plating. The project team needs to be sensitive to gold plating, as this is a major source of requirements creep, can impact the cost and schedule of the project adding to the risk of project failure. Refer also to Section 14.2.5.1 for more details concerning controlling needs and requirements creep.

To help combat the uncontrolled growth of requirements that are not necessary, one approach is for the project team to adopt a zero-based approach to defining the set of design input requirements and maintain this zero-based approach throughout project execution. A zero-based approach means that all requirements are thoroughly evaluated by the project not only for feasibility but also for applicability and value regarding meeting the agreed to lifecycle concepts and integrated set of needs and thereby "earn" their way into the set of requirements. The use of the concepts of criticality and priority will aid in this evaluation. If the requirement is not critical or of high priority, why add the requirement to the set?

In addition to the feasibility evaluation, ask:

- *Why is the requirement needed?* What role does it play in the realization of the need from which it is being transformed, its allocated parent, or other source? If there is no rationale or a weak rationale for its existence, consider not adding it to the set or removing it from the set.
- *What value does the requirement add as compared to the other requirements?* If the value is low or questionable, consider not adding the requirement to the set or removing it from the set.

- *What would happen if this requirement was not included in the set; would the developing organization address this concern anyway?* If nothing negative would happen or the developing organization would have to address this concern anyway to meet other requirements, then why should it be included in the set?
- *Is the requirement communicating "how" to do something (implementation), i.e., it is a design output rather than a design input?* If not justified, consider not adding it to the set or removing it from the set.
- *Is there a trace from the requirement to a need, allocated parent requirement, or source?* If it cannot be traced to a need, allocated parent requirement, or source, why should it be included in the set? *Note: There may be cases during analysis in which the project team determines the requirement is necessary even if it cannot be traced to a need, allocated parent requirement, or source. In this case, the issue may be a missing need, parent requirement, or source which the project team will need to resolve.*
- *Is the requirement appropriate to the level of architecture at which the system, subsystem, or system element it is being stated?* The requirement may be valid, but not appropriate for the system, subsystem, or system element at that level. If the requirement is not being stated at the appropriate level, move it to the appropriate level to which it applies.

Adopting a zero-based approach to defining design input requirements, assessing feasibility, and asking these questions will help avoid gold plating and specifying requirements that are not needed and defining a set of design input requirements that is necessary and feasible at a risk that is appropriate for the lifecycle stage of development.

6.2.6.3.3 Managing Change, Feasibility, and Risk The requirement feasibility bucket is also useful for managing change. Once the bucket is full and the set of requirements is baselined, change will happen. If any of the requirements change, or a new requirement is proposed to be added, a feasibility/risk analysis must be made as part of the change impact assessment activities.

When a stakeholder wants to add a new requirement to the already full bucket, responses include:

1. *Remove a lower priority, non-critical requirement* to make room in the bucket for the requirement being added without adding risk or eating into the management reserves. (This is a key reason for assigning priority and critically to each requirement statement.)
2. *Reject the addition of the new requirement.* However, politically, this may not be advisable. A better, response would be "Yes, but… " and explain the possible consequences of implementing the added requirement. If management still wants the requirement added, then accept the addition.
3. *Accept the addition of the new requirement.* Often management will want to accept the new requirement without removing any of the existing needs. In this case, either or both the management reserves could be impacted as well as the risk to the project potentially increased.

Note: Any proposed change to a requirement will involve an assessment of the sources from which the requirement was transformed as well as any dependent peer requirements. For a more mature project, the project team will also need to look down the trace chain to assess the potential impacts on the design and design characteristics that implement that requirement.

Refer to Section 14.2.5 for a more detailed discussion concerning managing change and assessing the impacts of a change.

6.3 BASELINE AND MANAGE DESIGN INPUT REQUIREMENTS

The outputs of the *Design Input Requirements Definition* activities, as shown in Figure 6.1, are listed below. This list is not inclusive as it does not include the PM work products that are developed concurrently with the SE artifacts.

- System and allocated baselined sets of design input requirements.
- Updated feasible lifecycle concepts.
- Updated system architecture, analytical, and behavioral models.
- Physical architecture.
- Product Breakdown Structure (PBS).
- Definitions of interactions across interface boundaries.
- Design and system verification planning artifacts.
- For each design input requirement (as applicable depending on level):
  - Traceability to need(s), parent requirement(s), or source.
  - Traceability to peer requirements.
  - Traceability to model elements in the architecture, analytical, and behavioral models.
  - System verification attributes defined.
  - Traceability to system verification planning artifacts.
  - Allocation (flow down) of design input requirements to the next level of the system architecture.
  - Budgeting from design input requirements to the next level of the system architecture.

Prior to baselining, the project team must verify and validate the set of design input requirements as discussed in Chapter 7. The requirement statements are verified against the requirement statements quality standards defined in the GtWR (or similar organizational requirement's quality definition document) and validated against the needs, parent requirements, and other sources from which they were derived.

Once the set of design input requirements has been verified and validated, they will be baselined, along with other PM and SE work products and artifacts, at a gate review such as an SRR, in accordance with the activities discussed in Section 14.2.1.

Applications within the project toolset should allow tracking of the status of the *Design Input Requirements Definition* activities using a dashboard that communicates key metrics related to these activities. These metrics can be obtained using the requirement attributes discussed earlier. Making this information accessible will enable stakeholders to have the same view of the maturity of the design input requirement definition activities. This information also represents a single ASoT for the project.

A major activity in managing requirements definition activities is managing change. Establishing traceability and keeping that information current is key to being able to manage the impacts of change throughout the system development lifecycle. The details of requirements management are included in Chapter 14, *Needs, Requirements, Verification, and Validation Management.*

This activity is critical to ensuring the resulting individual requirements have the GtWR characteristics *C1—Necessary, C3—Unambiguous, C4—Complete,* C6—Feasible, and *C8—Correct,* and sets of requirements have the GtWR characteristics *C10—Complete, C11—Consistent, C12—Feasible, C13—Comprehensible, C14—Able to be Validated,* and *C15—Correct.*

Example Set of Design Input Requirements defined by the LIR Project Team

The following represents an initial minimum set of design input requirements for the LIR. Not shown are the attributes that must be defined and combined with the requirement statements to form requirement expressions. Each requirement statement has a unique identifier to enable traceability. Each requirement is traced to the need from which it was transformed N0xx].

Acronyms used: LIS: Lid Installation System; LIR: Lid Installing Robot; FCR: Facility Control Room; LDS: Lid Distribution System; JPS: Jar Processing System. Terms with the form Xxx_Yyy are defined in the project glossary or data dictionary.

Functional/Performance

R001: The LIR shall obtain Lids from the LDS as described in the LIR/LDS ICD xyz, Section 123. [N001]

R002: The LIR shall install on the Jars, Lids having the characteristics defined in Lid Specification LID1234. [N002]

R003: The LIR shall install Lids on Jars having the characteristics defined in Jar Specification JAR1235. [N002]

R004: The LIR shall install Lids on the Jars with a torque within the torque ranges specified in Table 6 for the Jar size and material. [N003]

R005: The LIR shall install Lids on Jars per the timing defined in JPS ICD xyz, Section 123. [N004]

R006: The LIR shall install a Lid on a Jar within 10 sec of receiving from the FCR C&M software a Jar_in_Place message having the characteristics defined in the FCR ICD xyz, Section 123. [N005]

R007: The LIR shall report within 1 sec the completion of Lid installation to the FCR C&M Software as defined in the FCR ICD xyz, Section 123. [N005]

R008: The LIR shall install Lids on Jars positioned on the JPS CBS conveyor belt as defined in the JPS ICD xyz, Section 123. [N006]

R009: The LIR shall maintain the Jar position within ± 0.1 inches on the conveyor belt during Lid installation. [N006]

R010: The LIR shall accept from the FCR C&M software Valid_Commands defined in the FCR ICD xyz, Section 123. [N007]

R011: The LIR shall execute each Valid_Command defined in the FCR ICD xyz, Section 123 within 1 sec of receipt from the FCR C&M software. [N007]

R012: The LIR shall perform the Power_Up_Sequence within 1 sec of receipt of a Power_Up_Command from the FCR C&M Software having the characteristics defined in the FCR ICD xyz, Section 123. [N007]

R013: The LIR shall report the completion of each Valid_Command to the FCR C&M software as defined in FCR ICD xyz, Section 123 within 1 sec. [N008]

R014: When sending Messages to the FCR, The LIR shall use the Message format defined in FCR ICD xyz, Section 123 paragraph 1.2.5. [N008]

R015: The LIR shall perform a Self_Test within 1 sec after Power_Up_Sequence is complete. [N009]

R016: The LIR shall report LIR_Health_and_Status to the FCR C&M Software as defined in the FCR ICD xyz, Section 123 within 1 sec of the completion of a Self_Test. [N009]

R017: The LIR shall install Lids on at least 99% of the Jars presented as defined in the JPS ICD xyz, Section 123, during each Shift. [N010]

R018: The LIR shall update LIR_Health_and_Status every [TBD minutes ± 0.1 min] when the LIR is powered. [N011]

R019: The LIR shall report LIR_Health_and_Status data every [TBD minutes ± 0.1 min] to the FCR C&M software as defined in FCR ICD xyz, Section 123. [N012]

R020: The LIR shall report a Critical_System_Failure within 1 sec to the FCR C&M software per the FCR ICD xyz, Section 123. [N013]

R021: The LIR shall execute the End_of_Shift_Shutdown_Sequence within 1 sec whenever a Critical_System_Failure is identified. [N014]

R022: The LIR shall report changes in System_State to the FCR C&M software as defined in the FCR ICD xyz, Section 123 within 1 sec of occurrence. [N015]

R023: The LIR shall execute the End_of_Shift_Shutdown_Sequence within 1 sec of End_of_Shift. [N016]

R024: The LIR shall update LIR_Current_Time to the FCR_Current_Time received from the FCR C&M Software as defined in the FCR ICD xyz, Section 123. [N016]

R025: If the FCR_Current_Time fails to be updated at least once a minute, the LIR shall execute the End_of_Shift_Shutdown_Sequence within 1 sec of when the fault is discovered. [N013]

Fit (Operational)

R026: The LIR shall Operate using 110–120 VAC, 60 Hz, 30-amp Facility_Power having the characteristics defined in Facility Drawing xyz. [N017]

R027: The LIR shall receive Facility_Power using an EIC 320-compatible electrical 3 prong connector defined in Facility Drawing xyz, Sheet 11. [N017]

R028: The LIR shall use power connectors for obtaining facility power that meet UL standards. [N046]

R029: If an Over_Voltage_Condition, the LIR shall remove power from all internal systems within 1 sec. [N017]

R030: The LIR shall meet the LIR Design Input Requirements while Operating in the Operating_Environment defined in Table 1. [N018]

R031: The LIR shall meet the LIR Design Input Requirements while Operating in the presence of electromagnetic emissions up to and including those shown in Diagram 2. [N019]

R032: The LIR shall meet the LIR Design Input Requirements after being shipped in shipping crates exposed to the Shipping_Environment defined in Table 3 for up to 24 hours. [N020]

R033: The LIR shall meet the LIR Design Input Requirements after being stored in shipping crates exposed to the Unconditioned_Storage_Environment defined in Table 5 for up to three years. [N021]

R034: The LIR shall meet the LIR Design Input Requirements after being sprayed with a 15 ± 2 gal/min water spray for up to 5 min AND allowed to dry for at least 24 hours. [N022]

Form (Physical Characteristics)

R035: The LIR shall limit Operations to within the 4-foot-wide x 16-foot-long x 8-foot-high volume within the LIS_Operations_Area as shown in Figure 1. [N023]

R036: The LIR shall fit through a 3-foot-wide x 7-foot-high door opening when fully assembled AND on the Supplier_Provided_Dolly, without having to remove the door. [N024]

R037: The LIR shall limit the force needed to move the LIR to less than 10 pounds when the LIR is loaded on the Supplier_Furnished_Dolly. [N025]

R038: The LIR shall install Lids on the Jars located on the JPS CBS conveyor belt through the 3 ft by 3 ft inch opening located between the LIS and CBS as shown in Figure 2. [N026]

Security

R039: The LIR shall only accept Valid_Commands from the FCR C&M Software that meet the Security_Protocols defined in the FCR ICD xyz, Section 567. [N027]

Safety

R040: The LIR shall enter Safe_State within 1 sec of the receipt of a Enter_Safe_State Command from the FCR C&M software as defined in the FCR ICD xyz, Section 123. [N028]

R041: The LIR shall execute the End_of_Shift_Shutdown_Sequence within 1 sec of receipt from the FCR software a Shutdown_Command as defined in the FCR ICD xyz, Section 123. [N029]

R042: The LIR shall set the Motion_in_the_Room parameter to TRUE when a Motion_Detected message is received from the LIS MMS having the characteristics defined in the LIS ICD xyz, Section 123. [N030]

R043: The LIR shall enter Safe_State within 1 sec of the MMS Motion_in_the_Room parameter being set to TRUE. [N030]

R044: The LIR shall set Door_Open to TRUE when a KAS Keypad_Accessed message is received from the LIS KAS having the characteristics defined in the LIS ICD xyz, Section 123. [N030]

R045: The LIR shall enter Safe_State within 1 sec of the KAS Door_Open parameter being set to TRUE. [N030]

R046: During Operations, the LIR shall limit sound levels such that the time-weighted average sound level is less than 85 decibels measured on the A scale across the frequency range of 20–20,000 Hz, when measured at 1 ± 0.1 feet from at least 10 equally spaced locations surrounding the LIR shown in Figure 8. [N031]

Quality (-ilities)

Maintainability

R047: The LIR shall have a MTTR not to exceed 15 min. [N032]

R048: The LIR shall identify each Critical_System_Failure to the Line Replaceable Unit (LRU) level. [N033]

R049: The LIR shall provide LIR_Health_and_Status data to the Maintenance Technician when the LIR is not in communication with the FCR C&M Software. [N033]

R050: The LIR shall have Parts, Components, and Assemblies that can be disassembled, removed, repaired, and maintained using those tools listed in [TBD] document. [N034]

R051: The LIR shall have Parts, Components, and Assemblies that are Interchangeable with the same parts, components, and assemblies contained within other LIRs. [N035]

Reliability

R052: The LIR shall have an Operational_Lifetime of at least 5 years when operating per the Shift defined in the JPS ICD 355 days/year AND when maintained per Maintenance Procedure xyz AND operating in the Operational_Environment defined in Table 1. [N036]

R053: The LIR shall have an MTTF of less than or equal to 720 operational hours, when maintained per Maintenance procedure xyz AND operating in the Operating_Environment defined in Table 1. [N037]

Availability

R054: The LIR shall install Lids on Jars during the Shift defined in the JPS ICD, 355 days/year. [N038]

Transportability

R055: The LIR shall fit within shipping packaging that meets the crated size and weight limits for the shipping methods defined in Table 4, when either assembled or unassembled. [N039]

R056: The LIR shall meet its Design Input Requirements after being dropped 4 ± 0.1 feet onto a concrete surface while unpowered AND in its shipping crate or crates. [N040]

Compliance (Standards/Regulations)

R057: The LIR shall comply with [TBD] EPA Regulation xyz concerning the use and handling of hazardous materials. [N041]

R058: The LIR shall have labels that communicate hazards defined in OSHA Cautions and Warnings Regulation xyz that may cause harm to people physically interacting with the LIR. [N041]

R059: The LIR shall have labels that communicate hazards formatted as defined in OSHA Cautions and Warnings Regulation xyz. [N041]

R060: The LIR shall comply with OSHA Regulation xyz, Section 4.6, concerning human/machine interfaces. [N042]

R061: The LIR shall comply with OSHA Regulation xyz, Section 7.8, concerning the safety of humans working with robots. [N043]

R062: The LIR shall comply with FDA Regulation xyz, Section 3.4, concerning the handling of food. [N044]

R063: The LIR shall have Parts that come in contact with the Lids and Jars that are non-porous meeting the requirements defined in FDA Standard xyz, Section 7.9. [N044]

R064: The LIR shall have Parts that come in contact with the Lids and Jars that can be sanitized using sanitizing liquids or materials approved in FDA Standard xyz, Section 8.1. [N044]

R065: The LIR shall meet the LIR Design Input Requirements when lubricants that meet non-contamination lubricant FDA Standard xyz, Section a.b, are used for lubrication of the parts that come in contact with the Lids and Jars. [N044]

R066: The LIR shall limit radiated emissions to be no greater than allowable per MIL-STD-461E, Section 8.9, Table 8.5. [N045]
R067: The LIR shall meet the LIR Design Input Requirements when Lubricants, Fluids, and Fuels that meet UL 546xx non-flammability standard are used for lubrication. [N047]
R068: The LIR shall have Labels that communicate using the English language. [N048]
R069: The LIR shall have Labels having the characteristics defined in Label Standard xyz. [N048]
R070: The LIR shall have Parts, Components, and Assemblies whose physical characteristics are defined using English units of measure. [N049]

6.4 ARCHITECTURAL LEVELS, FLOW DOWN, ALLOCATION, AND BUDGETING

Note: If the reader has not done so, it is highly advised that the reader review Section 2.3.2 concerning levels prior to proceeding.

Unless an SOI requires no further elaboration, once its set of design input requirements has been defined, verified, validated, and baselined, the project team will flow the requirements down to the subsystems and system elements at the next level of the architecture via allocation and budgeting.

For some, architecting is part of the *Design Definition Process*, when in fact they are two vastly different processes. The *Architecture Definition Process* defines the architecture of the subsystems and system elements that make up the integrated product or system. The focus is on function, organization, interactions, and relationships between the subsystems and system elements that make up the SOI. Each subsystem and system element within the architecture is defined by its own set of lifecycle CNR in the context of the system of which they are a part.

Architecture definition and design input requirement definition go together iteratively and recursively as the project team defines the levels of the architecture [52]. The project team identifies subsystems and system elements at one level of the architecture, defines the lifecycle CNR for those subsystems and system elements, flows the requirements down to the subsystems and system elements at the next level, and identifies and defines the interactions (interfaces) between the SOI and its external systems. Because SE is iterative and recursive, these actions are repeated for each entity in the system architecture as the project team moves down the levels of the architecture until all system elements have been defined and no further elaboration is needed for the project team to make a build, make, or reuse decision. The system elements can then be realized via the customer's *Acquisition Process* or internal to the developing organization using the *Design Definition Process*.

If the system element is to be developed by a supplier, the system element becomes that supplier's SOI, and the supplier can develop lifecycle CNR using the customer-supplied system requirements as inputs, as discussed previously. Using their lifecycle CNR for the SOI, the supplier is responsible for the design and definition of the set of design output specifications. If the customer has completed the design and has developed the design output specifications, the supplier is responsible for the production (manufacturing and/or coding) of the system element in accordance with those design output specifications.

The *Design Definition Process* focus is on the physical realization of the system elements based on the set of lifecycle CNR for the SOI. The outputs of the *Design Definition Process* are a set of design output artifacts and specifications to which a system element is produced along with design verification and design validation artifacts.

Once the system elements have been produced, they can be verified against their set of design output specifications (production verification) and set of design input requirements (system verification), validated against their integrated set of needs (system validation), and integrated into the higher-level system element, subsystem, or system of which they are a part.

For more details on the *Architectural Definition Process* and *Design Definition Process*, refer to the INCOSE SE HB.

6.4.1 Moving Between Levels of the Architecture

When talking about moving between levels, what is being referred to is defining and baselining a set of lifecycle CNR for an SOI, defining the next level of the architecture, and then flowing those requirements down to the next level subsystems and system elements for implementation.

For brownfield systems, some elements of the SOI physical architecture will already be defined. For greenfield systems, as part of the feasibility analysis, a functional architecture is developed and mapped to a physical architecture. This enables the project team to identify and assess the maturity of critical technologies needed based on the stakeholder's needs for SOI performance, quality, safety, security, and interactions with external systems. Also, it is the subsystems and system elements within the physical architecture that will be procured or designed, manufactured/coded, integrated, verified, validated, and delivered.

From a design input requirements perspective, it is this architecture that is the focus of requirements flow down from one level of the architecture to another.

Unless the SOI is a system element and needs no further elaboration, for each subsystem and system element at the next level, the *Lifecycle Concepts and Needs Definition* activities *discussed in Chapter 4* and *Design Input Requirement Definition* activities *discussed in this Chapter* will repeat.

6.4.2 Product Breakdown Structure and Document Tree

From a PM perspective, the physical architecture can be mapped to a PBS. The PBS is similar in concept to a work breakdown structure (WBS). The system, along with each subsystem and system element within the architecture, represents an entity within the PBS. Each of these entities is represented by a budget, schedule, development concept, procurement concept, and a set of both project work products and engineering artifacts.

Key engineering artifacts for each entity within the PBS include:

- A set of lifecycle concepts.
- An architectural description.
- Analytical/behavior/architectural models (contained within the integrated system models).
- An integrated set of needs.
- Traceability (may be included with the integrated set of needs).
- System validation planning artifacts.
- A set of design input requirements.
- System verification planning artifacts.
- Interface definition documentation (such as ICDs or IDDs).
- Traceability (may be included with the set of design input requirements).
- Allocation and budgeting (may be included with the set of design input requirements).

In addition to these artifacts, system elements that require no additional elaboration the PBS will also include:

- A design description.
- Design verification and design validation records.
- A set of design output specifications.
- Production verification records.
- System integration records.
- System verification and validation records.
- A maintenance plan.
- Instructions for use.
- A disposal plan.

These sets of artifacts for the system and the lower levels in the system architecture are shown in Figure 6.18 in the context of the PBS and physical architecture.

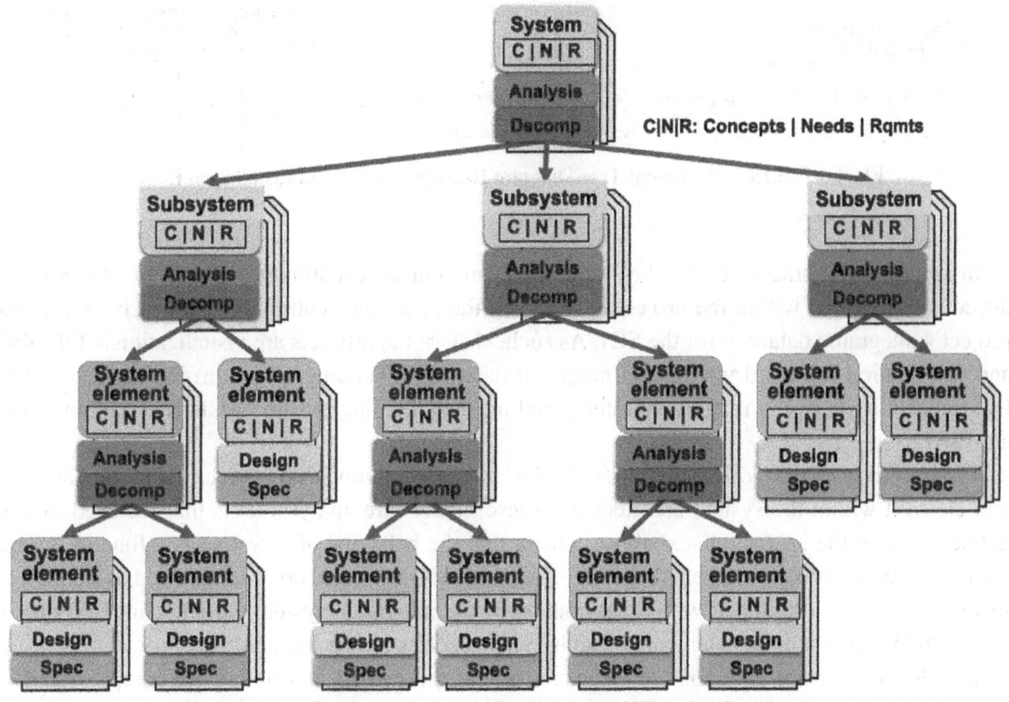

FIGURE 6.18 PBS and Associated Sets of Engineering Artifacts.

The artifacts shown in Figure 6.18 are organized by levels into what has been referred to historically as a "document tree." An example for the LIR is shown in Figure 6.19. This hierarchical view includes the above mentioned artifacts associated with each entity within the PBS that is defined as part of project planning and development of the SEMP. Each artifact is included in the project's

WBS, and the development and delivery of each is included in the project budget, master schedule, and acquisition plan. For subsystems or system elements that are contracted to a supplier, these sets of artifacts are addressed in an SOW and associated Contract Deliverables Requirements List (CDRL) or in an SA as a list of contract deliverables.

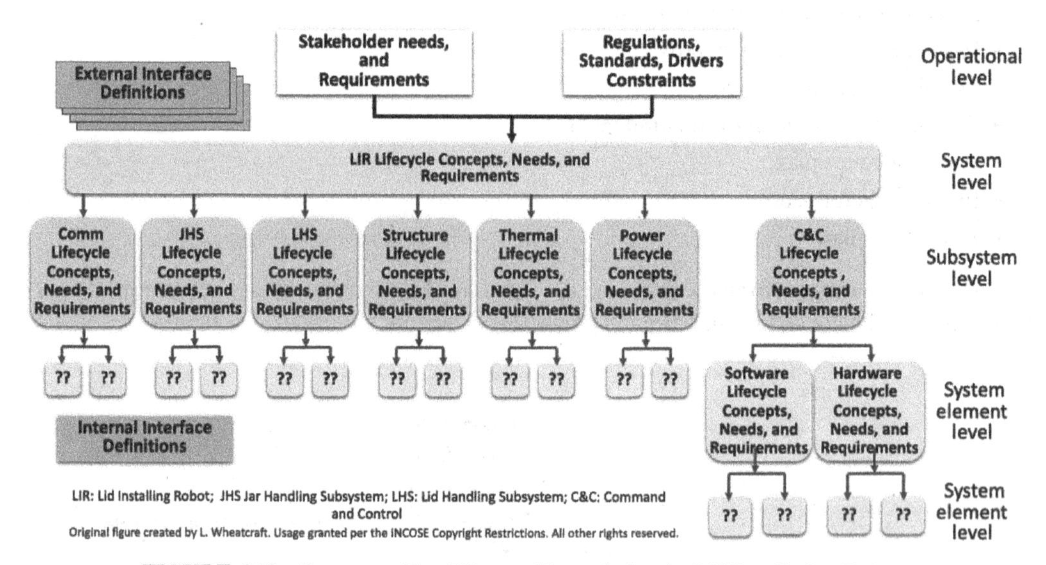

FIGURE 6.19 Document Tree Diagram Example for the Lid Installation Robot.

In a data-centric practice of SE, the data and information representing the various artifacts are produced and managed within the project toolset, and the data representing these artifacts exist in the project's integrated database for the SOI. As such, each of the artifacts are visualizations of the data and information managed within the integrated database. This enables the project team to establish traceability between related data items and artifacts helping ensure consistency and establish an ASoT.

The project team develops analytical/behavior/architectural model(s) for each subsystem and system element within the system architecture. These models are analysis tools that are used to help define many of the artifacts listed above. Given that the behavior of a system is a function of the interaction of its parts, subsystem and system element models should not be developed as standalone models, but rather developed within the context of the integrated system model. Thus, the PBS is represented within a model element hierarchy, such as class or block diagrams.

In a document-centric practice of SE, each form of these artifacts listed above is represented by an actual document in either printed or electronic form. From a needs and requirements perspective, there are separate needs and requirements documents for each entity in the PBS.

To completely describe the SOI, lifecycle CNR sets for each entity within the PBS are needed. As discussed earlier, system, subsystem, and system element lifecycle CNR sets are highly interrelated and dependent. These relationships and dependencies must be considered to ensure the integrated and verified system can be validated to meet its integrated set of needs. The result is an integrated family of lifecycle CNR that exists at multiple levels of the system architecture, linked together to form a data and information model of the integrated system.

Other items to consider when developing the family of lifecycle CNR for the system are the inclusion of other sources of requirements concerning industry standards, regulations, constraints, interface definition documentation, operational environments, enabling systems, and test equipment that may be referred to within the integrated set of needs and set of design input requirements. Some projects will invoke these within the various requirements for subsystems and system elements at each level, others may generate standards and interface definitions that are levied across the system, subsystem, and system elements at each level of the physical architecture or across levels (such as common structural design standard or an ICD) as discussed previously.

For some, the system-level set of lifecycle CNR and associated ICDs or IDDs are referred to as the "functional baseline," and the family of lifecycle CNR and interface definition artifacts for the SOI are referred to as the "allocated baseline." At the PDR, the design team addresses how their design will result in the implementation of the allocated baseline. The sets of design output specifications for the subsystems and system elements that are part of the SOI physical architecture are referred to as the "product baseline" to which the subsystems and system elements will be manufactured or coded.

6.4.3 Allocation—Flow Down of Requirements

Allocation is an activity by which the design input requirements defined for an entity at one level of the architecture are assigned (flow down) to the entities at the next lower level of the architecture that has a role in the implementation of the allocated requirement (shown in Figure 6.18). The arrows shown in Figure 6.20 indicate the flow down of requirements from an entity at one level of the architecture to entities at the next level.

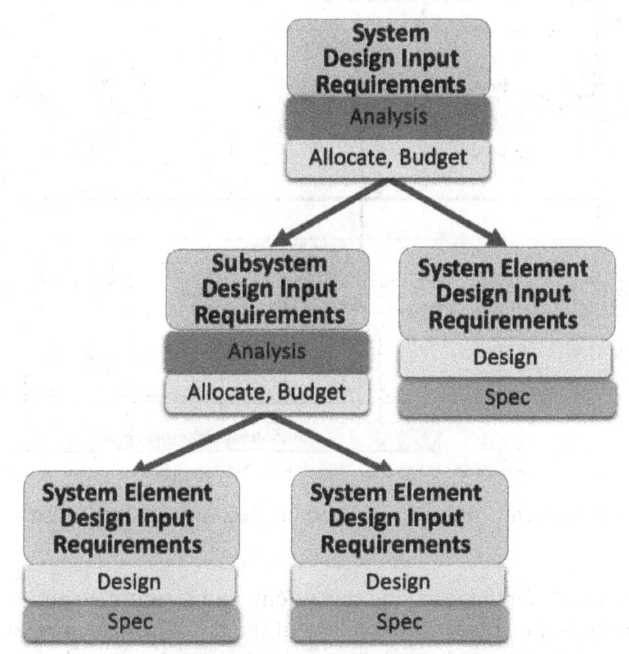

FIGURE 6.20 Flow down of Design Input Requirements via Allocation and Budgeting.

Based on analysis of the design input requirements and the function of the system whose requirements are being allocated, the *Architecture Definition Process* decomposes the system into subsystems and system elements, resulting in the next level of the architecture.

As part of this analysis, the project team determines what "role," if any, each subsystem and system element at the next level of the architecture has in the implementation of each design input requirement being allocated.

For each subsystem or system element that has a role, the requirement will be allocated to those subsystems or system elements. As shown in Figure 6.21, the allocated requirements are drivers for the receiving subsystems or system elements; as such, they are inputs to the *Lifecycle Concepts and Needs Definition* activities. The resulting set of needs is transformed into a set of design input requirements for the subsystem or system element the higher-level parent requirements were allocated to.

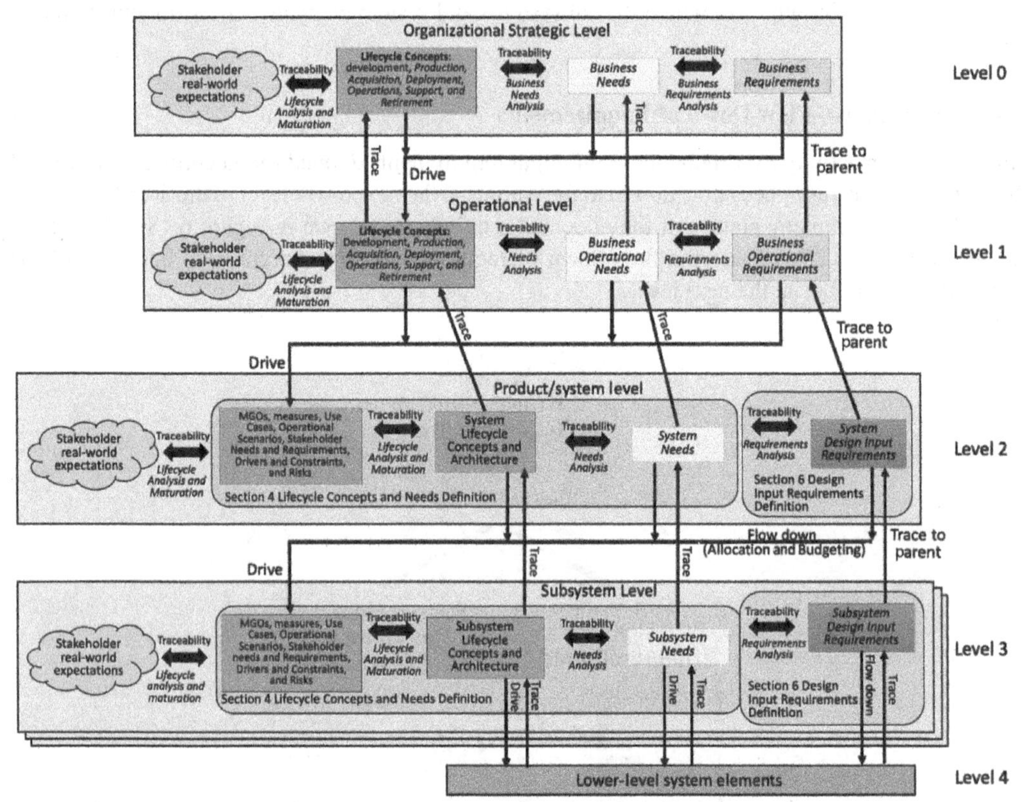

FIGURE 6.21 Requirements At One Level Drive the Requirements at the Next Level.

The design input requirements for the receiving subsystems and system elements will trace to the needs from which they are transformed. In addition, any of the resulting design input requirements that address the allocated parent will also have a trace to that parent. As such, they are referred to a child requirement for the allocated parent. As a set, the child requirements contained in all the sets of design input requirements for each subsystem and system element the parent requirement was allocated to must be necessary and sufficient to meet the intent of the allocated parent requirement. *See Section 6.4.4 for a more detailed discussion on defining child requirements.*

It is important to understand the distinction between allocation and traceability in the context of design input requirements. While requirement traceability is establishing a link between two requirements (parent/child, child/parent, peer/peer, and requirement/source), allocation is linking a parent requirement to a subsystem or system element at the next level of the architecture that has a "role" in meeting that requirement.

Theoretically for greenfield systems, at the time of allocation, the design input requirements for the subsystems and system elements at the next level of the architecture have not yet been defined.

Once the child requirements have been defined for each of these subsystems and system elements, they will be recorded and traceability to the allocated parent requirement will be established. With the concept of bi-directional traceability, once the trace from child to parent is established, a reverse trace from parent to child is also established.

If using the data-centric approach advocated in Chapter 3, role analysis and allocation are done within a single integrated model. As such, the role analysis and allocation are like "pealing back layers of an onion" increasing the level of abstraction as the project team moves down the levels of architecture within the model. Out of this analysis comes both the definition of the next level of the architecture as well as the allocations of requirements to the applicable subsystems and system elements at that next level.

In a document-centric practice of SE, the role analysis and allocation are more reliant on the knowledge and experience of the members of the project team. Often spreadsheets are included as appendices to show allocation as well as parent/child relationships.

6.4.3.1 Software-Intensive System Architecture It is common to define physical architectures in terms of "system," "subsystem," and "system element" designations. While these naming conventions work for electrical/mechanical-centric systems, they may not work well for software-intensive systems that tend to have a "flatter" architecture organized by function or feature.

In the past, when a subsystem included software, the software was listed under the subsystem along with other hardware or mechanical parts that made up that subsystem (as shown in the LIR example presented earlier). As a result, the software was shown several levels below the system level at either the subsystem or system element levels. An issue with this approach is that software development is then done in isolation (silos) by the organization responsible for developing that subsystem or system element. When there are multiple subsystems and system elements containing software, the result is often a fragmented approach to software development, leading to system integration issues and failed system validation.

In today's software-intensive systems, it is common for aircraft, spacecraft, and even automobiles to have over a million lines of code. Core functionality and performance are part of the software rather than the hardware. However, while discrete functionality is assigned to software, software performance is dependent on hardware: processor, memory, power, communication buses, sensors, and actuators.

Because of this, team members responsible for developing the embedded software must work closely and collaboratively with the team members responsible for the hardware within which the software modules run and which provides the inputs to the software and are commanded by the software.

Unfortunately, many organizations are still using the same traditional hardware-centric architecture when developing software-intensive systems, resulting in multiple software segments not only being developed in a silo but also going through independent software segment verification "testing" with emulators or simulators of external hardware and software systems rather than actual system hardware and software. To save time and money, integrated system software and hardware verification with the software running within the actual integrated hardware/software operational environment with the intended users is often skipped, resulting in the integrated system failing system validation.

A different approach to systems engineering architecture is needed to accommodate software-intensive systems. **Rather than developing subsystems that contain software, the view should change that the project team is developing software-intensive systems where hardware and mechanical systems are enabling systems for the software.**

For the purpose of this discussion, hardware is defined as system elements that involve electricity in any form; mechanical system elements have no electricity, but could have fluids (hydraulic, cooling, or fuel for example.). The mechanical system elements hold the hardware together providing structure and the "skin" of the system. The software runs on hardware microprocessors, using hardware memory, communicating over hardware-supplied buses, gets data from hardware sensors, interacts with the user via hardware-supplied I/O devices, and commands hardware systems.

In many cases, the hardware components include electronic control units (ECUs) containing "firmware" that communicates with the central system software over communication buses. Given that the firmware is really software as well, the result is software distributed across the system. For example, today's automobiles may have over 150 software modules and associated hardware microprocessors and multiple communication buses.

Rather than standalone applications, the software is "embedded" in the hardware and "dependent" on that hardware. In this context, the software is "constrained" by the hardware. However, it is the software that provides the core functionality and performance as viewed by the users.

Before dividing a system directly into traditional subsystems and decomposing the subsystems into system elements, an additional layer needs to be acknowledged that exists between the top system level and traditional subsystems. An alternate approach is to first divide the system into hardware, mechanical, and software, as shown in Figure 6.22, and then decompose the hardware, mechanical, and software systems into lower-level subsystems and system elements.

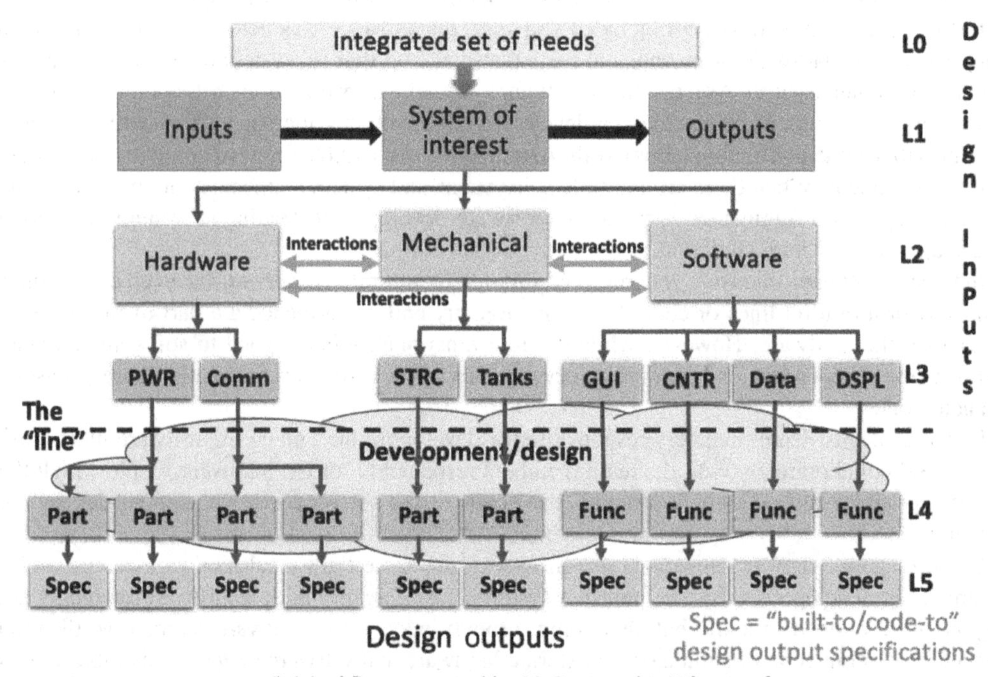

FIGURE 6.22 Alternate Architecture for Software-Intensive Systems.

The top-level SOI requirements are first allocated and budgeted to the hardware, mechanical, and software involved in meeting a system-level requirement. There will be dependencies and interactions between the hardware, mechanical, and software, thus the increased importance of identifying interface boundaries, defining the interactions across those boundaries, defining interface requirements, and identifying constraints for each interaction.

Using this approach, the hardware, mechanical, and software systems are much more tightly coupled. Because of this, the software modules must not be disconnected from the hardware architecture in which it is embedded. The intent is to define a single integrated architecture with multiple views—a hardware/mechanical view and a software view. With this approach, the software architecture view is overlaid with the hardware/mechanical architecture view. This is a major difference between developing standalone software applications and software embedded within the SOI.

Rather than the hardware, mechanical, and software decomposition shown in Figure 6.22, some organizations choose a different breakdown. For example, railroad systems are commonly first divided into "onboard systems," "off-board systems," and "control systems."

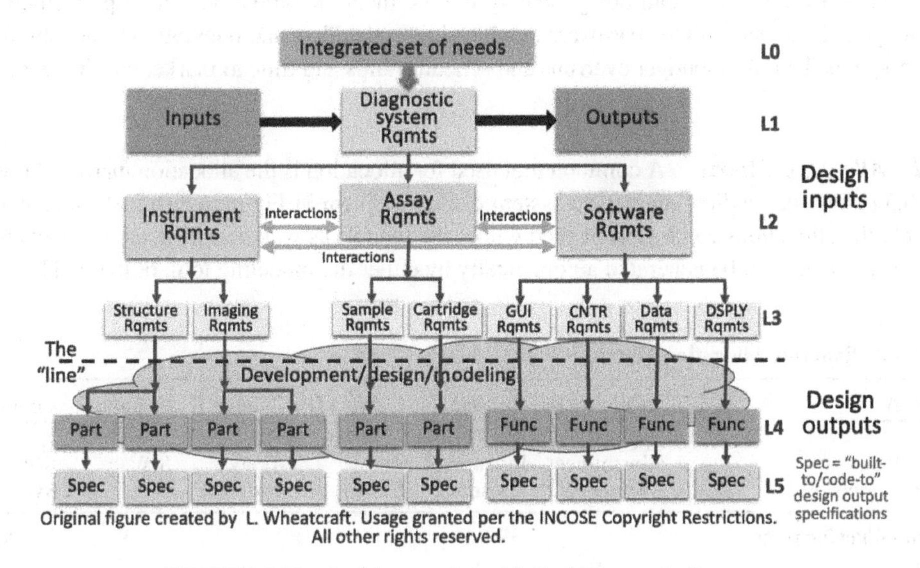

FIGURE 6.23 Architecture for a Medical Diagnostic System.

As shown in Figure 6.23, a medical diagnostic system may be divided into "instrument," "assay," and "software," where the assay consists of a biological sample, slide or cartridge, and reagents. The assay is inserted into the instrument that collects data in several forms (for example, high-definition images at different wavelengths of light) and then the data are analyzed by the software to detect whatever it is intended to detect. The software also has modules involved in the user interface with the instrument, control of hardware components, and reporting the results of the analysis.

For modern software-intensive systems to be successful, the project team members must work much more closely as an integrated, collaborative, multidiscipline team to develop integrated models of the SOI and focus on the integrated system behavior. As an example, for the medical diagnostic system discussed above, system-level requirements associated with accuracy, precision, total allowable error, and total time to get a result are allocated to the instrument, assay, and software, as shown later in Table 6.5. The engineers and scientists responsible for each must work together to be able to ensure each of the system-level requirements are met.

It is also important to note that the traditional hardware-centric approach and the traditional software-centric approaches to system development will need to be combined into one integrated approach where the organizational silos are minimized and there is a single data and information model of the integrated system.

Project team members will need to have a better understanding of the culture associated with the other team members. For example, software architectures tend to be "flatter" with fewer levels of decomposition and tend to be organized more based on features and capabilities (with implicit assumptions about enabling hardware performance) rather than the traditional hierarchical view of hardware/mechanical systems. In addition, the trend is toward a more "Agile" approach to software development that will need to be acknowledged by the SE approach commonly used by hardware/mechanical focused systems engineers. This can result in an approach to SE taking advantage of the strengths of both approaches.

Using this approach, software can be developed under a common management structure, using a common ontology and definition of interactions in terms of messages, data, telemetry, and commands. The software testing and design and system verification validation can be performed in a more integrated manner with the hardware in which it is embedded, and integration issues should be much less, thus the risk of budget overruns and schedule slips and time to market will be reduced.

6.4.3.2 Allocation Matrix A common tool used for allocation is the allocation matrix, shown in Table 6.5 (using the medical diagnostic system example shown in Figure 6.23). For a data-centric approach, the allocations can be established within the models as well as within the RMT enabling the allocation matrix to be generated automatically by either the modeling tool, or the RMT.

TABLE 6.5 Example Allocation Matrix.

Column A	Column B	Column C	Column D	Column E	Column F	Column G
	Instrument Systems		Assay Systems		Software Systems	
System A Design Input Requirements	Instrument System 1	Instrument System 2	Assay System 1	Assay System 2	S/W System 1	S/W System 2
Functional/performance					X	X
Rqmt 1	X	X			X	X
Rqmt 2		X	X	X	X	X
Operational (fit)						
Rqmt 3	X	X			X	
Rqmt 4			X	X		X
Physical Characteristics (form)						
Rqmt 5	X	X	X	X	X	X
Rqmt 6	X	X		X		
Quality (-ilities)						
Rqmt 7	X	X	X	X	X	X
Rqmt 8	X	X	X	X	X	X
Compliance (standards/regulations)						
Rqmt 9	X	X		X	X	X
Rqmt 10	X	X	X	X	X	X

While similar, the analysis associated with allocation is slightly different for each of the types of requirements discussed earlier: functional/performance, operational (fit), physical characteristics (form), quality, and compliance. The functional/performance and operational requirements will be allocated to the specific subsystems and system elements that have a role in meeting that parent requirement. Operational requirements dealing with power will only be allocated to those systems that use power, but not to a software or mechanical system. Physical characteristics like mass will be allocated to all applicable physical subsystems and system elements, but not software. Requirements dealing with quality tend to apply to all subsystems and system elements, so they would be allocated accordingly. Standards and regulations would be allocated to all subsystems and system elements to which a specific standard or regulation applies.

Generating the allocation matrix is important as it allows the project team to review and assess the allocations. If allocation is addressed properly within an RMT or model, the allocation matrix should be able to be generated as a report. For each set of design inputs for entities within the system architecture that require further elaboration, this assessment should address the following questions:

- Has every requirement been allocated? *If a system element requires no further decomposition and a buy, make, or reuse decision can be made, further allocation of that system element's requirements is not applicable.*
- Are the allocations made in the RMT and within the models consistent?
- Does the owner of the allocated parent requirement agree with the allocation?
- Do the owners of each receiving entity agree with the allocation? Is their role in meeting the requirement allocated to them clearly understood?
- If a requirement is allocated to a single entity, should it be also allocated to another entity at the same level, or is the parent requirement stated at the wrong level? *If a parent requirement only applies to a single entity, then it could be it is at the wrong level and a proper parent requirement is missing.*
- For the requirements that are allocated to multiple entities, does that involve a dependency or an interaction (interface) between those entities?

While a report can be generated from the RMT or model to determine if each requirement is allocated, the report cannot address the correctness of the allocation—was the requirement allocated to the right entities and all the entities that have a role in meeting the intent of the allocated requirement? Correctness questions will have to be answered by members of the project team.

In a document-centric practice of SE, the allocation matrix is often included as an appendix to the design input requirements document. In some cases, rather than an allocation matrix, the project includes a parent/child traceability matrix, with the assumption that if the project team responsible for a system, subsystem, or system element derives child requirements for a parent requirement, that parent must have been allocated to that system, subsystem, or system element. Conversely, if there are no child requirements for a system, subsystem, or system element, then the parent requirement must not have been allocated to that system, subsystem, or system element. *In both cases, these assumptions could be incorrect.* It is best to have separate matrices for allocation and parent/child traceability. As discussed in Sections 14.2.6 and 14.2.7, the project team can do the analysis needed to assess the "correctness" of both the allocations and traceability.

This activity helps establish individual requirement statements that have the GtWR characteristic *C2—Appropriate*, and the set of design input requirements has the GtWR characteristics *C10—Complete* and *C11—Consistent*.

6.4.4 Defining Child Requirements That Meet the Intent of the Allocated Parents

As shown in Figure 6.21, once the parent design input requirements have been allocated, they become drivers for each of the receiving subsystems or system elements along with all the other drivers and constraints identified during the *Lifecycle Concepts and Needs Definition* activities discussed in Chapter 4. As such, the project team for the receiving entity will include in their lifecycle concepts analysis and maturation activities, the specific "role" they have in the implementation of the intent of each of the allocated parent requirements.

This cannot be done in a silo. The "role" a system, subsystem, or system element has must take into consideration the "role" of the other subsystems or system elements at the same level that were allocated the same parent requirement. Because they were allocated the same parent requirement, there will often be, to some degree, a dependency and/or interaction.

For example, if the parent requirement is a functional/performance requirement, then each receiving system, subsystem, or system element needs to define one or more subfunctions and performance child requirements, which together will result in the parent requirement's function and performance to be achieved. This may involve an interaction between subfunctions and thus an internal interface whose interaction must be defined and implemented. Each of the child requirements for each of the receiving subsystems or system elements contributes to the performance associated with the parent requirement.

This is especially true for software-intensive systems. It will be common to have a functional/performance requirement allocated to both hardware and software. Each of the project teams will need to determine their specific role and how each will interact such that the intent of the parent requirement is met. (*Note: this may be a new concept for software engineers experienced in developing standalone software applications, but who do not have experience in developing embedded software dependent on hardware systems.*)

These dependencies are much easier to determine and assess within the functional, analytical, and behavioral models of the integrated system. If using the data-centric approach advocated in this Manual, dependencies will be discovered and managed within the models helping to ensure consistency of the requirements within a set as well as consistency with requirements that have a dependency with requirement sets for other subsystems and system elements.

Performing the lifecycle concepts and needs definition activities discussed in Chapter 4, there will likely exist needs that address the allocated parent requirements. For example, referring to the example Allocation Matrix shown in Table 6.5, there could be a Subsystem 3 need statement that says: "*The stakeholders need Subsystem 3 to implement the allocated requirements from System A shown in Column D of the System A Allocation Matrix contained/recorded in [TBD location].*" Based on this, the project team would transform this need statement into the set of derived Subsystem 3 child requirements that address Subsystem 3's role in meeting the allocated parent requirements in accordance with the activities discussed earlier in this section. The project team must determine whether the resulting child requirements for their SOI, as well as the other subsystems or system elements to which the allocated parent requirement was allocated, are both necessary and sufficient to meet the intent of the allocated parent requirement.

Caution: When defining the child requirements, DO NOT just copy and paste the parent requirement and change the noun. For example, if there is a parent requirement: "System A shall do X with a performance of Y." Do not write child requirements "Subsystem 3 shall do X with a performance of Y." and "Subsystem 4 shall do X with a performance of Y." that are just copies of the parent requirement with a noun change.

Rather, the project team needs to do the analysis and derive child requirements based on the specific role that each Subsystem 3 and Subsystem 4 have that contributes to the allocated parent requirement being met. It is the architecture definition activity and analysis that drives and controls the decomposition and allocation of requirements at the next lower level of the architecture.

This activity helps establish the set of design input requirements has the GtWR characteristics *C10—Complete* and *C11—Consistent.*

6.4.5 Budgeting of Performance, Resource, and Quality Requirements

Allocation involves more than just "flowing down" requirements from one level to another.

There are two types of allocation:

1. One type of allocation is for responsibility, where the receiving entity has some role in meeting the intent of the allocated parent requirements.
2. The other type of allocation involves the allocation of some quantity such as resource production or utilization, performance, quality, or some physical attribute. Physical attributes include mass, volume, etc. Performance is associated with functional requirements in terms of how well, how fast, and how many (for example, accuracy, precision, time, bandwidth, consumption of a consumable, or power use).

Budgets need to be managed and controlled at the system level. In a data-centric practice of SE, budgets are managed and controlled within an integrated model of the system. In a document-centric practice of SE, budgets are commonly managed and controlled either within the RMT or a spreadsheet referenced by the requirements.

A critical concept associated with budgeting is that the budgeted quantities result in requirements that have a dependency—a change in one will result in the need to change another, as shown in Figure 6.24. Because of these dependencies, establishing traceability between the child requirements and their allocated parent as well as between peer requirements is critical. The RMT being used should allow management of the allocations to the entities separately from the traceability.

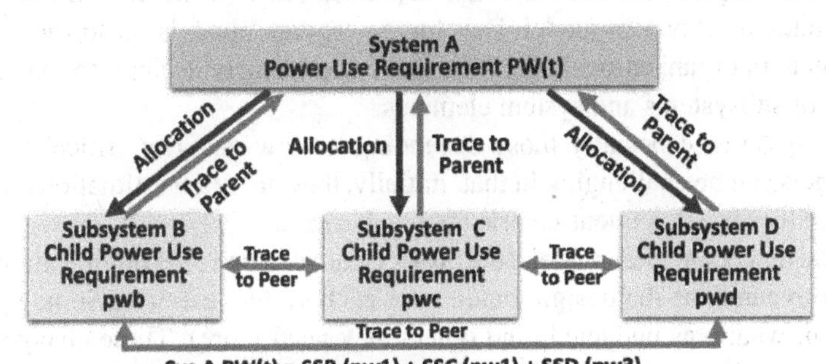

FIGURE 6.24 Allocation and Budgeting Example.

As a result of these dependencies, it is useful to view allocation of resources as an equation.

$$\text{Sys A PW(t)} = \text{SSB (pw1)} + \text{SSC (pw2)} + \text{SSD (pw3)}$$

where PW(t) is the total power available for use, SSx = subsystem, and pw$x$ = the allocated power use values for Subsystems B, C, and D. In this context, changes to any variables on the right side of the equation will require a change in the other variables to keep the equation balanced. It is important that the project team identifies and understands these dependencies and relationships.

The project team can use models or spreadsheets to establish and manage these dependencies and relationships. However, each of the subsystems may also suballocate/budget their values to the system elements that make up the subsystem, creating second-order allocations/budgets. Each of the system elements at the lower level may, in turn, suballocate/budget their values, resulting in third-order allocations/budgeting. There will be cases where there may be a dependency between second- or third-order allocated/budget values across subsystems or system elements, as shown in Figure 6.25.

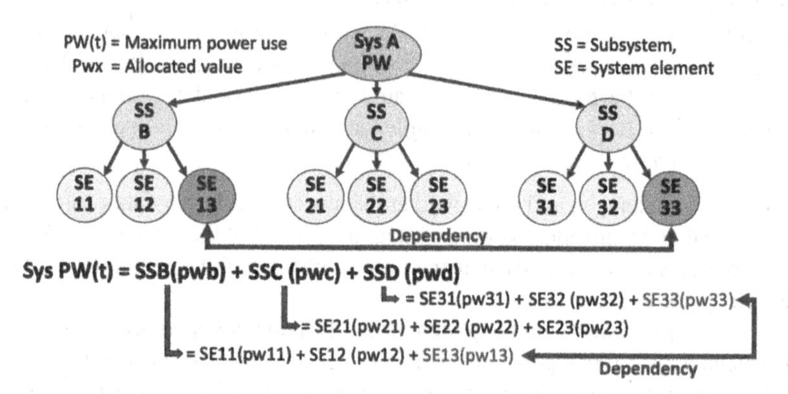

FIGURE 6.25 Allocation and Budgeting Dependencies.

Identifying and managing these lower-order dependencies is difficult to do unless the project team is using an integrated system model. Developing separate models for lower-level subsystems and system elements in organizational silos will make it extremely difficult to manage the budgets at multiple levels of subsystems and system elements.

For greenfield systems, especially those depending on low-maturity critical technologies, the managing of budgets can be challenging in that, initially, they are often estimations with a minimum of analysis, especially for a document-centric approach to SE.

Using analytical and behavioral models, the project team can do some optimization. However, the budgets are often dynamic as the design matures for each system element. Some system elements may need less than what was budgeted (and others may need more). These changes have a ripple effect for all dependent budgets up and down the levels of the system architecture. Because of this, it is critical for the project team to manage all budgets using a single integrated model of the SOI.

Due to the dynamic nature of budgeted qualities, many organizations do not include the actual budgeted values within the requirement statements. The budgets are managed separately in a configuration-managed form (database, spreadsheet, or model). With this approach, the individual design input requirements need to include a pointer to their source. For example, the project team may export the allocations into a spreadsheet, where the requirements would point to the spreadsheet.

For subsystems or system elements that are contracted to a supplier, special provisions need to be made in the contract to manage the dynamic nature of budgeting. To help manage the risk associated with the uncertainties associated with budgeted values, the customer and supplier will need to define and agree to margins for the budgeted quantities, as discussed in Section 6.4.6.

For spacecraft, mass is a critical allocated quantity as is power utilization. The spacecraft would have a maximum allowable mass and then allocate that mass to the hardware and mechanical systems. *"The Propulsion Subsystem shall have a mass equal to or less than the Propulsion Subsystem mass allocation in [spacecraft mass management spreadsheet xyz or database]."* As lower-level subsystems and system elements are defined for the *Propulsion Subsystem*, the mass would be suballocated and managed within the same spreadsheet or database.

From an electrical power perspective, the power used by subsystems and system elements must be kept in balance with generation and storage capabilities of the spacecraft. Similar to mass, the spacecraft would have a maximum power generation and storage capability that would be allocated to all subsystems that use electrical power.

6.4.5.1 Allocations and Budgets Across Interface Boundaries A common occurrence is when a system-level performance requirement is allocated to multiple subsystems and system elements that have a role in meeting that requirement.

A classic example shown in Figure 6.26 is an observation spacecraft that has an instrument with sensors to obtain observation data (for example, the Earth's sea level, temperature, or an image) and transmits that data to observers on Earth for display and analysis. There is an instrument on the spacecraft that collects the observation data, processes, stores, and provides the observation data to the spacecraft communications subsystem. The communications subsystem integrates the observation data into its downlink data stream and transmits that data stream to a communications satellite that transmits the data stream to a ground communications network, which provides the data stream to the spacecraft control center. The spacecraft control center strips off the instrument observation data from the spacecraft data stream and sends that observation data to the instrument control center, which processes and stores the observation data. These data are then made available for download by observers who need to receive, store, process, display, and analyze the data.

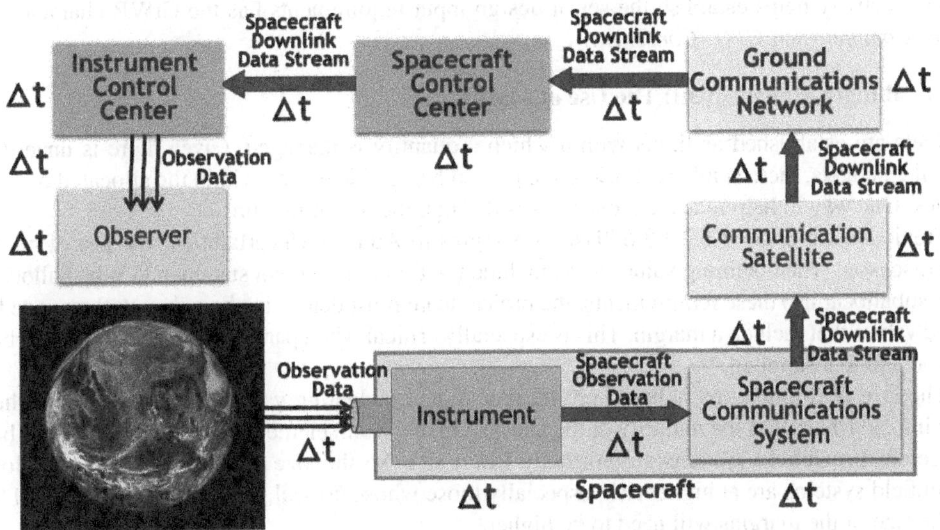

FIGURE 6.26 Allocation and Budgeting Across Interface Boundaries.

There are many systems, both internal to the instrument and spacecraft and external, that are involved; several of which are enabling systems owned and operated by external organizations (space and ground communications systems). There are interface boundaries between each of these systems and interactions across those boundaries. Each step of this process involves some form of data manipulation as well as the time of each interaction across the interface boundaries. Each instance of data manipulation and each instance of interaction across and interface boundary takes some amount of time (delta T). In the example shown in Figure 6.26, there are 13 delta-Ts involved. *(There will be internal delta-Ts internal to each system as well.)*

At the operational level, there may be a stakeholder need for the maximum time allowed from the time an observation event occurs on the surface until the time that the observation data are available for display at the instrument control center by an observer. The implementing system-level time requirement would be allocated to the various systems that make up the overall architecture. Each of those systems would then do an analysis of their allocated time to 1) define a concept for achieving that time and 2) determine if that time is feasible. In some cases, feasibility will be an issue for one or more of the systems involved. The project team will then need to address these issues, find a solution, and reallocate the times based on this analysis—assuming they can define an end-to-end concept that is feasible. If not feasible, they will need to push back to the customer and renegotiate a value that is feasible.

For systems in this chain that are existing, the delta-T will be known (often fixed), leaving the remaining time to be divided up between the developing systems. For the developing systems, the actual answer as to what is feasible will be a function of the physical design and the maturity of critical technologies. For cases when the SOIs are being developed by different suppliers external to the project team's organization, managing these allocations will be a challenge unless addressed in the supplier's contract.

This example illustrates the dynamic nature of allocation and budgeting, which gets more complex with the number of subsystems and system elements involved. This complexity also results in increased uncertainty concerning feasibility, which may not be known until the design is complete. To help manage this uncertainty, the concept of defining margins as discussed below can be used by the project teams for each of the subsystems or system elements being developed.

This activity helps establish the set of design input requirements has the GtWR characteristics *C10—Complete* and *C11—Consistent.*

6.4.6 Budget Management: The Use of Margins

Budgets are established as limits within which a quantity is managed. Given there is uncertainty with the budgets, there is inherent risk to the project being able to stay within the allocated budgeted values. One way to help manage those risks is through the use of margins.

Previously, in Section 6.2.1.3.6 "Use of Margins to Address Uncertainties" the use of margins was discussed when defining values to be included within requirement statements. When allocating (and suballocating) these requirements, the project team must determine how they will manage budgeted values that include a margin. This is especially critical when parts of the architecture are being outsourced to a supplier.

The size of the margins is based on the risk. When budgeting values, the project team should take into consideration the maturity of the subsystem or system element for which the value is being budgeted. Brownfield systems are normally lower risk, so the size of the margins can be lower. Greenfield systems are at higher risk, especially those whose critical technologies are at low TRLs, so the size of the margins will need to be higher.

The early establishment of adequate margins and the effective management of them throughout the project's lifecycle play a critical role in the ability of the project to deliver a winning system.

During allocation and budgeting, the project team is strongly encouraged to define adequate margins, especially for parts of the architecture with significant complexity and high risk. Failing to define these margins places the project at great risk of cost overruns and schedule slips.

6.4.7 The SE Vee Model and Moving Between Levels

In Figure 6.27 on the next page, a hierarchical view of the system architecture is mapped to the SE Vee Model showing what subsystems and system elements are part of the system physical architecture and how their needs, design input requirements, and design output specifications relate to the SE Vee Model.

This view shows the iterative and recursive nature of SE as a system is decomposed into a hierarchical architecture of individual subsystems and system elements, realized by design, built, made, or reused, and integrated with other parts of the system.

The left side of the SE Vee shows a top-down series of activities where the definition of lifecycle concepts, integrated sets of needs, sets of design input requirements, architecture, and design take place. The larger dark arrows point to the needs and design input requirements for the system, subsystems, and system elements to where they are represented in the SE Vee Model. The dashed arrows point from the system element design output specifications in the hierarchical view to where they are represented in the SE Vee Model.

As discussed previously, parent/child relationships are important concepts during design input requirements definition as they deal with the flow down (allocation and budgeting) of requirements from the system at one level of the physical architecture to subsystems and system elements at the next lower level of the physical architecture.

As the project moves down the system architectural hierarchy, the determination of the proper parent/child relationships and whether the child design input requirements are both necessary and sufficient are part of requirements verification, requirements validation, design verification, and design validation activities during development as the project team moves down the left side of the SE Vee.

6.4.8 Combining Traceability and Allocation to Manage Requirements

Combining the concept of allocation with the concept of traceability provides the project team is a powerful way to manage the design input requirements, especially across levels and across subsystems and system elements within a specific level.

Referring to the example in Figure 6.18, there are 16 subsystems or system elements within the SOI architecture. Assuming an average of 100 requirements per system, subsystem, or system element, this results in 1600 requirements across the SOI architecture. In many organizations, the average number of requirements per system, subsystem, or system element is more like 400 requirements—for a total of 6400 requirements for the spacecraft example shown in Figure 6.16.

As requirements are developed and flow down from one level to another, it is critical that allocation and traceability are assessed for completeness, correctness, and consistency. These assessments are not only needed while defining the sets of design input requirements but are also a major function of managing the sets of needs and sets of design input requirements, especially when assessing changes.

While the use of the traceability and allocation matrices discussed above is useful for smaller sets of requirements with fewer levels, as the number of subsystems and system elements and their requirements increase, the usefulness of these matrices, or any other visualizations, is limited as a tool to assess the correctness, consistency, and completeness between and within the sets of design input requirements. For this reason, a data-centric approach is recommended for complex systems with many levels and requirements, enabling the ability to do this assessment.

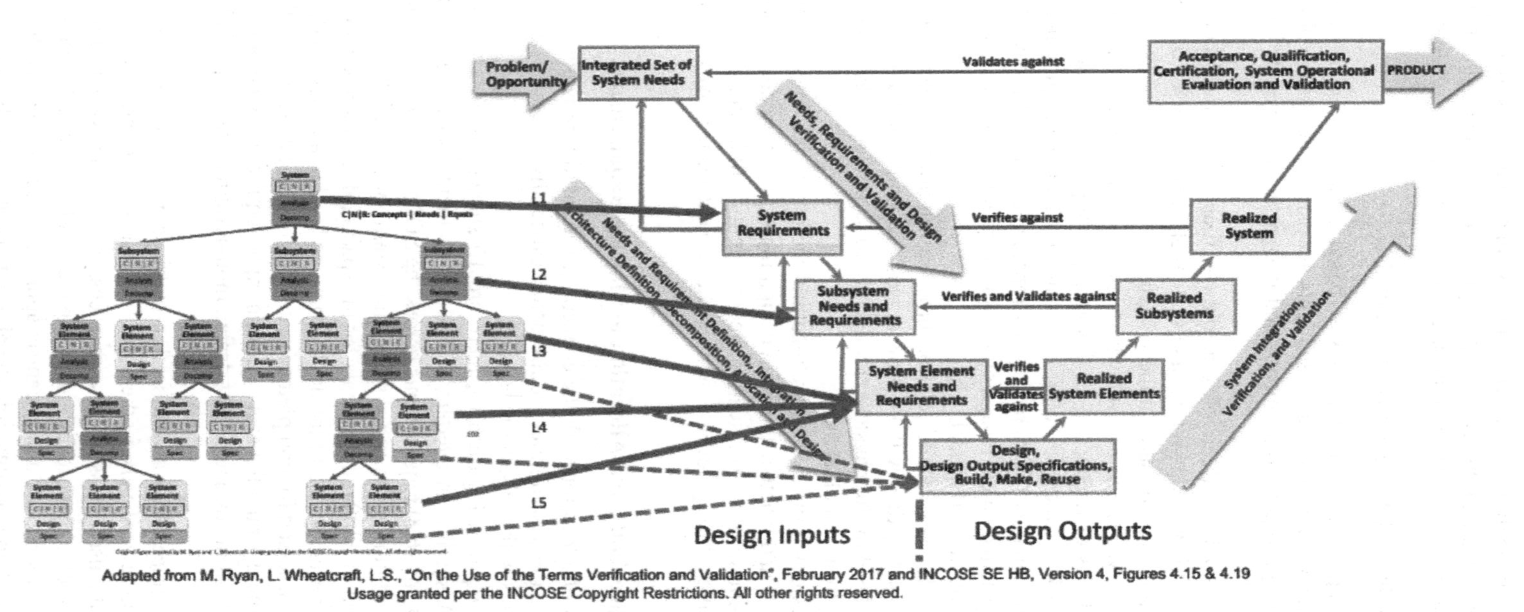

FIGURE 6.27 System Architecture Hierarchy Mapped to the SE Vee Model.

Refer to Section 14.2.7 for a more detailed discussion concerning the use of allocation and traceability to manage the sets of design input requirements.

This activity helps establish each design input requirement has the GtWR characteristics *C1—Necessary* and *C8—Correct*. It also helps establish the set of design input requirements has the GtWR characteristics *C10—Complete, C11—Consistent,* and *C14—Able to be Validated.*

6.5 SUMMARY OF DESIGN INPUT REQUIREMENTS DEFINITION

The focus of the *Design Input Requirements Definition* activities is to define well-formed sets of design input requirements for the integrated system as well as each subsystem and system element within the system architecture. These sets of design input requirements represent the allocated baseline to which the *Design Definition Process* will be implemented.

"Well-formed" refers to the quality of the sets of design input requirements in terms of content as well as the structure as defined in the INCOSE GtWR. The goal is to have well-formed sets of design input requirements that clearly communicate the intent of the needs to users of the requirements. These users include those responsible for the design, design verification, and design validation; developing the design output specifications; production, integration, system verification, system validation, and management of the design input requirements.

Ensuring the sets of design input requirements have the characteristics of well-formed design input requirements as defined in the INCOSE GtWR is necessary to ensure high-quality, requirements that will not be as volatile as many organizations have historically experienced.

Projects often ask the question: "How do we know the requirements are 'done' enough to proceed with design?" There is always a trade-off between "better" and "good-enough." One definition of "good enough" is *the point where the cost of potential changes is less than the effort needed to define every requirement.*

There really is not a simple indicator, the decision should be knowledge driven and not schedule driven. Baselining poorly formed sets of design input requirements often results in the accumulation of technical debt that will be more costly in terms of both schedule and budget rather than spending the time and effort that will result in well-formed sets of design input requirements, as discussed in thisu guide.

7

REQUIREMENTS VERIFICATION AND VALIDATION

This chapter provides a detailed discussion on the planning, activities, and artifacts associated with design input requirements verification and requirements validation, as illustrated in Figure 7.1.

In this chapter, "requirements verification" and "requirements validation" are about the design input requirement expressions themselves, not about verification that the system meets a requirement, as discussed in Section 2.3.4.2 and described in Chapters 10 and 11. It is also distinct from design verification and design validation, as discussed in Chapter 8.

Requirements verification and validation consist of a series of activities assessing the quality of the individual requirements, and sets of requirements, to determine whether they correctly represent the needs, parent requirements, and other sources from which they were transformed (validation) and whether the requirement statements and sets of requirements have the characteristics defined in the Guide to Writing Requirements (GtWR) (verification).

If the integrated set of needs was developed as defined in this Manual, the needs should already have traces to these other sources; however, if the organization has not documented an integrated set of needs, then the requirements will need to trace to the sources shown in Figure 4.13 from which the needs would have been derived.

Requirements verification and validation activities should be done both continuously, as the requirements are defined, and as well as part of the baseline activities during an System Requirements Review (SRR) or similar type of gate review. Using a data-centric approach to requirements definition, rather than waiting to do requirements verification and validation on the completed sets of design input requirements, they can be verified and validated individually during requirements definition activities.

The status of requirements verification and validation can be tracked using the attributes (discussed in Section 6.2.1.5). These attributes allow the project team to track the status

INCOSE Needs and Requirements Manual: Needs, Requirements, Verification, Validation Across the Lifecycle,
First Edition. Louis S. Wheatcraft, Michael J. Ryan, and Tami Edner Katz.

of the requirements verification and validation activities. Once verification and validation of each individual requirement in the set are complete, the set itself can be verified to have the characteristics of well-formed sets of requirements defined in the GtWR.

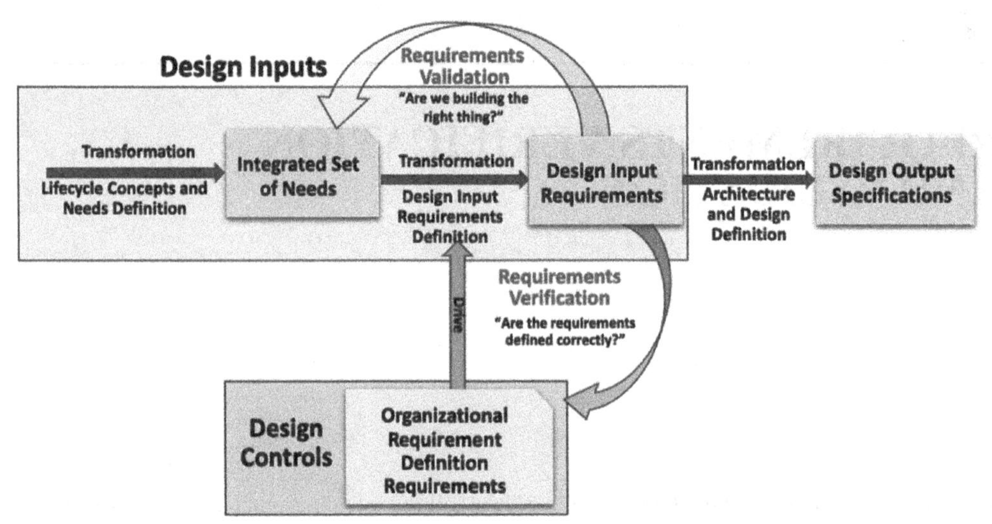

FIGURE 7.1 Design Input Requirements Verification and Validation.

While requirements verification and validation are discussed separately, the activities will often be performed concurrently. If a Natural Language Processing (NLP)/Artificial Intelligence (AI) tool is used for some of the requirement verification activities, that analysis should be completed prior to requirement validation, enabling the reviewer to focus on content and intent, rather than on structure.

Requirements verification and validation are not a discrete set of activities performed once. While the set of design input requirements will be baselined and put under configuration control, they will most certainly change over time. During physical architecture definition and design analysis and maturation activities, as well as the definition of the resulting design output specifications, it is common to discover issues with the baselined set of design input requirements, requiring them to be updated when the physical architecture, design, and design output specifications are baselined.

7.1 REQUIREMENTS VERIFICATION

A summary of the design input requirements verification activities is shown in Figure 7.2.

The requirements verification activities will review the requirement expressions and the sets of requirements to verify they have the expected characteristics and adhere to the rules that result in those characteristics as defined in guidance such as the INCOSE GtWR, or the developing and customer organization's guide to writing well-formed requirements.

Requirements verification activities include inspection of traceability records from the sources from which the requirements were transformed to ensure traceability with those sources, as well as an audit of the interface requirements, as discussed in Chapter 6.

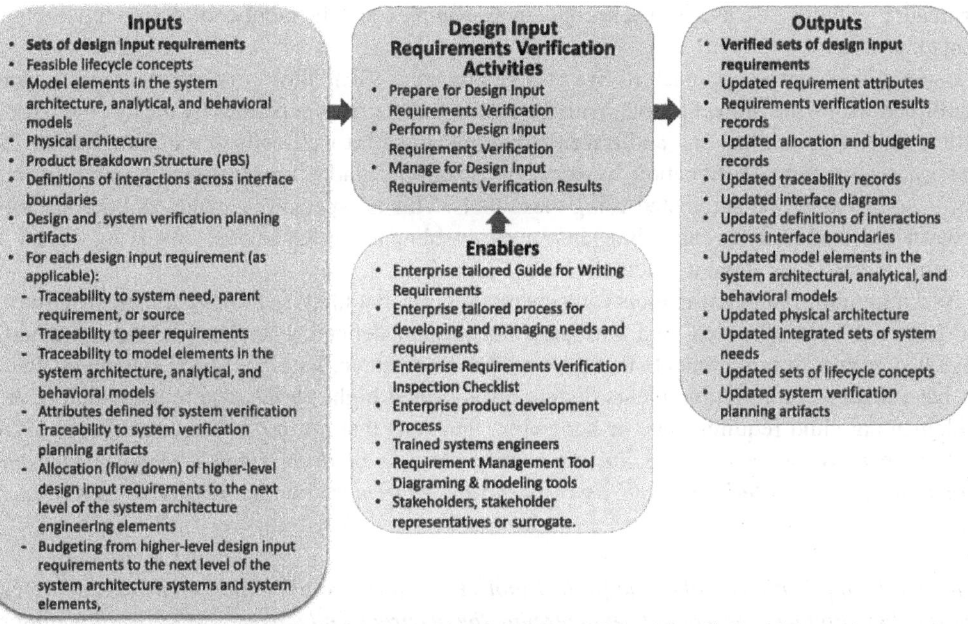

Inputs
• Sets of design input requirements
• Feasible lifecycle concepts
• Model elements in the system architecture, analytical, and behavioral models
• Physical architecture
• Product Breakdown Structure (PBS)
• Definitions of interactions across interface boundaries
• Design and system verification planning artifacts
• For each design input requirement (as applicable):
- Traceability to system need, parent requirement, or source
- Traceability to peer requirements
- Traceability to model elements in the system architecture, analytical, and behavioral models
- Attributes defined for system verification
- Traceability to system verification planning artifacts
- Allocation (flow down) of higher-level design input requirements to the next level of the system architecture engineering elements
- Budgeting from higher-level design input requirements to the next level of the system architecture systems and system elements,
Design Input Requirements Verification Activities
• Prepare for Design Input Requirements Verification
• Perform for Design Input Requirements Verification
• Manage for Design Input Requirements Verification Results
Enablers
• Enterprise tailored Guide for Writing Requirements
• Enterprise tailored process for developing and managing needs and requirements
• Enterprise Requirements Verification Inspection Checklist
• Enterprise product development Process
• Trained systems engineers
• Requirement Management Tool
• Diagraming & modeling tools
• Stakeholders, stakeholder representatives or surrogate.
Outputs
• Verified sets of design input requirements
• Updated requirement attributes
• Requirements verification results records
• Updated allocation and budgeting records
• Updated traceability records
• Updated interface diagrams
• Updated definitions of interactions across interface boundaries
• Updated model elements in the system architectural, analytical, and behavioral models
• Updated physical architecture
• Updated integrated sets of system needs
• Updated sets of lifecycle concepts
• Updated system verification planning artifacts

FIGURE 7.2 Requirements Verification IPO Diagram.

7.1.1 Prepare for Requirements Verification

The first step in preparing for requirements verification is to ensure the enablers shown in Figure 7.2 are in place. Assuming the enablers are in place and tools are in use, the artifacts listed as inputs should have been produced and matured during the design input requirements definition activities to the point where they are ready for the requirements' verification activities. Using a data-centric approach as defined in this Manual, these artifacts should have been recorded within the project toolset.

Requirements verification is enabled with the establishment of defined requirements templates for all associated requirement types so that the resulting requirements can be compared (and verified) with respect to the defined standard. Standardized templates for requirements expedite the requirements verification activities.

While some of the requirements verification activities must be done manually, others may be able to be automated depending on the capabilities of the applications in the project toolset. NLP/AI applications provide a capability that allows requirements verification to be automated to some extent. Many of these tools use the characteristics and rules defined in the INCOSE GtWR as a basis for assessing the quality of the requirement statements and sets of requirements. These tools can be used as both a "digital assistant" to aid in the writing of requirement statements as well as to assess the quality of individual requirement statements and the set of design input requirements. Many of the NLP/AI applications provide a "score" concerning the quality of the requirement statements based on criteria defined by the project team, as well as identify specific defects on which the score was based. The project team will determine how this score will be used in the management of the requirements and requirement verification.

To what extent the project team can rely on automated requirements verification depends on a tradeoff between effort and risk. The need for and importance of using automation for requirements verification will increase as systems become more complex and the number of design input requirements increases.

Using a data-centric approach to Systems Engineering (SE), allocation and traceability can be established within the project toolset, with the associated matrices generated by the tools as a report. With a well-defined allocation and traceability strategy and a well-defined requirements attribute strategy, requirements verification in terms of allocation, budgeting, traceability, and attributes can, to some extent, be automated using these tools. This is especially enabled as interoperability between tools and analysis capabilities improves, enabling advanced analysis involving traceability of requirements to the input artifacts shown in Figure 7.1.

As a minimum, data-centric tools can generate reports that identify missing traces, requirements that have not been allocated, and attributes that are not defined. For example, reports could be generated listing all requirements that do not trace to a source, listing all need statements that do not have implementing requirements, listing all allocated higher-level requirements that have no implementing child requirements, or budgeting equations that are not balanced. *Refer to Section 14.4 for a more detailed discussion concerning the use of these kinds of reports to manage allocation and traceability as well as to perform requirement verification as discussed in this section.*

Note: While applications within the project toolset can produce allocation and trace matrices or reports, these applications cannot yet determine the accuracy and correctness of the allocations or traces between entities. This assessment must be done manually. Deriving requirements from the behavioral and architectural models [48] *significantly reduces the risk of errors in the definition and allocation of design input requirements.*

A key preparation activity for requirements verification is the creation of a Requirements Verification Inspection Checklist if one does not already exist. If the organization has a generic checklist, then the project team can tailor the checklist to the System of Interest (SOI) being developed and the project's processes. This checklist should be part of the organization's requirements definition process requirements. In the medical device industries, these requirements are referred to as "design controls," as shown in Figure 7.1, which are part of a larger Quality Management System (QMS).

This checklist serves as a standard to measure the requirements against defined criteria and will help guide all the requirements verification activities. Addressing each of the areas and questions within the checklist will lead to successful completion of the requirements definition and verification activities. (*An example Requirements Verification Inspection Checklist is included in the GtNR, Appendix D.*)

7.1.2 Perform Requirements Verification

There are several activities that should be performed to ensure the individual requirement expressions and integrated set of design input requirements are well-formed and provide objective evidence they have been verified. Using the Requirements Verification Inspection Checklist as a guide, the project team should:

- Manually verify that individual requirement expressions and the sets of requirements have the characteristics in accordance with the rules defined in the GtWR or similar guide. This could be done by using the Requirements Verification Inspection Checklist as a guide to inspect each requirement statement by individuals or as part of a tabletop or peer review. Given the number of characteristics and the number of rules to help ensure the requirements have those

characteristics, this task is difficult to do manually, especially for large sets of design input requirements.
This activity helps establish that each design input requirement has the GtWR characteristic C9—Conforming.

- Verify that the set of requirements contains individual requirements that are unique, do not conflict with or overlap with other requirements in the set, and that the units and measurement systems they use are homogeneous.
This activity helps establish that each need has the GtWR characteristics C9—Conforming and C11—Consistent.
- Verify that the language used within the set of requirements is consistent (i.e., the same words are used throughout the set to mean the same thing) and that all terms used within the requirement statements are consistent with the architectural model and project data dictionary.
This activity helps establish that each need has the GtWR characteristics C9—Conforming and C11—Consistent.
- Alternatively, if in the project toolset, use an NLP/AI application that provides the capability to automate the verification of the requirement statements in terms of how well they adhere to the rules for writing requirements and sets of requirements, as well as checking for consistent use of terminology.
  - For requirement statements with defects, members of the project team will need to examine each defective requirement statement and fix the defects the NLP/AI application identified.
- Verify that individual requirement expressions have the set of attributes agreed to by the project team defined. The project toolset should be able to produce a report concerning whether any of the attributes are "null"—that is, no values have been defined for a given attribute. While a report can tell if an attribute has been defined, it cannot assess the quality or accuracy of the information in the attribute—that assessment will have to be done manually. For example, does the text in the rationale statement include the information expected to be in the rationale? Does the rationale state why the requirement is necessary? Is the source of any numbers identified?
This activity helps establish that each design input requirement expression has the GtWR characteristic C4—Complete.
- Verify that individual requirement expressions have the set of system verification attributes agreed to by the project team defined. The project toolset should be able to produce a report concerning whether any of the system verification attributes are "null"—that is, no values have been defined. While a report can tell if a system verification statement has been defined, it cannot assess the quality or accuracy of the information in the statement—that assessment will have to be done manually. For example, does the text in the verification statement include the information expected?
This activity helps establish that each requirement expression has the GtWR characteristic C4—Complete.
- Use the project toolset to generate traceability reports to verify each requirement traces to the need, an allocated parent requirement, or a source from which it was derived. In addition, use the project toolset to generate a report that lists all requirements that do not trace to a need, parent, or source. In a document-based approach, trace matrices are often managed manually, requiring considerable time and effort to ensure their completeness and correctness, and are often defective. The focus of this verification activity is to ensure that traceability has been established. The correctness of the traceability is addressed in the Section 7.2.2.
This activity helps establish that each design input requirement has the GtWR characteristic C1—Necessary.

- Use the project toolset to generate traceability reports to confirm each SOI need, source, or parent requirement allocated to the SOI has at least one child requirement that addresses that need, parent, or source. In addition, use the project toolset to generate a report that lists all needs, parent requirements, and sources that do not trace to an implementing design input requirement. The focus of this verification activity is to ensure that traceability has been established. The correctness of the traceability is addressed in the Section 7.2.2.
  This activity helps establish that each design input requirement has the GtWR characteristic C1—Necessary, and the set of design input requirements has the GtWR characteristic C10—Complete.
- Perform the interface audit discussed in Section 6.2.3.8.
  - Using the interface diagrams and architectural diagrams developed during lifecycle concepts analysis and maturation activities and refined as part of the transformation of the needs into the design input requirements as a guide, verify that all interfaces have been addressed and the associated interface requirements are included in the requirement set.

  This activity helps establish the set of design input requirements has the GtWR characteristics C10—Complete and C11—Consistent.
  - For each interface requirement, verify it is clear what the specific interaction is between the SOI and the external system and that the requirement includes a pointer to where that interaction is defined, recorded, and agreed to.

  This activity helps establish that each interface requirement has the GtWR characteristic C4—Complete and C7—Verifiable.
  - For each interface requirement, verify the external system referred to has a corresponding interface requirement or has included the interaction with the SOI being developed in its interface control documentation.

  This activity helps establish that the set of design input requirements has the GtWR characteristics C10—Complete and C11—Consistent.
- For the set of design input requirements, verify there are requirements included that address function/performance, operational (fit), physical characteristics (form), quality (-ilities), and compliance.
  This activity helps establish that the set of design input requirements has the GtWR characteristics C10—Complete, C14—Able to be Validated, and C15—Correct.
- Assess allocation to the next level of architecture. As discussed in Section 6.4.2, a system will have a hierarchy of requirements based on the architecture. Unless no further elaboration of the system, subsystem, or system element is needed and is ready for a buy, make, or reuse determination, the SOI requirements will be allocated to the next-level subsystems and system elements contained within the system's architecture. Using the project toolset to generate allocation and trace matrices, perform an analysis that verifies:
  - Each of the SOI requirements has been allocated to the next level of the architecture.
  - Each allocation is correct and complete, i.e., the requirements were allocated to all applicable subsystems and system elements at the next level of the architecture and each of the allocations was to the correct subsystems and system elements.
  - The resulting dependent child requirements in response to allocations of performance, quality, or resources are linked to manage changes to the allocated/budgeted values.

  The focus of this verification activity is to ensure that allocation has been established. The correctness of the allocation is addressed in the Section 7.2 on requirement validation.
  This activity helps establish that the set of design input requirements has the GtWR characteristics C10—Complete and C11—Consistent.

Note: Allocation assessments will be easier if the project has developed architectural, analytical, and behavioral models during the lifecycle concepts analysis and maturation activities and refined those models during the transformation from needs to design input requirements. These models will be refined and elaborated for each system, subsystem, and system element within the SOI architecture until all the design input requirements for each system elements have been defined.

7.2 REQUIREMENTS VALIDATION

The GtWR includes the characteristic C14—"*Able to be Validated*" for sets of requirements. In the context of requirements validation, "*Able to be Validated*" means the project team or customer will be able to validate that the SOI was designed and built/coded in accordance with the set of design input requirements.

The requirements validation activities will review the requirement expressions and the set of requirements to confirm each requirement statement and the set of design input requirements clearly communicate the intent of the needs, parent requirements, and sources, from which they were transformed, in a language understandable by all project team members, customers, and other key stakeholders.

A summary of the requirements validation activities is shown in Figure 7.3.

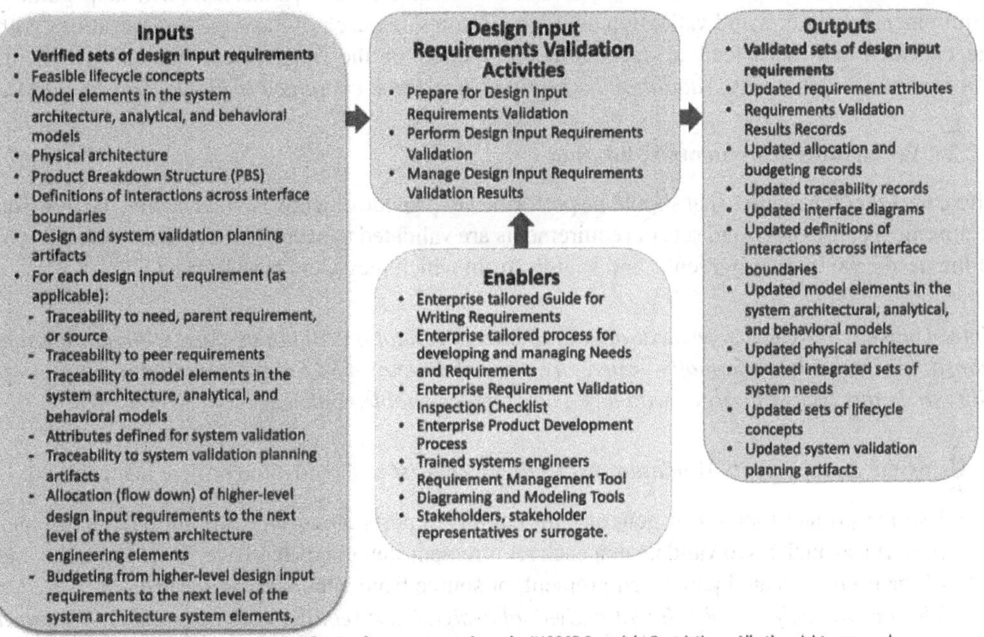
Inputs
• Verified sets of design input requirements
• Feasible lifecycle concepts
• Model elements in the system architecture, analytical, and behavioral models
• Physical architecture
• Product Breakdown Structure (PBS)
• Definitions of interactions across interface boundaries
• Design and system validation planning artifacts
• For each design input requirement (as applicable):
- Traceability to need, parent requirement, or source
- Traceability to peer requirements
- Traceability to model elements in the system architecture, analytical, and behavioral models
- Attributes defined for system validation
- Traceability to system validation planning artifacts
- Allocation (flow down) of higher-level design input requirements to the next level of the system architecture engineering elements
- Budgeting from higher-level design input requirements to the next level of the system architecture system elements,
Design Input Requirements Validation Activities
• Prepare for Design Input Requirements Validation
• Perform Design Input Requirements Validation
• Manage Design Input Requirements Validation Results
Enablers
• Enterprise tailored Guide for Writing Requirements
• Enterprise tailored process for developing and managing Needs and Requirements
• Enterprise Requirement Validation Inspection Checklist
• Enterprise Product Development Process
• Trained systems engineers
• Requirement Management Tool
• Diagraming and Modeling Tools
• Stakeholders, stakeholder representatives or surrogate.
Outputs
• Validated sets of design input requirements
• Updated requirement attributes
• Requirements Validation Results Records
• Updated allocation and budgeting records
• Updated traceability records
• Updated interface diagrams
• Updated definitions of interactions across interface boundaries
• Updated model elements in the system architectural, analytical, and behavioral models
• Updated physical architecture
• Updated integrated sets of system needs
• Updated sets of lifecycle concepts
• Updated system validation planning artifacts

FIGURE 7.3 Requirements Validation IPO Diagram.

7.2.1 Prepare for Requirements Validation

The first step in preparing for requirements validation is to ensure the enablers shown in Figure 7.3 are in place, and those project team members doing the requirements validation have access to the inputs.

Because the focus of requirements validation is on content and intent concerning what is being communicated, requirements validation is difficult to automate using NLP/AI algorithms. However, if the requirements are *derived* from a model, then validation can have lower risk of missing critical needs from which they were transformed.

Even though individual requirements and sets of requirements may be well-formed and traceability has been verified, the message they are communicating may not be as intended. It is important that organizations do both requirements verification and validation. Failing to do so results in technical debt adding risk to the project that the SOI being developed will fail system validation and will not be approved for its intended use.

Assuming the requirements definition activities were completed as defined in this Manual and tools are in use, the artifacts listed as inputs should have been produced and matured to the point where they are ready for the requirements validation activities. Using a data-centric approach, these artifacts should exist within the project toolset.

The requirements definition activities should have resulted in analysis records, diagrams, and models that provide rationale for the transformation that resulted in each of the requirement expressions. Each requirement expression should include rationale to help those validating and implementing the requirement, so they have knowledge of the source and intent of what the requirement statement is communicating.

A key preparation activity for requirement validation is the creation of a Design Input Requirements Validation Inspection Checklist if one does not already exist. If the organization has a generic checklist, then tailor the checklist to the system being developed and the project's processes. This checklist serves as a standard to measure the requirements against and will help guide the requirements definition and validation activities. Addressing each of the areas and questions within the checklist will help lead to successful completion of the requirement validation activities. (*An example Requirements Validation Inspection Checklist is contained in GtNR, Appendix D.*)

7.2.2 Perform Requirements Validation

There are several activities that should be performed to provide objective evidence that the individual requirement expressions and sets of requirements are validated to accurately communicate the intent of the needs, parent requirements, and sources from which they were transformed.

Note: Unlike requirements verification, requirements validation cannot be done without the project team doing the analysis manually—currently, none of the known NLP/AI applications have the capability to do this type of analysis with any degree of acceptable confidence.

Using the *Requirements Validation Inspection Checklist* as a guide, the project team should:

- Use the project tool set to generate traceability reports along with the rationale attribute to perform an analysis to validate that each requirement statement clearly communicates the intent of the need, allocated parent requirement, or source from which it was derived.
  This activity helps establish that individual design input requirements have the GtWR characteristic C8—Correct.
- Assess the correctness of allocation, budgeting, and traceability within the set of requirements. A system will have a hierarchy of sets of requirements for the system and each subsystem or system element within the system architecture. Assuming the allocation and traces exist as previously assessed during requirements verification, use the traceability, allocation, and budgeting reports to perform an engineering analysis to validate that:
  - The allocations from onc level to the next were completed for each requirement and the allocations are correct. The focus of this validation activity is to ensure that allocation was

done correctly. The fact that allocation was done at all was addressed during requirement verification discussed previously.

This activity helps establish that individual design input requirements have the GtWR characteristic C8—Correct, and the set of design input requirements has the GtWR characteristic C10—Complete.

- Confirm budgeted values (physical characteristics, performance, and quality) are managed such that changes can be made as the design matures, changes are within the budgeted allocation, and dependent requirements are identified and changed appropriately.

This activity helps establish that individual design input requirements have the GtWR characteristic C8—Correct, and the set of design input requirements has the GtWR characteristic C11—Consistent.

- For each need, allocated parent requirement, and source, assess whether the derived requirement(s) are necessary and sufficient to meet the intent of the need(s), allocated parent requirement(s), or source they are traced to and from which they were derived. Given that an allocated parent requirement is often allocated to more than one subsystem or system element at the next lower level of the architecture, this means evaluating all child requirements that trace back to the allocated parent requirements for all subsystems and system elements to which the common parent requirement was allocated.

This activity helps establish that the set of design input requirements has the GtWR characteristics C10—Complete, C14—Able to be Validated, and C15—Correct.

- For each requirement that references a capability, performance value, or physical characteristic dependent on a critical technology, confirm the risk attribute has been defined indicating this dependency. Also, confirm a TRL has been assigned to the technology and there is a plan to mature this technology.
  This analysis helps establish that each design input requirement has the GtWR characteristic C6—Feasible.
- Confirm the project documented their assessment that the set of requirements is feasible in terms of cost, schedule, and technology maturation and has included key product development activities in their cost and schedule estimates, including use of enabling systems, lifecycle concepts analysis and maturation, flow down of requirements, design verification, design validation, system integration, system verification, system validation, and procurement.
  This analysis helps establish that the set of design input requirements has the GtWR characteristic C12—Feasible.
- To reduce the risk of unknowns, using the priority, critically, and Key Driving Requirement (KDR) attributes (reference Chapter 15), confirm the project has included adequate budget and schedule margins and reserves as well as defined a descope plan.
  This analysis helps establish that the set of design input requirements has the GtWR characteristic C12—Feasible.
- Confirm with the stakeholders associated with each need, allocated parent requirement, or source that the message being communicated by the resulting requirement expression is correct and acceptable to ensure the set of requirements are the right requirements, i.e., do they accurately represent the intent of the agreed-to needs, parent requirements, or other sources from which they were transformed? It is critical that the project team has confirmation from the stakeholders because it is the set of requirements that represent what is necessary for acceptance against which the built or coded system will be verified.

This activity helps establish that each design input requirement has the GtWR characteristics C1—Necessary, C3—Unambiguous, and C8—Correct and that the sets of design input requirements have the GtWR characteristics C10—Complete, C11—Consistent, C13—Comprehensible, C14—Able to be Validated, and C15—Correct.

- Ensure system verification attributes have been included with each requirements expression addressing the verification *Method*, *Strategy*, and *Success Criteria* for each requirement. This information must be defined and documented before the set of requirements is baselined. This helps ensure each requirement statement is worded such that the design and the system can be verified to meet the requirement.
  This activity helps establish that each design input requirement has the GtWR characteristics C4—Complete and C7—Verifiable, and that the set of design input requirements has the GtWR characteristic C14—Able to be Validated.
- Confirm with the *Approval Authority* that the defined verification attributes represent what is necessary for acceptance.
  This activity helps establish that each design input requirement has the GtWR characteristic C7—Verifiable, and that the set of design input requirements has the GtWR characteristic C14—Able to be Validated and C15—Correct.

7.3 MANAGE REQUIREMENTS VERIFICATION AND VALIDATION RESULTS

Requirements verification and validation records are created documenting the results and outputs of the requirement verification and validation activities. The outputs include the updated input artifacts and Project Management (PM) work products shown in Figures 7.2 and 7.3.

The attributes listed in Section 7.4 should be updated for each requirement expression. Applications within the project toolset should allow tracking of the status of the requirements verification and validation activities using a dashboard that communicates key metrics related to attributes showing results of the requirements validation. Making this information accessible will enable all project team members and key stakeholders to have a shared view of the maturity and status of the requirement verification and validation activities.

Defects found in the set of design input requirements, as well as any defects found in any of the input SE artifacts or PM work products, must be addressed and corrected before they are baselined.

Changes to any of the system verification information contained in the system verification attributes should be updated in the system verification planning artifacts discussed in Chapter 10.

If there is insufficient data to complete requirements verification and validation, a lack of an integrated set of needs from which the set of design input requirements was transformed, or a lack of traceability records within the project toolset, the issue should be recognized as a risk that could cause both design and system verification to be unsuccessful.

7.4 USE OF ATTRIBUTES TO MANAGE REQUIREMENTS VERIFICATION AND VALIDATION

Requirement attributes that aid in requirements verification and validation management include:

A1: *Rationale*: intent of the requirement.

A2: *Trace to Parent*: the allocated parent from which the requirement was formed.

A3: *Trace to Source:* the need from which the requirement was transformed.

A5: Allocation/Budgeting: link to the subsystems or system elements at the next level to which the requirement was allocated.

A26: *Stability:* stable, likely to change, and incomplete.

A28: *Requirement Verification Status*: binary—true/false, yes/no, or incremental—not started, in work, complete, and approved. (A28 is a prerequisite for A29 and A30)

A29: *Requirement Validation Status*: binary—true/false, yes/no, or incremental—not started, in work, complete, and approved.

A30: *Status of the Requirement* (in terms of maturity): draft, in development, ready for review, in review, and approved.

A32: *Trace to Interface Definition*: helps to ensure completeness of interface requirements.

A33: *Trace to Peer Requirements*: manage dependencies when requirements are related to each other, to help to ensure consistency and completeness.

A36: *Risk (of implementation)*: indicating the risk of the need or requirement not being met. This could be a result of the instability (A26) of the requirement or due to feasibility in terms of the TRL of the technology needed for a requirement to be jet, cost, or schedule.

Refer to Chapter 15, for a more detailed discussion on attributes and definitions of each.

8

DESIGN VERIFICATION AND DESIGN VALIDATION

This chapter provides a discussion on the planning, activities, and artifacts associated with design verification and design validation. *Refer to the International Council on System Engineering Systems Engineering Handbook (INCOSE SE HB), Section 2.3.5.5, for a more detailed discussion on the Design Definition Process.*

The goal of successfully completing design verification and validation is to improve the project team's (and customer's) confidence that the design and resulting set of design output specifications, when implemented, will result in a system that meets its intended use in the operational environment, when operated by the intended users, as defined by the integrated set of needs.

Performing design verification and validation enables the project team to find and correct issues before production, where the cost and time to address these issues are much less than waiting to discover these issues during system integration, system verification, and system validation of the actual physical subsystems and system elements that make up the integrated system.

As shown in Figure 8.1, design verification and validation help ensure the design is the "right" design as communicated by the design output specifications.

For each system element, it is assumed there exists a verified and validated integrated set of needs to validate the design against, a verified and validated set of design input requirements to verify the design against, and organizational architecture/design definition criteria to verify that the process of design and resulting design output specifications were done "right" by the project team.

A major outcome of design verification and validation, and early system verification and validation, is a mature physical architecture, design, and design characteristics communicated via a well-formed set of design output specifications, lowering the risk of issues during system integration, verification, and validation. Successful integration, system verification, and system validation are dependent on a proper design and system elements that are built in accordance with the design output specifications.

INCOSE Needs and Requirements Manual: Needs, Requirements, Verification, Validation Across the Lifecycle,
First Edition. Louis S. Wheatcraft, Michael J. Ryan, and Tami Edner Katz.

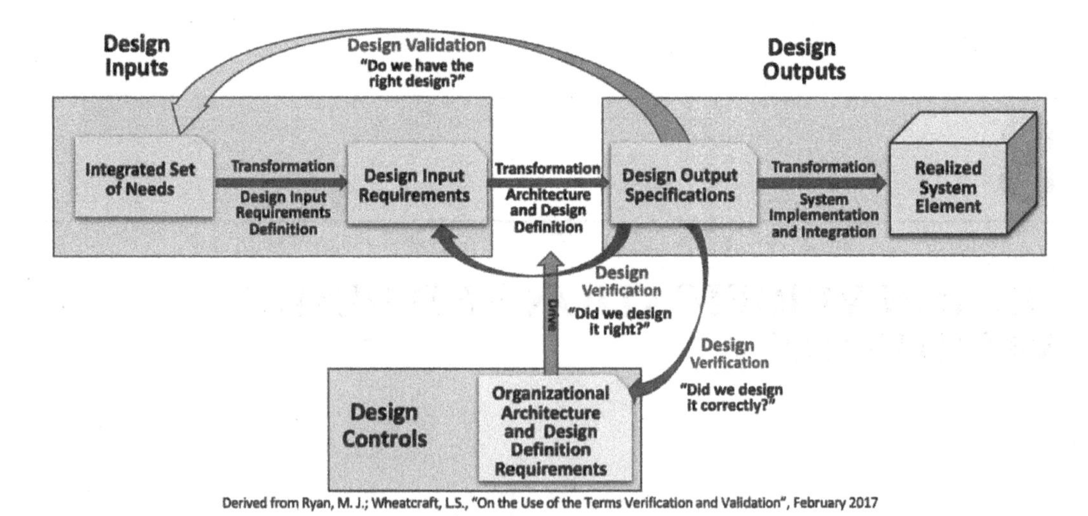

FIGURE 8.1 Design Verification and Validation.

8.1 DESIGN DEFINITION PROCESS OVERVIEW

As shown in Figure 8.1, during the *Design Definition Process* activities, the baselined set of *design input requirements* is transformed into a design that is communicated to the builders/suppliers via a set of *design output specifications*.

A powerful tool in performing design verification and validation is the creation of integrated system architectural, analytical, behavioral models, and simulations of the SOI. These models and simulations can be crucial in confirming with some level of confidence that the design will result in an SOI that can be validated to meet its integrated sets of needs and verified to meet its set of design input requirements.

The architectural and analytical/behavioral models will mature as the design concepts mature through the development lifecycle activities. These models help ensure completeness, correctness, consistency, and feasibility of the integrated set of needs as well as the set of design input requirements that are transformed from these needs. As the integrated set of needs is defined, they are linked to the models, as are the resulting design input requirements. Traceability is established and managed across the lifecycle. The resulting end-to-end traceability across all lifecycle artifacts is key for successful design verification, design validation, system verification, and system validation.

As shown in Figure 8.2, design cycles are used to incrementally mature the system design concepts, models, and associated logical and physical architectures zeroing in on a feasible design that implements the SOI's set of design input requirements and meets the intent of the integrated set of needs.

This process starts with the integrated set of needs and set of design input requirements for the SOI, as well as a preliminary physical architecture, models, prototypes, budgets, and schedules.

Candidate design solutions developed during the lifecycle concepts analysis and maturation activities (discussed in Chapter 4) are refined. Trade studies are done based on feasibility analysis of the candidate design concepts. This analysis includes functional, performance, and behavioral analysis; simulations; and technology readiness and risk assessments.

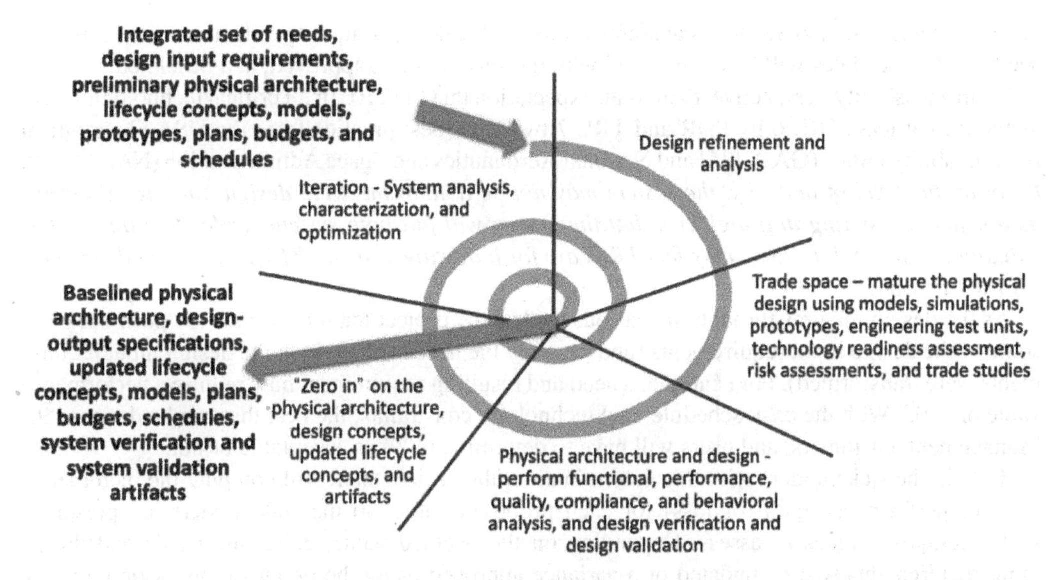

FIGURE 8.2 Zeroing in On a Feasible Design and Physical Architecture.

The level of detail for the initial physical architecture is matured by expanding it to include lower levels of the architecture detail such that all assemblies, subassemblies, parts, and components can be defined in terms of design output specifications.

During design, it is common to use prototypes and engineering units to mature the physical architecture and design. The performance of these prototypes and engineering units is examined in more detail to ensure the design concept will result in a system that can be verified to meet its design input requirements and validated to meet its integrated set of needs.

Throughout each cycle, the project team must involve key stakeholders, including the customer(s) and users/operators, as part of continuous validation. Coordination with the key stakeholders is similar to the approach used in Agile methodologies in which "sprints" are defined, implemented, and reviewed periodically with the stakeholders leading up to one of the formal gate reviews. Based on stakeholder feedback, the resolution is increased for the next iteration and the project team enters the next cycle. For each iteration, the physical architecture, design concept, artifacts, and associated models and prototypes are revised based on the increased level of knowledge.

The number of cycles that are needed would ideally be specified in the project System Engineering Management Plan (SEMP); in other cases, management may have dictated the fixed number of cycles based on cost and schedule, and the role of the project team is to perform as many design cycles as needed within these constraints.

Progress through the cycles is tracked and assessed both continuously and at formal discrete gate reviews (such as the System Design Review [SDR], Preliminary Design Review [PDR], Critical Design Review [CDR], or similar reviews). By PDR, there is often an expectation for physical system elements that ~20% of the design output specifications are completed, and by CDR, there is an expectation that ~80–90% of the design output specifications are completed.

As the design concepts mature and design verification and design validation activities are performed, issues are identified and resolved. The functional architectural and analytical/behavioral models are updated, and the physical architecture and the design output specifications are revised as needed. These updates will be coordinated with the stakeholders, approved, and baselined.

From a feasibility perspective, there is an expectation that the TRL of all critical technologies have matured to at least TRL 6 by PDR and TRL 7 by CDR [best practice from the USA Government Accountability Office (GAO) [35] and National Aeronautics and Space Administration (NASA) [53]. *If not at this level of maturity, the system may not be able to meet the design input requirements and/or needs resulting in a high risk that the system will fail both system verification and system validation. Note: While some may feel TRLs are for hardware systems, TRLs have been defined for software as well.*

As the design matures through the various cycles, the project team may have to "push back" on some of the design input requirements (and possibly the needs from which the design input requirements were transformed). For example, a need and resulting requirement may require a performance value of "10." With the cost, schedule, and technology constraints, the best that can be done is "9." Management, customers, and users will have to determine if "9" is acceptable or not.

To help the stakeholders with this determination, the project team will compute the incremental cost for performance gain (or loss) for each proposed value, so the stakeholders are presented with alternative values to assess. Depending on the decided value, either the needs and design input requirements will be updated or a variance approved using the organization's Configuration Management (CM) process.

In other cases, the project team may discover missing needs (and their corresponding requirements), resulting in changes that must go through the CM process. For cases where there is a customer/supplier relationship, these changes will involve a contract change as well. If the activities discussed in Chapters 4 and 6 were completed in the definition of the customer-supplied set of design input requirements, the likelihood of missing requirements is reduced, avoiding expensive contract changes.

Without this early warning of a missing (or unfeasible) need or design input requirement and the negotiations that follow, the result could be a failure to pass system verification and system validation. If the project team learns of an issue that could result in the SOI failing system verification and system validation, the issue must be brought to the attention of management immediately and resolved. As discussed previously, assigning priority and criticality to each need and requirement and addressing margins when quantifying performance requirements are valuable tools to use when these types of issues are uncovered during design maturation activities.

Change is a normal part of SE and the *Design Definition Process*. From a needs and requirements perspective, changes (updates) to the integrated sets of needs and sets of design input requirements are indicative of the interactive and recursive nature of SE resulting in a maturation of the needs and requirements such that individual needs and requirements have the Guide to Writing Requirements (GtWR) characteristics: *C3—Unambiguous, C4—Complete, C6—Feasible, C7—Verifiable, and C8—Correct*, and the sets of needs and requirements have the GtWR characteristics: *C10—Complete, C11—Consistent, C12—Feasible, C13—Comprehensible*, and *C14—Able to be Validated.*

Note: Any changes made may have a ripple effect that could impact meeting other dependent requirements, satisfying the needs from which they were transformed or satisfying higher level needs or requirements. This ripple effect could also impact meeting lower-level needs and requirements. Changes could also cause substantial updates to the system verification and validation planning artifacts. All changes must be assessed as to the possible impacts to all SE artifacts, including cost, schedule, and risk. Traceability will need to be updated as needed. Refer to Section 14.2.5.3 concerning change management across the lifecycle and Section 14.2.5.4 on change impact analysis.

Once the design concept and implementing design output specifications have passed design verification and validation, they will be ready to be baselined at a formal gate review, such as CDR or other similar review.

The project team can decide to either buy, make, or reuse each mechanical, hardware, and software system elements and develop a set of design output specifications for each.

If the decision is to make (build or code) internally, the design output specifications will be provided to those responsible for manufacturing or coding the system element. If the decision is to buy, the design output specifications will be provided to a supplier as part of a procurement action.

The procurement action could be to obtain an Off-the-Shelf (OTS) system element or obtaining the services of a supplier to manufacture the system element or code the software (if the decision is to reuse an existing system element, they will need to map the design output specifications to the as-built specifications for the reused system element). *Refer to Chapter 12, for a more detailed discussion for procuring OTS system elements and reusing existing parts.*

Assuming the project team completes the above-mentioned activities, the risk of issues will be minimized during physical integration of the physical system elements, system verification, and system validation. While no approach completely guarantees success, this approach minimizes risk during system integration, system verification, and system validation of the subsystems and system elements because it (in part) provides the project team opportunities to find issues and defects early during the *Design Definition Process.*

8.2 EARLY SYSTEM VERIFICATION AND VALIDATION

There is sometimes a misconception concerning when system verification and validation may occur. System verification and validation is not a set of activities limited to the right side of the SE "Vee." In nearly all cases, it may be possible to begin some system verification and validation activities early during development before production, substantially reducing risk of technical debt. In this context, because these system verification and validation activities occur before system integration, system verification, and system validation of the manufactured physical system, they are sometimes referred to as "early" system verification and validation.

Note: There are some who define "early verification and validation" as any verification or validation activity that occurs prior to formal system verification and validation of the manufactured physical system. This would include need verification and need validation, requirement verification and requirement validation, and design verification and design validation. In this section, the focus is on early system verification and validation that occurs as part of or concurrently with design verification and design validation before the physical system is built or coded.

Early system verification and validation can be performed as part of, or concurrently with, design verification and validation before the physical SOI is manufactured (or software written) using the behavioral models, simulations, prototypes or engineering development units or test articles that were developed to mature the design and refine the design output specifications.

Each verification and validation activity should consider the reduction of risk associated with identifying errors early and not waiting until a system is fully built or coded before verifying that it meets a requirement or validating that it meets a need.

This is a crucial activity, as manufacturing and writing code takes time and costs money—sometimes a large percentage of the overall development budget. The intent of early system verification and validation is to postpone committing these resources on a design that has not been properly verified and validated. Doing these activities can help to avoid expensive rework if these issues are discovered later, after the physical system has been built or coded during more expensive system integration, system verification, and system validation activities.

It is a best practice to perform early system verification and validation progressively during design, obtaining objective evidence with the required level of confidence that the architecture and design, as communicated by the design output specifications, will result in a system that will pass system verification and system validation.

Simulations, prototypes, and engineering test units allow early system verification and validation of the system or parts of the system to help uncover issues with the architecture and design.

Below are examples of early system verification and validation activities performed as part of design verification and validation:

- Develop an enabling facility and test fixtures for testing jet engines and the series of disposable jet engines that are tested to destruction during the design program. Each facility and test stand has a development cycle of its own, defining needs and requirements, defining a design, building it, and completing its own system verification and validation prior for use during jet engine design and test activities.
- Perform bench testing of a component or system element by integrating it into an assembly consisting of the control system (software), other components, and a test rig (test facility). This assembly will never be part of the final product; its existence and use are for design maturation and design verification and validation only.
- Conduct tests of the control system connected to individual subsystems or system elements it controls for design maturation and design verification and validation. For example, tests of a control system connected to a fuel pump or thrust vector control system for a launch vehicle. Doing so helps to mitigate the risk concerning the ability of the control system to control individual subsystems or system elements it controls before system integration, system verification, and system validation.
- Perform lifetime testing over extended periods to gain sufficient confidence that the SOI can meet its End of Life (EOL) performance and quality requirements.

Concurrently with the early system verification and validation activities, the project team is updating the draft set of design output specifications used to build further prototypes and engineering verification units or proceed to produce the SOI. The architecture and design cycle activities are repeated using the physical prototypes and engineering test units, as shown in Figure 8.2. Based on the results, the draft design output specifications are updated and verified to correctly represent a system that can be verified to meet its design output specifications, the set of design input requirements, and its integrated set of needs.

Using this approach, the project team is continuously doing early system verification and validation during design verification and validation, never losing sight of both meeting the integrated set of needs and set of design input requirements. Continuous design verification and validation can be performed in support of design concept maturation activities for any SOI—no matter the level of the architecture.

Early system verification and validation helps ensure that individual needs and requirements have the GtWR characteristics: *C3—Unambiguous*, *C4—Complete*, *C6—Feasible*, *C7—Verifiable, and C8—Correct*, and the sets of needs and requirements have the GtWR characteristics: *C10—Complete*, *C11—Consistent*, *C12—Feasible*, *C13—Comprehensible*, and *C14—Able to be Validated*.

8.2.1 Modeling and Simulation Challenges

A major challenge when planning for and doing early system verification and validation is the ability to run simulations of the *integrated system*, including models of all the subsystems and system elements within the system architecture as well as interactions with external systems and the operational environment.

As discussed in Section 2.3.2.6, a major tenet of systems thinking is that a system will have characteristics and behaviors that are not directly relatable to the parts that make it up, resulting in emergent behaviors—both good and bad—that are not observable until all the parts are integrated into the whole. While there are some emergent behaviors that are out of the reach of even the most sophisticated modeling tools, an end-to-end integrated system model, created and executed early, can still uncover some of these behaviors.

Another key tenet of systems thinking is that the behavior of the system is a function of the interaction of the parts that make up the system as well as interactions with the external systems and the operational environment. These interactions include the classic interactions across interface boundaries as well as interactions with other systems within the macrosystem of which it is a part. The SOI may also impact these other systems, including the human operators. To be of the maximum benefit, the models and simulations must simulate both the classic interactions across interface boundaries, the operational environment, as well as the induced environments (impacts the SOI has on the operational environment of the other systems within the architecture as well as external systems in the operational environment).

Just because it has been verified that an SOI meets its requirements during standalone system verification and validation activities does not mean it will do so when integrated into the macrosystem it is a part of, in its actual operational environment, over a sustained period.

Developing models and simulations of an SOI in a silo, independently from the integrated system of which it is apart, is of limited benefit without the knowledge of how the SOI interacts with the other subsystems and system elements within the integrated system in its actual operational environment.

A major challenge is being able to simulate the actual operational environment in the physical world, which is governed by the laws of physics, thermodynamics, chemistry, biology, and human factors. Given that validation is concerned about the system doing what is intended in its operational environment when operated by its intended users, early system verification and validation can consider either the actual operational environment without expensive environmental chambers or a simulated operational environment that is as close to the actual operational environment as is reasonably achievable.

Fortunately, many modern tools are sufficiently powerful to not only model the SOI but also the operational environment and interactions across interface boundaries. Doing so provides an opportunity to detect interface or environmental issues early before the production and integration of the physical parts of the architecture, and both natural and induced environments (such as heat, acoustics, friction, vibrations, and Electromagnetic Interference and Electromagnetic Compatibility [EMI/EMC]) must be considered.

8.3 UPDATING SYSTEM VERIFICATION AND VALIDATION ARTIFACTS

The focus of design verification and validation activities is on the architecture and behavioral models, prototypes, engineering units, and simulations used to mature the design, and implementing design output specifications. However, maturation and validation of the system verification and validation planning artifacts discussed in Chapters 4 and 6, and system verification planning artifacts (test cases) discussed in Chapter 10, is also a major consideration.

System validation planning starts when developing the system validation attributes that are part of each need expression addressing question, "How will we validate the SOI fulfills this need?" Likewise, system verification planning starts when developing the system verification attributes that are part of each design input requirement expression addressing question, "How will we verify the SOI fulfills this design input requirement?"

As part of this planning, the system verification and validation *Success Criteria, Strategy, Method, Activities,* and *Procedure Requirements* should be defined as described in Chapter 10. As much as practical, the project team should define and use these artifacts during design verification and validation, as well as early system verification and validation, to both mature and validate these artifacts. Any issues discovered with these artifacts should be corrected so there are fewer issues during system integration, system verification, and system validation.

Note: Any changes to the Success Criteria will most likely result in a change to the need or requirement expression from which the Success Criteria was defined. These changes will need to be submitted to a CCB for approval, as discussed in Chapter 14.

The reader is urged to consider an automated tool to manage the data associated with all the system verification and validation artifacts.

8.4 DESIGN VERIFICATION

Design verification is a confirmation that the design clearly implements an agreed-to transformation of the design input requirements into a design as communicated to the builders or coders within the sets of design output specifications.

In some organizations, design verification is referred to as a functional configuration audit (FCA) where the design is verified to meet its functional baseline approved at the PDR and CDR. Other organizations refer to this as a compliance assessment, where the design is assessed against its design input requirements with an assessment of compliance captured in a compliance matrix. These assessments typically leverage similarity to heritage, analysis techniques, process compliance, and, potentially, development unit inspection and test activities.

In addition to obtaining objective evidence that the SOI's design input requirements have been met by the proposed design, design verification also provides objective evidence that the design definition activities were performed by the project team per the organization's design definition guidelines and requirements. These design guidelines and requirements are commonly developed at the business management and operational levels as a result of lessons learned and best practices and are part of the design controls shown in Figure 8.1 that are part of the organization's Quality Management System (QMS).

Design verification involves both assessing the "correctness" of the design in the context on how well the design will result in a system that can be verified to meet its set of design input requirements as well as the quality of the design and resulting design output specifications that communicate the design to builders and coders. Two distinctive designs could satisfy the design input requirements equally—but may differ in quality and adherence to organizational or industry-specific design standards and processes, especially for software development processes.

Design verification also involves inspection of traceability records from the design input requirements to both of the following:

- To the design itself and resulting design output specifications.
- To the analytical/behavior models of the system elements that make up the system as well as the integrated system modeled in its macro environment.

As discussed previously, design verification should also include verification that prototypes, engineering verification units (mechanical, hardware, or software), and simulations also meet the design input requirements. The design verification activities will include a combination of testing or demonstrations (if prototypes, engineering verification units, or simulations are used) and analysis.

Note: Some projects may choose not to map individual design input requirements to physical system elements—such as a cam arm or choice of metal for the exterior "skin." In this case, prototype testing may be mapped back directly to the design input requirements. While this is not the most rigorous approach, the project team may choose it as long as there is sufficient objective evidence with a sufficient level of confidence that the design, if realized, would result in an SOI that will meet its design input requirements.

The practice of SE is intended to give the designers freedom to use their creativity in a design solution, as long as the baselined set of design input requirements and integrated set of needs are met.

A summary of the design verification activities is shown in Figure 8.3. *Note that it is assumed the draft design output specifications are reasonably mature and at least preliminary traceability of both the design input requirements to the design artifacts and the design output specifications has been completed.*

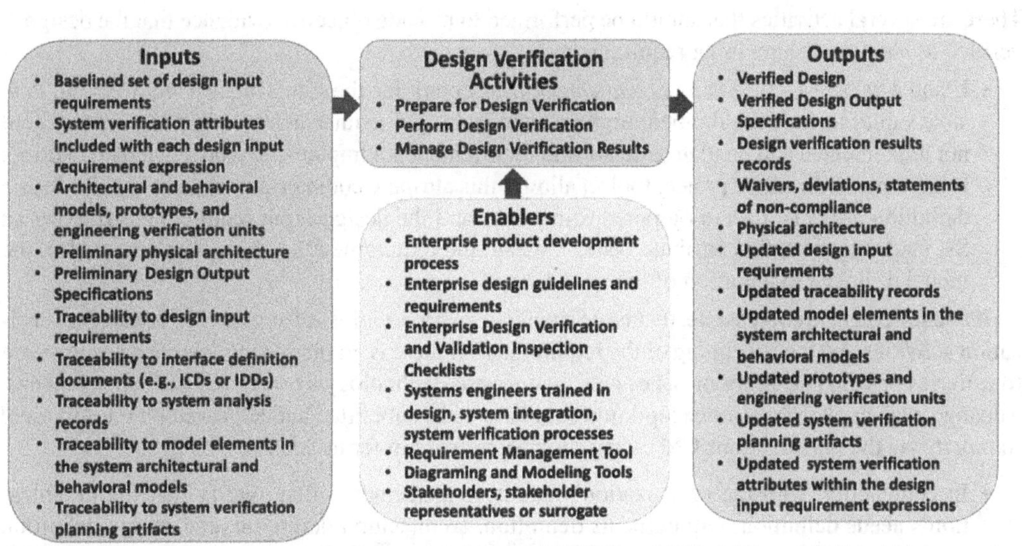

FIGURE 8.3 Design Verification IPO Diagram.

Design verification activities also help ensure individual needs and requirements have the GtWR characteristics: *C3—Unambiguous*, *C4—Complete*, *C6—Feasible*, *C7—Verifiable*, *and C8—Correct.*

8.4.1 Prepare for Design Verification

As shown in Figure 8.3, there are several enablers to successful design verification. These include a process for developing and managing sets of needs and sets of design input requirements, an enterprise product development process and *Design Definition Process*, and systems engineers trained in and knowledgeable on how to perform the design verification activities.

Inputs and enablers to design verification include a verified and validated set of design input requirements along with the enterprise's design guidelines and requirements and the architectural and behavioral models of the SOI. Of particular importance are the engineering design applications and modeling/diagramming tools within the project's toolset used to produce the initial concepts and artifacts of the design.

Traceability of the design with respect to the needs, requirements, interface definition documents, system analysis records, model elements, and finally to the system verification planning artifacts is essential to successful design verification.

An important input to design verification is the system verification attributes that (ideally) are part of each design input requirement expression. During design verification, the project team must ensure the design output specifications will result in an SOI that can pass system verification as communicated within these attributes.

Assuming the design verification processes are in place and tools are in use, the artifacts listed as inputs shown in Figure 8.3 should have been produced and matured to the point where they are ready for the design verification activities. Preparing for design verification consists of gathering or obtaining access to these input artifacts. On many projects, design verification can begin on one set of system elements, while artifacts on another part of the design are still being defined.

8.4.2 Perform Design Verification

There are several activities that should be performed to provide objective evidence that the design is verified to meet the design input requirements.

- Using *Attribute A31—Status of implementation,* determine the status of each requirement. Possible values include requirement implemented in design; requirement not feasible, requirement not implemented and no plan to do so; and requirement not implemented but have plan to do so. If implemented and the project toolset allows, this attribute could be a trace between the design definition artifacts, design output specification, and the design input requirement. If using an SE modeling tool, this attribute could indicate the requirement has been implemented in the model or linked to a portion of the model.

If a requirement is not feasible, that issue should have been addressed as part of requirement verification activities and the baselining of the requirement. If there is no plan to implement a requirement (due to cost or schedule issues or issues with maturing the technology a requirement is dependent on), a change, waiver, deviation, or exception would have to be submitted and approved by the *Approval Authority* via the organizations CM change control process (refer to Section 14.2.5).

- In conjunction with the organization's quality management office, verify that the organization's needs definition, requirements definition, architectural definition, and design definition guidelines and requirements were followed and the required traceability records are available.

If not followed, the required input artifacts may not be available including the verified and validated sets of design input requirements.

- Use the project's toolset to generate traceability records that show traceability of the design descriptions and other design artifacts to the design input requirements. If traceability records are not available, it will be difficult to do effective design verification. This traceability is extremely important for requirements that are addressing a risk mitigation action. Safety and quality personnel will use this information to close out this risk. *This activity helps show that the design is complete.*
- Using traceability records, verify the design and resulting design output specifications adequately address each of the design input requirements. During gate reviews (such as PDR and

CDR), it is a customary practice to first list specific subsets of design input requirements and then provide objective evidence that the design will result in a system that will meet those requirements. *This activity helps show that the design is complete and correct.*

- Using the information in the system verification attributes included within each design input requirement expression and associated system verification artifacts, verify the proposed design and resulting design output specifications will result in a system that can be verified to meet the design input requirements.

  To do this, the project team can define system verification activities (test cases) that will determine if the SOI, once built, will meet the system verification *Success Criteria* in accordance with the *Strategy* and *Method* defined in the verification attributes. These activities can first be performed using the models and then using the prototypes and engineering verification units.

  As part of this effort, the project team may need to update the verification *Success Criteria, Strategy,* and *Method* defined in the system verification attributes. The updated information will be reflected in system verification artifacts discussed in Chapter 10 and will be used for system verification. Updating the verification plan or strategy at this point is far easier than midway through system verification.

  If the design verification activities have resulted in changes to the existing design input requirements, these changes will need to be managed via the project's configuration control process, and updates to the design must be incorporated into the design output requirements.

- Perform a TRA. Based on the project's or applicable corporate or government design guidance and requirements, verify the required TRL for critical technologies has been achieved appropriate for this lifecycle stage as defined in the project's Technology Maturation Plan.
- Using approaches similar to those discussed for design input requirements verification, verify the quality of the design output specifications and other design artifacts. How this is done will be organization specific. If the organization has an internal Quality Management (QM) process that is certified to a quality standard (such as ISO 9000 and AS 9100), the project team needs to verify conformance to the QM process requirements and guidelines. The design output specifications and other design artifacts need to have characteristics consistent with the QM requirements and guidelines and verified to be correct, complete, consistent, feasible, understandable, and verifiable. Smaller projects may not adopt such standards, although different corporate requirements for quality still apply.

8.5 DESIGN VALIDATION

Design validation is a confirmation that the physical architecture and design, as communicated in the design output specifications, will result in an SOI that meets its intended purpose in its operational environment, when operated by its intended users and does not enable unintended users to negatively impact the intended use of the system, as defined by the baselined integrated set of needs.

Closely related to design validation is the concept of System Effectiveness Analysis (SEA) to determine if criteria defining the effectiveness of the system, such as measures (*refer to Section 4.3.2.4*), are addressed and implemented in the design and will result in a realized system that will meet these criteria.

Design validation also involves assessing whether the defined validation *Success Criteria, Strategy,* and *Method* as documented in validation attributes that are part of each need expression will be able to be executed by the project team and achieved by an SOI built to the set of design output specifications.

Even though the design has been verified that the resulting system will meet the design input requirements, there is still the possibility that the resulting SOI will fail system validation. This can happen if the design input requirements were defective or were not a correct transformation from the baselined integrated set of needs.

This can also happen if the designers did not address all the design input requirements or ignored one or more design input requirements. Even worse, some organizations fail to define, verify, validate, and baseline an integrated set of needs before defining the set of design input requirements. *Without a baselined integrated set of needs, what will the design and system be validated against?*

The design validation activities will include a combination of testing, demonstrations, and analysis using the models, prototypes, and engineering validation units.

A summary of the design validation activities is shown in Figure 8.4.

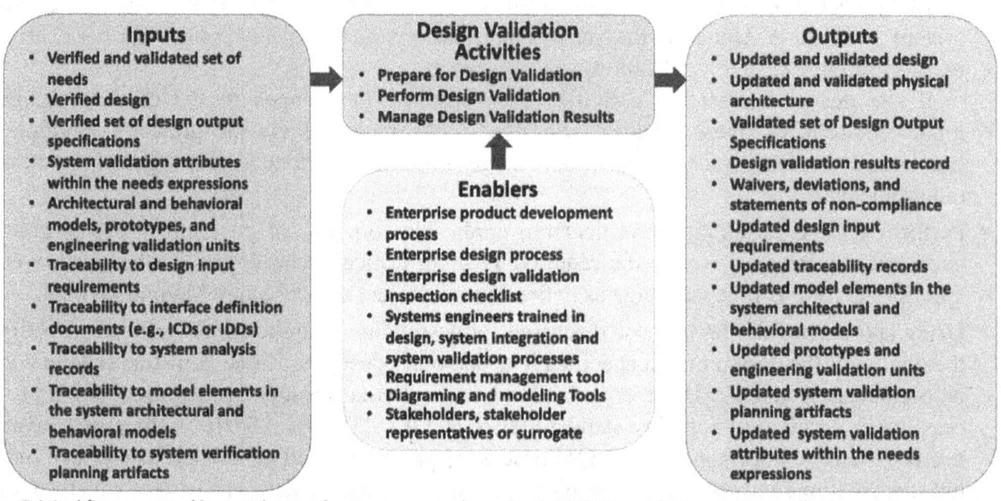

FIGURE 8.4 Design Validation IPO Diagram.

The following discussion on design validation assumes there is a baselined integrated set of needs to validate the SOI against. If the set of needs has not been defined, verified, validated, and baselined as described in this Manual, the project is at risk of failing system validation.

Design validation helps ensure the sets of needs and requirements have the GtWR characteristics: *C10—Complete*, *C11—Consistent*, *C12—Feasible*, *C13—Comprehensible*, and *C14—Able to be Validated.*

8.5.1 Prepare for Design Validation

As shown in Figure 8.4, there are several enablers to successful design validation. These include a process for developing and managing the integrated sets of needs and sets of design input requirements, an enterprise-level design definition process, and systems engineers trained and knowledgeable on how to perform the design validation process activities.

Inputs and enablers to design validation include a verified and validated integrated set of needs along with the enterprise's design guidelines and requirements and the architectural and behavioral models of the SOI. Of particular importance are the applications and modeling/diagramming tools within the project's toolset used to produce the input artifacts from which the design was transformed.

Traceability of the design with respect to the needs and their sources, to interface definition documents, system analysis records, to model elements in the architectural and analytical/behavioral models, and to the system validation planning artifacts is essential to successful design validation.

An important input to design validation is the system validation attributes that (ideally) are part of each need expression. During design validation, the project team must ensure their design and resulting design output specifications will result in an SOI that can pass system validation as communicated within these attributes.

Assuming the processes are in place and tools are in use, the artifacts listed as inputs in Figure 8.4 should have been produced and matured to the point where they are ready for design validation. Preparing for design validation includes accessing these input artifacts. The project team may be able to start design validation on some areas and system elements of the SOI, while artifacts on another part of the design are still being defined.

The process for developing and managing sets of needs and sets of requirements should result in analysis records to provide sufficient rationale for the transformation that resulted in the design and resulting set of design output specifications.

Ideally, rationale should be one of the attributes maintained with the need statements to help understand the intent of what a need statement is communicating. For whatever is being communicated in the need statement, there must be rationale that includes the assumptions made during the transformation, as well as traceability to the sources from which the need was transformed.

8.5.2 Perform Design Validation

There are several activities that should be performed to provide objective evidence that the design is validated to meet the integrated set of needs; some of these may already have been completed concurrently during design verification.

- Confirm that the organization's needs, requirements, and design definition processes were followed, resulting in input artifacts shown in Figure 8.3, and that the required traceability records are available. If the processes were not followed, the required input artifacts may not be available, including the verified and validated integrated set of needs and traceability matrices. If not available, it will be difficult to do effective design validation.
- For each need, perform a design analysis to confirm that the design, as communicated within the design output specifications, will result in a system that will meet the intent of the need. Often to understand the *intent* of the need, the rationale (if it exists) may not be sufficient. The use of traces from a need statement to its source is a good way to better understand the intent. Once the intent is understood, the project team can determine whether the design solution will result in a system that will meet the intent of the need.
- Using the information in the system validation attributes included within each need expression and associated system validation artifacts, validate the proposed design and resulting design output specifications will result in a system that can be validated to meet the baselined integrated set of needs.

  To do this, the project team can define validation activities (test cases) that will determine whether the SOI, once built, will meet the system validation *Success Criteria* in accordance with the *Strategy* and *Method* defined in the system validation attributes. These activities can first be done using the models and then using the prototypes and engineering validation units.

  As part of this effort, the project team may need to update a need statement or the system validation *Success Criteria*, *Strategy*, and *Method* defined in the system validation attributes for each need statement. The updated validation attributes will be reflected in system validation artifacts that will be used for system validation. Updating the validation plan or strategy at this point is far easier (less time, cost, and resources) than midway through system validation.

If the design validation activities have resulted in changes to the existing integrated set of needs, these changes will need to be managed via the project's CM process. The project team is encouraged to put any new knowledge on history or intent of the need, into the need's rationale statement/attribute.

- If a system simulation capability was developed for the integrated system within a simulated operational environment, the project team should run simulations using representative inputs to determine if the needed outcomes result. If a range of inputs is possible, the simulations should include the range of all possible inputs. If multiple inputs are involved, then all combinations of inputs across each of their ranges must be exercised along with across the ranges of each of the operational environmental factors.

  Having a simulation capability enables the discovery of issues that may not be discovered until actual system verification or worse yet, after the system has been released for operations. This simulation capability is also an integral part of the SEA discussed at the beginning of this section.
- If prototypes and engineering design validation systems have been developed, the project team will use those as part of design validation. If possible, it is a best practice to involve the actual users in the actual operational environment to do these activities.

8.6 MANAGE DESIGN VERIFICATION AND VALIDATION RESULTS

The data involved in, and resulting from, the design verification and validation activities must be recorded and managed. First, the project team should create design verification and validation records documenting the results and outputs of these activities. These records contain the data and information needed to determine whether the design and associated design output specifications will result in a system that meets the baselined integrated set of needs. These records address two things: 1) the ability of the design to implement the design input requirements and meet the integrated set of needs, and 2) traceability to the objective evidence that the set of design output specifications will result in a system that can be verified to meet the design input requirements and validated to meet the integrated set of needs. This information will be helpful during system integration, system verification, and system validation, as well as during final acceptance.

If it is identified that there is insufficient data to complete design validation, such as a lack of a baselined integrated set of needs, the issue should be recognized as a risk that could cause system validation to be deemed unsuccessful. *In cases where a formal integrated set of needs has not been defined, it is even more important to involve the customer and other stakeholders during the design process.*

It is common to have issues with the design not being able to meet a requirement or need; this is part of the evolutionary, knowledge-based practice of SE and a prime reason for the design cycles discussed previously.

For all nonconformances, the project team must either address the design deficiency, request the subject requirement or need be changed, or request a variance from management and the *Approval Authority*. Any issues with the baselined integrated set of needs, or set of design input requirements, discovered during design verification and validation activities need to be addressed and updates made to the needs and requirements. *Refer to Section 10.3.3 for a more detailed discussion concerning variances, concessions, waivers, and deviations.*

An updated need should drive an impact analysis to see if any design input requirements, design output specifications, system verification attributes, or verification activity descriptions should be updated. All changes to design input requirements must be assessed concerning the impacts on other related artifacts in the trace chain. Often the change of one requirement in the chain can have a ripple effect that must be managed. Failing to do so could result in other design or system verification and validation activity failures, or a consequential failure of the system during operations.

Note that even updating one need can have large domino effect, rippling across many of the development lifecycle artifacts, especially at the higher levels of the architecture. A change to the integrated set of needs at the system level may result in updates to the set of design input requirements, architecture, design concept, models and simulations used for design verification and design validation, design output specifications, prototypes and engineering test units, and design verification and design validation activities redone. Failing to address the impacts of a change to a need could also result in other design or system validation activity failures.

Project teams are highly encouraged to use a documented change impact analysis process as discussed in Section 14.2.5, including CCBs if the needs or requirements and associated artifacts need to be changed. Doing so results in a change history of the evolution of the needs and requirements.

Practicing SE from a data-centric perspective, Requirement Management Tools (RMTs) that allow needs and requirements to be managed across the lifecycle along with the use of modeling tools can also be useful in addressing and managing changes due to issues found during design verification and validation, helping maintain consistency and correctness of all artifacts developed across the system lifecycle.

The project's toolset should allow tracking of the status of the design verification and validation activities using a dashboard that communicates key metrics related to the design verification and validation activities. Many of these metrics can be obtained using the need and requirement attributes discussed in Section 8.7 within the project's toolset. Making this information accessible will enable all members of the project team to have the same view of the maturity and stability of the design, status of the design, CCB activity, and status of the design verification and validation activities. *To ensure this information is current, the project team must keep Attribute A31—Status of implementation current.*

8.7 USE OF ATTRIBUTES TO MANAGE DESIGN VERIFICATION AND DESIGN VALIDATION

Requirement attributes that can aid in design verification and validation management include:

- A1: *Rationale*: intent of the need or requirement.
- A26: *Stability of a need or requirement*: values could include stable, likely to change, and incomplete.
- A31: *Status of implementation*: an indicator of the status of the implementation or realization of the requirement in the design outputs. Possible values include requirement implemented in design; requirement not implemented and no plan to do so; and requirement not implemented but have plan to do so. *Note: It is a good practice to establish trace between design artifacts and the requirements being implemented by the design. Many SE tools and modeling tools enable this capability.*

Refer to Chapter 15, for a more detailed discussion on attributes and definitions of each.

9

PRODUCTION VERIFICATION

Note: While production verification is normally performed by a quality organization and not the responsibility of systems engineers, those involved in manufacturing are key stakeholders that have needs and requirements concerning the system being built or coded. Also, the manufacturing system is developed using systems engineering processes and activities. In terms of verification, the same principles apply in that the manufacturing system is verified against its own design input requirements and design output specifications and validated against its own integrated set of needs.

Once the design, as communicated by the design output specifications, has passed design verification and design validation and has been baselined at a gate review, the System of Interest (SOI) can be realized (manufactured or coded). In this Manual, the manufacturing and coding are referred to as "production"; in the International Council on System Engineering Systems Engineering Handbook (INCOSE SE HB), it is referred to as "implementation."

The focus of this chapter is on "production verification" sometimes referred to as a Physical Configuration Audit (PCA). While system verification is performed once against the design input requirements, *production verification is done continuously as a quality function for each system element that is manufactured or coded.*

9.1 PRODUCTION VERIFICATION DEFINED

Referring to Figure 9.1, production verification is concerned with both verifying the system was "built right" as defined by the design output specifications as well as verifying that it was "built correctly" per the organizational design controls (guidelines, best practices, and requirements) concerning the equipment and processes used to manufacture and code the system element were followed. These guidelines and requirements are part of the organization's design controls that are part of the organization's Quality Management System (QMS).

INCOSE Needs and Requirements Manual: Needs, Requirements, Verification, Validation Across the Lifecycle,
First Edition. Louis S. Wheatcraft, Michael J. Ryan, and Tami Edner Katz.

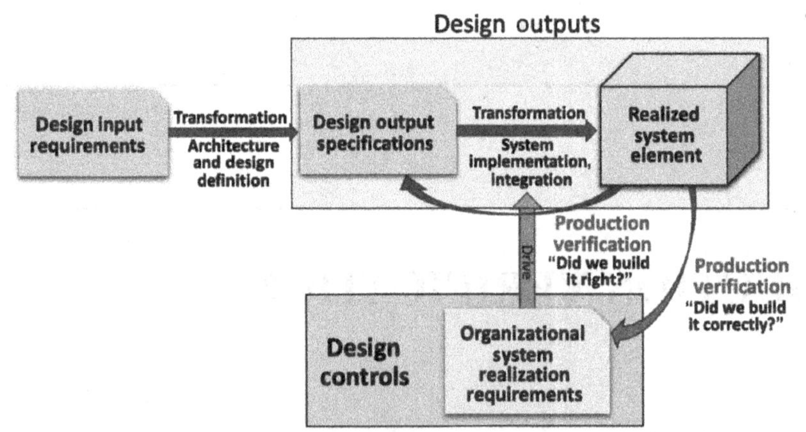

FIGURE 9.1 Production Verification.

In addition, production verification ensures any additional requirements concerning the manufacturing/coding processes defined by the customer, standards organizations, and regulatory agencies are followed. The focus of these requirements is ensuring the quality of the system being produced as defined by work instructions and standards. For example, soldering or welding standards, or for food and pharmaceuticals, contamination standards.

These standards can present unique challenges as to which production verification activities need to occur routinely during each step of the manufacturing process. This may drive the need for requirements concerning test or inspection points for components, subassemblies, assemblies, and subsystems that are only used during manufacturing, but not used, nor accessible, once the system elements are integrated to form the system. (*In many cases, these test points may also need to be accessible during system verification as well.*) These requirements for accessibility may be included as part of both the design input requirements and the design output specifications or may be included within the organization's design controls for design and manufacturing.

9.2 SCALABILITY, REPEATABILITY, AND YIELD

Another challenge for production is the *scalability* and *repeatability* of the manufacturing and coding processes to consistently produce a safe, secure, and high-quality product.

Scalability is an important concept. Being able to build a working prototype or engineering unit in a laboratory may not carry through to the manufacturing of multiple versions of the SOI at the volume and frequency required.

One of the goals of production verification is to verify the resulting manufactured/coded versions can meet the design output specifications that were based on the laboratory prototype or engineering unit that will result in an SOI that can be verified to meet its set of design input requirements and validated to meet its set of needs. However, there are cases where a design that passed design verification and design validation in a laboratory or development setting, results in a realized system that fails system verification and validation after production in a production facility.

For example, medical devices that involve instruments to assess biological samples. An "assay" is developed consisting of a special "slide" or "cartridge" along with chemical "reagents." These are provided in a special test kit for use by technicians to prepare a biological sample for insertion into the instrument. An issue often encountered is that while the chemistry or formulation has been demonstrated to work in the laboratory, there are often scalability issues concerning preparing 10,000 kits during manufacturing that consistently work the same way to produce the intended results when used by the intended users in the intended operational environment. In this case, the manufacturing organization must develop and validate a process that will produce test kits that will consistently meet the design input requirements for the kit.

Repeatability is another important concept. Building one copy of a system is a lot different from production line manufacturing that is turning out hundreds or thousands of copies. This is especially important when manufacturing safety-critical systems. For example, the Food and Drug Administration (FDA) addresses the manufacturing process in Title 21, Part 820, defining [manufacturing] process validation as "*establishing by objective evidence that a process consistently produces a result or product meeting its predetermined specifications.*"

Verifying repeatability would mean the need for obtaining objective evidence that a manufacturing process used during production consistently produces an SOI meeting its design output specifications, compliant with both organizational and regulatory guidelines, requirements, and best practices for manufacturing and coding.

Closely related to repeatability is the concept of "yield," where "yield" is defined as the percentage of copies of a system that pass production verification. While yield does not directly impact product design, if yield is too low, it could indicate other issues such as a hard-to-manufacture design or a poorly planned-for and implemented manufacturing concept. In either case, the organization may need to consider a redesign, either to the SOI being produced or the equipment used to produce the SOI. Any redesign would need to be carefully analyzed and assessed in terms of cost, schedule, risk, scalability, repeatability, and yield.

9.3 PRODUCTION VERIFICATION VERSUS SYSTEM VERIFICATION

It is common for production verification against design output specifications and production standards to be performed separately from system verification, which is against the design input requirements (as discussed in Chapters 10 and 11). In many organizations, production verification is overseen by the quality engineering or quality control department; quality assurance would then monitor production, including assessment and management of any production drift. Tools such as lean six-sigma, process controls, and Pareto charts are often used in this effort.

When a manufactured product fails system verification or system validation, it may not be clear if the cause was a production fault or a design flaw. To reduce the risk of a production fault, production verification is a formal and important activity in the product development lifecycle.

However, this approach may not always be the case. For parts and components at lower levels of the physical architecture, some projects may accept the production verification outcome as objective evidence that a part or component meets its design output specifications and skip the system verification and validation activities needed to provide objective evidence that it also meets its set of design input requirements and set of needs. This approach assumes that if a system's design verification and design validation was performed successfully, a system built to the resulting design output specifications will pass system verification and system validation. This is often a budget and schedule decision, where the project is willing to accept the risk of this approach.

For cases in which the project procures a Commercial Off-the-Shelf (COTS) system element or contracts out the production of the SOI to a supplier, the project may elect to accept the supplier's certification that the system element or SOI passed production verification by using appropriate, pre-negotiated set of documentation from the vendor. In other cases, the customer may have their own acceptance process or may make provisions in the contract to have their own quality inspectors be involved in the production and production verification activities in the supplier's facility. *Refer to Chapter 12, for a more detailed discussion concerning the use of Off-the-Shelf (OTS), Modified Off-the-Shelf (MOTS), and COTS system elements and Chapter 13, concerning supplier-developed systems.*

System verification is normally performed once against the design input requirements; however, on multi-build system elements to be used in the future, organizations may do a subset of verification based on risk of poor workmanship (such as grounding requirements that support Electromagnetic interference [EMI] controls). In addition, when larger quantities of a part or component are being produced, some aspects of system verification may be performed as part of production verification on a periodic basis (lot testing). Even with lot testing, once production begins, the verification activities are intended to detect production drift—*not* repeat development testing that has already been performed.

9.4 VERIFICATION AND VALIDATION OF THE MANUFACTURING SYSTEM

The manufacturing system is a system in its own right. Because of this, the manufacturing facility, equipment, and processes will need to be treated as enabling systems undergoing their own system development, successfully passing their own system verification and validation, and being certified to be ready for production of the SOI. This is sometimes called, "qualifying the (production) line."

Engineering a manufacturing line is a technical task that requires knowledge of not only part design but also design input requirements and overall needs. Defining concepts for production and production verification are part of the lifecycle concepts that are defined and matured during *Lifecycle Concepts and Needs Definition* discussed in Chapter 4.

Like Technology Readiness Levels (TRLs) discussed in Chapters 4 and 6, Manufacturing Readiness Levels (MRLs) can be used to define and develop the organization's manufacturing system and associated processes to the point where a quality and safe SOI can be produced consistently, which results in a product that meets both the design output specifications (production verification) as well as its design input requirements (system verification). MRLs are defined using a scale of 1–9, similar to those used for TRLs. The lower the number, the more risk there is concerning the ability to manufacture a quality system that meets its design output specifications.

9.5 SPECIAL CONSIDERATIONS

Special considerations during production verification include:

- *Building systems that either cannot be tested as an integrated system in its operational environment, or it is not practical to do so.* This could include a space launch vehicle or a weapons system. In these cases, system verification and validation will consist of a combination of production verification, design verification, and design validation activities. The assumption is that if design verification and validation were successful AND the system was manufactured or coded in accordance with the verified and validated design output specifications using organizational best practices and guidelines, the result should be an integrated system that will meet its intended purpose in the operational environment.

- *Consumer products that are built in large quantities.* In this case, classical system verification and validation, as discussed in Chapters 10 and 11, will be completed once per production unit, but to ensure quality and repeatability, production verification will need to be performed periodically to assure quality. It is common that for some consumer products, the first time they are put into use by a customer is the first time electrical power will have been applied.
- *Production verification of software applications.* Production verification for software would involve verifying that the code set, or executable delivered for production or insertion into a hardware assembly, is the same exact executable obtained from the repository. Example: When a digital optical disk, i.e., DVD is "burned" with the software, there is an error-checking step to ensure the disk was "produced" right—i.e., no bits were flipped in the disk production process. *Note that the hardware that receives the software would be tested in the normal way: production verification, followed by system verification and system validation.*
- *Embedded software will also use different approaches.* Software production verification would be performed as described in the previous item; after this, the embedded software would need to pass its "operational" system verification and system validation. *This means the firmware would need to be verified and validated after being installed in the destination hardware and that hardware has been integrated into the macro system it is a part.* Just because firmware worked (passed system verification and system validation) in a simulation or on development hardware does not mean it will work in the actual SOI or once the SOI is integrated into the macro system it is a part.
- *Projects that have multiple suppliers, each of which has their own unique production facilities and processes.* The lead system engineer, or product engineer, will need to review each supplier's processes, procedures, and documentation to determine if their production processes are acceptable, and whether the received items were truly built to the design output specifications. This may involve several vendor visits and audits. In some cases, the customer may define mandatory inspection points in the production process witnessed by a customer quality control person.
- *Quality of raw materials, parts, and components used as part of the manufacturing process.* The quality of the raw materials, parts, and components, either as part of the manufacturing system or used to produce the SOI, has a direct effect on the quality of the manufactured SOI. Periodic auditing of these raw materials, parts, and components in terms of quality is important to ensure they continue to meet their own acceptance criteria over time. Of particular concern is when counterfeit versions are provided by a less than reputable supplier rather than the contracted versions, this can result in security, safety, and quality issues.
- *Manufacturing equipment degradation over time.* The mechanical parts and equipment that make up the manufacturing system degrade over time. In some cases, this could result in hazardous contaminants being introduced into a product. This is a major concern for food and pharmaceuticals. While contamination requirements are initially met, due to degradation of the manufacturing requirement, these requirements are no longer compliant; this can result in expensive recalls and impact a manufacturing organization's reputation.

Addressing each of these considerations is beyond the scope of this Manual. For more information concerning production and production verification, refer to the INCOSE SE HB v5 Sections 3.1.7, Manufacturability/Producibility Analysis, 3.1.10, Sustainability Engineering, and 2.3.5.7, Implementation Process.

9.6 PRODUCTION VERIFICATION

A summary of the production verification activities is shown in Figure 9.2.

Preparing, performing, and managing production verification results would be performed according to an organization's production readiness and manufacturing procedures. This is outside the scope of this Manual.

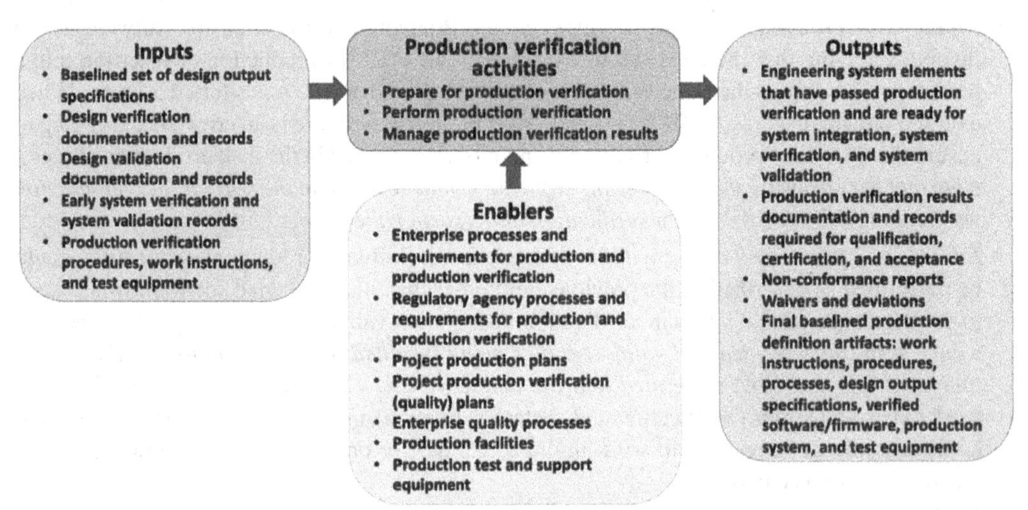

FIGURE 9.2 Production Verification IPO Diagram.

10

SYSTEM VERIFICATION AND VALIDATION COMMON PRINCIPLES

While system verification and system validation are two distinct processes, they share many common principles. Before discussing these process areas individually in Chapter 11, this chapter covers principles each has in common.

The definition, development, and management of the system verification and validation artifacts described in this chapter are concurrent and crosscutting activities that begin early in the system lifecycle during *Lifecycle Concepts and Needs Definition*, continue to be developed and matured during *Design Input Requirement Definition*, and are further matured and validated during design, design verification and validation, and early system verification and validation. Conducting early system validation and verification, as discussed in Chapter 8, will help reduce the number of issues during system integration, system verification, and system validation activities on the actual physical System of Interest (SOI), further reducing the risk of potential budget overruns and schedule slips.

Some organizations use the phrase "Test Phase" where the word "Test" refers to a series of activities being conducted across the lifecycle. These activities could be part of system verification or system validation as discussed in this section. When this is the case, it is important to note that activities performed to address system verification and validation are not only accomplished by Test as a *Method* but can also be accomplished using other methods such as *Analysis, Inspection,* or *Demonstration.* Additionally, organizations conduct test activities as part of a design cycle, maturing the design, or doing design verification and validation, and not necessarily tests in support of system verification or validation. In this context, an organization may conduct acceptance testing, qualification testing, certification testing, functional testing, and environmental testing as examples. These may be a subset of system verification or validation, or may be standalone tests.

Another common issue is the use of the phrase "requirement verification" when what is really meant is "system verification," as discussed in Chapter 2.

INCOSE Needs and Requirements Manual: Needs, Requirements, Verification, Validation Across the Lifecycle,
First Edition. Louis S. Wheatcraft, Michael J. Ryan, and Tami Edner Katz.

To avoid these ambiguities, it should be made clear what is being verified or validated and against what. System verification refers to verifying the SOI against its design input requirements; system validation refers to validating the SOI against its integrated set of needs.

System Verification and Validation Stages

Both the system verification and validation can be divided into five stages: Planning, Defining, Execution, Reporting, and Approval, as shown in Figure 10.1. These stages are an iterative process, where feedback from later stages may prompt updates of data from the earlier stages (ideally by using a data-centric practice of Systems Engineering (SE), where these artifacts are included within the project's data and information model to help ensure consistency, correctness, and traceability resulting in an authoritative source of truth (ASoT)—as discussed in Section 10.7).

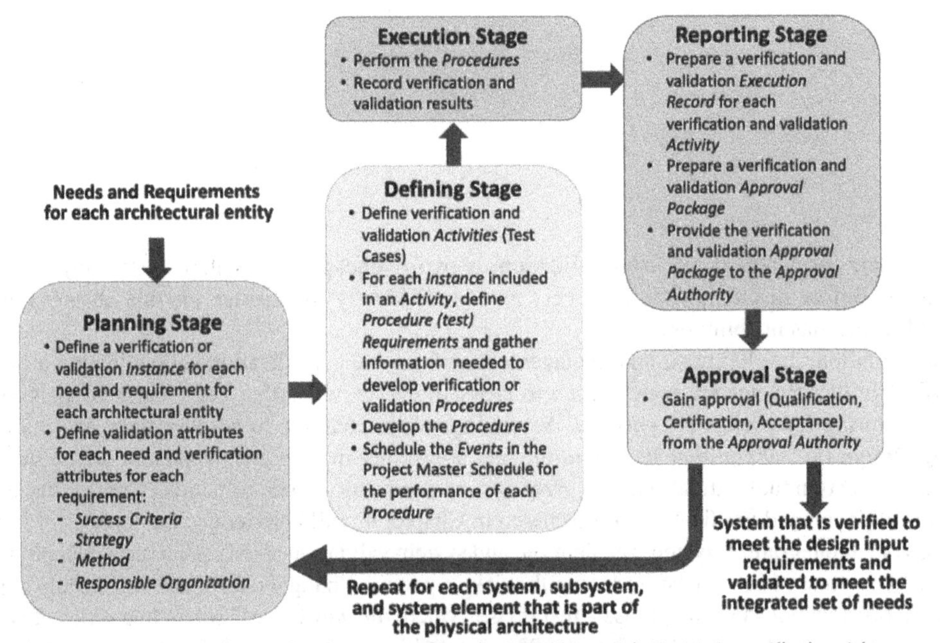

FIGURE 10.1 System Verification and Validation Stages.

Given system integration, system verification, and system validation are a series of bottom-up activities, system verification and validation are repeated for each subsystem and system element within the system's physical architecture.

The five stages may overlap in the sense that some system verification and validation activities for the subsystems and system elements at one level of the architecture may be completed before other system verification or validation activities for other subsystems and system elements have been fully defined at the next higher level of integration:

- *Planning Stage*. Each need and design input requirement represents a system verification or system validation *Instance*. For each *Instance*, system verification or validation attributes concerning the *Success Criteria*, *Strategy*, *Method*, and *Responsible Organization* are defined. These attributes are included within the need or design input requirement expressions when they are defined. *Refer to Section 10.1 for a detailed discussion on the Planning Stage*.

- *Defining Stage*. When sufficient architecture and design detail are known, a system verification or validation *Activity* is defined that addresses one or more *Instances*. For each *Instance* included in an *Activity*, a set of *Procedure Requirements* are defined that will result in the *Strategy* to be realized using the selected *Method*. Based on these *Procedure Requirements* and other key information gathered, a *Procedure* is developed consisting of a sequence of steps or actions whose performance will result in the objective evidence whether the SOI meets the *Success Criteria* for each *Activity* based on the *Strategy* and *Method* for that *Activity*. An *Activity* can address more than one *Instance* and a *Procedure* can address more than one *Activity*. The project team schedules system verification or validation *Events*, which can include the execution of one or more *Procedures* per *Event*. *Refer to Section 10.2 for a detailed discussion on the Defining Stage.*

  Note: What is defined in this Manual as a system verification or system validation activity is often referred to as a "Test Case" in some domains even if the Method is not "Test." In a similar manner to that discussed above for "Test Phase," it is important to understand that the word "test" is being used to refer to an action rather than a Method of verification or validation (Test, Demonstration, Inspection, and Analysis) as discussed in Section 10.1.3.

- *Execution Stage*. In accordance with the *Events* scheduled in the project's master schedule, the *Procedures* are performed to obtain the data needed to provide objective evidence that the SOI meets the *Success Criteria* defined for each *Instance* included in that *Procedure*. The data resulting from the performance of each step or action within the *Procedure* is recorded. *Refer to Section 10.3 for a detailed discussion on the Execution stage.*
- *Reporting Stage*. A system verification or validation *Execution Record* for each *Activity* is created after the completion of each *Procedure*. The *Execution Record* will state the status and outcome of each *Activity* in terms of whether the resulting objective evidence obtained can be used to determine whether the *Success Criteria* for each *Instance* within the *Activity* were met. The family of *Execution Records* are combined into a system verification or validation *Approval Package*. *Refer to Section 10.4 for a detailed discussion on the Reporting stage.*
- *Approval Stage*. The *Approval Package* is submitted to the appropriate *Approval Authority* for approval (qualification, certification, or acceptance). *Refer to Section 10.5 for a detailed discussion on the Approval stage.*

When different subsystems or system elements of an SOI are procured from a supplier, system verification and system validation for those elements are part of the qualification, certification, and acceptance criteria specified as part of the procurement contract.

In some cases, the acceptance of a procured system element will be based on production verification against the design output specifications (quality-control-based acceptance criteria); in other cases, acceptance will be based on acceptance of the objective evidence obtained as a result of the supplier's completion of system verification and validation against the system element's integrated set of needs and set of design input requirements. *Refer to Chapter 9 for a more detailed discussion on production verification and Chapter 13 for a more detailed discussion on customer/supplier system verification and validation considerations.*

Successful system validation showing that an SOI meets its integrated set of needs is dependent upon successful system verification that that SOI meets its design input requirements. At the integrated system or product level, the execution, reporting, and approval stages for system verification will normally happen prior to the execution, reporting, and approval stages for system validation, especially when system verification is done by a different organization than the organization responsible for system validation (discussed in Chapter 13). *Note: For some government projects, federal regulations require that system validation be done by a different organization than the one that performed system verification.*

However, for some needs and the resulting requirements, both system verification and validation may be accomplished concurrently when the same organization is responsible for doing both, as is common for systems developed and integrated "in-house" and with many consumer products. When cost-effective, and warranted by analysis, the expense of system validation activities separately can be mitigated by combining them and performing system verification and validation concurrently.

There will be some cases where objective evidence that an SOI has been validated to have met a need can be based on the objective evidence from verifying that the SOI has met its design input requirements, which have been transformed from that need. In other cases, verification that the SOI has met its design input requirements traced to that need is a prerequisite to the system validation activity associated with that need.

Which case applies will be determined by the project during the planning stage when the system validation attributes are defined for each of the needs expressions. A key consideration is that system validation is intended to be completed in the actual (or analogous) operational environment by the intended users (or surrogates); for system verification, that is not always the case.

For each stage shown in Figure 10.1, specific system verification and validation artifacts are generated and managed. Figure 10.2 shows the relationships of each of these artifacts.

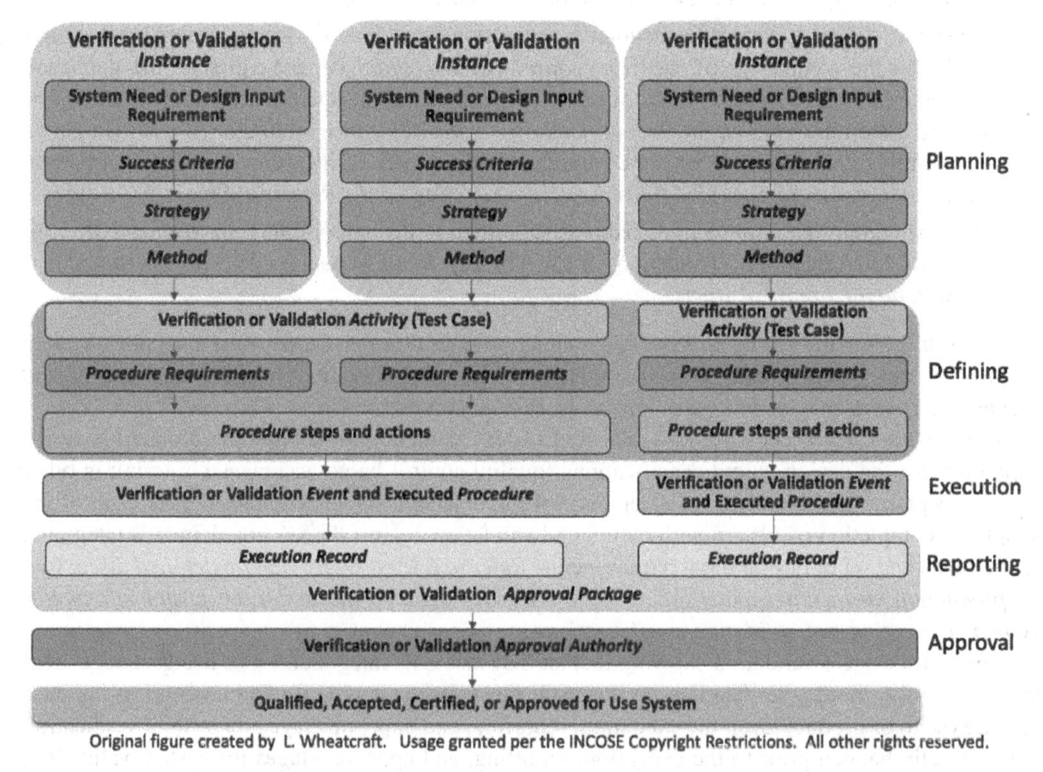

FIGURE 10.2 System Verification and Validation Process Artifacts.

10.1 PLANNING STAGE

Thorough system verification and validation planning from the beginning of the project [54] is critical to ensuring a timely and smooth system integration, system verification, and system validation program. Because of this, the planning and definition of the system verification and validation activities and expected artifacts must be completed before implementing these activities.

Planning begins during *Lifecycle Concepts and Needs Definition* to mitigate the risk of budget overruns and product launch delays, which are often caused by incomplete system verification and validation planning. These activities include defining lifecycle concepts for procurement, design, design verification, design validation, early system verification, early system validation, production, production verification, system integration, system verification, and system validation.

The planning activities should be documented in the project's Program/Project Management Plan (PMP) from a Project Management (PM) perspective and elaborated within the project team's System Engineering Management Plan (SEMP) from a SE perspective. The PMP and SEMP provide an overview of the project's verification and validation program across the lifecycle, highlighting the details toward implementing these activities, i.e., how each of the subsystems and system elements that are part of the SOI physical architecture will be validated to meet their needs and verified to meet their design input requirements. *Refer to Chapter 14 for a more detailed discussion concerning the PMP and SEMP.*

Because system integration, system verification, and system validation are formal development lifecycle processes, project teams develop dedicated plans for how they will address these processes. These plans include the Master Integration, Verification, and Validation (MIVV) and System Integration, Verification, and Validation (SIVV) plans. *Refer to Chapter 14 for a more detailed discussion concerning the MIVV and SIVV plans.*

Given that system verification and validation are key cost and schedule drivers, early planning is especially important. *Early identification and maturation of the Success Criteria, Strategy, Method, and Responsible Organization helps to ensure the project team has the budget, schedule, and resources needed to successfully execute and complete its system integration, system verification, and validation program.* This information also helps to ensure the project has included in their Work Breakdown Structure (WBS), schedule, and budget adequate time and resources for all system integration, system verification, and system validation activities that need to be completed before the SOI is accepted and approved for use.

Developing these concepts early is also critical when contracting the development efforts of an SOI to a supplier. The information from the maturation of the design verification, design validation, system verification, and system validation concepts and implementing plans will be used to define the requirements and deliverables in the Statement of Work (SOW), Purchase Order (PO), or Supplier Agreement (SA) (as discussed in Chapter 13).

System verification and validation planning begin as each need and design input requirement is being defined. Attributes containing this information are included within the needs and design input requirements expressions; the expressions are not considered complete until these attributes are defined.

For each system verification and validation *Instance*, key information is defined to start system verification and validation planning.

- Well-defined *Success Criteria* (*Section 10.1.1*), which must be met by the SOI to which the need or requirement applies.
- The *Strategy* (*Section 10.1.2*) that must be followed to obtain objective evidence that the *Success Criteria* have been met.

- The *Method* (*Section 10.1.3*) (Test, Demonstration, Inspection, or Analysis) that is part of the *Strategy* to obtain data that provides objective evidence the *Success Criteria* were met.

Ultimately, it is the responsibility of the project team responsible for the integrated system to integrate all system verification and validation plans, attributes, activities, and results into a coherent set of artifacts for eventual *Approval Authority* acceptance of the SOI. This includes finalizing (and updating) the budget and schedule based on the number of needs and associated system validation attributes and the number of design input requirements and associated verification attributes.

To avoid technical debt, planning for system integration, system verification, and system validation must be addressed for each system, subsystem, and system element within the SOI's physical architecture.

Note: Because budgets and schedules of the system verification and validation activities evolve during development of the SOI, initial budgets are only estimations with a margin of error. Once all the sets of needs and implementing design input requirements for each system, subsystem, and system element at each level of the physical architecture are defined, a more accurate budget and schedule can be defined. This frequently happens when the project conducts the Preliminary Design Review (PDR). If contracts must be established early, these uncertainties must be addressed when contracting out an SOI to a supplier and should be documented in the contract.

For projects using system modeling and tools, initial planning can begin with the creation of a system verification or system validation *Instance* in the SOI system model for each need and design input requirement. Each *Instance* is a unique entity within the system model used to manage system verification and validation actions associated with a given need or design input requirement.

Even non-modeling tools that are table driven can define system verification and validation *Instances*; the important thing is that for each design input requirement and need, the above-mentioned artifacts be defined and managed within the project toolset to identify each system verification and validation *Instance*.

10.1.1 Define the System Verification and Validation Success Criteria

For each system verification and validation *Instance*, the *Success Criteria* are defined and recorded as an attribute within each need or design input requirement expression. The *Success Criteria* are defined by the customers, regulatory agency, or developing organization that must be shown to have been met, with some level of confidence, before the *Approval Authority* will qualify, certify, or accept the SOI to be used as intended in the operational environment when operated by its intended users. The *Success Criteria* represent the expected result of a system verification or system validation *Activity* for a specific *Instance*.

10.1.1.1 Formulating the Success Criteria Statement The *Success Criteria* is not a restatement of a need or design input requirement, nor is it intended to change the scope or intent of the associated need or design input requirement. By defining and recording the *Success Criteria* concurrently when the need statements and design input requirement statements are formed, the quality of the statements will improve, as well as help define the objective evidence needed to determine whether the SOI will be able to meet them.

For example, as tolerances and ranges are examined, the need for margins is assessed to ensure the *Success Criteria* can be achieved. In addition, the operational environment in which an *Activity* must take place is determined, as well as the quality expectations (for example, reliability). The feasibility to meet a need or design input requirement will be reassessed as the project team moves through the SOI development lifecycle stages.

The wording of the need or design input requirement statements provides the foundation for defining the *Success Criteria*. For example, given the following requirements for a pressure vessel:

- "The pressure vessel shall maintain an operating pressure of 200 ± 5 psi when operating within an external temperature range of xx degrees F to yy degrees F and an internal temperature range of xx degrees F to yy degrees F."
- "The pressure vessel shall not burst when pressurized to a maximum pressure of 300 ± 5 psi when operating within an external temperature range of xx degrees F to yy degrees F and an internal temperature range of xx degrees F to yy degrees F."
- "The pressure vessel shall have a burst pressure safety margin of xx % when operating within an external temperature range of xx degrees F to yy degrees F and an internal temperature range of xx degrees F to yy degrees F."

The value "200 ± 5 psi" is the value the *pressure vessel* must be verified to operate at, "300 ± 5 psi" is what the *pressure vessel* must be verified to achieve without bursting, and xx% is the required safety margin the pressure vessel must be verified to have. The external and internal temperature ranges address "under what conditions."

At this point, to define the *Success Criteria*, the requirements are examined collectively to determine what data must be obtained that will provide objective evidence that the pressure vessel can meet these requirements. The system can be verified to meet these requirements by conducting a series of tests across the expected range of performance and thermal ranges. Some tests could be conducted concerning the 200 and 300 psi requirements, and other tests concerning the safety margin would be run at the pressure that represents the safety margin. Multiple tests would be needed at the temperature extremes. *Note: A destructive test on an engineering unit would be conducted first as part of design and early verification in that the pressure would be increased until the vessel bursts to establish what the burst pressure is, which is then compared to the required safety margin.*

Assuming the system verification or system validation *Method* is Test, the resulting *Success Criteria* statement for the operating pressure requirement could be: "*This requirement will be shown to have been met when the pressure vessel has been pressurized to the operating pressure of 200 ± 5 psi for at least 4 hours for each combination of the internal and external thermal extremes.*"

For the maximum pressure requirement, the *Success Criteria* could be: "*This requirement will be shown to have been met when the pressure vessel has been pressurized to the maximum pressure of 300 ± 5 psi for at least 1 hour AND the pressure vessel has not fractured for each combination of the internal and external thermal extremes with a 99% confidence level.*"

For the safety margin requirement, the *Success Criteria* could be: "*This requirement will be shown to have been met when the pressure vessel has been pressurized to safety margin of xx% greater than the maximum operating pressure of 300 ± 5 psi for at least 4 hours for each combination of the internal and external thermal extremes, AND the pressure vessel has not fractured, with a 99% confidence level.*"

Note: The actual time periods and confidence level shown above would be computed based on a hazard analysis done by the organization's safety office. They may also be specified in standards or regulations for this type of system.

The pressure vessel would be verified to meet its requirements by obtaining the data needed to provide objective evidence the *Success Criteria* were met via the steps and actions in the Verification *Procedure* with the stated level of confidence.

In is important to understand that the *Success Criteria* must not contain any hidden requirements that are not included in the set of design input requirements. For example, the need for qualification safety margins specified by a regulatory agency must be addressed in the set of design input requirements along with the requirements for normal and maximum operating pressure, as all these requirements must be implemented by the design and not specified using the *Success Criteria.*

As shown in Figure 10.3, the *Success Criteria* are driven by at least three sources: the need or design input requirement; organizational design guidelines and requirements; and qualification, acceptance, or certification requirements.

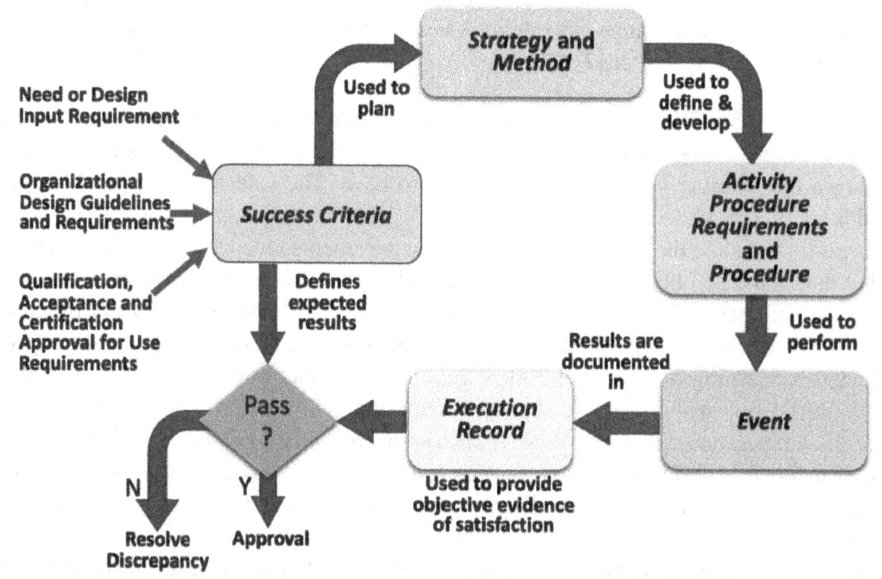

FIGURE 10.3 Success Criteria Influence on System Verification and Validation Planning and Implementation.

The *Success Criteria* are used to plan the *Strategy*, select the *Method*, plan associated system verification and validation *Activities*, and define the system verification and validation *Procedure Requirements*, which are implemented via the resulting *Procedures* which collects the objective evidence needed to determine that the SOI met the need or design input requirement. The objective evidence will be compared to the expected results as defined by the *Success Criteria* to determine whether the SOI has met the need or design input requirement.

Questions that need to be addressed when defining the *Success Criteria* include:

- What objective evidence must be gathered to determine if the need or design input requirement is met?
- Beginning of Life (BOL) or End of Life (EOL)? *See discussion below.*
- Under what operational environment?
- What users/operators need to be involved?

- What level of confidence is needed consistent with the priority and criticality of the need or requirement? *See discussion below.*

The *Success Criteria* will evolve during development of the SOI and will be finalized after the completion of design verification and design validation at a formal gate review (such as the CDR or Integration Readiness Review [IRR]) prior to system verification, system validation, and system integration. A major outcome of design verification and validation and early system verification and validation is updated *Success Criteria, Strategy, Method, Activity* definition, and *Procedure Requirements,* based on the results of these activities as discussed in Chapter 8. This updated information will be reflected in the *Procedures*.

Note: Any changes to the Success Criteria will most likely result in a change to the need or requirement expression from which the Success Criteria was defined. These changes will need to be submitted to a change authority for approval, as discussed in Chapter 14.

In some cases, the customers or regulatory agencies will specify the *Success Criteria, Strategy,* and *Method* that must be used to provide objective evidence that the sets of needs and sets of design input requirements have been met to their satisfaction. In other cases, the customers or regulatory agencies may specify the *Success Criteria* and leave the *Strategy* and *Method* up to the supplier.

Not specifying the *Strategy* and *Method* may appear to save cost and schedule for the project but must be traded against the need to specify this to the suppliers to ensure the objective evidence conforms to the needs of the *Approval Authority* and the needs of the customers and other relevant stakeholders to ensure the supplier's approach will be accepted at their levels. As discussed in Chapter 4, it is the *Approval Authority* that determines what is necessary for acceptance as defined by the *Success Criteria* for each need and design input requirement.

Whichever approach is used, the project team must clearly document in the SOW, PO, or SA the supplier's system verification and validation roles and responsibilities with required activities, deliverables, and contents clearly specified. *Refer to Chapter 13 for a more detailed discussion concerning customers and supplier roles and requirements for system verification and validation activities.*

10.1.1.2 Considerations for Defining Success Criteria When defining the need and design input requirement statements and associated *Success Criteria*, there are several things that must be considered. These considerations are major reasons why the system verification and validation attributes must be defined and included for each need and design input requirement expression and approved when the needs and design input requirements are baselined.

10.1.1.2.1 Confidence Level The confidence level concept is rooted in statistics. When making measurements of performance values or other quantitative parameters, not only the correctness (accuracy and precision) of the data collected needs to be considered but also the degree of confidence in the correctness and repeatability of that data. The confidence level is expressed as a percentage within a range of 0 – 100% on the confidence that subsequent tests will yield the same result within the accuracy and precision requirements, for example 90%, 95%, or 99% confidence level. What is the confidence level that subsequent testing will result in the same "correctness" of the data? If 100 tests are conducted and the system meets the *Success Criteria* for the subject need or requirement just once, the degree of confidence would be low. What about a 50/50 pass/fail rate? What about a 95 out of 100? What pass/fail rate is acceptable to the business stakeholders, customers, and regulatory agencies who are the *Approval Authority*? Keep in mind Six Sigma represents 1.3 defects per million.

For example, if the system under development is an instrument that measures some quantity, along with the sample being measured, there will be requirements on the accuracy and precision of the measurements over the life of the instrument. During system verification, there would be a calibration sample with a known, baseline value of the thing being measured. Each time the instrument is used to take a measurement, the result will probably be a slightly different result. Accuracy is a measure of how close an average of measures is to the baseline value. Precision is a function of the distribution of the measurements taken. A Test Case could be defined where 10 measures of the baseline sample are taken and averaged and the results compared to the accuracy and precision requirements. If a single test is conducted, what is the confidence level that a subsequent run of the test will result in the same result? The test after that? For a weapon system, how many live-fire tests need to be conducted such that the customers and users will have a high confidence level the weapon will meet its accuracy and precision requirements when firing on a target? What about BOL versus EOL? (*See discussion below concerning BOL and EOL considerations.*)

Confidence levels can be addressed in either the requirement statement or in the *Success Criteria* definition. When stated as part of the requirement, the *Strategy* will need to address how that confidence level will be achieved. For example, "*The instrument shall measure [something] in the range of xxx-yyyy [units] with an accuracy of +/− z [units] with a 95% confidence level.*" (*Note: both range and accuracy are stated in the same requirement statement because, in this case, it is assumed they are dependent variables, and the system will be verified to meet each value in the same verification Activity.)*

When stated as part of the *Success Criteria*, the issue is confidence that the system verification or validation *Activity* will result in objective evidence that the system meets the need or requirement as defined within the *Success Criteria* with the desired degree of confidence. The confidence level stated in the *Success Criteria* will influence the *Strategy* and *Method* chosen, the size of the population of the entities being verified, as well as the number of verification actions. Generally, the degree of confidence will be greater when the *Method* is Test versus Analysis.

When considering priority, critically, or risk, not all requirements are of such stature to need a confidence level to be stated. For example, "*The system shall be the color blue as defined in xxxxx standard.*" or "*The system shall be x ± yy inches high.*" Safety and security needs and requirements would justify a confidence level to be stated, as would higher priority or critical needs and design input requirements.

When defining the confidence level, considerations of the impacts on the cost and schedule of the design, system verification activities, and system validation activities must be considered. The higher the required confidence level, the more expensive it will be to achieve that level of confidence both in design and during system verification and system validation. The population size (number of samples tested) and number of tests (tests per sample) will have a direct impact on budget and schedule. Because of this, tradeoffs are often made as to what confidence level is practical and feasible within the project constraints.

10.1.1.2.2 System Lifetime and Expected Performance Another important consideration is the stated lifetime versus needed performance of the system. For a design input requirement like the requirement above for the instrument, does that requirement represent the design input requirements for BOL of the instrument or EOL? All too often, this question is not considered when defining the sets of needs and sets of design input requirements.

This concept complicates system verification and validation planning in that to address the issue, for any given performance requirement, the members of the project team responsible for design will have to understand which parts of the system could degrade over the operational life of the SOI that would impact the ability of the system to meet the performance requirements at EOL; these considerations must be taken into account when defining the *Strategy* and *Method* to be followed during system verification and system validation. For example, the *Method* for a BOL requirement may be Test, but because of a long lifetime (for example, 14 years or more), the *Method* for EOL requirement would be Analysis. In cases like this, it is common to procure or build key parts that are subject to degradation and conduct lifetime testing in a laboratory to collect the data needed for the *Method* Analysis. This is often done as part of design verification and validation as well as early system verification and validation, as discussed in Chapter 8.

10.1.2 Determine the Strategy

For each system verification and validation *Instance*, once the *Success Criteria* have been defined, a *Strategy* must be determined which will result in the data needed that can be used as objective evidence the *Success Criteria* is met.

While this Manual discusses *Success Criteria*, *Strategy*, and *Method* separately, each are closely related and dependent. As shown in Figure 10.4, defining the *Success Criteria*, determining the *Strategy*, and selecting the *Method* are iterative. Selecting a *Method* is part of defining the *Strategy*. Using the knowledge gained in defining the *Strategy* using a specific *Method*, the project team will assess the resulting costs, schedule, risk, confidence level, and enabling systems needed.

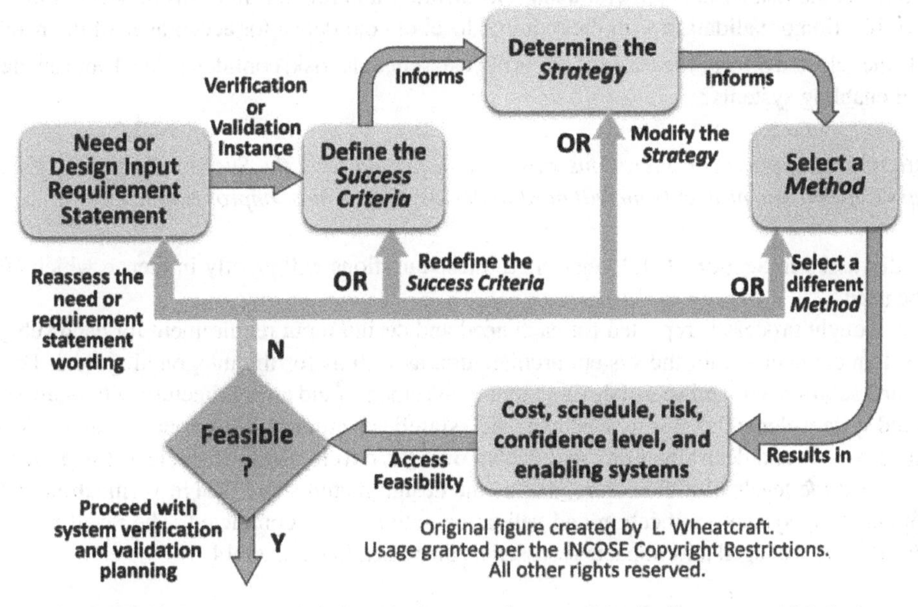

FIGURE 10.4 Iterative Relationship Between Success Criteria, Strategy, and Method.

By defining the *Strategy*, the project team provides an overview of the methodology to be implemented for a system verification or validation *Activity* which will result in the data needed to provide objective evidence the SOI has met the defined *Success Criteria*. The *Strategy* is tailored to the defined *Success Criteria* based on cost, schedule, risk considerations, and required confidence level. The *Strategy* could also be influenced by the *Strategy* selected for the level below or above the level the SOI being verified or validated. An effective *Strategy* reduces technical debt while also preventing gaps and duplication in efforts across the lifecycle.

Based on this assessment, the project team will assess the feasibility of the *Strategy*. If feasibility is an issue, the project team will need to either reassess the wording of the need or design input requirement, redefine the *Success Criteria*, modify the *Strategy*, or select an alternate *Method* such that the *Strategy* is feasible given the constraints levied on the project.

If a need or design input requirement can be changed, that provides several options; if not, the challenge is to find an approach (*Success Criteria, Strategy, Method*) that is feasible while able to obtain the data needed to provide objective evidence the SOI meets the subject need or design input requirement.

Questions that need to be addressed when defining the *Strategy* for each *Instance* include the following:

- What specific actions? How many times will the actions be repeated? On how many samples?
- What will these actions cost? How long will it take to complete each action?
- Are the intended users/operators available or will surrogates be used?
- Will the *Activity* be conducted in the actual operational environment or a simulated or analog operational environment?
- What enabling systems are needed? Do they exist? Will they be available when needed? If they exist, do any need to be modified? If any do not exist, how will they be acquired and at what cost? If any need to be modified, at what cost and timeframe?
- What is the risk associated with using this *Strategy* and *Method* in terms of successful system verification or validation with the required level of confidence for acceptance of the results?
- Is the selected *Strategy* feasible in terms of cost, schedule, risk, confidence level, and availability of enabling systems?

Note that some of these considerations may have been defined in the Success Criteria. If not in the Success Criteria, the project team will need to decide what is most appropriate.

As discussed in Section 10.1.3, answers to these questions will greatly influence which *Method* will be used.

This thought process is repeated for each need and design input requirement for each subsystem and system element within the system architecture, as well as for the integrated system. For large integrated sets of needs and sets of design input requirements (and an architecture with many subsystems and system elements), this process can take a significant amount of time and resources. Also, the initial selections and determinations will evolve over the development lifecycle as the project team moves down the levels of architecture, and as the design matures and design verification and validation and early system verification and validation activities are completed. The resulting changes will need to be managed and impacts assessed as discussed in Chapter 14.

10.1.2.1 Considerations When Determining the Strategy

10.1.2.1.1 Operational Environment Ranges versus Ranges of Performance versus Quality

There is an expectation that system verification and validation *Activities* are conducted in the operational environment or representative equivalent; for example, for a spacecraft, the use of a thermal-vacuum chamber to simulate the operational space environment.

In addition, for system validation, the expectation is that the intended users/operators will be involved. To help ensure success, the classes of users/operators (personas/user classes) should have been defined and addressed in the integrated set of needs and resulting design input requirements.

The operational environment includes natural and induced environments that exist within the macro system containing the SOI. The project team should have addressed the operational environment within the sets of needs and design input requirements.

By their nature, requirements dealing with environmental factors, such as temperature, pressure, Electromagnetic interference (EMI), Electromagnetic compatibility (EMC), humidity, mechanical loads, vibrations, and acoustics, are stated as a range of values. Thus, there are multiple environmental factors—each having a range of values that must be considered when defining the *Strategy* and selecting a *Method*. In some cases, the environmental factor range is stated in a three-dimensional graph of frequency, magnitude, and range. As a result, there are multiple environmental factors each with their own range of values. Even if not explicitly stated in the *Success Criteria*, the expectation is that the system elements, subsystems, and the integrated system meet all needs and requirements in any combination of extremes across the ranges for each environmental factor.

Another complication to be addressed is that system performance values are usually defined within a range. In some cases, the defined performance across a range of a given environmental factor is not linear. The performance may be better in the middle of the range and degrades closer to the extremes. In other cases, the time the system operates at the extremes of performance combined with the environmental factor extremes affects performance; for example, while the system can operate at the nominal operating temperature for an extended time, it may not be able to do so at the extremes of temperature, pressure, or another environmental factor.

From a stakeholder perspective, the expectation is that the SOI performs predictably and reliability across all the stated ranges of performance as well as the stated range of environmental factors as defined in the integrated set of needs (system validation) and the set of design input requirements (system verification). Quality factors (-ilities) needs and requirements must also be considered.

In addition, needs and requirements have dependencies that must be considered when defining the *Success Criteria* and *Strategy*.

The main point of this discussion is that, in many cases, system verification and validation activities will involve the system verification or system validation of multiple needs or requirements concurrently within the same system verification or system validation *Activity*.

Because of the number of dependencies, it is important to have developed detailed models of the SOI as well as models of its operational environment and external systems it interacts with that can be used to run simulations during design verification and validation activities (as well as early system verification and validation activities), as discussed in Chapter 8.

With this capability, all the combinations and permutations of dependent requirements across each of their ranges of performance, as well as ranges of environmental factors, are possible, allowing evaluation of "stacked" ranges or tolerances and possible failures. "Stacked" is used when the system is tested when all performance ranges and all environmental ranges are at an extreme simultaneously. There are many cases where a system works as intended within the middle of the combined ranges but fails when tested at the extremes, especially if all the dependent requirements are at their extremes (all high or all low ends of their ranges.)

10.1.2.1.2 Level of Integration At what level of integration should system verification and validation activities be performed? As part of system verification and validation planning, *when*, *where*, and at *what level* are primary concerns. In general, doing system verification and validation at the lowest possible level is usually the lowest risk and most cost-effective approach. It is lower risk as it enables issues to be discovered sooner and lower cost as the efforts of verification for system elements can often be less complex than for a complete system (of course, there are exceptions).

A general rule of thumb is that a need or design input requirement for a system, subsystem, or system element should be recorded and managed at the level the system verification and validation activities will be conducted (i.e., a verification activity expressed in terms of a subsystem should not be associated with a system-level need or design input requirement; it should be associated with the need or design input requirement stated for that subsystem).

As a bottom-up activity, system integration, system verification, and system validation should start with the system elements, then work up to the subsystems they are a part, then to the top integrated-system level. However, for some projects that may not be the case. Some projects may choose to only require production verification, showing the system element meets its design output specifications, accepting risk and skipping the system verification and system validation activities which would also show that it meets its design input requirements/needs.

10.1.2.1.3 Budget and Schedule Common considerations when developing the project's MIVV and SIVV plans are cost, resources, and time required to do the verification and validation *Activities*. Due to budget and schedule constraints, the project team often is forced to re-evaluate their initial concepts for system verification and system validation. There are several factors on a project that may drive the amount and complexity of the system verification and validation efforts, including:

- New system compared to a system with heritage.
- Amount of regulatory criteria for the system.
- Large production effort compared to single build effort.
- High visibility/high consequence of failure system compared to lower risk system.

A minimal amount of system verification and validation activity may be appropriate when the developer is working with a known system for a known customer in a known operational environment with known users. In this case, the likelihood of the system itself not meeting its needs or requirements is lower. However, when developing a new system for a new customer in a unique environment for a newly defined intended use incorporating a new technology, the probability of not meeting a need or requirement and system non-acceptance could be higher.

For products in highly regulated industries, the likelihood of not meeting a need or requirement may be low, but the consequence may be high (for example, the system is rejected for use). In this case, additional resources may be necessary for system verification and validation, regardless of need or requirement type or complexity. A new product that is heavily regulated may need to provide additional objective evidence that can be used to prove compliance with standards and regulations or there is a risk that it may not be approved for use.

A large production project is more likely to need a more robust system verification and validation plan than a small, one-time project. Large projects involve more resources, larger budgets, and more time; thus, the consequences of failure are higher based (in part) on the size of the investment. More complex projects can increase the likelihood of failure—especially when the development extends over longer periods of time and there are multiple supplier organizations involved.

The consequences of failure of highly visible projects could be greater than smaller, less visible projects. The inverse is also true: for projects with high consequences of failure, internal visibility may increase as stakeholders recognize their share of risk and invest attention in ensuring the project completes successfully.

For example, a major system that is highly visible to the public such as a launch vehicle or spacecraft, especially when humans are on-board, a medical diagnostic device whose results can be trusted, or a vaccine that needs to be effective without long-term negative effects on the health of the recipients—all may require additional system verification and validation to buy down the overall risk. In the inverse, a project with high levels of company resources invested may commit to additional verification and validation activities to ensure the invested resources are not lost.

10.1.2.1.4 Earlier Detection of Defects Defects are more costly and time consuming to correct later in the lifecycle—especially when they are not discovered until verification and validation activities at integrated-system level.

The project may be tempted to reduce the budget and shorten the schedule for system verification and validation by doing only subsystem or integrated-system level system verification or validation against their design input requirements and needs—and only using production verification for lower-level system elements. The logic of using this strategy, though faulty, is that once objective evidence has been obtained that a subsystem or integrated system has met its set of needs and set of design input requirements, then the system elements within its architecture must have met their set of needs and design input requirements.

However, this strategy could be risky; a critical problem may not be discovered until later in the system integration process or during integrated system validation or customer's or regulatory agency acceptance. It may be difficult (time consuming and expensive) to trace the problem to its source for a large, complex system.

This is a primary example of a project incurring technical debt. Once the source of the problem is discovered, it will be more difficult to resolve, especially if more than one part, component, or system element involved, resulting in disassembly of the subsystem/system as part of the anomaly resolution.

A single defect discovered during integrated-system level system verification or system validation may cancel all savings projected from eliminating lower-level system element verification and validation activities. Worse yet, a defect may go undetected until after the integrated system has passed system verification or system validation and has been released for use. In the case of safety critical consumer products (such as an automobile, a crewed spacecraft, an aircraft, a medical device, or pharmaceuticals) the consequences could be a loss of life.

It is also risky to assume an SOI was built or coded to the design output specifications without proof. If there were production quality issues, then it will be more difficult to trace the nonconformance to its source—was the design incorrect, were the resulting design output specifications wrong, or was the subsystem or system element built or coded incorrectly?

If an organization completes design verification and validation as well as early verification and validation, as discussed in Chapter 8, the risk of failing system verification and validation during system integration is reduced. Doing so should result in any issues with the design being addressed before production. As a result of completing these activities, the source of a defect discovered during system verification and validation will be easier to trace to either a production issue or an integration issue. As discussed in Chapter 9, production issues should be identified during production verification. A robust production verification process should help quality engineers isolate any production faults.

Another consideration is defects at the system element or subsystem level, which are not discoverable after being integrated into the parent system due to latent issues, lack of perceptibility at an integrated level, or system verification/validation *Activities* at the integrated level, which do not address a lower-level requirement/need.

Theoretically, as discussed in Chapter 2, if allocation and budgeting were done properly, the resulting necessary and sufficient set of child requirements for the subsystems and system elements should adequately address the quality and compliance requirements. However, without objective evidence obtained during system verification and validation of the subsystem or system element against its own needs and requirements, it is risky to assume the subsystem or system element was designed and built such that it meets those needs and requirements.

10.1.2.1.5 Using a Risk-based Approach to Reduce Cost and Schedule Budget and schedule constraints may make system verification or validation of every subsystem or system element in the SOI architecture against every one of their needs and design input requirements cost prohibitive.

If system verification or validation activities must be reduced due to cost and schedule constraints, this can be done using a risk-based approach. If the system, subsystem, or system element has needs and design input requirements that are high priority, critical, or associated with safety or security, or risk mitigation, system verification and validation must be done to provide objective evidence the system, subsystem, or system element meets those needs and requirements prior to integration.

For system verification and validation activities dealing with noncritical, lower priority needs and design input requirements that are low risk, i.e., the risk of a need or requirement not being met or the consequences of not meeting the need or requirement are low, the system verification or system validation *Activity* may be a candidate for removal.

For Off-the-Shelf (OTS) or Modified Off-the-Shelf (MOTS) systems, subsystems, or system elements (discussed in Chapter 12) or in a contracted development or production setting (discussed in Chapter 13), it may be that the suppliers are only required to show the customer that the delivered system, subsystem, or system element meets its design input requirements. Lower-level system verification and validation of internal subsystems and system elements is treated less formally as a supplier "in-house" activity. This approach may be risky, but it could result in significant reductions in cost and delivery times; this is why including production quality requirements in the supplier contract is important.

Projects must do a risk assessment and determine which needs and corresponding design input requirements represent the highest risk if not implemented. Inadequate system verification and validation represent risk to the project.

The following types of needs and requirements warrant additional assurance that system verification and validation is well planned and executed. Types of needs and requirements that represent significant risk to the project if not met include:

- *Critical needs and requirements* that must be met; otherwise, there is a high risk the SOI will be unable to provide critical functions and performance, i.e., will not be able to be used as intended in the intended operational environment by the intended users.
- *Safety and security needs and requirements* that must be met; otherwise, there is a risk of a safety, security, or loss issue, i.e., a requirement that ensures safety of people, a requirement that addresses the risk of unintended users negatively impacting the intended use of the system or a requirement whose implementation will prevent misuse of the system or a loss as addressed in Chapter 4 when defining the misuse cases and loss scenarios during elicitation.
- *Needs and requirements associated with risk mitigation* that must have objective evidence that they have been implemented to determine that the risk has been addressed.
- *High-priority needs and requirements* that must be met; otherwise, there is a risk the SOI will be unable to provide functions and performances key stakeholders view as high priority. Even if not critical, safety, or security related, the key stakeholders expect their high-priority requirements to be met.
- *Interface requirements*. Failure to meet interface requirements can 1) impact the system integration, system verification, and system validation activities, 2) result in a failure of the system to operate once integrated into the macro system it is a part, and 3) may enable unintended users to negatively impact the intended use of the system.
- *Regulatory needs and requirements* the project must show objective evidence of compliance. For safety or security-critical systems, the regulatory agency may specify the *Strategy* and

Method to be used that will provide them with the objective evidence they need to approve and certify the system for its intended use when operated by the intended users.

10.1.2.2 System Verification and Validation Using Models Caution is advised if planning to use design verification/validation results based on models, simulations, or prototypes as evidence for system verification/validation in place of objective evidence using the actual physical SOI hardware and software. Doing so reduces the confidence level of the system verification and validation effort.

Showing that a model of an SOI in a simulated operational environment with simulated interfaces meets its set of design input requirements (system verification) or a set of needs (system validation) *does not mean* the actual physical system interacting with actual external systems in its actual operational environment when operated by the intended users will meet its set of design input requirements or set of needs.

While some organizations may need to save time and money on system verification or system validation by substituting the use of models in place of the actual physical components and actual operational environment, they are adding and accepting additional risk as well as technical debt. Such use of simulations to verify the system model meets the design input requirements or validate the system model meets its set of needs in lieu of system verification/validation against the actual physical system should be done with caution and with full knowledge of the project concerning technical and operational risk.

There are several reasons for this:

- It is difficult to model the actual physical and functional interactions between system components in their actual operational environment.
- A goal of system verification is to detect errors, faults, and defects within the realized physical integrated system [1]. Using models and simulations of the SOI will not always uncover errors, faults, and defects in the actual physical subsystems and system elements.
- The behavior of a system is a function of the interaction of its parts as well as interactions with the external systems and environment of which it is a part. Thus, a major goal of systems validation is assessing the behavior of the integrated physical system and identifying emergent properties not specifically addressed in the needs or requirements. Emerging properties may be positive or negative. For example, cascading failures across multiple interface boundaries between the subsystem and system elements that are part of the SOI's architecture. Relying on models and simulations of the SOI and operational environment may not uncover all the emerging properties and issues that may occur in the physical realm.
- Another goal of system validation is to evaluate that the product or service performance is predictable in its intended operational environment when operated by its intended users and does not enable unintended users to negatively impact the intended use of the system. Relying on models and simulations of the SOI and operational environment is not a reliable substitute for uncovering issues when the actual physical system is operated by its intended users as well as showing that unintended users (for example, hackers) are unable to negatively impact the intended use of the system.

While design verification/validation using models and simulations allows a theoretical determination that the modeled system will meet its needs or requirements in the operational environment once realized, determination that the actual physical integrated system meets its needs and requirements when operated by its intended users/operators, as well as addresses the above goals of system verification and system validation, must be done whenever possible in the physical realm with the actual hardware and software integrated into the SOI in the actual operational environment by the intended users [1].

As long as the physical system is not completely integrated and/or has not been validated in the actual operational environment by the intended users and does not enable unintended users to negatively impact the intended use of the system, no result must be regarded as definitive until the acceptable degree of confidence is realized.

See also Section 10.1.3.4 discussion on "Analysis by Analogy or Similarity" *and* "Analysis by Certification" *for more information concerning using models as part of Analysis as a Method of system verification or system validation.*

10.1.2.3 Strategy Impacts on Design The SOI may need specific features dedicated to support production verification, system verification, or system validation activities. The need to provide objective evidence that an SOI performs as defined by its integrated set of needs and set of design input requirements and has been built correctly as defined by its set of design output specifications results in design considerations concerning how test or diagnostic data will be obtained.

In the definition of "Test" as a *Method* of system verification or system validation, the need for special instrumentation and test support equipment, is included in the definition. In many cases, this will require that the SOI design includes dedicated interfaces, which may not be needed for operations, to connect this instrumentation and test support equipment to collect the data defined in the *Success Criteria* for a need or design input requirement.

For example, additional interfaces may be required to provide diagnostic information, access to test data, or provide commands for test operations. This could result in the need for additional connectors, test points, and wiring harnesses to connect to special test instrumentation or external power, extra data to display or record, or a database to give visibility into an internal process, an inspection portal, or a bracket to hold the SOI in a test fixture.

When conducting system verification or validation of software or hardware with test code or test points inserted into the SOI, do these test points and extra code remain in the SOI after delivery? If so, how is the extra code disabled—how is that test code made unreachable with assurance that it will not affect system performance? What if the customer stipulates no unreachable code? Or, that all code be exercised during system verification and system validation?

Alternatively, are the test points and/or test code only in engineering test units and not in the production items? If the test points are only in the engineering test units, then the actual SOI being produced and delivered is not the same exact SOI undergoing system verification or system validation. How is that managed? These are questions that the project team must address.

This is one of the advantages of early system verification and validation planning, especially addressing the *Strategy* and *Method*, because it enables system requirements to be derived to support system verification and validation activities long before the actual activities.

10.1.2.4 Addressing Enabling Systems When defining the *Strategy* for system verification or validation, it is important to pay particular attention to other systems that "enable" system verification or validation. Unique support or test equipment, special facilities, or models may be needed. Of particular importance is what is needed to perform system verification or system validation at the interfaces: the actual system, a simulator, or an emulator? In some cases, the system going through system verification or validation is dependent on external systems to supply something as an input. Will that other system (or a simulator or emulator) be available when needed to support the system verification and validation activities?

These systems are referred to as *enabling systems*. They need to be identified, defined, designed, modified, built, or procured, verified, and validated in time to meet the system verification or system validation schedule for the SOI under development.

Each facility, support equipment, and other enabling systems will have its own development cycle where needs and requirements will be defined, the enabling systems designed and built, and will go through their own system verification and validation prior to being certified for use for the SOI's system verification or validation activities.

Even OTS support equipment will need to be exercised and at least validated as certified for use. For any enabling system that needs to be developed, there could be a long lead time before it will be ready for use. For existing enabling systems, there may be modifications needed to meet a project's specific needs—these modifications take time and money and must be included in the project budget and schedule.

Even existing facilities, equipment, and personnel must be scheduled so they can be certified and are available when needed. Facilities, support equipment, and enabling systems may be needed by other projects, and schedule conflicts could arise that impact the system integration, system verification, and system validation activities.

Because of this, it is critical to begin system verification and validation planning early in the lifecycle so the enabling systems are identified, certified, and available when needed. In some organizations, there are dedicated departments that build and furnish custom support equipment; once these groups have requirements, they can work concurrently with SOI development. If this is the case, schedule alignment will be important as well as technical alignment. For technical alignment, a major consideration is at the interface boundaries and the interactions across those boundaries.

If sufficient planning was conducted early in the project, these enabling systems should be ready to support the SOI's system integration, system verification, and validation activities when needed.

10.1.2.5 System Integration Approach System integration, system verification, and system validation are bottom-up activities. The way a system is assembled during integration is primarily driven by the design, the physical architecture, procurement, and the manufacturing and coding sequence, sometimes influenced by the need to package and transport sub-assemblies when manufacturing or coding is geographically distributed.

For mass produced products, the concepts of "design for assembly" or "design for manufacturing" are common to reduce the time and expense to build and assemble a system. Both design for assembly and design for manufacturing are needs, the satisfaction of which will have to be validated. These needs also may result in specific design input requirements.

The sequence of integration is influenced by the defined system integration concepts reflected in the SEMP and MIVV plan. Prior to integrating a component into the physical assembly that it is a part it must be approved for integration into the assembly. Sometimes, this can be done by inspection, or a "bench test" of the component is performed by *integrating* the component into a test assembly consisting of the control system (software), power supply, other support components, test instrumentation, and a test stand (mechanical support) that mimics the larger system. Interfaces with external components may be simulated or emulated as needed.

This specific system verification or system validation configuration may never be part of the final system. It is a temporary configuration solely to support subsystem or system element verification or validation activities so they can be certified for acceptance prior to integration into the next higher assembly. Such a setup can also enable electrical continuity and input-source checks to ensure power supply currents are not exceeded, or a random short is not present in the given subsystem or system element that is being verified.

There may also be cases where there is a need to do a subset of integration and system verification/validation for a set of system elements that have a dependency and interact (interface) with each other prior to their integration into the final physical system they are a part.

- For example, to mitigate a particular risk for a launch vehicle, there may be a need to conduct tests of the control subsystem connected to the fuel pumps prior to integrating these systems into the launch vehicle. In this case, an operational environment would need to be provided to allow this testing to be completed.
- Another example from the Oil and Gas industry is when hydrocarbons are released into the system during commissioning. Prior to releasing hazardous materials into the system, partial commissioning is done to mitigate the risk. These approaches allow the project team to verify the SOI against high-priority or critical requirements concerning the interaction of these two systems prior to their final integration into the overall system.

These considerations drive home the need to plan how the system will be integrated for system verification and validation purposes. As a system verification or system validation *Activity* is planned, the corresponding integration of components for that *Activity* is also planned, even if the assembly of such parts, components, or subsystems cut across the system architecture. While this may seem tedious, the payoff is avoiding unforeseen behaviors at the top-level of integration due to a defect or simply an unanticipated interaction that could have been detected earlier.

When preparing for system integration, system verification, and system validation everything must be included that is part of the operational environment, including all maintenance and support equipment, software, and all facilities into which the system is to be deployed. When developing a new system, all support and enabling systems needed to maintain, handle, power, support, test, configure, and transport the system being integrated must be developed and go through their own system verification and validation prior to using those items to integrate and conduct system verification and validation for the SOI under development.

For example, for a new launch vehicle, the Assembly, Integration, and Check-Out Facility must be "proofed," including verifying that the overhead crane, handling equipment, access stands, and ground support equipment meet their requirements, including regulatory, security, and safety requirements prior to being certified for system integration, system verification, and system validation activities.

10.1.2.6 Preliminary versus Final System Verification and Validation In addition to the early system verification and validation conducted concurrently with the design verification and design validation activities discussed in Chapter 8, some system verification and validation may be performed in phases, as shown in Table 10.1.

Preliminary system verification and validation may be done on an engineering version of the SOI, and final system verification and validation will be done on the final version of the SOI. This is often the case where access to key data needed as part of the *Success Criteria* is not accessible for the end system, so additional test code or test points are included within the engineering version of the SOI but not in the actual end system. In software, this is called a "debug" version of the code—where the final production executable would have the debug code removed before compilation.

Because the actual end item is not being used during these system verification and validation activities, the project team will need to determine which system verification and validation activities will need to be done on the engineering version of the SOI and which will need to be done on the actual end item.

TABLE 10.1 Tracking Preliminary and Final System Verification and Validation Status.

Phase	Component	Subsystem (System Elements)	Integrated System
Preliminary system verification and system validation	All subsystem parts/components developed in-house have successfully completed production verification and preliminary system verification and validation by the project in a non-operational environment.	The subsystem has successfully completed subsystem verification and subsystem validation by the project in a non-operational environment using simulated interfaces with external systems.	The integrated system has successfully completed system verification and validation by the project in an operational "like" environment with emulators or simulators for external systems and operated by surrogate users.
	OR	*OR*	*OR*
	All subsystem parts/components developed by a supplier have passed production verification and have been successfully verified and validated by the supplier in a non-operational environment.	The subsystem has successfully completed subsystem verification and subsystem validation by the supplier in a non-operational environment with or without using simulated interfaces with external systems.	The integrated system has successfully completed system verification and validation by the supplier in an operational "like" environment with emulators or simulators for external systems and operated by surrogate users.
Final system verification and system validation	All subsystem parts/components developed in-house have successfully completed preliminary system verification and validation by the project in an operational "like" environment.	The subsystem has successfully completed subsystem verification and subsystem validation by the project in an operational "like" environment using simulated interfaces with external systems.	The integrated system has successfully completed system verification and validation by the project in the operational environment when operated by the intended users or trained surrogates and actual systems.
	OR	*OR*	*OR*
	All subsystem parts/components developed by a vendor have been verified and validated by the supplier in an operational "like" environment AND the vendor results have been accepted by the project.	The subsystem has successfully completed subsystem verification and subsystem validation by the supplier in an operational "like" environment AND the supplier results have been accepted by the project using simulated interfaces with external systems.	The integrated system has successfully completed system verification and validation by the supplier in the operational environment when operated by the intended users and actual interfaces with external systems AND the supplier results have been accepted by the project.

As discussed in Chapter 13 for supplier-developed SOIs, the supplier may be responsible for system verification and the customer responsible for system validation. In other cases, the customer may accept as preliminary the results of the supplier system verification and validation in an "operational-like" environment, interacting with simulators or emulators that represent external systems, using contractor personnel as the users/operators. Then the customer (or another organization that represents the customer) will conduct final system verification and validation of the SOI in the actual operational environment, interacting with the actual external systems and operated by the intended users.

There may also be a mixed approach. For lower priority, non-critical, or non-safety, non-security sensitive requirements for the supplier-developed SOI, the customer may accept both the supplier's system verification and validation results. However, for high-priority, critical, safety, or security-related requirements, the customer may elect to do their own or additional system verification and system validation.

The strategy used comes down to acceptance of risk versus the cost to buy down the risk. Supplier history and track record can come into play making such a decision. These strategies must be addressed in the MIVV and SIVV plans.

Often, a mix of both preliminary and final system verification and validation activities can be the most efficient strategy. There are some details to consider when such a mix is used:

- In some cases, the first round of system verification and validation activities will be conducted without establishing the actual operational environment or actual external systems, and then a second round of system verification and validation activities are performed that only addresses needs and design input requirements that are related to the operational environment and interactions with the actual external systems.
- In other cases, especially for system verification and validation activities associated with quality, safety, security, and lifetime requirements, preliminary system verification and validation activities may be warranted to verify the system meets these types of requirements in the operational environment prior to being accepted for final integration.
- If the operational environment is not available, then extensive environmental testing in specialized chambers may be substituted at the system element level.
- Another reason for preliminary system verification and validation (at an Integration and Test (I&T) facility, test stand, or on the workbench) of a part or component is to gather data prior to system verification and validation by *Analysis* of the next higher level of architecture that part will be integrated into. Here, data is collected at the sub-component or system element level—and that data is used a part of the *Analysis* of the next-higher level assembly. The preliminary system verification and validation activities will directly affect the confidence level of the final system verification and validation results once that system element is integrated into the next higher level of the system architecture.

10.1.2.7 Embedded Software System Verification and Validation Challenges From a system integration, system verification, and system validation perspective of hybrid hardware/software systems, many standalone software development and system verification/validation strategies are not well suited to embedded software. In this context "operational environment" has a different meaning.

Embedded software exists within a hardware system element. This software has interactions with other software modules that are embedded within other hardware system elements and communicates across multiple communication buses. In addition, the embedded software has interactions with external sensors that provide data needed by the embedded software, as well as interactions with physical system elements (for example, actuators) that are controlled by the embedded software.

In this context, standalone testing of this software is not a substitute for system verification and validation of the software once embedded into the hardware, and that module is then integrated into the macro system it is part.

Many of today's automobiles contain over 150 electronic control units (ECUs) [55]. This is especially an issue for integration given that there could be multiple suppliers, each with their own development approach, operating systems, and programming languages. Each of the suppliers has limited insight into the ECUs being developed by the other suppliers, or how their ECUs will be integrated into the larger system. Because of this, it is not always clear how changes to the embedded software in one ECU will affect the performance of other ECUs.

The result is a much more complex system that presents additional challenges for system integration, system verification, and system validation than the traditional hardware/mechanical-centric systems of the past. There may be fault modes and the possibility of cascade failures where a failure of a hardware component (for example, a sensor) could ripple through the entire system, resulting in system failure.

To summarize this sub-section, a major part of defining the *Strategy* is deciding on how the project will conduct system integration, system verification, and system validation at each of the levels of the physical architecture. These activities also allow tailoring of the system verification and validation effort in terms of cost, schedule, and risk.

10.1.3 Select a System Verification or System Validation Method

As each need or design input requirement is defined, the *Success Criteria* defined, and the *Strategy* determined, a *Method* (Test, Demonstration, Inspection, Analysis) is selected that will be used to verify the system meets a design input requirement or used to validate the system meets a need.

Selecting the *Method* is based on the *Success Criteria* and *Strategy* defined for each need and requirement. The *Method* selected affects cost, schedule, risk, and confidence level.

No matter which *Method* is selected, the project team should document the rationale for why a specific *Method* is chosen. If Test is possible, but Analysis was chosen as the *Method*, why? Section 10.1.3.5 provides more detailed guidance on selecting the *Method* most appropriate for a given need or design input requirement against which the SOI is being verified or validated.

10.1.3.1 Test Test is a *Method* that leverages data taken from one or more actions to provide objective evidence that the SOI satisfies a need or requirement. The *Method* Test, uses controlled, predefined sets of inputs, data, or stimuli to exercise the SOI. Using special test equipment or instrumentation to obtain the required data, results are compared against expected outcomes as defined by the *Success Criteria*.

- Test as a *Method* for system verification is often conducted in a controlled operational environment (or simulation of) in accordance with predefined use cases or operational scenarios and is typically conducted by engineers, technicians, or test personnel using baselined system verification or validation *Procedures*.
- Test as a *Method* for system validation should be conducted in the actual operational environment by the intended users using actual operating procedures or instructions for use.

Using Test, one or more tests (actions) can be performed using final versions of the end item or dedicated design or engineering units or platforms of varying levels of fidelity.

Emulators or simulators of external systems may be used to verify or validate interactions across interface boundaries prior to integration of subsystems and system elements into the next higher level of the physical architecture.

Test is a preferred *Method* as it provides direct, measurable objective evidence. However, due to the expense relative to other methods, tests must be used strategically and consistently within the cost, schedule, and risk constraints as discussed previously. Test is generally selected as the *Method* when other methods are incapable of providing results of sufficient fidelity or high enough degree of confidence or when a need or requirement is high priority, critical, or of particular importance to the system's ability to meet needs or requirements concerning safety, security, or risk mitigation.

For the *Method* Test, the following questions should be considered; in addition to helping to describe the *Strategy*, the answers to these questions will also be used to define the system verification and validation operational environment, special conditions for operations, the *Activity*, *Procedure Requirements*, and resulting *Procedures*.

- Where will the test activities be conducted?
- What objective evidence will be measured?
- Who will be conducting the tests? The project? The supplier? An independent contractor? The intended users or representative surrogate users?
- At what lifecycle stage or level of integration?
- What fidelity of SOI is needed? Engineering test unit? Production unit?
- What enabling equipment, facilities, and tools are needed?
- What special equipment is needed to command the SOI, power the SOI, physically support the SOI, and collect and store the data collected?
- Have the equipment, facilities, tools, and special equipment been verified, validated, and qualified or certified for this specific use?
- Are additional measurement points needed to obtain the needed data?
- Are there additional interfaces needed to connect special hardware/software?
- Which interfaces with external systems will the SOI under test need to interact with during operations that will need to be emulated or simulated versus using the actual systems to support the test?
- Have the emulators or simulators for external systems been verified, validated (especially at the interface boundaries), and qualified or certified for use?
- If using the actual external systems associated with the interaction, will they be available per the project schedule to support the testing?
- What is the operational environment needed to conduct the test? Normal facility environment, simulated operational "like" environment, or the actual operational environment?
- Does the facility have that capability to vary the operational environment characteristics across a range of values?
- Are the conditions for use clearly stated and understood?
- How many tests (actions) need to be successfully completed to obtain the stated confidence level?
- If multiple copies of the SOI are being manufactured, how many copies need to be tested (lot size) and what number of tests per unit (sample size) are needed to obtain the stated confidence level? *See the discussion in Section 10.1.1 for a more detailed discussion on confidence levels.*
- What analysis tools are needed to analyze the data collected as part of the test activities?

10.1.3.2 Demonstration Demonstration is a *Method* that involves the operation of an SOI to provide objective evidence that the required functions were accomplished under specific scenarios and operational environments. Demonstration is the manipulation of a system, subsystem, or system element as it is intended to be used to verify that the results are as expected as defined by the *Success Criteria*. Demonstration consists of a *qualitative* determination made through observation, with or without the use of special test equipment or instrumentation, to verify characteristics such as human engineering features, services, access features, and physical accommodation (compared with Test, which uses a *quantitative* approach).

If the result of an action or stimulus is needed to be observed (for example, a parameter exists), then demonstration could be used. Demonstration may be appropriate for event-triggered requirements. For example, issuing a command and observing the immediate result. Using Demonstration is often less complicated and less expensive than Test as it is usually performed at the extremes of range of performance rather than multiple actions across the entire range of performance.

For the *Method* Demonstration, the following questions should be considered; in addition to helping to describe the *Strategy*, the answers to these questions will also be used to define the system verification and validation operational environment, special conditions, system verification and validation *Activity*, system verification and validation Procedure Requirements, and resulting *Procedures*.

- Where will the demonstration test activities be conducted?
- What objective evidence needs to be observed?
- Who will be conducting the demonstration? The project? The vendor? An independent contractor? The intended users, or representative users, or surrogate users?
- At what stage or level of integration?
- What enabling equipment, facilities, tools, etc. are needed to perform the demonstration test?
- Which interfaces with external systems that the system interacts with during operations will need to be emulated or simulated versus using the actual systems to support the demonstration test?
- Have the emulators or simulators been verified and validated, especially at the interface boundaries and qualified or certified for this specific use?
- If using the actual systems, will they be and qualified or certified for this specific use and available to support the demonstration test?
- What is the operational environment needed to conduct the demonstration test? Normal facility environment, simulated operational "like" environment, or the actual operational environment?
- Are the conditions for use clearly stated and understood?
- For the SOI, how many demonstration actions need to be completed to obtain the stated confidence level? *(See the discussion in Section 10.1.1 for a more detailed discussion on confidence levels.)*
- If multiple copies of the system are being manufactured, how many copies need to be demonstrated (lot size), and what number of demonstrations per system (sample size) are needed to obtain the stated confidence level?

10.1.3.3 Inspection or Examination Inspection or Examination is a *Method* that is nondestructive and typically includes the use of sight, hearing, smell, touch, and taste; simple physical manipulation; or mechanical and electrical gauging and measurement. Inspection could be inspection of the end-item-system (code, physical attributes) or artifacts that represent the SOI, such as drawings, parts listings, and part specifications.

Usually, when using Inspection, a checklist is developed using the *Success Criteria* defined for each need or requirement that the system is being verified against. This checklist would be included within the system verification or validation *Procedure*. For physical inspections, the criteria may also be included in quality inspection records or production build documentation.

If a person can observe or use a simple measurement to determine if the requirement is satisfied, then this is an appropriate *Method*. This tends to be applicable when:

- There is a binary decision.
- Can involve a simple judgment, one where no special competency is necessary.
- May be a simple measurement (torque, length, weight, electrical resistance, etc.).
- Can be seen to exist (note on a drawing, feature is present, label present).
- Nondestructive and executed on an actual system or system documentation (drawing, physical feature, line of code, record from a supplier).

Inspection tends to be the least expensive *Method* and can be used to verify design details communicated in the design output specifications, e.g., built as specified in a drawing or coded per an algorithm or logic diagram, built or coded to specification determination.

For the *Method* Inspection or Examination, the following questions should be considered; in addition to helping to describe the Strategy, the answers to these questions will also be used to define the system verification or system validation environment, special conditions of operations, the system verification or system validation Activity, system verification or system validation procedure Requirements, and resulting Procedures.

- Where will the inspections be conducted?
- What objective evidence will be observed?
- Who will be conducting the inspections? The project? The vendor? An independent contractor?
- At what stage or level of integration?
- What specifically is being inspected?
- What special access is needed to perform the inspection?
- What equipment, facilities, tools, etc. are needed to perform the inspection?
- Are the equipment, facilities, tools, etc., qualified or certified for this specific use?
- Under what conditions (lighting, environment, etc.) will the inspections need to take place?

10.1.3.4 Analysis Analysis is a *Method* that uses established mathematical models, simulations, algorithms, charts, graphs, circuit diagrams, or other scientific principles and procedures to provide objective evidence that design input requirements or needs were met. Analysis also includes engineering assessments and logic usage based on previously recorded evidence obtained during design verification, design validation, early system verification, early system validation, and production verification activities, as well as results of lower-level system, subsystem, or system element system verification or system validation activities.

Analysis allows the organization to make predictive statements about the expected performance of an SOI based on the confirmed test results of a sample set or by combining the outcome of individual tests, demonstrations, or inspections performed on lower-level subsystems or system elements to conclude the SOI meets its needs or design input requirements.

Analysis is often used to predict the breaking point or failure of a product or system by using nondestructive tests to extrapolate the failure point. Analysis is also the primary *Method* for the system verification or system validation of many quality (-ilities) needs and requirements.

For example, for lifetime or reliability requirements or needs, the project will conduct a series of tests for an extended time on parts of the actual system or on developmental versions of the physical system in its operational environment to collect data needed as dictated by statistical analysis, for example, [xxxx] number of data sets (samples) to have a confidence level of [yyyy]. *Refer to Section 10.1.1 for a more detailed discussion on confidence levels.*

There are several subcategories of the *Method* Analysis; these are further described in the following sub-sections:

- Analysis by Analogy or Similarity
- Analysis by Certification
- Analysis by Model or Simulation
- Analysis by Engineering Assessment
- Analysis by Sampling

10.1.3.4.1 Analysis by Analogy or Similarity Analysis by Analogy or Similarity is considered a type of an Analysis rather than a separate *Method.* Analysis by analogy or similarity is normally performed by comparing the SOI under development to another similar SOI that has successfully been verified to meet a similar set of design input requirements or validated to meet a similar set of needs in a similar operational environment and operated by similar users/operators for a similar use. "Similar" is defined in terms of form, fit, function, quality, compliance, and operational environment. This approach is useful for a design upgrade or building of a similar design using the same components.

Analysis by analogy or similarity consists of assessment and review of hardware/software configuration, hardware/software application, function, performance, operational environment, external interfaces, and prior test data, including a comparison against prior design or early system verification and validation results as compared to the "similar" system.

Differences in configuration, application or use, operational environment, external interfaces, or test conditions usually require "enveloping" by the original item qualification, or performance of additional analyses and testing (activities) to complete system verification or system validation. In this context, enveloping is a determination that the characteristics communicated within the requirements for the system under verification fall within the characteristics of the "similar" system.

Using Analysis by analogy or similarity can be risky for interfaces. The above analysis makes a critical assumption that the SOI that is undergoing system verification or system validation was "built to spec" and there are no defects at the physical interface boundary. However, experience has shown that a major source of system integration issues involves issues concerning connectors, such as bent pins, shorts, grounding issues, etc.

10.1.3.4.2 Analysis by Certification Analysis by Certification is sometimes used for government-sponsored development projects where the supplier is directed to reuse certain components that are from an existing system (or a system that was fielded but has been decommissioned). In certain cases, the customer (government) may direct the supplier to use the part "as-is" without a rigorous physical testing regime as part of Test. This may happen when the technical risk of reusing the part is low, the part cannot easily or cost-effectively be remanufactured, or the customer has a limited supply of this part, and a testing program would consume too many reusable parts that cannot be replaced. With this approach, the customer would assume the technical risk.

The technical maturity of a system is based on a specific use in a specific operational environment. If the existing component or system was qualified for a different use in a different operational environment, there is a risk that it may not perform as needed when integrated into the new system.

For a supplier-produced system with a well-established supplier system verification and validation program, the customer may accept the Supplier's *Execution Records* as objective evidence rather than doing the system verification and validation activities themselves. For this to be an acceptable *Strategy,* provisions need to be made for the supplier to do the system verification and validation activities and provide the deliverables needed for Analysis by Certification. In addition, the customers may include in the contract provisions for customer participation (insight or oversight) in the Supplier's system verification and validation activities. *Refer to Chapter 13 for more detailed discussion on Customer/Supplier considerations for system verification and system validation.*

10.1.3.4.3 Analysis by Model or Simulation *Analysis by Model or Simulation* is performed using analytical/behavioral models or simulations that approximate the physical system and its interacting external systems as well as simulate the operational environment.

There are cases were the actual/physical system may not be able to be verified to meet a requirement or validated to meet a need directly by test of the actual system in the actual operational environment by the intended users. An example is a system that is to go into space or installed on the ocean floor. Another example is a weapons system where an actual test of the weapon results in the destruction of that weapon.

There are both natural and induced environmental requirements that will be encountered while operating a system within which the SOI must work during and/or after being subjected to those environmental conditions. In cases like this, obtaining objective evidence for system verification/validation showing the system can meet a subset of the system's operational, performance, and quality needs/requirements may require Analysis by Model or Simulation. For today's complex, software-intensive systems, Analysis by Model or Simulation is a major *Method* used for design verification and design validation (discussed in Chapter 8).

Analysis by Model or Simulation may be more difficult than Test as a *Method* as it could involve multiple runs (such as Monte Carlo simulations), and the approach and assumptions need to have a basis in reality. From an expense perspective, these models and simulations will need to first be developed, verified, and validated before being qualified (or certified) for use.

Model verification and model validation

For most customers, models and simulations are expected to be developed per specified methods and complete "model verification" and "model validation" to be considered certified for use as part of the SOI system verification and validation activities.

Model verification and model validation are done using measured data obtained from partial tests of similar systems, allowing a comparison of the inputs/outputs from predicted and actual data to confirm the model provides a realistic and accurate reflection of the actual physical system, subsystem, or system element it is representing. This activity is sometimes referred to as "model correlation".

Using modeling and simulation ignores the fact that one of the goals of system verification is to detect faults in the actual physical system that could be a result of either a poor design or a production issue that resulted in defective parts.

If an organization completes design verification and design validation as described in Chapter 8, then system verification or validation by model or simulation is really accepting the results of that activity in combination with production verification as evidence, rather than accepting results of an activity with the actual physical system that a specific design input requirement (system verification) or need (system validation) was met.

Going back to the space system example for a launch vehicle, one approach is to build a physical "mass model" of the launch vehicle that is well instrumented and then stimulate the mass model with forces at various vectors, magnitudes, and frequencies. The resulting data is then used to develop an analytical model which can be used for both design and system verification/validation. Once the launch vehicle is built, each segment can go through loads and vibration testing. The resulting data will then be fed into the analytical model. The results can then be used to verify the launch vehicle meets its design input requirements and validate the launch vehicle will meet the needs concerning loads and vibrations.

Final vehicle system verification and validation will occur during the first launch. Because of this, projects responsible for the development of launch vehicles (and other space systems) will often include additional instrumentation in the first set of vehicles to collect data to validate their models with actual "flight" data.

For example, the Space Shuttle Program included special "Development Flight Instrumentation (DFI)" on the first space shuttle to collect this information over several flights.

Considerations on the use of models and simulations as a *Method* of Analysis.

There are several things that must be considered when using models and simulations as a method of Analysis:

- *How credible is the model?* Was the model developed for a specific purpose that is applicable to the system verification and validation activity for which it is being used? Has the model itself been verified against its design input requirements and validated against its intended use?
- *How complete is the model or simulation?* A model or simulation approximates the actual physical SOI, addressing a subset of characteristics. It is difficult to model every aspect of the SOI under development. How complete is the model or simulation, i.e., what aspects of the SOI are included in the model or simulation? Logic? Conceptual behavior? Data, commands, message flow between software components? Is the model of the system or artifacts generated as part of defining the system? Is the model primarily of the software included in the system but not the physical attributes and characteristics of the system? How well are physical aspects of the system modeled, for example, physics, chemistry, biology, thermodynamics, or electrodynamics? How well were the actual operational environment, external systems, and interfaces modeled or simulated? If not modeled or not modeled well, the confidence level would be low that the actual physical system will work as intended in the actual operational environment when operated by the intended users solely based on the design verification and design validation results based on the model.
- *Developing models of parts of the architecture within the integrated-system model.* A basic tenet of systems thinking is that the behavior of a system is a function of the interaction of its parts as well as the interaction with the external systems within the macro system the SOI is a part. To accurately assess the behavior of the SOI undergoing system verification or system validation, an integrated model of the macro system it is a part must be used when using models or simulations for Analysis.
- *Combining models developed separately from the integrated-system model.* Assessing the behavior of a SOI when it has been modeled separately from the integrated-system model is problematic. For example, a model of a single pendulum can be developed. However, if there is a need for a system with two connected pendulums, the model of the single pendulum cannot be added to another copy of the model to produce a model of the two-pendulum system

because it is a different system, and the interaction of the two pendulums must be included in the model to accurately model the behavior of the two-pendulum system.

Likewise, a model can be developed for an automobile and a model of a trailer, but the two models cannot be added together to form a model of the integrated system consisting of an automobile attached to the trailer. Again, the integrated system is a different system and the interaction between the automobile and the trailer determines the behavior of the integrated–system—for example, the integrated behavior when the operational environment includes icy roads.

Note: With the advent of Model-Based Systems Engineering (MBSE), model-based design (MBD), and the increasing use of modeling and simulation, some practitioners of MBSE have proposed the use of modeling and simulation as a fifth system verification and validation Method. While this change is well underway, this Manual uses the four classical methods discussed in this chapter.

10.1.3.4.4 Analysis by Engineering Assessment Analysis by Engineering Assessment uses engineering judgment for a collection of data and information, along with knowledge of the design, to provide a logical outcome of results. This can include an assessment of a set of design features to show that a safety failure tolerance is in place (such as a hazard analysis or Failure Modes and Effects Analysis [FMEA]), a set of processes to show that a desired protocol will be enacted (such as a ground operation analysis), and a set of observations to show capabilities for a service.

This approach is less rigorous than the prior approach using models and simulation as it relies more on engineering judgment and logic; however, it is still a useful approach in certain applications. For validation that the system meets its needs, analysis by engineering assessment may be used when it can be determined that the intent of a given need was met by successful verification that the system met all the child requirements that were transformed from that need. For example, needs dealing with compliance and with quality (-ilities), a standard, or a regulation.

10.1.3.4.5 Analysis by Sampling Analysis by Sampling can be used when multiple copies of a system are being produced, and there may be issues concerning repeatability and consistency between the copies of the system. In other cases, multiple copies are produced in "batches." Just because one copy of the system has been shown to meet a need, design input requirement, or design output specification, can it be assumed all copies will do so - over time? It is often cost prohibitive to do system verification or system validation against all copies within a batch or multiple batches. Based on statistical methods, a sample of a given size will be selected for periodic verification. Analysis by sampling is often done as a quality function during production verification activities (discussed in Chapter 9) concerning quality of workmanship during manufacturing.

10.1.3.4.6 Factors to Be Considered for Analysis as a Method For Analysis, the following questions should be considered; in addition to helping to describe the *Strategy*, the answers to these questions will also be used to define the system verification and validation environment, special conditions, the system verification and validation *Activity*, *Procedure Requirements*, and resulting *Procedures*.

- Where will the analysis (actions) be conducted?
- What objective evidence is needed as a result of the Analysis)?
- Who will be conducting the system verification or validation by Analysis? The project? The supplier? An independent contractor? What are their qualifications to do this Analysis?

- At what stage or level of integration?
- What category of Analysis (method) will be used?
- If using models or simulations for Analysis, has the integrated model or simulation for the SOI been verified, validated, and certified or qualified for this specific use?
  - Has the operational environment (natural and induced) been included in the model?
  - Have the interfaces with external systems been included in the model?
  - How well have the physical aspects of the SOI been modeled?
- What specific data and information is needed for the Analysis?
- How will the data and information be obtained? What testing is needed on what fidelity of the system (physics model, engineering test unit, production unit) to obtain the data needed for the Analysis? How much data is needed? How many samples?
- What level of analysis (actions) is needed that will result in the required confidence level?
- What tools are needed to analyze the data?
- Have the tools to be used been verified and validated?
- Are there changes or differences in the SOI configuration or modeled operational environment that may impact the Analysis?

10.1.3.5 Questions Associated with Each Method As part of determining the *Strategy* and selecting a *Method*, Project managers need to know the answers to the above list of questions for each *Method* to plan for the resources, cost, and time needed to implement the project's system verification and validation programs as defined in their MIVV and SIVV plans.

It is best to address these questions as the project is maturing the lifecycle concepts, defining the integrated set of needs, and defining the set of design input requirements. These questions must be addressed before the project unknowingly spends resources on designing or building a system based on a design input requirement that the SOI cannot be verified to meet, or a need that the SOI cannot be validated to meet, or performing a system verification or system validation *Activity* that cannot be accomplished within the project's budget, schedule, and resources.

System verification and validation is a knowledge-based set of activities that evolve over the system lifecycle. The answers to the above questions for each *Method* may not be known until later in the development lifecycle, needs and design input requirements may change, or the status and availability of enabling systems may change. Based on these changes, the project team will need to actively manage the system verification and validation planning and resulting artifacts throughout the system lifecycle.

10.1.3.6 Considerations When Selecting a Method

10.1.3.6.1 Risk, Cost, Schedule Considerations When selecting a *Method,* it is best to consider factors of risk, cost, schedule, accuracy and precision of the needed data, confidence level, and consequences of not satisfying the need or requirement.

Risk is a major consideration when choosing the *Method*. This is an important issue from a PM standpoint. Test provides the highest confidence level that the system meets a design input requirement or need. While it is preferable to use Test as the *Method*, it may not be practical for all needs or design input requirements. In some cases, Test may be banned by treaty (for example, the Nuclear Test Ban Treaty). In many cases the project cannot afford the cost or does not have the time or resources to use Test for all design input requirements and needs at all levels of the system architecture. In other cases, a special facility to simulate the operational environment or equipment to assess the system's interactions with other subsystems or external systems may not be available nor cost effective to develop.

When a need or design input requirement poses a lower risk to the project, and a lower confidence level is acceptable, one of the other *Methods* may be more appropriate. However, Analysis as a *Method* can be misused or overused, resulting in a higher risk to the project when good rationale for using Analysis is not defined. Reducing the number of system verification or system validation Activities or selecting a lower-cost *Method* due to cost overruns or schedule slips can also add risk to the project. These considerations need to be addressed when defining the *Success Criteria*, *Strategy*, and *Method* when each need or design input requirement is defined.

Because choosing the *Method* is really a cost, schedule, and risk decision, the Project Manager and Lead Systems Engineer for in-house system development efforts and customers for contracted development need to be involved in the decision and agree on the *Success Criteria* (including level of confidence), *Strategy*, *Method*, and organizational roles and responsibilities. This is especially important when contracting development to a supplier. *Refer to Chapter 13 for a more detailed discussion concerning contractor/supplier considerations for system verification and system validation.*

10.1.3.6.2 Analysis Instead of Test There may be a temptation to select Analysis instead of Test because it may seem to be a less expensive alternative. However, while Analysis may appear to be less expensive than Test—that is not always true. Some form of measured test data on parts of the actual system or on developmental versions of the physical system is still required to generate the appropriate analysis correlation to actual expected outcomes. In addition, the number of hours required to do the analysis and develop the level of technical documentation that is required, can drive up the cost of the effort.

In the end, the pass/fail determination for Analysis is based on the knowledge and experience of the engineer, and the type and fidelity of the models, simulations, and data used to verify and validate the model or simulation. A major consideration is how well the physical system, the actual operational environment, as well as external systems and interfaces are modeled.

The form of the analysis (actions) performed results in data used to make the pass/fail determination. Because the confidence level may be lower when using Analysis, the risk is higher that the system will fail to perform as expected as compared to actual operation of the physical system in the actual operational environment when operated by the intended users.

10.1.3.6.3 Need or Requirement Type The type of need or requirement can also influence which *Method* is the most appropriate. Functional/Performance needs and requirements are best verified by Test whenever possible. Needs or requirements whose action is binary (it either does an action or not based on a defined trigger) may be best suited for Demonstration. A need or requirement that deals with a physical characteristic that can be observed by sight may be best suited for Inspection. Quality (-ility) type needs and requirements are most suited for Analysis or Inspection. While there are many -ilities are verified by Analysis (such as maintainability, reliability, and availability), some organizations may verify requirements associated with testability, manufacturability, and quality control using Inspection. Organizations should have guidelines as to which *Method* should be used based on the type of need or requirement. These guidelines should be included in the project's MIVV and SIVV plans.

10.1.3.6.4 The Macro System of Which the SOI Is a Part What information is needed to verify or validate the SOI in the context of the next higher level "macro" system which it is a part, in its intended operational environment when both are being operated by their intended users? The type and quality of the information needed is dependent on the *Method* used.

Failing to verify and validate that the system performs as needed across its external interfaces and in its operational environment, when operated by its intended users, can result in major issues when the delivered system fails to perform as expected when integrated within the macro system it is a part.

Even though every system element and subsystem that is part of the system's physical architecture has been verified to meet their design input requirements and validated to meet their needs, the project must ensure they work together as a whole within the integrated system before the system is delivered to the customers (or released to the public for use). Once integrated, can the integrated system be verified to meet its design input requirements and validated to meet its set of needs?

System verification and validation of both internal and external interfaces often require Test as the preferred *Method* or Inspection if the interface is mechanical. *(Using any of categories of Analysis can be risky for interactions across an interface boundary, as discussed earlier.)* While the cost can be high, failure to successfully verify the SOI interacts with external systems as required across all interface boundaries can lead to major, and even catastrophic, problems during operations.

10.1.4 Defining Each Instance

Each need expression and each requirement expression represent a system validation or system verification *Instance*.

Key information needed to define each *Instance* includes the following information. *Note: The specific information to be defined will be specified in the project's MIVV and SIVV Plans.*

- *The need or requirement identifier and the need or requirement statement.*
- *Success Criteria.*
- *Strategy.*
- *Method.*
- *Priority.* Relative importance of the need or requirement.
- *Criticality.* Whether the need or requirement is critical in terms of functionality, performance, safety, security, or compliance.
- *Operational environment.* The environmental conditions under which the SOI will be verified or validated to meet the subject need or requirement. This includes environmental conditions such as temperature, pressure, altitude, humidity, or vibration loads as well as the operational environment needed to perform the system verification or system validation Activities. This could be environmental characteristics such as mountainous terrain, underwater, specific rainfall amounts, electromagnetic interference conditions, or other such environmental constraints necessary to determine whether the operation of the system is per the intent of the design criteria. Also included are the external interfaces as well as the users (intended or surrogate) when doing system verification or system validation activities. The operational environment should be defined when the *Strategy* is defined.
- *Special conditions.* The special conditions (such as item configuration) necessary to determine that the SOI verified or validated against not only nominal operating conditions but worst-case conditions or boundary conditions, if possible. The angle at which a line replaceable unit must be positioned relative to rainfall would be considered a special condition; for example, place the line replaceable unit at 45° angle to the rain path and conduct functional testing periodically

(at a minimum of 5 minutes) throughout the course of the test. Again, special conditions should be defined when the Strategy is defined.

- *Rationale*. The rationale for why the *Strategy* and *Method* were chosen. For example, why a *Method* of Analysis versus Test? Why a specific configuration for the system verification or system validation *Activity*?
- *Activity* in which the *Instance* will be included (once defined).
- *Status*. The system verification or validation status of the *Instance* and the *Activity* it is a part—in work, passed/failed, *Execution Package* approved/disapproved, open/closed. *Note: "closed" refers to the fact that the organization has approved the results of the Activity the Instance is a part and the Activity has been successfully completed.*

In a data-centric practice of SE, this information is defined and managed electronically within the SOI's integrated or federated database, as discussed in Section 10.7.

It is important to maintain traceability throughout the system development lifecycle linking all the artifacts associated with system verification and validation planning, defining, execution, reporting, and approval for each system verification and validation *Instance*. This should be able to be done within the project's SE toolset.

10.1.5 System Verification and Validation Matrices

Historically, in a document-based practice of SE, organizations developed System Verification Matrices (SVM) and System Validation Matrices (SVaM) which are a matrix/table visualization of the information defined above for each *Instance*. An example partial SVaM is shown in Table 10.2.

TABLE 10.2 Example System Validation Matrix (SVaM).

Need ID	Need	Validation Method	System Validation Success Criteria	Criticality
N001	The stakeholders need the LIR to obtain Lids from the LDS.	Demonstration	The intend of this need will be validated to be met when it is demonstrated that the LIR can reliably obtain Lids from the LDS.	Yes Capability
N043	The stakeholders need the LIR to comply with [TBD] OSHA requirements concerning the safety of humans working with robots.	Analysis	The intent of this need will be validated to be met when an analysis of all requirements transformed from this need have been verified to have been met.	Yes Compliance
N048	The stakeholders need the LIR to be labeled in English.	Inspection	The intent of this need will be validated to be met when inspection of the LIR shows each LIR label is worded in English.	Yes Safety

An example partial SVM is shown in Table 10.3.

TABLE 10.3 Example System Verification Matrix (SVM).

Rqmt ID	Requirement	Verification Method	System Verification Success Criteria	Criticality
R001	The LIR shall obtain Lids from the LDS as described in the LIR/LDS ICD xyz, Section 123.	Demonstration	The LIR will be verified to have met this requirement when the demonstration test provides objective evidence that the LIR obtains Lids from the LIR as described in the LIR/LDS ICD xyz, Section 123.	Yes Capability
R010	The LIR shall accept from the FCR C&M software Valid_Commands defined in the FCR ICD xyz, Section 123.	Test	The LIR will be verified to have met this requirement when the test provides objective evidence that the LIR will accept only valid commands from the FCR as defined in the FCR ICD xyz, Section 123.	Yes Safety, and Security
R061	The LIR shall comply with OSHA Regulation xyz, Section 7.8, concerning the safety of humans working with robots.	Analysis	The LIR will be verified to have met this requirement when the analysis provides objective evidence that the LIR complies with each requirement in OSHA Regulation xyz, Section 7.8, concerning the safety of humans working with robots.	Yes Compliance

These matrices can be created for each subsystem and system element within the system architecture as well as for the integrated system. The SVaM will contain a listing of all the needs in an SOI's integrated set of needs. The SVM will contain a listing of all the requirements within the set of design input requirements. Each row of the matrix represents an *Instance*.

Note: In the above examples, only a sample of the possible fields (columns) are shown due to space. Organizations will often include more columns based on their needs.

In the past, using a document-centric approach to SE, these matrices were created manually using a common office word processing tool or spreadsheet. However, in a data-centric practice of SE, the information displayed in the matrices is included in the SOI's integrated or federated dataset. As such, the project's toolset should be able to generate these matrices as reports, which could be imported into a spreadsheet if desired.

If the organization has a separate tool within their project toolset dedicated to defining and managing the system verification and validation information and artifacts, this information would be defined and managed within that tool with traceability to the needs and requirements that are defined and managed within other tools within the project toolset.

10.2 DEFINING STAGE

During the Defining Stage, when sufficient design detail is available and the integrated set of needs and set of design input requirements are mature enough, the organization responsible for executing the system verification and validation program can begin defining specific *Activities* and *Procedure Requirements* and gathering information that will enable *Procedure* development.

"Mature enough" is subjective; if the set of needs or design input requirements are not complete and are evolving, defining *Activities* may result in unnecessary rework. However, waiting too long to define the *Activities* could result in resource, budget, and schedule issues. Defining the details associated with each *Activity* early in the system lifecycle minimizes the risk of cost overruns and schedule delays.

For example, to complete a given *Activity* or set of *Activities*, a test facility, test fixture, or other enabling system many need to be modified or constructed, a thermal-vacuum chamber may be needed for environmental testing, or special test and support equipment may be needed to collect the data needed as objective evidence to determine whether the *Success Criteria* has been met for each *Instance*.

In some cases, the resources or enabling systems may be shared with or owned by another organization. The project team will need to budget for their use and possible modifications as well as ensure the availability and readiness of these resources or enabling systems are consistent with the project's system verification and validation schedule and budget.

10.2.1 Defining Each *Activity*

An *Activity* is a specific set of actions that are defined for the purpose of verifying the SOI meets one or more design input requirements or validating the SOI meets one or more needs. In some domains, such as software engineering, an *Activity* is referred to as a "Test Case" even if the verification *Method* is not Test.

An *Activity* may address more than one *Instance*, for example, dependent functional performance requirements along with the requirements dealing with conditions of use and operational environment. There are often cases where multiple needs can be the focus of a single system validation *Activity*, or multiple design input requirements can be the focus of a single system verification *Activity*. There are also examples where a single *Activity* could address both system verification and system validation for a given need and design input requirement transformed from that need.

Considerations for combining multiple *Instances* within a single *Activity* include:

- A function may have multiple performance requirements associated with that function, as well as related requirements dealing with conditions of use, the operational environment, and quality. In such cases, it may be possible to verify each of these requirements within the same system verification Activity, often concurrently.
- The *Method* is the same.
- Doing multiple inspections, tests, or demonstrations on the same part or component in the same location.
- The *Strategy* is similar in terms of the operational environment, configuration of the SOI, conditions of use, facility, support equipment, command and control, data acquisition, and personnel.

For each *Activity*, the responsible project team members will use information defined for each *Instance* as well as answers to the questions addressed when defining the *Strategy* and selecting the *Method* to collect key information needed to define the *Activity*.

Below is an example of the types of information that could be collected as part of the definition of each *Activity* and stored and managed within the project's toolset:

- *Activity title*. A title that clearly communicates the nature of the *Activity*.
- *Activity number*. A unique identification number for the *Activity*.
- *Type of Activity*. System verification or system validation.
- *Instance(s)*. List of each *Instance* (need or design input requirement) included within this *Activity*.

- *Integration level* within the physical architecture at which *Activity* will take place.
- Project phase in which the *Activity* will take place (for example, design, purchasing, manufacturing, assembly, integration, commissioning).
- *Organization responsible* for the *Activity, Procedure*, and creation of the *Execution Record*.
- *Organization responsible* for submitting the *Execution Record* to the *Approval Authority*.
- *Activity location*. Where will the *Activity* be performed?
- *Hazardous operation*. Is this a hazardous operation? Is a hazard analysis needed?
- *Procedure*. Identifier for the implementing *Procedure—when known*.
- *Event name*. Name of the *Event* associated with the performance of the *Procedure the Activity is a part (once the Event has been scheduled. (Refer to Section 10.3.)*
- *Predecessor activity(s)*. Are there any prerequisite *Activity*(s) that must be completed before this *Activity* can be performed?
- *Successor activity(s)*. Are there any *Activity*(s) that cannot be executed before this *Activity* has been completed? Is this *Activity* part of a larger (parent) *Activity*? If so, which *Activity*(s)?
- *Constraints*. Are there any constraints to the performance of the *Activity*? Include anything that limits or constraints the performance of the *Activity*. Also include impacts/constraints on other *Activities* during the performance of this *Activity*. For example, no other work is allowed in test area while this *Activity* is being performed.
- *Workmanship codes/standards*. List any workmanship codes or standards with which this *Activity* must be compliant.
- *Resources needed*. A description of the equipment/components/assemblies/subsystem/systems involved in the *Activity*. List of any resources needed to perform this *Activity* (utilities: steam, power, GN2, air) as well as any other systems, portable equipment, leak test unit, or power supply that are needed during the performance of the *Activity*. Do any of the resources need certifications for their intended use?
- *Documentation*. Do new *Procedures* need to be developed to perform this *Activity*? Or are there existing *Procedures* that can be used? Are there any existing supplier *Procedures* that can be used to perform parts or all the *Activity*?
- *Organizations/personnel*. What key organizations need to be involved in the planning and/or performance of the *Activity*? (Project Engineer, Test Director, Supplier, Safety, Security, Quality Assurance (QA), Technicians, Test Operations, others?)
- *Training/certifications*. Are there any training/certifications required by personnel involved in the *Activity*?
- *Status*. The status of the *Activity—Procedure* in work, *Event* scheduled, *Procedure* completed, *Procedure* in the process of being closed out (issues are being resolved), *Execution Package* approved/disapproved, *Procedure* open/closed, *Activity* open/closed. *Note: "closed out" refers to the fact that the organization has approved the results of the Activity as defined and the Procedure that included that Activity has been successfully completed.*

Again, this information is recorded and managed electronically within the SOI's integrated or federated database, as discussed in Section 10.7.

10.2.2 Gathering the Information Needed to Develop the Procedures

Note: In the past, system verification or system validation "requirements" were used to state the Method, Strategy, and Success Criteria for each design input requirement or need [56]. *The approach used in this Manual is to define the Method, Strategy, and Success Criteria within the set of attributes for each design input requirement or need, as discussed in Chapter 15. What was missing before was the concept of defining specific requirements on the organization performing each Activity that would be implemented within a Procedure that would be used to address the steps and actions needed for each Activity. To address this, the concept of "Procedure Requirements" is used as described in this chapter. For those that use the phrase "Test Case," or conduct a series of test activities as discussed at the beginning of this Chapter, what is referred to as Procedure Requirements may be referred to as "Test Requirements."*

A project's *Activities* are implemented via *Procedures*. The steps and actions within a *Procedure* are defined based on *Procedure Requirements* that are implemented by the *Procedure*, along with other information gathered by the project team concerning the performance of the *Procedure*.

The *Procedure Requirements* consist of a set of requirements concerning the details involved in implementing the *Strategy* and *Method* defined for each *Instance* that will result in objective evidence that a need or design input requirement has been satisfied by the SOI as defined by the *Success Criteria* for each *Instance* included in the *Activity* or *Activities to be* addressed by the *Procedure*.

Procedure Requirements are written from the perspective of the organization and personnel responsible for planning and performing the steps and actions within a *Procedure*. For example,

"The [Operator] shall [stimulate the SOI in some way],"

"The [Operator] shall [record/document] [the result of the stimulation]."

"The [QA Representative] shall [certify] [the result of the stimulation] was recorded accurately."

When gathering the *Procedure Requirements* and other information needed to develop the *Procedure*, the project teams will use information related to each *Instance* and *Activity* from multiple sources. This information may or may not be communicated in the form of requirements (shall statements). No matter the form, this information must be considered when developing a *Procedure*.

Sources of information used to develop a *Procedure*.

- *Procedure or Test Requirements.*
- Other information gathered concerning the performance of the *Procedure*.
- The organization's design controls for system verification and validation. These design controls will contain requirements on the personnel developing and performing the procedures.
- The answers to the questions addressed when determining the *Strategy* and selecting a *Method* for each *Instance* discussed in Sections 10.1.2 and 10.1.3.
- The information defined for each *Instance* as discussed in Section 10.1.4.
- The information defined for each *Activity* as discussed in Section 10.2.1.
- Requirements associated with the facility being used.
- Requirements associated with the systems the SOI will be interacting with during the Activity.
- Requirements associated with the special support equipment being used.
- Answers to the following questions:
  - Where will the *Procedure* be performed?
  - What enabling systems are needed? What certifications or qualifications are needed to use the enabling systems for this specific use?

  - What power is needed? Will the facility supply power directly to the SOI or to a project-supplied power supply? Does the quality of the power and power connections need to be verified before being connected to the SOI? Is there an existing procedure or checklist for operating the facility power supply or project-supplied power supply?
  - What external command-and-control capability is needed and other support systems?
- How will the interactions with enabling systems be addressed—connecting with actual external systems or is an emulator or simulator needed?
  - What actions are required to configure and operate the enabling systems, emulators, or simulators? Is there an existing procedure or checklist for doing so?
  - What certifications or qualifications are needed to use the emulators or simulators for this specific use?
  - What power is needed for the enabling systems?
- For each *Instance*, what are the actions needed that will result in the data needed defined in the *Success Criteria*?
- What data is needed to be displayed to the operators conducting the *Procedure*?
- Has the required test code been inserted or access to test points established to allow the needed data to be collected?
- What data must be recorded? At what rate? At what bandwidth? What volume of data needs to be recorded? In what form does the data need to be recorded and format supplied to the organization responsible for the *Activity* being addressed by the *Procedure*?
- What collected data must be certified by a QA Representative?
- What environmental conditions need to be provided during the *Procedure*? Actual or simulated? Is the capability to adjust the environmental conditions across a range needed?
- Are there any special conditions or configurations of the SOI expected during the performance of the *Procedure*?
- Are there specific requirements specified with a standard or regulation that must be complied with during the performance of the *Procedure*?
- Who needs to be involved in the *Procedure*? Intended users/operators or surrogates? Customer or Customer representative? QA Representatives?
- Who must observe the performance of the *Procedure* and certify that specific steps and actions were performed?
- What training or certifications do the personnel need to have to participate in their assigned role?
- What safety and security considerations are there?
- What resources are needed to perform the *Procedure*?
- How long will it take to complete all steps and actions defined in the *Procedure*? How much will it cost?
- Who will be performing the analysis of the data resulting from the performance of the *Procedure* to determine if there is sufficient objective evidence that the *Success Criteria* have been met for each *Instance* addressed within the *Procedure*?
- Who is responsible for preparing the system verification or system validation *Execution Records* that formally document the result of performing the *Procedure*?
- Who is needed to signoff that the *Procedure* is ready to be used? Has it been completed?

The operators of the facility, external systems, and special support equipment may have their own set of requirements that drive and constrain the *Procedures* that are performed within the facility. There may be safety and security provisions that must be followed, especially for hazardous operations or operations involving classified or sensitive systems. Key personnel will be needed to conduct the steps and actions specified within the *Procedure*. Who supplies these personnel? There are cases where these personnel may need to be certified to conduct certain operations.

Procedure Requirements are based on a specific project. In some cases, the customer may perform system verification or system validation; in other cases, the customer or regulatory agency may require the supplier to provide objective evidence the system has successfully passed system verification and system validation in the operational environment when operated by the intended users. As discussed in Chapter 13, in the case of a customer/supplier contractual agreement, the specific roles, responsibilities, and deliverables must be clearly documented in a contract. The customer may levy requirements concerning system verification or system validation on a supplier in addition to the design input requirements as a means to impose contractual requirements on how the system will be verified or validated and to develop and provide the *Execution Records* as part of the contract deliverables.

Within a project, requirements for their in-house system verification and validation program are developed in accordance with the processes defined in their MIVV and SIVV plans. In the case of a regulatory agency, the specific roles, responsibilities, requirements, and deliverables will often be documented within a regulation, for example the Federal Aviation Administration (FAA) or Food and Drug Administration (FDA).

10.2.3 Procedure Development

A system verification or system validation *Procedure* consists of a sequence of steps or actions, which result in the collection of data and information that will be used as objective evidence that the system meets the *Success Criteria* defined for each *Instance* within the *Activity(s)* addressed by the *Procedure* in accordance with the set of *Procedure Requirements* defined for that *Procedure*.

For example, an *Activity* may require a site visit where the SOI will be verified to meet multiple requirements whose *Method* is Inspection. The set of inspections is divided into a series of steps or actions within the *Procedure*. Each series of steps or actions may focus on one or more specific inspections, such as witnessing the configuration of piping, orientation of pumps, or inspecting several aspects of a mechanical part to see if it was assembled as shown in the design output specifications.

When gathering the requirements and information needed to develop the procedure, the project team will need to coordinate with the organization responsible for integration and assembly to ensure the plans for assembly include the proper steps at the right time within the assembly sequence, as observed by the QA representatives.

While the specific form of *Procedure* will vary from one organization to another, a *Procedure* is not created in a vacuum. Developing a well-formed *Procedure* is a complex process. Before a *Procedure* can be written, there is a lot of preparation. The enablers shown in Figure 10.5 must be available, or the *Procedure* cannot be developed nor performed.

The organizational processes, procedures, and work instructions must be complied with; the system verification or system validation facility and support equipment must be "ready" to support the steps and actions as specified in the *Procedure*. "Ready" means that all test, support equipment, emulators, simulators, and models must have been identified, modified, or built, have gone through their own system verification and validation processes, and have been qualified or certified for use for this purpose.

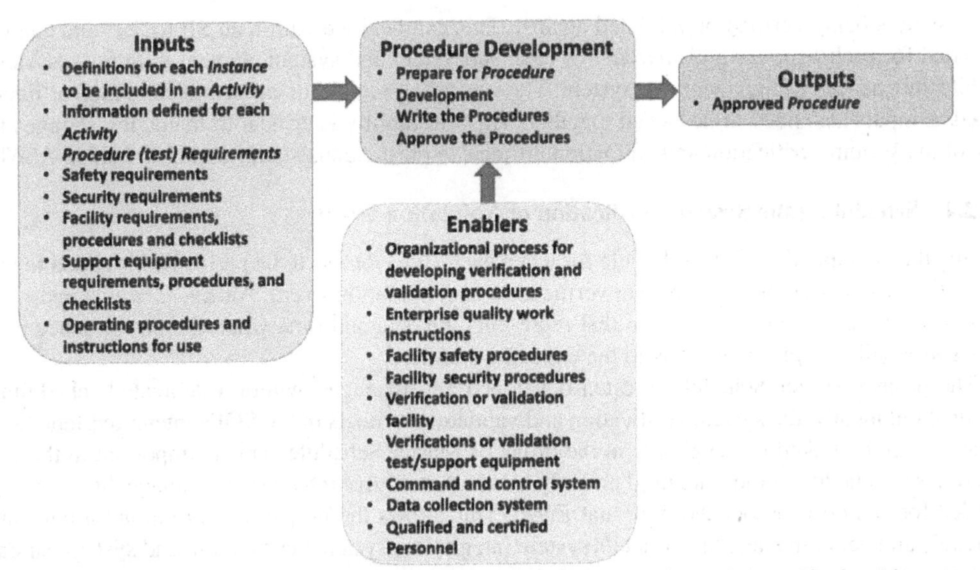

FIGURE 10.5 Procedure Development IPO Diagram.

A means to command and control (if applicable) the SOI must exist in the *Procedure*, along with the means to collect, display, and record the data that will be used as objective evidence that the SOI meets its needs, requirements, or specifications.

The preceding sections in this chapter walked the reader through the actions needed to define the inputs shown in Figure 10.5. In addition to this information, there will be organizational requirements dealing with quality, safety, and security. There will also be requirements concerning operations within the facility as well as requirements concerning the use and operation of the test and support equipment. All these inputs must be addressed within the *Procedure*.

Once the preparations are complete, the *Procedure* can be written, reviewed, and approved. Given that the *Procedure* is a work product (and could be a contract deliverable), a *Procedure* review is another verification and validation *Activity* applied to the *Procedure*, i.e., does the *Procedure* address all the *Procedure Requirements*, design controls, and drivers and constraints?

In many cases, procedures or checklists may already exist for common tasks such as operating facility systems or set up and operations of support equipment. In addition, operating procedures and instructions for use will have been developed as part of defining the conditions for use. It is a best practice to use these existing operating procedures and instructions for use when developing the *Procedure*. The organization should have validated the operating procedures, instructions for use, and checklists prior to using them during performance of the *Procedure*. When these are available, the *Procedure* will call out the execution of applicable steps and actions within the existing procedures, instructions for use, and checklists rather than including them within the *Procedure* itself.

The output of this process is an approved *Procedure* that is ready to be scheduled as an *Event* within the project's Master Schedule.

Once the *Procedure* identifier has been assigned, this identifier should be added to the definition information for the *Activity* which the *Procedure* is implementing.

It is important to maintain traceability throughout the system development lifecycle. The *Procedures* should be linked to the needs, design input requirements, or design output specifications the system is being verified or validated against. This can be done within an SE toolset and can be included for each *Instance* and *Activity* for each subsystem and system element within the system architecture as well as the integrated system. The various system verification and validation artifacts listed as inputs also need to be linked together. This traceability is critical to managing changes to any of the system verification and validation artifacts to ensure consistency and establish an ASoT.

10.2.4 Scheduling the System Verification or Validation *Event*

During the Defining Stage, the schedule for when each *Procedure* is to be performed is added to the project's Master Schedule as a system verification and validation *Event*. An *Event* is a named item within the project's Master Schedule that represents the date and time when the performance of a *Procedure* will take place, along with the expected duration.

The project Master Schedule is often developed and managed within a dedicated scheduling application. Ideally, the system verification and validation artifacts in the SOI's integrated/federated database can be linked to the *Events* in the project's Master Schedule. This is important in that, in many cases, scheduling tools are used not only to schedule an *Event* but also to manage the resources needed for and costs associated with that *Event*. This allows the project team to monitor cost and schedule and better manage the project's system integration, system verification, and system validation programs.

10.3 EXECUTION STAGE: PERFORMING THE *PROCEDURES*

For each *Event* scheduled within the Project Master Schedule, the steps and actions within the *Procedure(s)* are performed to obtain the data needed to provide the objective evidence that the SOI meets the *Success Criteria* defined for a specific *Instance* associated within each *Activity* addressed by the *Procedure*. As these steps and actions are completed, the results are recorded.

System verification and system validation are formal processes, often with legal and contractual implications. The integrity of the data collected as a result of performing a *Procedure* is of utmost importance. Because of this, organizations will define the process to be followed to ensure the trustworthiness of the data collected during the performance of a *Procedure*. This process is normally not a singular buy off, but rather may involve some sort of independent review. Considerations include the criticality of the system being developed, whether it is an in-house development effort versus developing the system for a customer, and whether the system is under tight regulatory control.

In many cases, the organization's QA office is involved in witnessing the steps and actions and certifying the data was recorded accurately. In some organizations, this involves a quality approval for each step or action that resulted in data being obtained and recorded. This process will involve a peer review of the completed procedure results and a formal signoff of the *Procedure* as discussed later. Depending on the regulatory authority or customer requirements, the certification of the data and final approval of the successful completion of the *Procedure* may be required to be done independently from the project.

For the performance of a *Procedure* by a supplier, the customer must clearly state in the contract the need for the QA function, as well as the roles of both supplier and customer QA representatives during *Procedure* execution, and the required deliverables. For some projects, the customer may specify "mandatory inspection points" where the customer's QA representatives are present to witness the execution of a particular step or action in the *Procedure* and certify the data collected or activity performed.

Many organizations will have a Test Readiness Review (TRR) or other similar gate review that begins the formal system integration, system verification, or system validation series of activities.

For cases where a supplier is the *Responsible Organization*, the customer may be present at this review as specified in the contract.

For each scheduled *Event*, it is common for the *Responsible Organization* to have an individual Pretest Review with all personnel that will be involved in the performance of the *Procedure*. During this review:

- The readiness will be assessed of the SOI and support equipment, facilities, and personnel that participate in the performance of the scheduled *Procedure*. Assuming the focus of the *Procedure* is the design input requirements, the readiness assessment could include a review of the results of the early system verification and system validation, design verification and design validation, and incremental system verification and validation activities as well as production verification and acceptance activities concerning whether the built or coded SOI is in compliance with the design output specifications and whether or not there were any nonconformance issues, waivers, or deviations.
- If the SOI to which the *Procedure* applies has interactions with external systems, emulators, or simulators, their readiness will be assessed concerning their ability to safely interface with the SOI during the execution of the Procedure.
- The readiness will be assessed of the facility, test/support equipment, the Command-and-Control System, and Data Collection System needed to support the performance of the *Procedure*.
- The approval of the *Procedure* will be confirmed.
- It will be confirmed that participating personnel are certified for their roles and will be available.

While the likelihood of a failure of the SOI "under test" is reduced if design verification, design validation, early system verification, early system validation, and production verification actions as described and advocated in this Manual have been completed, problems may still be uncovered.

When a problem occurs during performance of a *Procedure*, the *Responsible Organization* will troubleshoot and identify what the problem is. It could be an issue with manufacturing that was not uncovered during production verification against the design output specifications, an installation/integration problem, a failure of the SOI itself, or an incorrect performance of a step or action within the *Procedure*. Whatever the cause, the *Responsible Organization* will work with the project team to determine the course of action, correct the defect or anomaly, redesign the SOI, or get a waiver or deviation as discussed later. In the cases of either fixing or redesign, there will be a need to rerun the *Procedure*.

Need and design input requirement attributes, as discussed previously, allow the project to both track the status of the *Procedures* performance as well as the results.

Upon completion of the *Procedure*, it is common for the *Responsible Organization* to have some form of closeout process or post-test review. This involves a review of the completed procedures, any problems found during execution, problem resolution, quality "stamps" of applicable steps or actions and certification and approval of the Procedure's execution.

10.3.1 Formal System Verification and Validation Processes

Because qualification, certification, and acceptance of the SOI are based on the results of system integration, system verification, and system validation activities, a more formal approach is often used, especially for larger government-funded projects or for highly-regulated systems.

This formality often involves the participation of project managers, the customer, regulatory agencies, and other key stakeholders in the preparation and approval of qualification, certification, and acceptance plans and *Procedures*. The specific artifacts, events, reviews, and approval process will be documented in the project's MIVV and SIVV Plans.

If the customer contracts these series of system verification and validation activities to a supplier, the artifacts and events are formal contract actions and deliverables that will be specified in the contract as discussed in Chapter 13.

Whether the project team or their customer develops and/or performs the *Procedures,* usually QA representatives and customer representatives are involved along with the test conductor(s), facility personnel, safety, and security personnel. While the specific process and types of reviews may vary depending on organization, domain, and product type, a representative approach common to systems being developed for the government is outlined below. For in-house development projects, a less formal approach is often used.

- *Development of formal qualification, certification, and acceptance system verification and validation planning artifacts* defined earlier in this Manual. This involves review and iteration with the project manager, the customer, QA personnel, and/or regulatory agencies until the system verification and validation planning artifacts are agreed upon.
- *Dry run(s) and approval.* Given that system acceptance is a major milestone that precedes final payment for the system or approval for use and release to the public, it is critical that performance of the *Procedure* goes smoothly without unexpected issues occurring. Dry runs are often performed multiple times until the steps are error free, and the personnel involved in the performance of the *Procedure* know exactly how to perform the *Procedure*. Once a sufficient level of confidence in the *Procedure* has been achieved, the *Procedure* can be approved for use. *Procedure* review and approval is a prerequisite to the activities discussed below.

  Note: With the increased use of behavioral models and the concept of the "digital twin," early system verification and validation, as discussed previously, can be conducted as part of design verification and design validation activities. As part of these activities the Procedure developed to be used during system verification and validation should be used whenever possible. This will result in issues being resolved in both the design, as well as with the Procedure itself prior to starting formal system integration, system verification, and system validation activities. For example, a sequence diagram that matches the system verification or system validation Procedure will provide early insight into issues that can be resolved prior to execution of the Procedure for formal system verification or system validation.

- *Test Readiness Review (TRR).* A TRR often precedes each test phase (FQT, FAT, SAT). The TRR is often a formal gate review that must be passed prior to the execution of a *Procedure* or family of procedures. At the TRR, an overall assessment of readiness of the system, procedures, and personnel is conducted. A TRR will typically also address system configuration (including any simulators or emulators), risk, and other information pertinent to the test.
- *Formal Qualification Test (FQT).* Sometimes, especially when there are large numbers of design input requirements, an FQT phase may precede the performance of the system verification and validation acceptance *Procedures*. The FQT is a series of formal activities that involve the project manager and/or the customer that are designed to provide objective evidence that can be used to determine compliance (verification) with all contractual requirements. The FQT activities may use an environment and users/operators that is different than that expected during operations. This environment and test inputs are meant to stimulate the system in a particular way to verify the system meets each requirement. This is not a validation test.

- *Factory Acceptance Test (FAT)*. Often, but not always, the developing organization will conduct a series of FAT activities at their facility that are witnessed by the customer. The FAT series of activities, to the extent possible, given they are not usually in the operational environment, are designed to obtain objective evidence that can be used to determine that the SOI meets the customer's and other stakeholder's needs (system validation but in a more controlled (and often less expensive) environment. System-level SOI design input requirements may undergo final system verification during the FAT series of activities. While the focus of the FAT is more system validation against the integrated set of needs, system design input requirements may undergo final system verification during the FAT.
- *Site Acceptance Test (SAT)*. A system validation series of activities whose successful completion results in customer or *Approval Authority* acceptance of the SOI for use or release to the public. After the SOI is delivered to the customer's location or location of intended use, the SAT activities are executed in the operational environment with the intended users to validate system operability and to perform a final system verification of the interfaces with external systems.

  The SAT activities will often execute operational scenarios or use cases - usually "day in the life" type scenarios based on use cases or an operational scenario that exercises critical functionality and performance of the system. Depending on the contract type, this is equivalent to the final Developmental Test and Evaluation (DT&E) event. As an example, the US Department of Defense (DoD) requires that the DT&E acceptance be by the contracting branch of the government (for example, Air Force procurement), transferring the system from the contractor to the government (for example, Air Force). This is followed by an Operational Test and Evaluation (OT&E) event conducted by an independent agency [for example, Air Force Operational Test and Evaluation Center (AFOTEC)] prior to authorizing operation by military personnel.
- *Post-Test Review (PTR)*. A PTR often concludes each test phase (FQT, FAT, SAT) of the acceptance process. At the PTR, the result of each *Procedure* is presented, along with a list of any discrepancies and nonconformances. The nature of any issues is discussed to determine if they are acceptable or to agree on a course of action to resolve the issue. The severity and type of issues are reviewed. If the project manager, the customer, or regulatory agencies agree that the system is acceptable as-is, then the discrepancies and nonconformances may become liens that the system developer will need to work off. If the discrepancies and nonconformances are determined to be too severe, then the *Procedure* fails, and the development organization will need to resolve the issues and redeliver the updated system prior to final acceptance for use by the intended users in the operational environment.
- *Test Report (Execution Record)*. The system verification and validation *Execution Record* (reference Section 10.4) provides a formal written record of all formal tests. Usually, a separate report is done for the FQT, FAT and SAT. The result of the *Procedure* is documented, along with a complete list of deviations and waivers. The *Execution Records* are used to develop the system verification and validation *Approval Packages*.

10.3.2 Discrepancies and Nonconformances

As used in this section, a discrepancy is an inconsistency between facts, while nonconformance is failure to conform. In practice, the terms are often used interchangeably.

When the *Success Criteria* for a given *Instance* are not met, the discrepancy or nonconformance must be recorded in the applicable system verification or system validation *Execution Record* and reported to management for follow-up action and closure. Failure to meet the *Success Criteria* is a failure to provide objective evidence that the SOI complies with a need or design input requirement contained in the established configuration item (CI) baseline.

Follow-up actions could include one or more of the following:

- Repeat the system verification or system validation activity to confirm the discrepancy or nonconformance.
- Develop a troubleshooting plan to identify the possible cause of the discrepancy or nonconformance.
- Redesign of the part of the system that failed system verification or system validation and subsequent repeat of the system verification or system validation activity. (This could have a significant impact on cost and schedule.)

In some cases, when redesign is not feasible due to cost and schedule constraints, it becomes clear the project or its suppliers, cannot comply with a need or design input requirement contained in the established CI baseline. There may be any number of reasons for this occurring. Normally, these nonconformances would be rectified using the formal engineering change process as discussed in Chapter 14, such as:

- Request a change to the baselined need, design input requirement, or design output specification.
- Request a change to the system verification or system validation *Success Criteria.*
- Request a change to the *Strategy.*

A change to a need or a requirement could have a domino effect that could result in changes to other requirements, the architecture, design, and design output specifications associated with the changed need, requirement, or related artifacts.

The risk of nonconformance is reduced if the organization follows the needs definition, requirements definition, design verification, design validation, early system verification, and early system validation activities presented in this Manual.

10.3.3 Variances, Concessions, Waivers, and Deviations

There are situations when *nonconformance* with baselined needs, requirements, or design output specifications may be accepted by the customer or *Approval Authority* without making a change to the CI baseline.

Configuration Management (CM) calls these special cases *requests for variance* (previously called *deviations* or *waivers* [57–59]) or *concessions* [60]. Note that a variance differs from an engineering change in that an approved engineering change requires corresponding revision of the CI's current approved baseline, whereas a variance does not.

Note: There are various standards that use a range of terminology when talking about discrepancies and nonconformances. Although the terms "waivers" and "deviations" are still in common use, the US DoD [58] *has moved away from using "waivers" and uses "deviations" instead. Some CM standards do not use the terms "waivers" and "deviations" at all; rather calling these special cases either "requests for variance"* [57] *or "concessions"* [60] *yet acknowledge that the terms waivers and deviations are still in use.*

10.3.3.1 Deviations and Waivers In some organizations, the older terms of deviations and waivers may still be in use. Deviations and waivers are forms of request for variance—although, strictly speaking a deviation is the closest to a request for variance.

- A deviation is defined as "A specific written authorization, granted prior to the manufacture of an item, to depart from a particular requirement(s) of an item's current approved configuration documentation for a specific number of units or a specified period of time" [61].
- A waiver is defined as "A written authorization to accept an item, which during manufacture, or after having been submitted for Government inspection or acceptance, is found to depart from specified requirements, but nevertheless is considered suitable for use "as-is" or after repair by an approved method" [61].

The principal difference between these two types of variances relates to when the variance is authorized.

- Deviations are essential decisions made before the event (design, production, system integration, system verification, or system validation) to allow the project to continue without unnecessary delay.
- Waivers are granted after the event and, as tacit acceptance of failure to meet requirements, are becoming a less commonly accepted means of allowing a departure from the baseline.

Note again that deviations and waivers differ from an engineering change in that the CI's current approved baseline is not amended.

10.3.3.2 Variance Management When dealing with a discrepancy or nonconformance, it is important to know both the priority and criticality of the need, requirement, or design output specification the current SOI is unable to meet, as well as whether that need or requirement is associated with a risk mitigation determination.

When requesting a variance, the customer or other *Approval Authority* will want to see an analysis of the impacts of the discrepancy or nonconformance concerning the ability of the realized SOI to achieve its mission or mitigate a risk. For low-priority or noncritical needs or requirements, if it can be shown that the overall impact on the mission is minimal and the high-priority, critical, and risk mitigation needs and requirements can still be met, they are more likely to approve the variance. *Note: Depending on the type of contract (as discussed in Chapter 13), the customer themselves may be responsible for the direct submission of any variances to a regulatory Approval Authority and may utilize this data during that process.*

Once approved, the variance is included in the *Execution Record* and compliance matrix, where it is noted that the *Activity* was closed with a variance.

When there are waivers, often the waiver is a one-time occurrence, and the need or requirement is left unchanged. The waiver is approved based on the SOI being able to still meet the high priority, critical, and risk mitigation needs and requirements and achieve its mission. For future versions of the system, the customer's expectation is that the project will find a way to meet the baselined need or requirement that is temporarily being waived.

When there is a deviation, the developing organization may request from the customer or *Approval Authority* a permanent variance when it is determined the need or requirement, as written, is not feasible within current cost, schedule, and technology constraints or cannot ever be met. In these cases, the project or supplier requests from the customer or *Approval Authority* for a variance to tailor the baselined requirement to what is feasible using the organization's CM Process.

As in the case for a waiver, the customer or other *Approval Authority* may choose to keep the requirement as-is (there may be other suppliers that may be able to meet the requirement), or they assume the requirement will be able to be met in the future once critical technologies mature.

Either way, the acceptance of the discrepancy or nonconformance with variance is still considered closed, whether the SOI passed or failed system verification or validation of that specific *Instance*.

From a contracting perspective, it is best for the supplier to avoid accepting requirements that are not feasible. From a customer's perspective, they should do the necessary analysis to have some level of confidence as to what is feasible before issuing a contract. To do this, feasibility must be addressed during lifecycle concepts analysis and maturation as discussed in Chapter 4, resulting in needs and design input requirements and sets of needs and sets of requirements that have been assessed for feasibility. Failing to do so will result in significant technical debt in the form of expensive and time consuming contract changes.

For any contract, the supplier wants to have a clear definition of the problem, what is needed, and what objective evidence they need to provide for the customers to accept the product. Clearly defining this information in the contract helps to mitigate litigation. Best practice is to ensure what is required is feasible.

10.4 REPORTING STAGE: DOCUMENTING THE RESULTS

During the Reporting Stage, an *Execution Record* for each *Activity* is created after the completion of the *Procedure* associated with that *Activity*. The *Execution Record* will state the status and outcome of the *Activity* in terms of whether the results provide sufficient objective evidence that the *Success Criteria* for each *Instance* were met. The family of *Execution Records* are combined into *Approval Packages*.

Execution Records formally document the results of the performance of a *Procedure*. There will be one *Execution Record* for each *Activity*. The results of the performance of a *Procedure* will be recorded in one or more forms of objective evidence. Regardless of its form, each piece of objective evidence must be uniquely identifiable and stored in a configuration-managed document or database.

The *Execution Record* should contain the following types of information:

- A unique identifier. This identifier will be used to locate and access the *Execution Record*. If stored within the project's toolset, this identifier can be used as a pointer to the *Execution Record*.
- Date when the *Procedure* was completed that generated the data within the *Execution Record*.
- Location where the *Procedure* was executed.
- Name of the *Responsible Organization* that performed the *Procedure*.
- The Quality Control (QC) process that was used to manage the integrity of the data.
- The name of the QA person(s) who monitored the performance of the *Procedure* and who signed off on the integrity of the data collected.
- The *Success Criteria* that were to be met.
- The *Strategy* and *Method* used to collect the data.
- The environment and any special conditions in which the data was collected.
- Description and form of the data collected.
- Description of the analysis that was used to determine whether the data collected provided sufficient objective evidence that the *Success Criteria* were met.
- A statement of conformance or nonconformance by the *Responsible Organization* and person doing the analysis and making the pass/fail judgment.

The preparation of the *Execution Records* will often be done concurrently with the closeout activities for the *Procedure*. Once the *Execution Records* are complete, they are assembled, along with any other information needed by the *Approval Authority*, into an *Approval Package* for submittal to the *Approval Authority*.

In addition to the *Execution Records*, other records included in the *Approval Package* may include:

- A Certificate of Conformance to a standard or regulation, often applicable to design and manufacturing processes.
- Approved waivers, deviations, or exceptions.
- Objective evidence in component end item data packages (design output specifications), such as a materials and parts list or a test report.
- Objective evidence that demonstrates that the design was developed in accordance with the approved design plan and system design output specifications (design verification and design validation).
- Objective evidence that demonstrates that the device was manufactured in accordance with the approved processes and design output specifications for the system (production verification).

As an example, for medical devices developed under the oversight of the FDA, CFR Title 21, Part 820 defines the process organizations must follow and the set of records that must be developed, maintained, and submitted to the FDA to get a medical device approved for its intended use. The requirements in this regulation include key definitions for parts of the process, including design inputs, design outputs, specifications, design reviews, design verification, design validation, process validation, and many other definitions.

Concerning compliance and record keeping the regulation defines three key records and requirements for each:

- Design History File (DHF) is a compilation of records which describes the design history of a finished device. The DHF contains, or references, the records necessary to demonstrate that the design was developed in accordance with the approved design plan and the design input requirements. The results of the design verification, including identification of the design, method(s), the date, and the individual(s) performing the verification, are documented in the DHF.
- Device History Record (DHR) is a compilation of records containing the production history of a finished device. Each manufacturer is required to maintain DHRs. Each manufacturer is required to establish and maintain procedures to ensure that DHRs for each batch, lot, or unit are maintained to demonstrate that the device is manufactured in accordance with the DMR and the requirements of this part.
- Device Master Record (DMR) is a compilation of records containing the procedures and specifications for a finished device. The DMR for each type of device includes, or refers to the location of, the following information:
  - Device design output specifications, including appropriate drawings, composition, formulation, component specifications, and software specifications.
  - Production process specifications, including the appropriate equipment specifications, production methods, production procedures, and production environment specifications.
  - Quality assurance procedures and specifications, including acceptance criteria and the quality assurance equipment to be used.
  - Packaging and labeling specifications, including methods and processes used.
  - Installation, maintenance, and servicing procedures and methods.

The total finished design output consists of the device, its packaging and labeling, and the DMR.

In context of what is presented in this Manual, the DHF includes the design inputs, including the integrated set of needs and set of design input requirements. It would also include the *Execution Records* and *Approval Packages* that provide objective evidence the realized medical device was verified to have met the design input requirements and validated to have met the integrated set of needs.

The DHR includes all the records associated with production verification. The DHR includes all the design output specifications, which via production verification, the medical device was manufactured and coded to, the production procedures and processes used to manufacture and code the medical device as well as the quality assurance processes and procedures used during production verification. It would also include all the packaging, labeling, instructions for use, installation, maintenance, and servicing procedures.

10.4.1 Chain of Evidence Showing Conformance or Compliance

The project must maintain and manage records that provide a chain of evidence of conformance to and compliance with the integrated set of needs, set of design input requirements, and set of design output specifications. Of special importance is compliance with the customer's requirements, standards, regulations, and needs and requirements associated with the mitigation of safety and security risks.

The chain of evidence of conformance or compliance could be complex, given the various levels of needs, requirements, the design implementation of those requirements, the design output specifications, manufacturing/coding that implemented those specifications and all the system verification and validation artifacts developed and data gathered to determine compliance. There could easily be thousands of system verification and validation *Instances* and hundreds of *Execution Records* included in the *Approval Packages*, depending on the size and complexity of the SOI under development.

Because of this, the organization must address how they will record this chain of evidence of conformance/compliance in their SEMP, MIVV, and SIVV (or equivalent type plans.) These plans must specify the approach the project will use to compile records that provide objective evidence or to generate a summary document that builds on assessments and objective evidence to determine that the system is compliant with its needs, design input requirements, and design output specifications. It is important for an organization to clearly identify the processes and resulting records they will develop and maintain. Customers, regulatory agencies, and other *Approval Authority* must clearly define their requirements for records that are needed to determine compliance and acceptance.

Often, the organization will have an internal "compliance" or "quality" group responsible for both the CM of the sets of needs and requirements and resulting system integration, system verification, and system validation artifacts as well as to work with individual projects within the organization throughout the development lifecycles to help ensure compliance. This compliance group will also interact with the external regulatory agencies to make sure the intent of the customer requirements, standards, and regulations are being met as well as to ensure the objective evidence needed to determine compliance and obtain certification or qualification and acceptance is properly documented and submitted to the regulatory agency in accordance with the requirements defined by the agency.

Many regulatory agencies, as well as customer's, will want the project to show traceability between the requirements from the customer, relevant standards and regulations, concepts to meet them, needs that communicate the which specific standards the project will be compliant with, the resulting design input requirements that address what the must be done to meet the needs, the implementing architecture and design, design output specifications, production verification, the end system, and the project's system verification and validation activities that result in the objective evidence that proves that the intent of the relevant customer's requirements and standards and regulations have been met.

Using a medical device company as an example, as part of the development process for pharmaceuticals and medical devices, the FDA is concerned with quality and safety, especially during use. To address this, the project is expected to perform a detailed Use or User Failure Mode and Effects Analysis (UFMEA) as discussed in Chapter 4. During the UFMEA, all the things that could go wrong during manufacturing, shipping, use, and disposal are identified. The project must then assess the risk associated with each hazard and determine which ones they will mitigate and how. Mitigation could involve people (skills, certifications, training), processes [procedures, Instructions for Use (IFU)], and/or the medical device itself. If it is the medical device, the FDA requires traceability to be established from individual risks, concepts, needs, design input requirements, and the design outputs (design, specifications, manufacturing, system verification and system validation. This traceability must be documented in the DHF and DMR discussed previously.

10.4.2 Use of Compliance Matrices

The results of the system verification and validation *Activities* for each *Instance* should be tracked as soon as the *Procedure* has been completed. This can be done using the attributes for each need or requirement expression as defined earlier. This information can then be used to generate summary reports referred to as a System Verification Compliance Matrix (SVCM) and System Validation Compliance Matrix (SVaCM).

A SVCM or SVaCM is a matrix/table visualization of the system verification or system validation information that lists the needs or requirements, the system verification or system validation *Procedure*, the system verification or validation result in terms of compliance (pass/fail), and a summary of the objective evidence that was used to determine compliance or a pointer to where the evidence is located.

Table 10.4 shows an example of a simple SVCM, each *Instance* is listed as a row.

TABLE 10.4 Example System Verification Compliance Matrix (SVCM).

Req ID	Verification Procedure	Verification Result	Verification Evidence
R001	LIR Inspection Procedure xxxxx	Pass	Execution Record xxxxx
R005	LIR Demonstration Procedure xxxxx	Pass	Execution Record xxxxx
R061	LIR Analysis Procedure xxxxx	Pass	Execution Record xxxxx

Some organizations will add the information shown in Table 10.4 as part of their SVM and SVaM discussed in Section 10.1.4.

The resulting matrix provides a useful visualization of this information for each *Instance*, providing the project team with the ability to evaluate the status of the system verification or validation activities on the path to certification and approval for the SOI's intended use in its operational environment. For projects with a large number of needs and requirements, the use of this form may be of limited use. Modern SE tools allow the generation of dashboards that provide a status and summary of the project's system verification and validation activities using the attributes defined for each need and design input requirement, as discussed in Section 10.6.

10.5 APPROVAL STAGE

During the Approval Stage, the *Approval Packages* are submitted to the *Approval Authority* for approval (acceptance, certification, readiness for use, or qualification). Given system verification and validation are bottoms-up processes, they are repeated for each architectural entity (subsystem, system element, or integrated system).

The *Approval Authority* is any individual, group of individuals, or organization that has the authority to approve the system for use in its operational environment by its intended users. The *Approval Authority* could be a combination of internal or external customer(s), a regulatory agency, or a third-party certification organization. It is the *Approval Authority* that (a) *decides* what *constitutes necessary for acceptance* and (b) *determines* what *is necessary for acceptance.*

As part of the definition of what is necessary for acceptance, it must be made clear to what level of information is included in the Approval Package. Only system verification and system validation at the system level, or for each part of the system architecture? When there is a customer/supplier relationship, the answer to this question has significant cost and schedule implications and must be clearly specified in the Contract.

The requirements for the internal approval process within the developing organization may be different than for an external Approval Authority. Both sets of processes must be well defined within the organization's design controls that are part of the organization's Quality Management System (QMS). In some cases, the external Approval Authority may only require the system-level Approval Package along with assurance that the internal approval process for lower-level parts of the architecture was followed.

For regulatory agencies, the process is often defined as part of the regulations governing the system under development. Some organizations may conduct a formal System Acceptance Review (SAR) or Readiness for Use Review as part of the approval process.

Assuming the project follows the process defined by the *Approval Authority* and followed the system verification and validation processes for sets of needs, sets of design input requirements, design, and system verification and validation defined in this Manual, the risk of non-approval is low. Not following these processes increases the risk of non-approval. In the previous FDA examples, one of the major reasons for non-approval was the failure to follow the defined product development process and submit the required records to the *Approval Authority*—in this example the FDA.

For many products developed for use around the world, there will be multiple *Approval Authority*, each with different defined approval processes and associated requirements for documentation and records. For example, in US DoD programs, there is a DT&E process for system verification and OT&E process for system validation. United States Code (USC) Title 9 requires these processes to be conducted by different organizations. In this case, the project will have to be in conformance/compliance with each set of regulations and go through multiple system verification and validation approval processes.

There can be multiple outcomes involved with system approval [1], including:

- *Acceptance* is an activity or series of activities conducted before transition of the SOI from supplier to customer against predefined and agreed-to "necessary for acceptance" criteria such that the customer can determine whether the realized SOI is suitable to change ownership from supplier to the customer. Acceptance can include system verification, system validation, certification, and qualification activities, or a review of system verification, system validation, certification, and qualification results is completed in accordance with the necessary for acceptance criteria as agreed to in a contract.

- *Certification* is the result of some audit or a written assurance that the realized SOI has been developed and can perform its assigned functions in accordance with legal or workmanship, design, and construction standards (such as for an aircraft, consumer product, or medical device). The development reviews, system verification, and system validation results form the basis for certification. Often, certification is performed by a third party to determine compliance with the applicable standards. As such, a product could have multiple certifications for compliance by multiple certification organizations. Many standards include requirements for specific methods (test, demonstration, analysis, or inspection) required to be used for certification to that standard. For example, Conformité Européenne (CE) certification in Europe and Underwriters Laboratories (UL) certification in the United States and Canada. Certification activities are separate from the standard system verification, system validation, and qualification activities, but may use the results of these activities.
- *Qualification*. System qualification requires that all system verification, system validation, and required certification actions have been successfully performed. The qualification process demonstrates that each subsystem and system element that is part of the system physical architecture and the integrated system, meet the applicable needs, requirements, and design output specifications, including performance and safety margins. Because qualification includes testing at the extremes and testing to failure, special qualification versions of the SOI are developed as "qualification units" that are representative to what requirements and specifications are being qualified. Qualification activities can be done as a subset of design verification and design validation as well as system verification and system validation.
- *Readiness for Use*. A determination of readiness for use by *Approval Authority* based on an analysis of the system verification, system validation, certification, and qualification results. This may occur multiple times in the system lifecycle, including the first article delivery, completion of production (if more than a single system is produced), and following maintenance actions. In the field, particularly after maintenance, it is necessary to establish whether the system is ready for use. The readiness for use determination could be made by a supplier prior to submitting the realized SOI to the customer or regulatory agency for acceptance or by the customer prior to making the SOI available for operations or release to the public.

The system verification and validation approval process can be long, complex, and frustrating. Because of this, the focus of the project must be on both the approval process requirements what is "*necessary for approval*" in addition to verifying the SOI meets the design input requirements and validating the SOI meets its integrated set of needs.

10.6 USE OF ATTRIBUTES TO MANAGE SYSTEM VERIFICATION AND VALIDATION ACTIVITIES

As discussed in Chapter 15, there are needs and design input requirements attributes to help manage the project's system verification and validation programs across the system development lifecycle. Using these attributes, tools within the project's toolset can generate reports and dashboards that allow the project team to target various aspects of the project's system verification and validation programs based on these attributes.

Many current tools allow the user to define a dashboard display that includes a graphical representation of key metrics based on the selected attributes. This is only possible if the project includes in the tool database schema the attributes that contain the information needed to produce these reports and dashboards at the beginning of the project AND keeps the information within each attribute current.

This set of attributes can be used to help track and manage the system verification and validation programs and status to make sure the designed and built system meets the set of design input requirements (system verification), set of design output specifications (production verification), and the set of needs (system validation).

This set of attributes can also be used for developing many of the system verification and validation artifacts discussed in this section as well as provide metrics that management can use to track the status of the project's system verification and validation programs.

Below is a list of attributes that should be included in each need or design input expression for projects to be able to use attributes to help manage their system verification and validation programs. *Refer to Chapter 15 for a more detailed discussion on attributes and definitions of each.*

A1: *Rationale*: Intent of the need or requirement, reason for the existence of the need or requirements.

A6: *System verification and validation Success Criteria: Refer to Section 10.1.1 for a more detailed discussion on Success Criteria.*

A7: *System verification and validation Strategy*: *Refer to Section 10.1.2 for a more detailed discussion on Strategy.*

A8: *System verification and validation Method*: *Refer to Section 10.1.3 for a more detailed discussion on Methods.*

A9: *Responsible Organization*: The organization that is responsible for system verification and validation for the need or requirement.

A10: *System verification and validation level*: Architecture level at which the system verification or system validation activity will be performed to obtain objective evidence that the SOI meets the requirement (verification) or need (validation). Possible values could include: <system>, <subsystem>, <assembly>, <component>, <hardware>, <software>, or <integration> as examples.

A11: *System verification and validation phase*: Lifecycle phase in which the system verification or system validation activity will be performed to obtain objective evidence that the SOI meets the requirement (verification) or need (validation). Possible values could include design, coding/testing, proof testing, qualification, and acceptance.

A12: *Condition of Use*: Operational conditions of use expected in which a need or requirement applies. This information can be used to set up the environment in which system verification or system validation activities need to take place.

A13: *System verification and validation results*: The results of each system verification or system validation activity.

A14: *System verification and validation status*: Indicates the status of the system verification or system validation activities.

Attributes for priority, criticality, and risk could also be useful.

10.7 MAINTAINING THE SYSTEM VERIFICATION AND VALIDATION ARTIFACTS

This Manual uses a specific ontology of terms for the various types of data and work products needed to successfully execute and manage the project's system verification and validation programs. These include system verification and validation *Instance, Method, Strategy, Responsible Organization, Activity, Procedure Requirements, Procedure, Event, Execution Record, Approval Package*, and *Approval Authority*.

A major point made in the RWG whitepaper, "*Integrated Data as a Foundation of SE*" [6], is that all the SE and PM project data and information should be recorded and managed within an integrated/federated dataset. This allows each data item to exist in a single location helping to ensure consistency and the establishment of an ASoT.

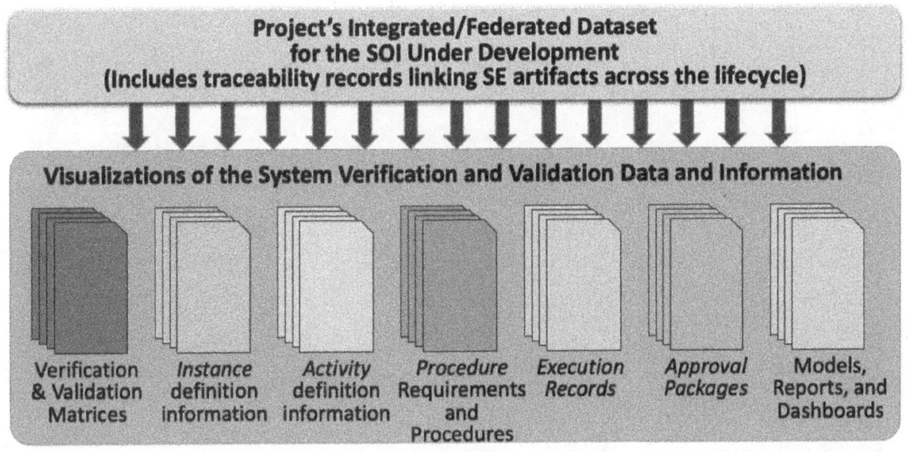

FIGURE 10.6 Visualizations of the Project's System Verification and Validation Data and Information.

As shown in Figure 10.6, the various system verification and validation artifacts (matrices, *Instance* and *Activity* definition information, *Procedure Requirements*, *Procedures*, *Execution Records*, and *Approval Packages* along with any models and reports), as well as dashboards used to manage the project's system verification and validation effort, should be thought of as different visualizations of this single set of data and information. The basic data included in each visualization is stored in a single location and then visualized as a report based on the needs of the project.

A key advantage of this data-centric approach is that a change in one piece of information will be reflected automatically no matter the visualization. If the *Method* for a specific *Instance* is changed, all matrices, reports, and dashboards will reflect that change. If the *Success Criteria* are updated, those updates will again be reflected across all the various visualizations.

From a CM perspective, it is the integrated set of data that is managed, rather than the individual visualizations as was done in the past using a document-centric approach to PM and SE.

11

SYSTEM VERIFICATION AND SYSTEM VALIDATION PROCESSES

This chapter, as well as Chapter 10, is an elaboration of the INCOSE SE HB verification and validation technical processes.

After the subsystems and system elements that are part of the System of Interest (SOI) physical architecture have been built or coded and passed production verification, the concepts of system verification and system validation take on a formal meaning.

Each of these processes represents a set of activities that cumulate with one or more gate reviews associated with the qualification, certification, and acceptance of the SOI by the customers or by regulatory agencies for release for use by the intended users. Because of the formality and legal aspects associated with the system verification and system validation processes and the need to provide documentation of the objective evidence obtained resulting from those processes with the required degree of confidence, system verification and system validation are separate and distinct processes within the system development lifecycle.

Some organizations focus only on system verification rather than both system verification and system validation. While the methods discussed in Chapter 10 are used for both system verification and system validation, as shown in Figure 11.1, the intent and focus of each process are much different.

System verification is used to obtain objective evidence of compliance with the system's set of design input requirements as well as to detect errors/defects/faults in the built/coded physical system. System verification process activities are the official "for the record" activities performed on an SOI to obtain objective evidence that is used to determine whether it meets its set of design input requirements.

System validation relates back to the integrated set of needs, use cases, user stories, operational scenarios, and lifecycle concepts defined during *Lifecycle Concepts and Needs Definition* activities discussed in Chapter 4. System validation determines whether the designed, built, and verified system, subsystem, or system element will result or has resulted in an SOI that meets its intended purpose in its operational environment when operated by its intended users and does not enable unintended users to negatively impact the intended use of the system.

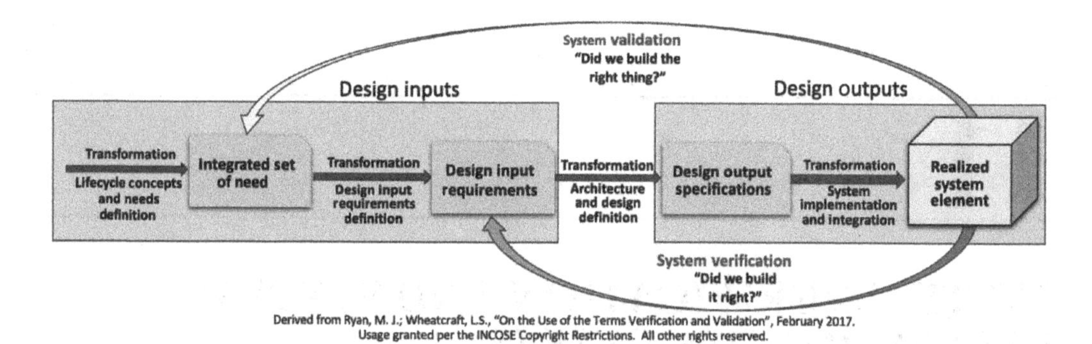

FIGURE 11.1 System Verification and Validation Processes.

To do this, whenever feasible, system validation is conducted under the actual operational conditions by the intended users. In addition to validating that the SOI meets its integrated set of needs, system validation is also used to access emergent properties (good and bad) and system behavior not expressively communicated via its set of needs or requirements. *To focus only on system verification and limiting system validation activities could result in a system that does not meet the stakeholders' real-world expectations.*

To be successful, objective evidence is required that the set of design input requirements and the integrated set of needs have been met by the realized SOI with the required degree of confidence.

While unusual, it is possible for an SOI to pass system verification and fail system validation; it is also possible for an SOI to fail system verification yet pass system validation. In either case, the root cause was most likely defective integrated set of needs or set of design input requirements.

To help prevent these failures, the project must adapt a continuous verification and validation approach across the lifecycle, as discussed in this Manual.

11.1 SYSTEM VERIFICATION AND VALIDATION PER LEVEL

A system will have several levels of subsystems and system elements within its physical architecture. Because of this, system integration, system verification, and system validation are a bottom-up series of activities. As discussed in Section 10.1.2.1 "Level of Integration", these activities begin with system verification and validation of lower-level system elements against their needs and requirements, integrating those into higher-level subsystems, doing system verification and validation of the higher-level subsystems against their needs and requirements, integrating those subsystems into the integrated system, and finally completing system verification and validation of the integrated system against its needs and requirements.

This bottom-up series of activities is shown in Figure 11.2, which shows a mapping of the hierarchical view of the system architecture to the Systems Engineering (SE) Vee Model. The thicker arrows point to the system, subsystems, and system elements to where they are represented in the SE Vee Model. The dashed arrows point from the system elements in the hierarchical view to where their design output specifications are represented in the SE Vee Model. The dashed arrows also represent production verification that the realized system elements meet their design output specifications. Successful production verification of each system element is a prerequisite to beginning final system integration, system verification, and system validation activities.

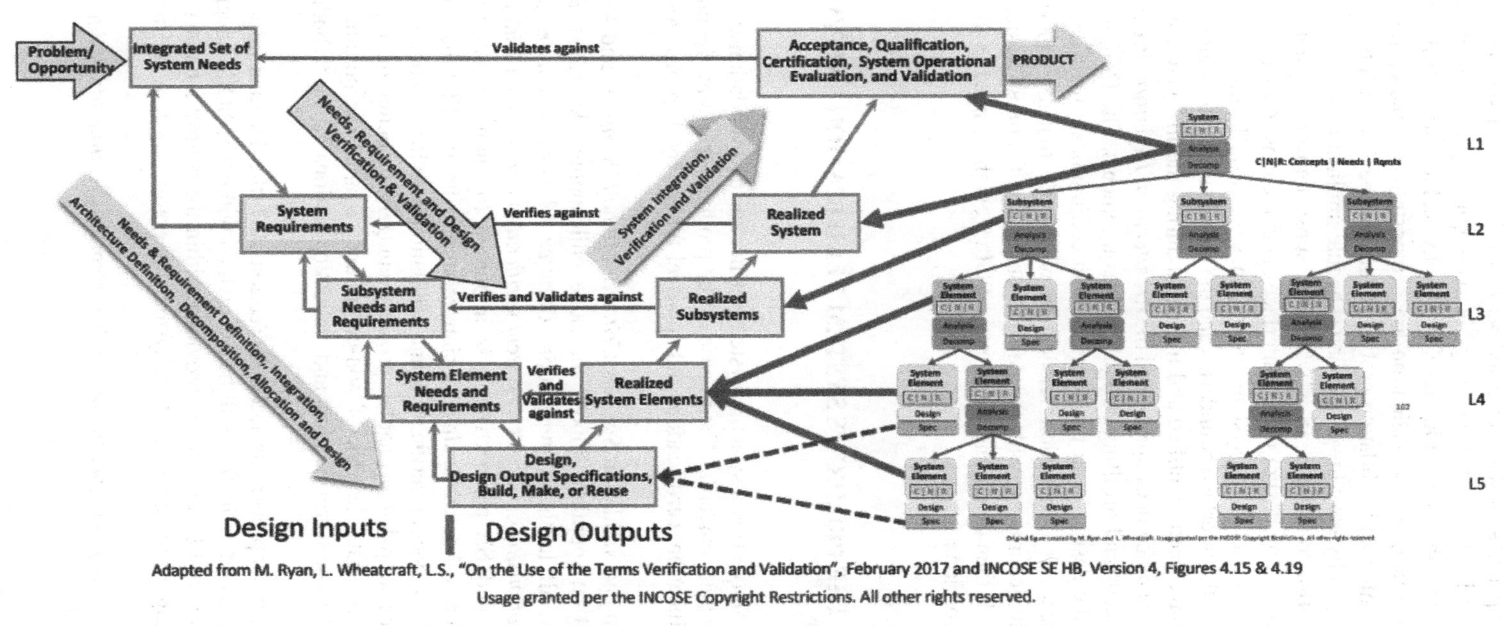

Adapted from M. Ryan, L. Wheatcraft, L.S., "On the Use of the Terms Verification and Validation", February 2017 and INCOSE SE HB, Version 4, Figures 4.15 & 4.19

FIGURE 11.2 Bottom-up View of System Verification and Validation.

Any issues or discrepancies must be corrected before a subsystem or system element is integrated into the next higher-level entity. A basic strategy is to complete system verification and validation activities at the lowest possible integration level of the physical architecture that makes sense. This ensures that system verification and validation are demonstrated at the earliest practical time to identify system verification and validation problems early in the system integration process when corrective measures typically have lower cost and schedule implications.

A major concern prior to integration is whether a subsystem or a system element meets its interface requirements and its design output interface specifications. Experience has shown that a major source of defects found during system integration, system verification, and system validation activities occurs both at the interface boundaries and the interactions across those boundaries. These types of risk should have been identified and mitigated during the lifecycle concepts analysis and maturation activities discussed in Chapter 4 and the *Architectural Definition Process* and reflected in the integrated set of needs and resulting design input requirements. The *Design Definition Process* will then implement the design input requirements and communicate their design via the sets of design output specifications. Production verification will verify that the subsystems and system elements were built in accordance with the design output specifications. System verification and validation activities will then determine whether the SOI meets these needs and requirements with the required level of confidence.

11.2 MANAGING THE PROJECT'S SYSTEM VERIFICATION AND VALIDATION PROGRAMS

For highly regulated safety-critical systems, system verification and validation are processes defined within the regulations and standards. The release of the product for use by the public is dependent upon successful qualification, certification, and approval for use by *Approval Authority* within the regulatory agency.

System verification and validation *Success Criteria*, *Method*, *Strategy*, and *Procedure Requirements* that are levied on suppliers impose the same philosophies described herein, ensuring a consistent *Strategy* at all levels of integration. *Chapter 13 goes into more detail concerning customer/supplier relationships in terms of system verification and system validation.*

From both a PM and SE perspective, it is important to know, at any time, which needs the SOI has not been validated against, which requirements the SOI has not been verified against, the status of all system verification and validation activities, anomalies discovered, and non-compliances. This knowledge enables project managers to better manage the budget and schedule, as well as estimate the risk of non-compliance against the possibility of eliminating some of the planned system verification and validation actions to meet budget and schedule constraints.

Using the data-centric approach discussed in Section 10.7, the data and information associated with system verification and validation will be maintained within the SOI's integrated dataset, and reports and dashboards will be able to be provided within the project's toolset. This enables the project team to have an accurate status of all system integration, system verification, and system validation activities.

More details concerning managing the project's system verification and validation programs are included in Chapter 14.

11.3 SYSTEM VERIFICATION ACTIVITIES

The purpose of system verification is to provide objective evidence with an acceptable degree of confidence that:

- The transformation to the SOI has been done "right" according to the design input requirements and design output specifications in accordance with the defined *Success Criteria.*
- No error/defect/fault has been introduced during the *Design Definition Process* transformation of design input requirements to design output specifications, or during the production process of transforming the design output specifications into the realized physical SOI.
- The selected verification *Strategy*, *Method*, and *Procedures* will yield objective evidence that if a fault were introduced, it would be detected.

The focus of system verification is on the physical SOI and subsystems and system elements that make up the SOI. As discussed in Section 2.3, during system verification, it is not the design input requirements being verified, it is verification that the SOI conforms, or is compliant, with its design input requirements. Any verification of the design input requirements and sets of design input requirements should have been done during the development and baseline of the design input requirements as discussed in Chapters 6, 7, and 14.

11.3.1 Prepare for System Verification—Planning and Defining

As shown in Figure 11.3, there are several enablers to successful system verification. These include an enterprise product development process that defines each development lifecycle stage, a project Master Integration, Verification, and Validation (MIVV) Plan, and specific System Integration, Verification, and Validation (SIVV) Plans for each system element and subsystem within the SOI physical architecture. It is critical that the project team members responsible for

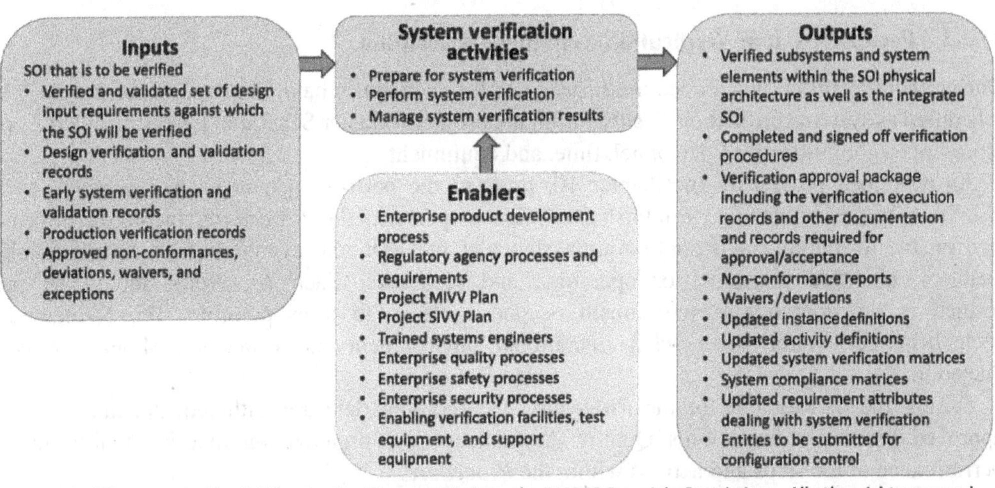

FIGURE 11.3 System Verification IPO Diagram.

planning for and implementing the project's system verification program are trained in these processes.

In addition, all system verification activities will need to adhere to the enterprise's Quality, Safety, and Security processes. If the SOI is governed by a regulatory agency, the system verification activities must be conducted in accordance with the requirements in the governing regulations and other guidance documents issued by the regulatory agency.

Finally, all enabling systems such as facilities, test equipment, and support equipment must be approved for use to support the project's system verification activities. The facilities need to have the ability to simulate the operational environment. If the actual external systems are not available, the capability to simulate or emulate interactions with external systems is required.

Inputs to the *System Verification Process* include:

- The SOI verified and validated set of design input requirements the SOI is being verified against.
- The completion of design verification and early system verification activities prior to baselining the sets of design output specifications for each entity and the production of each entity.
- Successful completion of production verification, where each entity was verified to have met its design output specifications, i.e., "built to spec."
- The design verification and design validation, early system verification and system validation, and production verification records for each entity. These records include information concerning any nonconformancc's, deviations, waivers, or exceptions.

As part of the Planning Stage, the system verification *Success Criteria*, *Strategy*, and *Method* are defined and included as attributes for each of the design input requirements for each entity. Information is gathered to define each verification *Instance*.

During the Defining Stage, for each entity, verification *Instances* are grouped into verification *Activities* and information gathered to define each *Activity*. System verification *Procedure Requirements* are defined along with other information needed to develop the *Procedures*, and each *Procedure* is developed, reviewed (verified and validated), and approved. *Reference Chapter 10 for information on the artifacts described.*

11.3.2 Perform System Verification Activities—Execution

During the Execution Stage, once the *Procedures* for each entity have been approved, they will be scheduled as system verification *Events* within the project's Master Schedule. Each scheduled *Event* represents a commitment of personnel, time, and equipment.

As described previously in Chapter 10, prior to the performance of each *Procedure*, there will be a pre-test review to establish readiness to perform the *Procedure*, including information on the availability and configuration status of the SOI, the availability of the verification enablers, qualified personnel or operators, and resources. Each *Procedure* should be performed in the operational environment or one as close to it as possible. The *Strategy* for each *Instance* should cover what operational environment the *Procedure* should be conducted in.

During the performance of the *Procedure*, the required data are gathered, documented, and approved by the organization's Quality Assurance (QA) organization that is monitoring the performance of the steps and actions within the *Procedure*.

After a *Procedure* has been completed, often there will be a Post-test Review (PTR) or similar review to review the status and results of the completed *Procedure* and any discrepancies and non-conformances that occurred. Discrepancies and non-conformances observed during the performance of the *Procedure* are recorded and resolved using the organization's QA process, as discussed earlier in Chapter 10. The discrepancies and non-conformances could be due to the verification *Strategy*, the enabling systems, execution of the *Procedure* steps and actions, faulty test equipment, incorrect input test data, a faulty design, poorly formed requirements, poorly formed design output specifications, or a manufacturing or coding defect. The *Procedure*, or portions of the *Procedure*, may be repeated as needed as part of anomaly resolution.

Once the discrepancies and non-conformances have been addressed, the completed *Procedure* will be signed-off in accordance with the organization's *Configuration Management Process*. Any discrepancies or non-conformances will be resolved prior to the final sign-off of each completed *Procedure*. *Note: "Signed-off" refers to key project team members to include their signatures certifying that the Procedure was completed, all data collected is accurate, and all discrepancies and non-conformances identified during the execution of the Procedure have been resolved.*

11.3.3 Manage System Verification Results—Reporting and Approval

During the Reporting Stage, for each system verification *Instance* covered by a *Procedure*, the results will be recorded in an *Execution Record* and maintained in accordance with organizational policy.

The recorded data will be analyzed against the *Success Criteria* defined for each design input requirement to determine whether the SOI being verified meets its design input requirements with an acceptable degree of confidence. Any discrepancies or non-conformances will be documented within the *Execution Record*.

Problem resolution and any subsequent changes will be managed through the organization's QA and *Project Assessment and Control Processes*. Any changes to the SOI definition artifacts (i.e., integrated set of needs, set of design input requirements, architecture, design, set of design output specifications, or interface definition artifacts, e.g., ICDs or IDDs), and associated engineering artifacts and PM work products are managed within the *Needs, Requirements, Verification, and Validation Management* activities discussed in Chapter 14.

Bidirectional traceability is established between the system verification artifacts and the design input requirements against which the SOI was being verified. How this traceability is managed should be defined in project's SIVV Plans or System Engineering Management Plan (SEMP).

During the Approval Stage, the individual *Execution Records* are combined into *Approval Packages* and submitted to the *Approval Authority*.

The project team will assess the success of the system verification activities. The evaluation of system verification results and follow-up corrective action(s) can vary depending on the purpose of the verification activity, the risk of the original requirements not being met, and the expected degree of confidence.

For lower-level subsystems or system elements within the SOI physical architecture, this could be a simple action to address a failed subsystem or system element followed by re-verification, or a more significant action such as a major project re-direction based on a failure to attain a key milestone.

As discussed in Chapter 10, a variance may be approved by the *Approval Authority* for system verification failures against non-critical, lower-priority design input requirements. Failed system verification can often lead to a wholesale project re-baseline.

The project team will provide baselined system verification artifacts and verified entities for Configuration Management (CM). The organization's *CM Process* is used to establish and maintain configuration items and baselines. The result will be a verified SOI ready to undergo system validation along with all the associated system verification artifacts.

11.4 SYSTEM VALIDATION PROCESS

System validation is performed as part of system integration, system verification, and system validation activities against the integrated set of needs for the SOI being validated. System validation is a process that normally occurs after system verification; however, there are cases where system validation and system verification activities can be performed concurrently within the same *Activity* [16].

Of particular importance is system validation of the integrated system. As advocated in this Manual, the project should manage the integrated system from the beginning of the project. The concept of integrated testing promotes a continuum of system integration, system verification, and system validation activities using collaborative planning and collaborative execution of these activities, providing shared data that can be used for independent analysis [62].

11.4.1 General Considerations

The integrated set of needs the SOI is being validated against is derived from the set of SOI lifecycle concepts. The lifecycle concepts include operational scenarios and use cases that are performed in a specific operational environment by the intended users for not only operations but also during other lifecycles including transportation, storage, installation, operations, maintenance, upgrades, and retirement. The integrated set of needs addresses all these lifecycle activities that represent the stakeholder needs for the SOI as well as expectations for function, form, fit, quality, and compliance.

11.4.1.1 Operational and Loss Scenarios and Use and Misuse Cases It is common for the operational scenarios and use cases for each part of the architecture to be exercised during conduction of the validation *Procedures*, using the operational environment and the intended users. The common aerospace saying "test as you fly, fly as you test" applies. In addition to just nominal operations, it is important to address loss scenarios as well as alternate nominal, off-nominal, and misuse cases that include attempts by intended users to misuse the system as well as unintended users to negatively impact the intended use of the system (as discussed in Chapter 4).

A positive system validation result obtained in a defined operational environment by specific users can turn noncompliant if the operational environment or class of user changes. These changes may not be immediately known by the developer; however, changing needs should be accommodated by the customer and developer's CM, PM, SE, and acquisition process activities.

During systems validation, especially for walkthroughs and similar activities, it is highly recommended to involve the intended users/operators under their own local operational environment.

When the system is validated at a supplier facility or development organization, often the customer will want to conduct some system validation activities in their own facility, operational environment, and with the actual users. The stakeholders who participated in defining the lifecycle concepts and integrated set of needs must be presented with the results of the system validation activities to ensure their needs and requirements have been met.

11.4.1.2 Assessing System Behavior and Emerging Properties Because the behavior of a system is a function of the interaction of its parts and interactions with the external systems and operational environment, a major goal of systems validation is assessing the behavior of the integrated physical system and identifying emergent properties not specifically addressed within the integrated set of needs or set of design input requirements.

Emergent properties may be positive or negative. An example of a negative property includes cascading failures across multiple interface boundaries between the system elements that are part of the SOI's architecture.

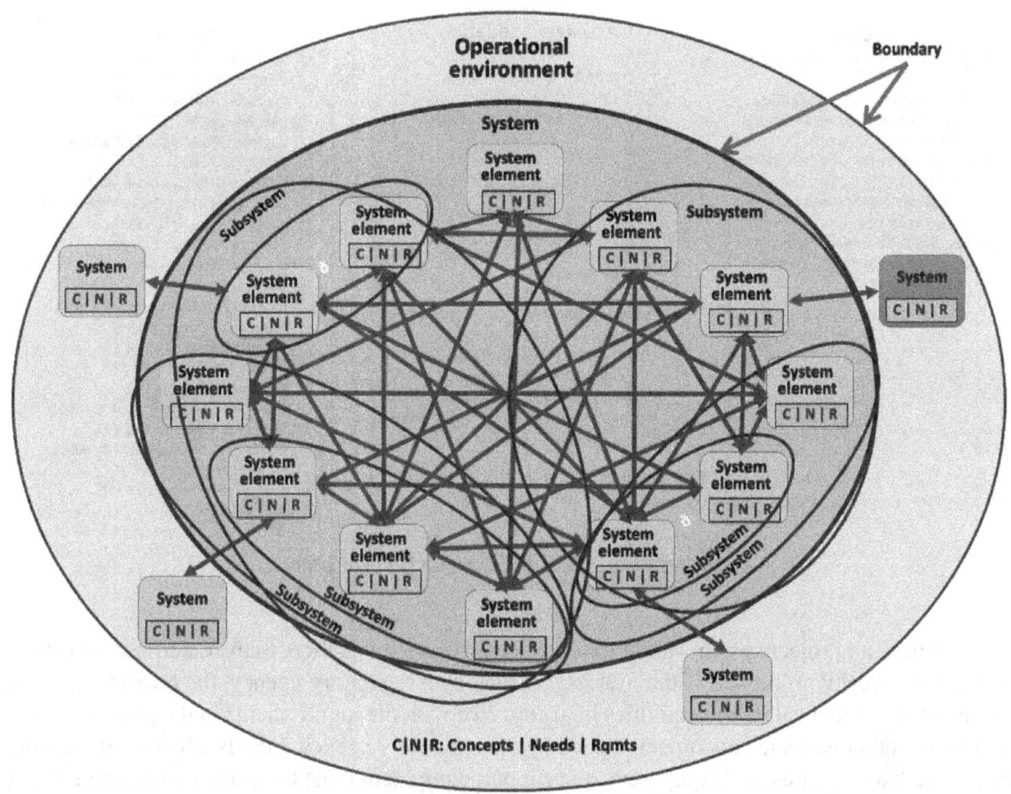

FIGURE 11.4 Holistic view of the SOI.

Figure 11.4 presents a holistic view of the integrated system that illustrates the many interactions between subsystems and system elements that are part of the system's physical architecture. It is the sum of these interactions that determines the overall behavior of the integrated system. Relying on models and simulations of the SOI and operational environment may not uncover all the emerging properties and issues that occur in the physical realm.

While validation using models and simulations allows a theoretical determination that the modeled system will meet its needs in the operational environment by the intended users once realized, the assessment of the actual physical system behavior (system validation) must be done, whenever possible, with the actual hardware and software subsystems and system elements integrated into the SOI in the actual operational environment by the intended users.

11.4.2 Prepare for System Validation—Planning and Defining

As shown in Figure 11.5, there are several enablers to successful system validation. These include an enterprise product development process that defines each development lifecycle stage, a project MIVV Plan, and specific SIVV Plans for each subsystem and system element within the SOI architecture. It is critical that the project team members responsible for planning and implementing the project's system verification and validation programs are trained in these processes.

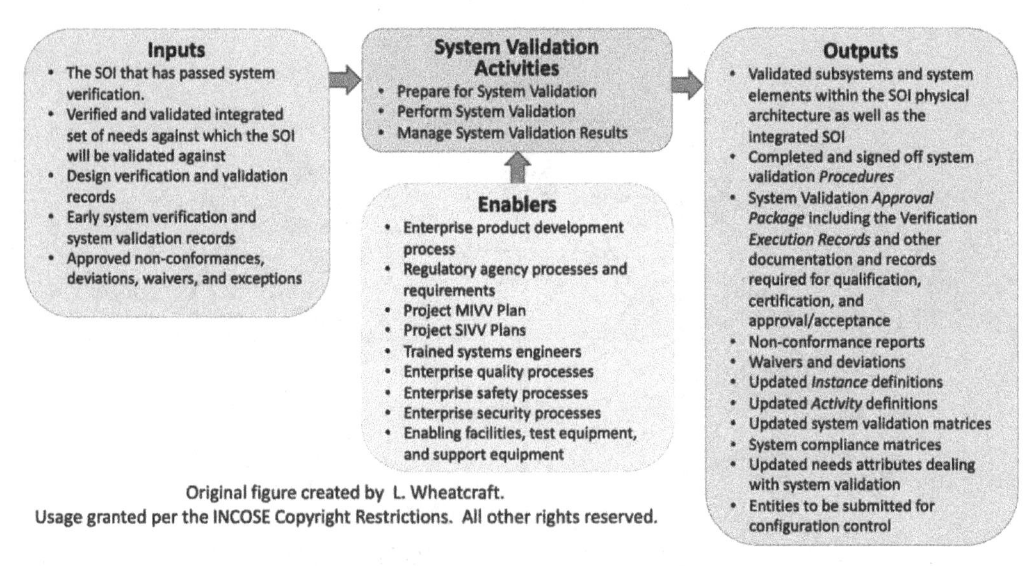

FIGURE 11.5 System Validation IPO Diagram.

In addition, all project system validation activities will need to adhere to the enterprise's Quality, Safety, and Security processes. If the SOI is governed by a regulatory agency, the project will need to conduct all system validation activities in accordance with the requirements in the governing regulations and other guidance documents issued by the regulatory agency. Finally, all needed enabling systems such as facilities, test equipment, and support equipment must be approved for the project's system validation activities.

Inputs to the *System Validation Process* include the SOI that has passed system verification and the verified and validated integrated set of needs against which the SOI will be validated.

As part of the Planning Stage, the validation *Success Criteria*, *Strategy*, and *Method* are defined and included as attributes for each of the needs. Using this information gathered to define each *Instance*.

During the Defining Stage, validation *Instances* are grouped into validation *Activities* and information gathered to define each *Activity*. Requirements for each validation *Procedure* are defined along with other information needed to develop the *Procedure*. The *Procedure* is developed, reviewed (verified and validated), and approved. *Refer to Chapter 10 for information on the artifacts described.*

11.4.3 Perform System Validation Activities—Execution

During the Execution Stage, once the *Procedures* have been approved, they will be scheduled for execution as system validation *Events* within the project's Master Schedule. Each scheduled *Event* represents a commitment of personnel, time, and equipment.

As described previously in Chapter 10, prior to the performance of the *Procedure*(s), there will be a pre-test review to establish readiness to perform the *Procedure*: availability and configuration status of the SOI, the availability of the validation enablers, qualified personnel or operators, and resources. Each *Procedure* should be performed in the operational environment or one as close to it as possible using the intended users; the *Strategy* for each *Instance* should define the operational environment the *Procedure* should be conducted in and which personnel should be involved.

During the performance of the *Procedure*, the required data are gathered, documented, and approved by the organization's QA organization that is monitoring the performance of the steps and actions within the Procedure.

Once a *Procedure* has been completed, often there will be a PTR or similar review to review the status and results of the completed *Procedure* and any discrepancies and non-conformances that occurred. Discrepancies and non-conformances observed during the performance of the *Procedure* are recorded and resolved using the organization's QA process, as discussed earlier in Chapter 10.

The discrepancies and non-conformances could be due to the validation *Strategy*, the enabling systems, execution of the *Procedure* steps and actions, faulty test equipment, faulty input test data, a faulty design, poorly formed needs, or a manufacturing or coding defect. The *Procedure*, or portions of the *Procedure*, may be repeated as needed as part of anomaly resolution.

Once all discrepancies and non-conformances have been addressed, the completed *Procedure* will be signed-off in accordance with the organization's *CM Process*. Any issues or non-conformances will be resolved prior to final sign-off of the completed *Procedure*. *Note: "Signed-off" refers to key project team members to include their signatures certifying that the Procedure was completed, all data collected is accurate, and all discrepancies and non-conformances identified during the execution of the Procedure have been resolved.*

11.4.4 Manage System Validation Results—Recording and Approval

During the Reporting Stage, for each system validation *Instance* covered by a *Procedure*, the results will be recorded in an *Execution Record* and maintained in accordance with organizational policy.

The recorded data will be analyzed against the *Success Criteria* to determine whether the SOI meets its needs with an acceptable degree of confidence. Any discrepancies or non-conformances will be documented within the *Execution Record*.

Problem resolution and any subsequent changes will be managed through the organization's QA and *Project Assessment and Control Processes*. Any changes to the SOI definition artifacts (i.e., integrated set of needs, set of design input requirements, architecture, design, set of design output specifications, or interfaces) and associated engineering artifacts and PM work products are managed within the *Needs, Requirements, Verification, and Validation Management* activities discussed in Chapter 14.

Bidirectional traceability is established between the validation artifacts and the needs against which the SOI is being validated. How this traceability is managed should be defined in project's system integration, system verification, and system validation plans or the SEMP.

During the Approval Stage, the individual *Execution Records* are combined into *Approval Packages* and submitted to *Approval Authority* for qualification, certification, and approval for use in the operational environment.

The project team will assess the success of the system validation activities. The evaluation of system validation results and follow-up corrective action can vary greatly depending on the purpose of the system validation activity, the risk of the baselined integrated sets of needs not being met, and the expected degree of confidence.

For lower-level subsystems or system elements, this could be a simple action to address a failed system element followed by re-validation, or a more significant action such as a major project re-direction based on a failure to attain a key milestone.

As discussed in Chapter 10, a variance may be approved by the *Approval Authority* for system validation failures against non-critical, lower needs. Failed system validation can result in the SOI failing to be qualified, certified, or approved for use in the operational environment by the intended users, which often leads to a wholesale project re-baseline or cancellation.

The project team will provide baselined system validation artifacts and validated entities for CM. The *CM Process* is used to establish and maintain configuration items and baselines.

The end goal is a successfully verified and validated SOI ready to be integrated into the next higher level of the system architecture or, if at the top level, the integrated system will be accepted, qualified, or certified and will be able to be released for use in the operational environment by the intended users.

12

THE USE OF OTS SYSTEM ELEMENTS

This chapter is a further elaboration of the INCOSE SE HB v5, Section 4.3.3, Commercial-off-the-Shelf (COTS)-Based Systems.

When the project team decides to either buy or reuse for a system element, one consideration is using an existing Off-the-Shelf (OTS) system element rather than developing the system element from scratch. This can be a commercially acquired product (COTS) obtained through procurement efforts, or an existing system element previously developed for use in other projects within the organization (such as a power supply, a display, a microprocessor, or an existing software application). In some cases, an "as-is" OTS system element may not be able to be used without minor modifications, resulting in a product referred to as Modified-off-the-Shelf (MOTS).

OTS system elements are produced to satisfy a capability needed within a specific market. These products are typically designed and built/coded to the supplier organization's lifecycle concepts, needs, and requirements (CNR), instead of being developed to meet the project's specific lifecycle CNR.

There are several reasons why an OTS system element may be an appropriate choice for use in a new development project:

- OTS system elements have a heritage of use and have been approved for a specific use in a specific operational environment.
- OTS system elements are often available sooner as compared to developing a custom system element from scratch.
- OTS system elements can be cheaper without the overhead associated with developing a custom system element (either internally or externally).
- OTS system elements are often built or coded to a standard interface (plug and play) based on an open architecture concept.

INCOSE Needs and Requirements Manual: Needs, Requirements, Verification, Validation Across the Lifecycle, First Edition. Louis S. Wheatcraft, Michael J. Ryan, and Tami Edner Katz.

Determination for OTS usage is made as the project moves down the levels of the physical architecture, as depicted in Figure 12.1, as part of the *Architecture Definition Process* and the *Design Definition Process*.

Most often the determination is made as a design decision at the lower levels of the architecture in response to design input requirements; however, the decision to use an OTS system element may be made by higher-level management stakeholders or the customer. In this case, the requirement to use the OTS system element is considered a constraint to which the project team must comply.

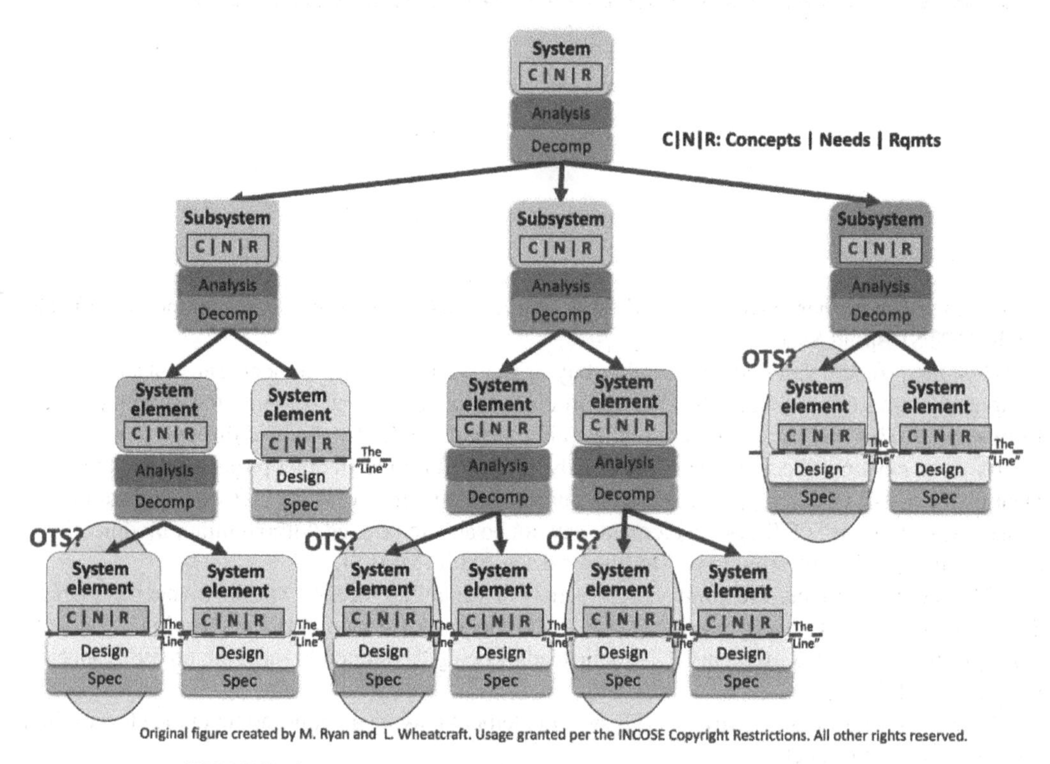

FIGURE 12.1 OTS Determination as Part of Architecture and Design.

As the project team moves down the levels of the physical architecture, the subsystems and system elements that make up the System of Interest (SOI) are decomposed through the various levels as discussed in Section 6.4. For each subsystem and system element in the physical architecture, design concepts are defined as part of the lifecycle concepts analysis and maturation activities in response to Mission, Goals, and Objectives (MGOs), measures, drivers and constraints, and risk for that system, subsystem, or system element (discussed in Chapter 4). As part of the lifecycle concepts analysis and maturation activities, the project team determines whether to buy, make, or reuse each of the subsystems and system elements in the SOI architecture.

For the cases when the buy or reuse determination is being considered, the project team evaluates existing OTS system elements (hardware, software, and assemblies) as part of their architecture and design concept definition activities. (*Note: The "line" shown in Figure* 12.1 *is when the project team has decided the system element needs no further decomposition or elaboration and they are ready to make the buy, make, or reuse determination.)*

As part of this determination, the acquisition or reuse of an OTS system element is traded against developing (building or coding) a custom system element (either internally or by an external supplier). Key considerations include:

- Can the project realize a reduction in cost and schedule (with acceptable risk) by using an OTS system element with required technical functionality, performance, quality, and compliance compared to a custom development effort?
- Can the OTS system element fulfill the needs and requirements for the intended use in the intended operational environment?

Choosing to use an OTS system element comes with some challenges; notably, there is still often an expectation of compliance to government regulations and standards applicable for the system being developed. This can be difficult considering the suppliers of an OTS system element may not provide the documentation required by the customer, and the supplier may not be willing to provide specific details of their product development and associated artifacts they consider proprietary. As an alternative, it is a customary practice for suppliers to develop the subsystems and system elements to industry standards and government regulations and supply the customer with a certificate of conformance or compliance (certification and qualification) against those standards and regulations rather than providing detailed Systems Engineering (SE) artifacts.

It is also common for suppliers to provide an "as-built specification," which states the physical characteristics, interface definitions for the interface boundaries and characteristics of the interactions across those boundaries, material listings including any hazardous materials, and operational environment (transportation, storage, and use) the OTS subsystem or system element was designed to operate within.

A major challenge for OTS system elements with embedded software, such as Electronic Control Units (ECUs), is that the customer has limited insight into the actual embedded software; while it may do what is specifically required, what else does it do that potentially could result in unintended behaviors?

The customer must address behaviors when the OTS system element is integrated within the SOI, including OTS performance in the operational environment, interface boundaries and interactions across those boundaries, and specific use for the OTS system element when integrated into the SOI. Conversely, the supplier often has little insight into how their OTS system element will be integrated into the customer's system, any modifications that may be made to result in MOTS, and any resulting behavior as a function of the interactions with other supplier-developed system elements.

OTS system elements are developed in response to the supplier-defined set of needs and set of design input requirements rather than specific needs and design input requirements defined for the SOI. These challenges represent risk when addressing system verification and validation of the OTS system element as well as the overall SOI.

When determining the needs and requirements for the OTS system element, the project team takes an external view of the OTS with a focus on form, fit, function, quality, and compliance addressing the following questions:

- What functionality and level of performance are needed when the OTS is integrated into the SOI architecture and is operating within the SOI's operational environment?
- What are the allowable induced environments the OTS system element is allowed to generate that could impact the performance of other system elements within the system architecture?
- Are the OTS system element interfaces compatible with the SOI it will be integrated within?

- What are the interface boundaries, where are they defined, and what are the characteristics of each interaction across those boundaries?
- What are the allowable envelope, mass, power requirements, and other physical attribute requirements for the OTS system element that enable it to be integrated into the system?
- What quality (-ilities) requirements need to be met by the OTS system element?
- What standards and regulatory requirements does the OTS system element need to comply?
- What supplier data and documentation are available to help address the above-mentioned questions and identify candidate OTS system elements for evaluation?

When addressing these questions, the project team must be careful to not over-specify the needs and requirements for the OTS system element, doing so will make it more difficult to find an acceptable OTS system element. It is also important to address specific rationale, priority, and criticality for each need and requirement.

Because the OTS candidates were not developed to meet the project's specific needs and requirements, not all needs and requirements defined for the OTS will be able to be met. Understanding the rationale, priority, and criticality of all needs and requirements allows the project team to make tradeoffs during their evaluation and selection process.

A framework [63] for evaluating the use of OTS subsystems and system elements is shown in Figure 12.2. This framework reconciles the needs and requirements for the SOI with the data available for the candidate OTS system elements from the suppliers.

For this proposed framework, the project team oversees the evaluation process, which includes the project manager, procurement personnel, systems engineers, design engineers, and subject matter experts. This framework involves assessing the needs and requirements for the OTS system element, the risk, budget, and schedule status of the project, and a set of project-developed criteria to assess the OTS system element capability to be used within the SOI.

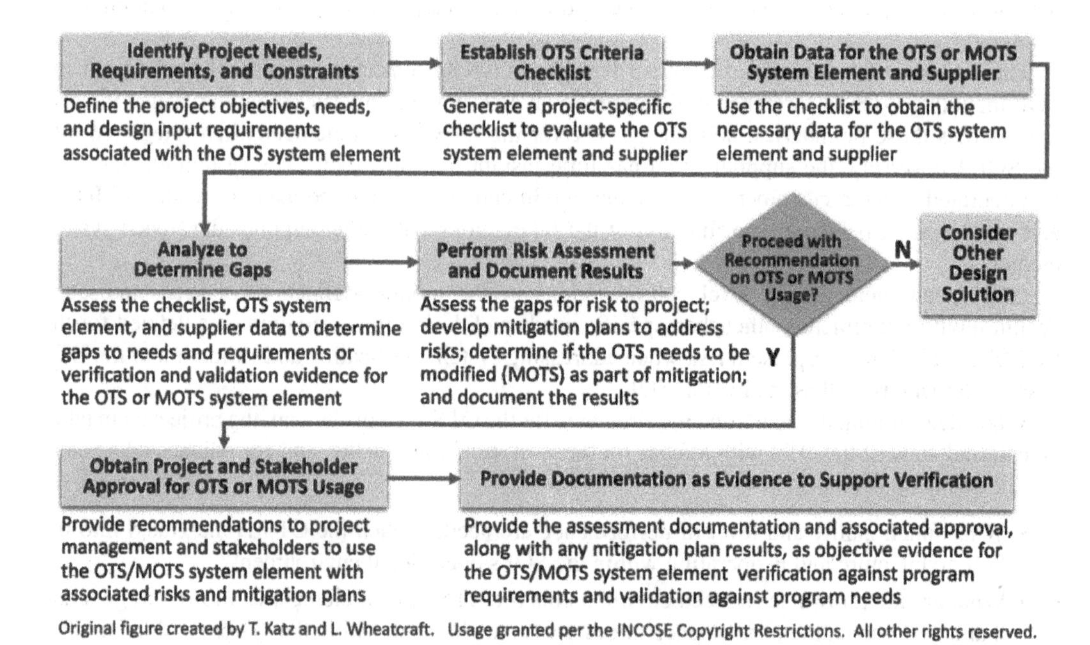

FIGURE 12.2 OTS Evaluation for Usage [63].

During this assessment, the OTS is compared against the needs and requirements from the project, and any gaps are identified. A risk assessment is performed to determine if the OTS is suitable for use in the SOI and if any modifications may be required to ensure acceptable risk (resulting in MOTS). If the OTS is unable to be accepted for use in any form, the project team moves to another design solution. If the OTS or MOTS is acceptable, then the results of the assessment are recorded, project and stakeholder approvals are obtained, and the evaluation records are used as objective evidence for system element verification and validation.

As part of this assessment, the project team will also need to address impacts to system verification and validation with the OTS or MOTS system element integrated within the SOI architecture:

- Have all interfaces and interactions been identified and assessed?
- Has the risk of any embedded software been identified and mitigated?
- Has the risk of any emergent behaviors been identified and mitigated?
- Has the operations of the OTS or MOTS in the integrated SOI in the operational environment been analyzed?
- Have system verification and validation approaches been captured to ensure SOI with OTS/MOTS can be verified to meet the design input requirements and validated to meet is integrated set of needs?

During the evaluation, it is critical to assess the overall cost. Data for costs associated with purchase of OTS system element and custom assemblies are available through various records; however, the hidden costs associated with OTS usage also need to be addressed. For example, the need to modify the OTS system element may offset any savings associated with using the OTS or MOTS system element. There have been cases where the costs and time associated with using a MOTS system element turned out to be greater than if the project would have designed and built its own custom system element; a prime example is the reuse of the US Space Shuttle Orbiter engines in subsequent launch vehicles.

Often a project will use the term "heritage" to refer to an existing system element that has a proven performance record. No matter the term used, heritage system elements are really OTS, and the above-mentioned discussion applies. The use of heritage OTS system elements frequently ends up driving system-level requirements.

When evaluating the use of any OTS or MOTS system element (hardware or software), it is important to assess the Technology Readiness Level (TRL) of the critical technologies used by that element. TRLs (as discussed in Chapter 4) consider the specific use in a specific operational environment. Thus, an OTS or MOTS system element that was designed and used for a specific purpose and operational environment is often assumed to have a TRL of 9; however, when being considered for use in a different manner or in a different operational environment, even if only slightly different, the TRL drops to at least TRL 5 until proven differently for the specific intended use and operational environment for which it is being considered. Advancing from TRL 5 to higher levels represents budget, schedule, and resource considerations that must be considered as part of the cost and risk evaluation. *In some cases, the cost and time associated with the TRL maturation activities pose a degree of difficulty that may counter any savings associated with using an OTS or MOTS system element.*

During the evaluation process of possible OTS or MOTS usage, all costs associated with assessments and risk mitigation plans (including design changes, additional testing, and TRL maturation activities) need to be included in the trade-off evaluation. OTS or MOTS usage becomes less favorable when the actions of mitigating the risks outweigh the cost of developing a custom system element.

While the steps shown in Figure 12.2 are basic, the true value of an OTS evaluation framework is in the process of establishing a checklist of criteria for the project that is optimized based on its parameters of concern. Going through the criteria development activity at least once at the project level will create an assessment approach that can be used for all OTS system elements evaluated for that project, addressing specific concerns of the project and its stakeholders. The result is an assessment approach for OTS or MOTS system element usage in a way that provides objective evidence that it meets the intent or is equivalent to the needs and requirements defined by the project team for the system element.

Note: For project needs and requirements for a system element not specifically met by the OTS or MOTS under consideration, the project team may need to get a variance approved by the Approving Authority prior to approval for use.

13

SUPPLIER-DEVELOPED SOI

This section provides considerations when a customer contracts out portions of an System of Interest (SOI's) development, production, system verification, or system validation activities to an external organization which may be referred to as a contractor, supplier, or vendor—for the rest of this discussion, the term "supplier" is used. *Refer also to the INCOSE SE HB v5 Section 2.3.2, Agreement Processes, Section 2.3.2.1, Acquisition Process, and Section 2.3.2.2, Supply Process.*

A primary goal of this Manual is to provide guidance and best practices that will help the customer avoid accumulating technical debt and enable projects and suppliers to deliver a winning product. For contracted system development, one of the best ways to reduce the risk of a system failing system validation (no matter who is responsible for system validation) is to ensure the processes and associated deliverables defined in this Manual are reflected in the supplier contracts.

13.1 CUSTOMER/SUPPLIER RELATIONSHIPS

All projects have customers, where a customer is defined as an organization or individual that has requested a work product. From a customer's perspective, the organization responsible for developing a work product may be either internal or external. Conversely, from a developer perspective, the customers may be internal or external to their organization.

The customer/supplier relationship will be formal in the form of a contract, Supplier Agreement (SA), or Purchase Order (PO). Contracts with a supplier for specific activities and deliverables often include a well-developed Statement of Work (SOW) that clearly defines the relationship between the customer and supplier and requirements for specific activities and deliverables.

For some organizations, the SA is a combination of contract items as well as an SOW, where the SOW is an appendix or attachment. The customer may also have a general contract with a supplier and issue a Task Order (TO) for specific tasks that are authorized by the contract.

Contracts for SOI development efforts will have a clear start date and delivery date with one or more deliverables. From a system verification and validation perspective, the customer's project plan, acquisition plan, System Engineering Management Plan (SEMP), Master Integration, Verification, and Validation (MIVV) Plan, and System Integration, Verification, and Validation (SIVV) Plans

INCOSE Needs and Requirements Manual: Needs, Requirements, Verification, Validation Across the Lifecycle,
First Edition. Louis S. Wheatcraft, Michael J. Ryan, and Tami Edner Katz.

will be used to develop an SOW that includes specific requirements on the supplier for the specific activities and deliverables expected from the supplier.

A key question for both the supplier and their customers is what their relationship across the lifecycle activities for the SOI that is being contracted for. The supplier's relationship with their external customer is dependent on contract type and approach. The contract type and approach could be:

- **Fixed Price (FP),** where the supplier is paid a fixed amount, no matter the supplier's expenses to provide a system. Fixed-price contracts are used to control costs and risks to the customer. The customer has no oversight of the supplier during the development of the SOI to be supplied.
- **Cost-Reimbursement (CR)** (sometimes referred to as cost-plus), where the customer pays the supplier for allowed expenses plus additional amount to allow for a profit.
- **Performance-based (PB)** [sometimes referred to as completion form (CF)] is an approach where the focus is on well-defined measures and outcomes rather than how the supplier does the work to produce the outcomes. With this method the customer may be provided some insight into the supplier activities as defined in the SOW, but no oversight (control) of those activities.
- **Level of Effort (LOE)/Term** is an approach to contracting that is a type of CR contract where supplier is paid for a specified LOE (time and materials) over a stated period on work that is stated in general terms. This is sometimes referred to as a "Term" contract as is has a defined start and end date specified in the contract. As contracted with PB contracts, the customer has both insight and oversight of the supplier activities.

As shown in Figure 13.1, the above-mentioned types and approaches can be combined in several ways. The following sections discuss these combinations in terms of verification and validation across the lifecycle of a supplier-provided SOI.

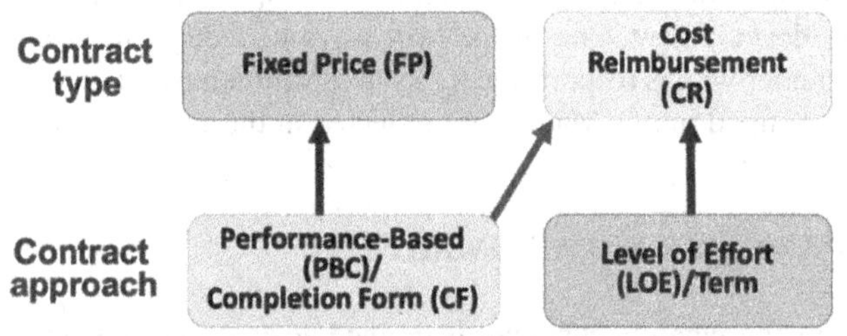

FIGURE 13.1 Contract Type and Approach.

13.1.1 Fixed-Price, Performance-Based Contract

For a FP, PB contract, there will typically be little interaction between the customer and supplier during the development of the SOI. The contracting mechanism is frequently a PO, and the SOI is frequently Off-the-Shelf (OTS) or Modified Off-the-Shelf (MOTS). The contractual obligation for the supplier is to deliver the SOI along with objective evidence it addresses what is *necessary for acceptance* and other deliverables as specified in the PO.

With this type of contract, the supplier's system verification and validation programs are what is normally followed. The form of the objective evidence that the SOI passed the supplier's system verification that the SOI meets the customer's requirements must be specified in the PO. For example, the *Approval Package* discussed in Chapter 10 would be included in the list of deliverables along with a description of its contents and organization.

In this case, often, the customer does their own system validation against their own set of needs in the actual operational environment with the intended users, and the supplier is not involved. This is especially true when the SOI developed by the supplier is a subsystem or system element, which the customer will be integrating with other subsystems and system elements. (*See also Section 13.2.*)

13.1.2 Cost-Reimbursement, Performance-Based Contract

For a CR, PB contract, there will be more interaction from the standpoint of the supplier providing more "insight" to the customer into the development lifecycle activities as defined in the contract. This interaction could be the supplier involving the customer in the development efforts of various artifacts and work products during the concept, development, production, integration, system verification, and system validation lifecycle activities. The integrated set of needs, design input requirements, architecture and design, design output specifications, production verification artifacts, system verification artifacts, and system validation artifacts could all be deliverables as specified in the contract.

Also included in the contract could be the customer's participation in major gate reviews. The customer may define specific deliverables in the SOW or SA to give them the insight they need, as well as may define key inspection points during production to ensure quality of the workmanship and as part of their participation in the supplier's production verification activities.

While the customer may participate in these activities, for performance-based contracting, they do not normally have an approval role. To do so would transfer the risk of noncompliance, cost overruns, and schedule slips from the supplier to the customer.

With this type of contract, the supplier's system verification and validation programs are what is normally followed. The form of the objective evidence that the SOI passed the supplier's system verification that the SOI meets the customer's requirements must be specified in the PO. For example, the *Approval Package* discussed in Chapter 10 would be included in the list of deliverables along with a description of its contents and organization.

Again, in this case, often the customer does their own system validation against their own set of needs in the actual operational environment with the intended users, especially when the SOI developed by the supplier is a system, subsystem, or system element, which the customer will be integrating with other subsystems and system elements. (*See also Section 13.2.*)

13.1.3 Cost-Reimbursement, Level of Effort/Term Contract

For a CR, LOE/Term contract, the customer is providing "oversight" to the supplier and will be involved in the development and approval of all SE artifacts developed across the SOI's development (for example, the integrated set of needs, set of design input requirements, architecture, design, system verification artifacts, and system validation artifacts).

The gate reviews result in customer approval of the intermediate work products and the realized SOI. Because of the customer's approval role, they are accepting the risk of noncompliance, cost overruns, and schedule slips. With this type of contract, the customer's system verification and validation programs are what is normally followed.

As discussed in more detail later in this section, for highly regulated systems, the customer may be responsible for the development of the *Approval Package* discussed in Chapter 10 that will be supplied to the regulatory *Approval Authority* for final approval for the system's intended use. The collection of the data and information to be included within the *Approval Package* is often specified in the contract list of deliverables and associated activities specified in the SOW.

13.1.4 Contracting for Specific Lifecycle Activities

One aspect of contracting is identification of which part of the development lifecycle activities the supplier is being contracted to support. From a lifecycle perspective, the customer must decide on specific portions of the development or production lifecycle they will issue a contract to a supplier. In some cases, the customer may develop a set of operational-level needs (sometimes referred to as Statement of Objectives [SOO]), and issue a contract to a supplier to develop system-level lifecycle concepts, integrated set needs, and a set of design input requirements, develop an architecture and design, produce a set of design output specifications, and produce the system element. In other cases, the design and/or manufacturing/coding may be done by separate suppliers.

From an architectural perspective, a customer may contract out the development of the entire SOI or a specific part or component. The supplier, in turn, may subcontract out system elements, parts, or components (creating multiple tiers of contracts). Some system elements could be OTS or MOTS, and some produced from a set of design output specifications levied in the contract. In cases like this, the customer/supplier relationships could be complex, especially if there are multiple suppliers and some of the suppliers are customers of their own lower-level suppliers.

13.1.5 Customer Versus Supplier Roles

A key area of consideration concerns identifying the organization responsible for integration of all SOI parts across all the lifecycle stages. Roles need to be defined as part of the contract concerning responsibilities such as:

- Who defines and manages the interface definitions?
- Who manages the flow down, allocation, and budgeting of requirements from subsystems and system elements at one level of the architecture to the system elements at the next level?
- Who defines the acceptance criteria for each subsystem or system element before it is accepted for system integration?
- Who is responsible for the system integration of all the subsystems and system elements that make up the integrated system?
- Who pays for the development and or use of special facilities, tests, or support equipment needed to do system integration, system verification, and system validation?
- What is the supplier versus customer role during system integration, system verification, and system validation?

No matter the contracting type and approach used, the customer must decide who is responsible for verification and validation activities across the lifecycle and clearly document the roles, responsibilities, and deliverables for each of these activities in the contract for each supplier.

- Who decides at which levels of the SOI architecture formal system verification and validation be conducted versus less formal (less documentation)? *(These decisions are cost, schedule, and risk decisions the customer may not want to delegate to a supplier.)*
- Who is the *Responsible Organization* for the system verification and deliverables versus system validation and deliverables?
- Who is the verification *Approval Authority*?
- Who is the validation *Approval Authority*?
- Who is responsible for preparing, managing, and submitting all records and documentation (*Approval Packages*) required by the *Approval Authority*?

All these decisions must be addressed in the supplier's contract, SOW, PO, or SA. In some cases, the *Responsible Organization* for system verification could be different from the *Responsible Organization* for system validation, as is required for many government-contracted system development projects. In many government- contracted systems, the customer is required to contract with a third party to undertake "independent verification and validation." *Refer to Section 13.2 for some challenges concerning who does system validation.*

Another major issue is whether a verified and validated integrated set of needs has been defined by the customer as defined in Chapter 4 of this Manual. A key tenet of the approach advocated in this Manual is that validation is a continuous activity across all lifecycle stages. Sadly, in many cases, there will not be a baselined set of integrated set of needs to validate the development artifacts and the realized SOI against. When this is the case, what is expected of the supplier in terms of continuous validation across the lifecycle and what are they expected to validate the artifacts generated across the lifecycle against?

Once the contracted SOI has been built and completed, the *Execution Records* will need to be prepared, assembled into *Approval Packages*, and submitted to the appropriate *Approval Authority*.

The requirements for the *Execution Records* and other documentation used by the *Approval Authority* vary across different companies, industries, and regulatory agencies. However, the end goal is the same: provide objective evidence that the SOI meets its set of design input requirements, set of design output specifications, and the integrated set of needs with the desired level of confidence. Without the required objective evidence, the *Approval Authority* will be unable to make this determination. It is, therefore, critical for the customer to articulate in the supplier contract what objective evidence is needed regarding production verification, system verification, and system validation, i.e., they must clearly define what is necessary for acceptance in the contract.

As part of defining the contract requirements for activities and deliverables, the customer must develop a plan concerning what activities the supplier is responsible for (or at least participates in) and the deliverables needed for qualification, certification, and acceptance. The customer must define their next steps once the supplier-delivered SOI has been qualified or certified and accepted by the customer, addressing factors such as

- Is the delivered SOI a system, subsystem, or system element that is going to be integrated into another system or SoS? If so, by whom?
- Is the *Approval Authority* in-house or a regulatory agency?
- For each case, what data, information, and *Execution Records* will the *Approval Authority* need from the supplier to be used as objective evidence that the system has passed system verification? System Validation?

Without this information, the system may not be approved for its intended use in the operational environment.

13.1.6 Consequences Resulting from Poorly Written Contracts

Customers do not always adequately define their requirements concerning system integration, system verification, and system validation in the contract, especially in terms of the various issues discussed above. As discussed in Chapter 2, failing to address these requirements in the contract can result in significant technical debt. There are many consequences that can result from poorly written contracts because of these missing requirements:

- The customer will lack insight into what will be involved in terms of cost, schedule, and resources needed to verify and validate the SOI.

- The customer will not have the opportunity to review and buy in as to the suppliers' system verification and validation programs as well as *Success Criteria*, *Strategy*, and *Method*, for each need and requirement.
- The customer frequently will be surprised by the actual cost and schedule needed to meet their undocumented requirements for system integration, system verification, and system validation.
- When issues occur during system integration, system verification, and system validation, the customer may not be kept apprised of these issues and the resulting impacts on cost, schedule, and risk.
- The customer's unstated requirements for quality and functionality may not be realized.
- What was expected by the customer and what was delivered by the supplier may be different. Perfection may not be achievable and there may be emerging properties (good and bad) that may not be discovered until after the SOI is approved for use and put into operations or in the market.
- When the supplier delivers the SOI, there may be a set of liens (technical debt) or open discrepancies accompanying the delivered SOI with requests from the supplier seeking waivers or deviations. If these are not approved, what will be the cost and schedule impacts?
- If not clearly specified in the contract as deliverables, the supplier's records and artifacts generated as part of their system verification and validation activities, and specifically during the execution of the system verification and validation procedures, may not be available, or their quality may not meet the customer's needs.
- Without adequate records, the customer may not have the objective evidence needed to be submitted to an external *Approval Authority* that can be used to determine the system meets its integrated set of needs, set of design input requirements, and set of design output specifications. This is often the case for suppliers who develop highly regulated products, such as medical devices; without adequate objective evidence, the regulatory agency may not approve the device for use.
- If not clearly specified in the contract as deliverables but later required by the customer, the resulting contract changes could be expensive and time consuming.
- If not clearly specified in the contract as deliverables, data and information gathered by the supplier as part of system verification and validation will not be available for use by the customer for sustaining engineering. They will not be able to use this data for analysis if there are failures of the SOI during later system verification and validation activities performed by the customer during integration, during operations, or after the system is made available to the public.
- If the supplier's roles concerning system verification and validation and associated deliverables are not clearly specified in the contract, suppliers operating on a very thin margin may be pressured to take liberal interpretation of any vagueness in the system verification and validation portions of the contract. Lack of agreement of the supplier's roles concerning system verification and validation and associated deliverables may also bankrupt the supplier if court action determines the supplier is responsible; this may be a short-term win for the customer, but a new supplier would then need to be sought at additional and unplanned expense and time before the SOI could be approved for use. If court action is in favor of the supplier, the result would also be additional and unplanned expenses and schedule slips for the customer.

To avoid these consequences, each of these issues must be addressed in the supplier contract, SOW, or SA.

Below is a case study that illustrates some of these issues.

Hubble Space Telescope Project—System Verification and Validation Lessons [64]

The Hubble Space Telescope project was originally funded for approximately $400 million with an anticipated launch date of 1983. However, it was not actually launched until 1990 with a cost of approximately $4.7 billion and was estimated to cost approximately $10 billion in the first 20 years of its life. With such a history, the project provides a rich source of PM case studies, including a particularly telling lesson regarding system verification and system validation.

The most critical component in the Hubble Space Telescope was a 2.4 m glass parabolic mirror. Only after the telescope had been launched into space in 1990 was it discovered that the telescope could not capture a clear image due to errors made in manufacturing the mirror, which had been polished with an aberration of some 2.2×10^{-6} m. The problem was able to be fixed with a corrective optics package that displaced the High-Speed Photometer (HSP)—a solution that cost many millions of dollars and required an unscheduled Space Shuttle mission in 1993.

The building of the telescope's mirror had been competitively contracted, with the successful supplier required to subcontract for a second backup mirror to be built by a losing bidder. The error in the launched mirror would have been discovered had each of the two companies been tasked to test the mirror of the other instead of relying on a single flawed test of the primary contractor's mirror. Late in the project, it had been too difficult to address the associated commercial issues, particularly since the project was already faced with significant cost and schedule pressure. Yet, had those managing the project been able to insist that this testing was to occur, or the requirement had been mandated by the original manufacturing contracts, the initial error would have been discovered before launch and then corrected by launching with the backup mirror.

When seeking corrective solutions to the defective mirror, it was discovered that the backup mirror was perfectly shaped. If the problem with the mirror had been detected during production or system verification, the backup mirror—which met the specifications—could have been substituted. Instead, a corrective optics package was developed and installed on the telescope during a service mission, which required transporting a new Corrective Optics Space Telescope Axial Replacement (COSTAR) system and installation during spacewalks by astronauts.

Lessons learned concerning verification and validation across the lifecycle would have avoided tens of millions of dollars in additional expenditure, several years of project delay, and significant loss of face for all responsible.

Source: Adapted from Smith [64].

From this case study, the following lessons are apparent:

- If more attention were placed on production verification (was the mirror built to its design output specifications), as discussed in Chapter 9, the manufacturing error with the primary mirror would have been discovered earlier in the project, before system integration, allowing the backup mirror to be installed before formal system verification and validation activities begin.
- When the customer developed the contract, provisions should have been made for a customer or other customer-delegated representative to participate in the production of the mirror, with mandatory inspection points during manufacturing to ensure the mirror was "built to spec" before system integration.
- Provisions should have been made in the contract for a pre-integration readiness review where the results of production verification activities were presented.

- After integration, specific system verification and validation activities should have been performed on the integrated system to check image clarity.
- Provisions within a contract must address verification and validation across the system lifecycle and not just at the end of development during system integration when the discovery and resolution of problems cause significant increases in cost and schedule slippages.
- Failing to learn from these lessons can result in problem discoveries after the system is put into operation, increasing the impact even more, as was demonstrated by the Hubble Space Telescope Project.

13.2 CUSTOMER/SUPPLIER VERIFICATION VERSUS VALIDATION CONSIDERATIONS

When a system is being developed by a supplier in response to a contract, some organizations may feel that they are only responsible for system verification and that the customer is responsible for system validation [16]. To help avoid issues concerning which organization is responsible for system validation, the customer should address the following considerations that are elaborated in the sections below.

13.2.1 Customer Is Responsible for System Validation

The SOI is contracted to a supplier for development according to the customer's design input requirements or production in accordance with customer developed design output specifications, but actual system validation is done by the customer or contracted out to a third party. In this case, the primary legal responsibility for the supplier is providing objective evidence that the contracted SOI meets the customer's design input requirements and/or design output specifications as defined in the contract.

In this arrangement, the SOI may have passed system verification and/or production verification activities performed by the supplier but fails during customer system validation. The customer determines whether their integrated set of needs, set of design input requirements, or set of design output specifications is defective and then makes a contract change to their needs, requirements, or specifications at their cost.

Often, for a supplier-developed SOI, which the customer will be integrating with other subsystems and system elements, final system validation of the supplier-developed SOI cannot be completed until it is part of the overall integrated system.

As discussed in Chapter 10, there are cases where the customer specifies in the contract that the supplier does some type of preliminary system validation in a simulated operational environment by surrogate users, and the customer will then undertake final system validation in the actual operational environment by the intended users once the supplier SOI is integrated into the macro system of which it is a part. In this case, the customer will have had to clearly define what the supplier will validate their SOI against (operational scenarios, use cases, and integrated set of needs) in the contract. This preliminary validation is done as a risk reduction effort.

For this type of contractual arrangement, the customer has accepted all the risk of an SOI failing final system validation, even if final system validation is contracted to a third party. If the system does not meet the customer's needs and does not meet its intended use in the intended operational environment when operated by the intended users, the impacts range from loss of reputation for the customer to program cancelation. Legal action is another possible outcome. As a minimum, often large sums of money and time will be spent fixing problems.

13.2.2 Supplier Is Responsible for System Validation

There are cases where, in addition to the supplier being responsible for system verification, the contract is written such that the supplier is responsible for final system validation. This approach only makes sense if the supplier can do system validation of their developed SOI in the actual operational environment by the intended users. *Note: The operational environment includes the actual external systems the SOI interacts with.*

For this case, the customer must clearly define, in the contract, what is *necessary for acceptance* in terms of the SOIs intended use in the operational environment when operated by the intended users. Using the approach advocated in this Manual, what is *necessary for acceptance* is not only the design input requirements the SOI will be verified against but also the set of operational-level needs and system-level set of needs the system will be validated against.

Contractually, the understanding is that when the supplier provides the customer objective evidence that both the design input requirements and the integrated set of needs are met, then the supplier-delivered SOI will be accepted by the customer and approved for use.

However, there is an issue with this approach as to who is accepting the risk if the system passes system verification but fails system validation. If the set of design input requirements provided by the customer is defective (not passing requirement validation as addressed in Chapter 7), then even if the system can be verified to meet these requirements, it could still fail system validation. Because of this, the customer has still accepted the risk of the SOI failing system validation even if system validation is performed by the supplier for the customer.

For the supplier to assume the risk associated with a system failing to pass system validation, the customer must provide the supplier with a well-formed integrated set of needs for the SOI being developed that have successfully passed need verification and need validation, as discussed in Chapter 5.

Assuming this is the case, the supplier would transform the integrated set of needs into a well-formed set of design input requirements, performing the activities associated with SOI development (Chapters 7–11) based upon these requirements. The supplier will also perform system validation against the customer's integrated set of needs. The result will be *Approval Packages* that would be provided to the customer for acceptance of the integrated system.

In this case, any issues associated with not meeting the customer-defined integrated set of needs would be the responsibility of the supplier.

Once the system is accepted by the customer, the system can be approved for its intended use in its operational environment by its intended users. If there are any issues with the system after that (for example, the integrated set of needs was defective), the issue resolution is the responsibility of the customer.

As stated before, an *Approval Package* is required to be submitted to a regulatory agency *Approval Authority*, the contract must include the deliverables necessary so that the *Approval Package* can be developed per the *Approval Authority* requirements.

13.2.3 Addressing the Amount of Supplier-Performed System Validation

Another issue concerns addressing how much system validation should be done by the SOI supplier.

For supplier-developed systems, some organizations tend to rely more on system verification than system validation, expending more time, money, and resources on system verification activities than on system validation. In this case, there is a question on the need for system validation, or how much system validation is needed. The answer to this question is really an issue of risk.

This issue should be addressed in the supplier contract. The need for system validation should be tied to the risk of negative consequences if the system does not meet the customer's needs in their operational environment by the intended users.

The basic approach to determine how much system validation is needed is to perform an analysis of the risk involved. A minimal amount of validation is probably appropriate when the supplier is developing a known SOI with existing technologies to be used in a known operational environment by known users and the customer has a good history of collaborating with that supplier.

However, a greater emphasis on system validation is warranted when contracting with a new supplier or when a supplier is developing a new system with new technologies to be used in an unfamiliar operational environment by intended users with whom the supplier is not familiar. In these situations, the probability of an SOI failing system validation is greater.

Having considered risk, customers can take action to collaborate with the suppliers to reduce the probability or magnitude of a system failing system validation even though the system has been verified to meet the design input requirements and/or the design output specifications.

13.2.4 Expecting System Validation Only

There is a trend, especially for standalone software applications developed using an Agile approach to product development, where the focus is on system validation rather than on system verification.

The logic used is that if the supplier can provide objective evidence that the SOI meets its intended use in its intended operational environment when operated by the intended users, and the customer is willing to accept that evidence, then the SOI is considered accepted for use. This negates any need for formally developing and managing a set of design input requirements and verifying the SOI meets those requirements. The proponents of this idea believe that it is not practical to try and develop a complete set of requirements at the beginning of a project, in that as they get deeper into the design, existing requirements will change, and new requirements will emerge. To address this issue, the Agile team is in frequent contact with the customer and users, continuously validating their work products with the customer and users.

While this approach is not suitable for all product types and domains, it may be the most effective way to engineer an SOI at reduced cost and a shorter development schedule when the approach is suitable. The primary cost and schedule savings are associated with informal definition and management of requirements rather than the more formal traditional definition and management of requirements and associated design and system verification activities defined in this Manual.

Using this paradigm, customers would focus on the lifecycle concepts and needs definition activities defined in Chapter 4 to ensure they have a well-formed integrated set of needs for the supplier to validate against.

If the SOI is being developed internally, the project team would elicit stakeholder needs, identify drivers and constraints and risk and mature the lifecycle concepts concurrently with the architecture and design definition activities. While there may be requirements associated with the needs, they would be communicated and recorded informally within the development team as appropriate to the level of communication needed by the various project team.

If development is being contracted out to a supplier, the customer would provide the supplier with the well-formed integrated set of needs against which the system would be validated against for acceptance for use. The customer/supplier roles concerning system validation would be defined within the contract as well as any other required contract deliverables.

This approach is similar to how a customer would select an OTS or MOTS SOI, as discussed in Chapter 12. The SOI would be described from the customer's external perspective of the SOI, communicating the customer needs for the SOI concerning the intended use, operational environment, constraints, quality, compliance, and risk. The supplier would be responsible for validating the delivered SOI meets the customer's integrated set of needs—how the supplier met those needs would be transparent to the customer. (*Note: This is how many consumer products are developed today.*)

13.3 ADDRESSING THE EVOLUTIONARY NATURE OF INTERFACE DEFINITIONS

As discussed in Chapter 6, the definition of interactions across interface boundaries evolves as the design matures for developing systems. In cases where development is contracted out to a supplier, the design input requirements provided to a supplier will include interface requirements dealing with what is crossing the interface boundary as part of an interaction. However, what the systems look like at the interface boundary and the media involved in the interaction are design dependent. As a result, these definitions may be To Be Determined (TBD) or To Be Resolved (TBR) at the time the contract is agreed to. Thus, the corresponding interface requirements will include a pointer to a TBD definition.

It is critical that the customer clearly define in the contract the process for dealing with the changes associated with these TBDs or TBRs as the design matures. In some cases, the supplier will be required to participate in the customer or integrating organization's Interface Control Working Group (ICWG) such that they can participate in decisions concerning the definitions, and assessing any impacts that may result from the specific definitions, especially the cost associated with one definition versus another.

For example, for the exchange of data, what preprocessing of the data is required? Should the sender of the data process the data in some way before sending the data across the interface boundary or is sending the raw, unprocessed data acceptable? Depending on the decision, costs will be incurred for the developer of either the sending or receiving system.

Additionally, consideration is needed regarding who is responsible for developing or supplying the media involved in the interactions across the interface boundary (such as the actual cables, pipes, or communication buses). Going even into more detail, who is responsible for the actual connectors? Experience has shown that it is risky for the suppliers involved to procure their own connectors independently from the other suppliers. While they may procure the same part number, often there are variants (dash numbers), and the variants may not be compatible. It is problematic to discover these incompatibilities during system integration (as is often the case).

14

NEEDS, REQUIREMENTS, VERIFICATION, AND VALIDATION MANAGEMENT

Needs, Requirements, Verification, and Validation Management (NRVVM) separates the "management" of Needs, Requirements, Verification, and Validation (NRVV) processes and data from the "execution" of NRVV activities.

NRVVM is a series of crosscutting activities that span all lifecycle process activities, including elicitation, lifecycle concepts analysis and maturation, needs definition, design input requirement definition, architecture, design, production, system integration, system verification, system validation, transportation, installation, operations, maintenance, and retirement.

The focus of NRVVM activities is to manage the artifacts that result from the NRVV activities (i.e., the integrated sets of needs, sets of design input requirements, sets of design output specifications, sets of system verification artifacts, and sets of system validation artifacts) to ensure alignment and consistency across the lifecycle. Included in these activities are the management of the interface definition documentation (such as Interface Control Documents [ICDs], Interface Definition Documents [IDDs], and similar interface definition repositories) and interface management in general.

NRVVM involves the following activities:

- Define, baseline, and oversee the organization's lifecycle Concepts, Needs, and Requirements (CNR) definition policies, requirements, processes, and work instructions.
- Define, baseline, and oversee the organization's verification and validation policies, plans, requirements, processes, and work instructions across the lifecycle.
- Define, baseline, and manage all artifacts concerning interfaces, both internal and external to the System of Interest (SOI).
- Define and manage the process to address the unknowns (for example, To Be Resolved [TBR] or To Be Determined [TBD]).

INCOSE Needs and Requirements Manual: Needs, Requirements, Verification, Validation Across the Lifecycle,
First Edition. Louis S. Wheatcraft, Michael J. Ryan, and Tami Edner Katz.

- Provide the enablers needed to implement these policies, requirements, processes, and work instructions, including the Project Management (PM) and Systems Engineering (SE) tools to be used to manage the project data and information.
- Ensure all PM and SE team members are trained in the use of these tools, processes, concepts, and work instructions associated with practicing PM and SE from a data-centric perspective.
- Receive, review, and baseline the NRVV artifacts.
- Manage changes to the baselined NRVV artifacts over the lifecycle.
- Manage and control the flow down, allocation, and budgeting of design input requirements between subsystems and system elements at all levels of the SOI architecture.
- Manage bidirectional traceability between parent/source and child requirements as well as traceability between dependent peer requirements.
- Ensure vertical alignment and consistency between lifecycle CNR for all subsystems and system elements at all levels of the architecture.
- Ensure horizontal alignment and consistency between NRVV artifacts and the project's plans, budgets, schedules, work products, and other SE and PM artifacts generated across the lifecycle.
- Monitor, control, track, and report the development and configuration status of NRVV artifacts across all levels of the SOI architecture and across the lifecycle.
- Manage and configuration control the work products and artifacts associated with NRVV activities.

A summary of the Needs, Requirements, Verification, and Validation Management activities is shown in Figure 14.1.

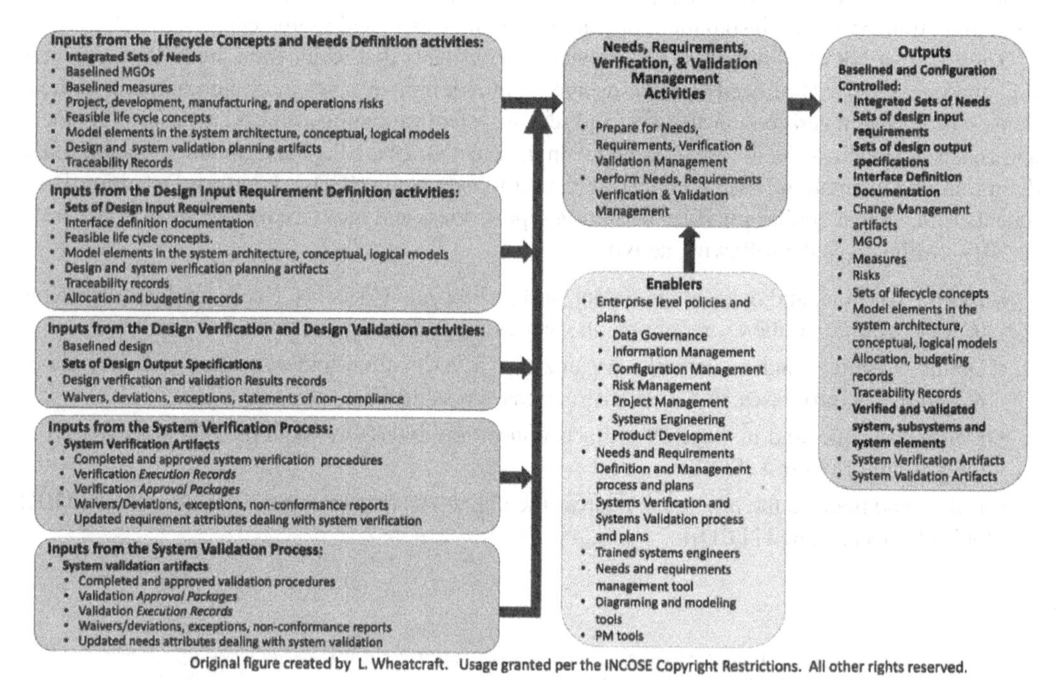

Original figure created by L. Wheatcraft. Usage granted per the INCOSE Copyright Restrictions.

FIGURE 14.1 Needs, Requirements, Verification, and Validation Management IPO Diagram.

14.1 PREPARE FOR NEEDS, REQUIREMENTS, VERIFICATION, AND VALIDATION MANAGEMENT

As shown in Figure 14.1, there are several enablers to successful NRVVM. These include enterprise and business-level policies and plans for data governance, information management, Configuration Management (CM), risk management, PM, and SE.

There should also be an organizational product development process, procedures, and work instructions, and experienced systems engineers trained in and knowledgeable in how to perform NRVVM activities according to these processes, procedures, and work instructions. Of particular importance are the Requirement Management Tools (RMTs), modeling/diagramming applications, and PM tools within the project's toolset used to develop and record the PM and SE artifacts.

Note: This Chapter assumes the project will be moving toward a data-centric approach to SE and recording and managing these artifacts using applications within the project's toolset that support the data-centric approach discussed in Chapter 3.

Preparing for NRVVM includes gathering or obtaining access to the required input artifacts shown in Figure 14.1 and preparing the project-level plans described in the following sections.

14.1.1 PM and SE Plans

NRVVM is a set of activities that involve those at the strategic and operational levels of the organization as well as the project/system level. The overall processes, procedures, and work instructions for NRVVM will be defined at the operational level of the organization.

The INCOSE SE HB Section 2.3.4.1, *Project Planning Process*, includes the development of a Program/Project Management Plan (PMP) and System Engineering Management Plan (SEMP) that establishes the direction and infrastructure necessary to enable the assessment and control of the project progress and identifies the details of the work as well as required personnel, skills, and facilities, schedule, and budget needed to produce the SOI. Based on this, each project will develop a PMP and an SEMP as part of their planning activities that are compliant with the operational-level lifecycle CNR concerning SE expectations.

The SEMP contains structured information describing how the SE effort, in the form of tailored processes and activities across all lifecycle stages, will be managed and conducted within the project. From an NRVVM perspective, the SEMP:

- Defines how the project will define and manage a project ontology and data dictionary.
- Includes key definitions of the needs and requirements definition, verification, and validation activities, and identification of all artifacts and work products generated as part of these activities, as well as the major deliverables of the project at each lifecycle stage.
- Provides a high-level description of the needs and requirements development and management process and how they fit into the overall system development lifecycle processes.
- Provides a high-level description of the verification and validation activities across the lifecycle and how they fit into the overall lifecycle process activities.
- Establishes how the project will develop, maintain, and manage the NRVV artifacts and the relationship of that information to the other SE artifacts and PM work products generated across the lifecycle.
- Defines how the project will manage interfaces, including identification, definition, interface requirements definition, and change control of all associated artifacts.

- Addresses the form of the work products (document versus database), the applications to be included in the project SE toolset to be used to generate, maintain, configure, and manage the artifacts, work products, their underlying data and information, and the IT infrastructure needed.
- Defines relationships between the needs and requirements definition and management activities defined in the Needs and Requirements Definition and Management (NRDM) Plan and the system verification and validation processes defined in the project Master Integration, Verification, and Validation (MIVV) Plan.
- Defines management of the work products and artifacts associated with the NRVV activities.

Both the PMP and SEMP identify the measures and reports that will be used to manage and track progress across the lifecycle. These reports help define the data and information needed to be managed within the SOI's integrated sets of data and information. Knowing which data and information will be included in the reports helps inform the formation of the project master schema to which individual sets of data and databases will conform.

14.1.2 Needs and Requirements Definition and Management Plan

The PMP and SEMP are supplemented by the NRDM Plan. The NRDM Plan focuses on the needs and requirements definition and management activities to be conducted, artifacts to be developed, and deliverables to be produced. Key provisions of the NRDM Plan can include, but are not limited to, the following:

- Provide details needed to implement NRDM activities, consistent with the project requirements defined in the PMP and SEMP, concerning how the NRDM activities of the project will be conducted.
- Define the oversight (control) and insight (monitoring) of the project's NRDM activities.
- Identify the relevant stakeholders who will be involved in the NRDM activities.
- Identify stakeholder engagement and peer review activities.
- Assign roles and responsibilities, authority, and resources needed to define the integrated set of needs and design input requirements, perform the NRDM activities, and develop the NRDM work products.
- Define the artifacts and work products that need to be developed as part of NRDM.
- Identify and manage the relationships and traceability within and between the integrated set of needs, sets of design input requirements, and sets of design output specifications and other project and systems engineering artifacts and work products throughout the system lifecycle.
- Provide a timeline for performing the NRDM activities and delivery of NRDM artifacts and work products.
- Define how the sets of needs, sets of design input requirements, and sets of design output specifications will be developed, maintained, tracked, managed, verified, validated, and reported across the lifecycle.
  - Define how and in what form the needs, design input requirements, and design output specifications will be recorded and managed.
  - Define the attributes that will be defined and maintained as part of each need and design input requirement expression.
  - Define the measures, metrics, and reports that will be used to monitor and control the development of the integrated set of needs, design input requirements, and design output specifications.

  - Define the review and approval processes for baselining the integrated sets of needs, sets of design input requirements, and sets of design output specifications.
  - Define the system verification and validation *Success Criteria*, *Strategy*, and *Method* for each need and requirement expression against which the SOI will be validated and verified (this may be included as part of the MIVV and SIVV plans).
  - Describe the process for the flow down (allocation) and budgeting of requirements from one level of subsystems and system elements within the architecture to another and the management of those allocations and budgets.
  - Describe the process for establishing bidirectional traceability between levels of requirements (parent/child, child/parent, source/child, child/source) as well as peer-to-peer traceability between dependent requirements at the same level.
- Define the activities associated with tracking and resolution of unknowns (TBDs and TBRs).
- Define the change management process for changes to needs, design input requirements, and design output specifications.
- Define the level of CM/data management control for all NRDM artifacts and work products.
- Identify the training for those who will be performing the NRDM activities.

NRDM activities do not exist in isolation from other PM and SE activities; a successful NRDM Plan requires integrated execution with the other activities defined in the PMP and SEMP. It is critical to get buy-in from all key stakeholders to reduce the risk of stakeholders failing to support the work described in the NRDM Plan within budget and schedule constraints. Getting this buy-in will result in an approved NRDM Plan.

14.1.3 System Verification and Validation Management Plans

The MIVV Plan expands and implements the project team's design verification, design validation, early system verification, early system validation, system integration, system verification, and system validation philosophy/concepts outlined in the PMP and SEMP. The MIVV Plan:

- Defines the system verification and validation *Success Criteria*, *Strategy*, and *Method* for each need and requirement against which the SOI will be validated and verified.
- Defines the more detailed approach and activities that will be used for design verification, design validation, early system verification, early system validation, system integration, system verification, and system validation.

Developing an MIVV Plan removes the need for the SEMP to contain this detailed information and can result in a much more manageable SEMP.

The MIVV does not have individual system, subsystem, or system element system verification and validation plans and schedules; these are addressed by individual SIVV plans developed for each subsystem and system element within the SOI physical architecture.

SIVV plans specific to each subsystem and system element are created concurrently by the project teams responsible for each subsystem and system element as individual sets of needs and sets of design input requirements are defined, containing detailed descriptions of the activities to be performed to implement the project team's MIVV Plan activities. Resources including test equipment, facilities, and personnel are addressed, as well as a detailed budget and schedule consistent with the MIVV Plan.

When an SOI is contracted out to a supplier, the SOW should require the supplier to develop a SIVV Plan consistent with the customer's MIVV Plan and submit the SIVV to the customer for approval. *See also Chapter 13 for supplier-developed SOIs.*

14.1.4 Configuration Management Plan

The Configuration Management Plan (CMP) supplements the PMP and SEMP to address the functions associated with CM. The CMP ensures SOI NRVV artifacts are defined, documented, validated, and verified to establish product integrity and that changes to these artifacts are identified, reviewed, approved, documented, implemented, and managed.

Not all artifacts and work products need to be put under configuration control. Artifacts or work products that are put under configuration control are identified in the PMP and SEMP and often represented via a "document tree" (Section 6.4.2). From a supplier contract perspective, contract deliverables are usually put under configuration control while intermediate work products may not be.

It is the set of configuration-managed artifacts and work products that represent the SOI baseline, representing the authoritative source of truth (ASoT) concerning what was agreed-to and is being developed and delivered in accordance with those agreements.

Historically, the configuration-managed artifacts are represented by documents in either a hardcopy or electronic form. A key issue with these documents is that they are often generated at different times and are valid only at the time they were baselined and put under configuration control. To make this clear, most organizations adhere to ISO 9001 standards and include a statement on the front page of these documents stating this fact. Because of different timing of the baselined documents, they are often out of sync and inconsistent; thus, the ASoT is not always clear.

A key benefit of a data-centric approach advocated in this Manual is that the artifacts are visualizations generated from the SOI's integrated dataset. When this is the case, changes to the data and information managed within the integrated dataset are propagated across all the various visualizations of artifacts which helps to ensure consistency across the lifecycle.

In the future, projects that have matured their CM process to be consistent with a data-centric approach may baseline the dataset from which the visualizations are generated rather than the individual artifacts. Using this perspective, all the artifacts should be consistent with the data and information in the baselined dataset. SE tools support this approach, enabling the creation of baselines by removing write permission from the dataset and freezing the data at the time of baseline. A copy of the data set can be made, and changes to the baseline data set are allowed only if they go through a formal CM change process.

Moving from a document-centric view of CM to a data-centric view of CM is a major challenge for many organizations. *Refer to the INCOSE SE HB Section 2.3.4.5 for more details concerning the Configuration Management Process.*

14.2 PERFORM NEEDS, REQUIREMENTS, VERIFICATION, AND VALIDATION MANAGEMENT

NRVVM involves the activities shown in Figure 14.2.

Key NRVVM activities include:

- Baseline needs, requirements, and specifications (Section 14.2.1).
- Monitor the status of needs, requirements, and specifications (Section 14.2.2).
- Control the needs, requirements, and specifications (Section 14.2.3).
- Manage unknowns (Section 14.2.4).
- Manage change (Section 14.2.5).

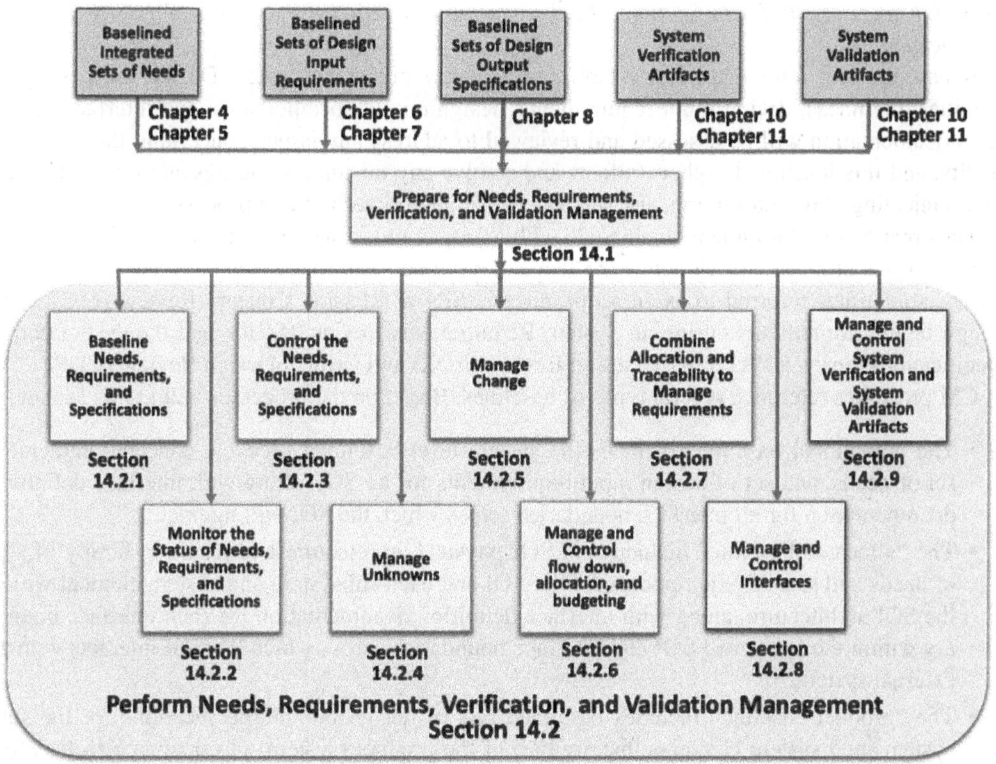

FIGURE 14.2 Needs, Requirements, Verification, and Validation Management Activities.

- Manage and control the flow down, allocation, and budgeting (Section 14.2.6).
- Combine allocation and traceability to manage requirements (Section 14.2.7).
- Manage and control interfaces (Section 14.2.8).
- Manage and control system verification and validation activities and associated artifacts (Section 14.2.9).

14.2.1 Baseline Needs, Requirements, and Specifications

A key part of system development is the establishment, control, and maintenance of baselines of the integrated sets of needs, sets of design input requirements, sets of design output specifications, and sets of interface definition documentation for the SOI under development. Changes to needs and requirements must be managed as they evolve, identifying inconsistencies that may occur among them, as well as with other project plans, work products, and engineering artifacts. Prior to baselining, needs and requirements will change as they are initially defined and matured.

When the sets of needs, design input requirements, design output specifications, and interface definition documentation for the SOI are completed and ready for baselining, they will undergo verification and validation activities to review their quality (discussed in Chapters 5, 7, and 8) and then baselined. *Note: The interface definitions will evolve during the SOI lifecycle based on design maturity. Refer to Section 14.2.8 for a more detailed discussion concerning interface management.*

A baseline is a reference point established as a key milestone "gate". During the baselining activities, the needs, design input requirements, design output specifications, and interface definition documentation will be assessed and reviewed to address any issues concerning their content, quality, and relationships to other artifacts and resolve any misunderstandings as to what they are communicating. After baselining, any suggested change initiates the CM process.

The creation of a baseline may coincide with a project milestone or decision gate. For example, integrated sets of needs, which define the scope of a project, are commonly baselined during a gate review, sometimes referred to as an scope review (SR) or Mission Concept Review (MCR); the design input requirements during an System Requirements Review (SRR); and the design output specifications during a Preliminary Design Review (PDR) and Critical Design Review (CDR).

CM processes reference several types of baselines. Based on the processes within this Manual:

- The "functional baseline" includes the system-level baselined lifecycle concepts, integrated set of needs, and set of design input requirements for an SOI, along with interface definition documentation for all interface boundaries across which the SOI will interact.
- The "allocated baseline" includes the SOI physical architecture and resulting family of sets of needs and sets of requirements for the SOI and each subsystem and system element within the SOI architecture, along with interface definition documentation for each interface boundary within the integrated SOI and interface boundaries across which the SOI interacts with an external system.
- The "product baseline" includes baselined sets of design output specifications for the subsystems and system elements that are part of the product (system) physical architecture (the realization of the allocated baseline).

Note: It is important to access the timing of when the needs and requirements are baselined. Baselining them too early could result in unnecessary work as the needs and requirements mature; to late could result in unnecessary rework as the architecture and design definition mature. When part of a contract, it is important that the set of design input requirements or design output specifications are well-formed and baseline prior to issuing the contract. Failing to do so could result in expensive contract changes.

The baselined integrated sets of needs, sets of design input requirements, and sets of design output specifications represent an agreement between the customer (internal or external) and the developing project team (internal or external) responsible for the SOI. Once baselined, they are put under CM in accordance with the process defined in the project's CMP–any changes proposed by the supplier would have to be approved by the customer via their CM process. If a contract deliverable, the contracting organization will accept and approve the work product as defined in the Contract Deliverables Requirements List (CDRL) that is part of the contract.

Baselining and CM of the NRVV artifacts is normally the responsibility of a configuration control board (CCB). A CCB is a formally chartered group of operational-level stakeholders responsible for reviewing, evaluating, approving, delaying, or rejecting changes to project SE artifacts and PM work products, in addition to recording and communicating such decisions. Not all projects require the use of a CCB. A project in a heavily regulated industry, or one with numerous components, interfaces, risks, and stakeholders, may require the use of a formal CCB more than a project without those characteristics.

The systems engineer, project manager, and other key personnel usually participate in the CCB approval and change management processes to assess the impact of a change (for example, cost, performance, quality, compliance, programmatic, security, and safety). In a data-centric practice of SE, this impact assessment is enabled by the traceability between the data being changed and related project data (such as verification *Activities* and architecture elements).

The CCB is often the ultimate source for approval of the integrated sets of needs, sets of design input requirements, and sets of design output specifications.

Note: For supplier-developed SOIs, there will be a CCB function for both the customer organization as well as the supplier organization. In this case, it is important that the customer clearly define in the SOW the relationship between the two CCBs and which changes must be submitted to the customer CCB versus which can be managed by the supplier CCB. One criterion could be whether the change would also trigger a change in the contract.

Note: The approach to CM and the baselining of the various artifacts is different depending on whether the organization is using a document-centric versus a data-centric approach to SE. If document-centric, the individual documents are baselined. As a result, it is a challenge to keep each document current and consistent with other documents; often resulting in there being no ASoT. If data-centric, then it is the databases that contain the underlying data and information that are baselined, rather than individual documents generated as reports from the data and information. As such, an ASoT can be established within the data and information generated and managed within the project's toolset.

This activity is critical to ensuring the resulting individual needs have the Guide for Writing Requirements (GtWR) characteristics *C1—Necessary, C3—Unambiguous, C4—Complete, C6*—Feasible, and *C8—Correct* and sets of needs have the GtWR characteristics *C10—Complete, C11—Consistent, C12—Feasible, C13—Comprehensible,* and *C14—Able to be Validated.*

Refer to Sections 14.3.3 and 14.3.4 for more details concerning needs and requirements creep and change management.

14.2.2 Monitoring the Status of Needs, Requirements, and Specifications

Monitoring the needs, requirements, and specifications provides project stakeholders *insight* into the status of needs and requirements definition process across the lifecycle activities helping to maintain a level of insight and collaboration. This is especially important with today's increasingly complex software-intensive systems.

The overall status of the needs and requirements definition and management activities, key metrics, and their implementation into design outputs via the *Design Definition Process* should be captured and communicated to stakeholders in accordance with the project's Stakeholder's Communications Management Plan (SCMP).

Associated metrics should be captured and maintained within the SOI's integrated data set and communicated via reports and dashboards shared with the project's stakeholders. A major use of the needs and requirements attributes as discussed in this Manual is to aid in the management and reporting on the status of the project in terms of the needs and requirements, their implementation, and system verification and validation progress. *Refer to Chapter 15 for a detailed discussion on the use of Attributes to manage the product development effort.*

14.2.3 Controlling the Needs, Requirements, and Specifications

Controlling the needs, requirements, and specifications provides project management *oversight* over their development, baseline, and implementation over the lifecycle.

Controlling involves the following activities:

- Controlling the number of needs and requirements as well as helping to ensure the sets of needs and requirements are feasible in terms of cost, schedule, technology, and risk as discussed in Sections 4.6.4.4 and 6.2.6.3.
- Controlling change as discussed in Section 14.2.5. Once baselined, any changes to the needs, requirements, and specifications are submitted through the CCB.
- Managing the flow down, allocation, and budgeting of requirements from one level to another as discussed in Section 14.2.6.
- Managing the traces between sets of needs and requirements for each subsystem and system element that is part of the SOI architecture. Maintaining the correctness and completeness of these traces is critical in support of change management as discussed in Section 14.2.7.
- Managing and controlling the interactions (interfaces) between parts of the architecture as well as the interactions of the SOI with external systems including users, operators, maintainers, and disposers as discussed in Section 14.2.8.
- Ensuring the design input requirements are implemented during the *Design Definition Process,* and the that making a SOI per the resulting design output specifications will result in a SOI that can be verified to meet its design input requirements and validated to meet its integrated set of needs. This is done as part of design verification and design validation activities participation in key gate reviews such as the System Design Review (SDR), Preliminary Design Review (PDR), and Critical Design Review (CDR), and participating in the system verification and system validation activities as discussed in Section 14.2.9.

It is important to note that controlling is a shared activity between project management, configuration control, systems engineering, procurement, and quality control personnel.

14.2.4 Manage Unknowns

During the definition of the needs and design input requirements, there may be unknowns, resulting in the project team having to make assumptions regarding criteria to allow the subsequent lifecycle activities to proceed.

This often happens when the project team skips the lifecycle concepts analysis and maturation activities prior to defining the integrated sets of needs and transforming them into sets of design input requirements. In other cases, further analysis or research is still required that would interrupt the overall workflow to develop and baseline a complete integrated set of needs and resulting set of design input requirements, for example, the maturation of critical technologies. While this work will continue, it is critical to capture the unknowns and resulting ongoing work to ensure the associated activities to address the unknowns are funded, tracked, and managed.

One common method to do this is to use the TBD or TBR indications in the need or requirement statement in place of (or in addition to) an actual value. When the actual value has not been determined, this can be represented as TBD, while a starting value that is not confirmed may be stated next to a TBR. This is highlighted in the example below. *Note: The term "**TBX**" is often used to generically refer to any of the "To-be- ???" unknown designations.*

In the example on the next page, the use of an TBX is an indication the requirement will undergo further analysis during transformation from the parent need into the implementing design input requirements to resolve what the stakeholder believes is "fast" (what is the minimum value and why) and align it with what is physically feasible based on the TRL of the critical technology needed to achieve that value. In this sense, an TBX is a place holder that indicates additional work needs to be done. As such, all TBXs need to be identified and managed formally as action items. A person within the organization should be assigned to manage the resolution of each TBX by a specific date.

Initial need statement: "The stakeholders need the SOI to [*process*] the input data *fast*."

In this need statement, "fast" is ambiguous and not verifiable. If this ambiguity was not able to be resolved during lifecycle concepts analysis and maturation activities, then a TBD can be used:

Updated need statement with TBD: "The stakeholders need the SOI to [*process*] the input data at a rate of TBD [*parameter*]/sec"

Given those that defined the integrated of needs did not provide a value for the TBD, the issue will need to be addressed during transformation of the need into one or more design input requirements.

Initial Design Input Requirement: "The SOI shall [*process*] the input data at a rate of at least TBD [*parameter*]/sec".

There may be cases where the value indicated by the TBD has been defined; however, there is disagreement by the stakeholders about what the specific value should be. In other cases, achievability of the value may still be in question. In this case, it is common to put square brackets "[xxx]" around the value that is uncertain and assign a TBR, as shown in the example below.

Updated requirement statement with TBR: "The SOI shall process the input data at a rate of at least [1000 TBR] [*parameter*]/sec."

In the case where the customer is going to contract out the transformation of the integrated set of needs into the set of design input requirements, they must decide who is responsible for resolving the TBXs. Resolution being assigned to the supplier for resolution must be addressed in the SOW along with a requirement for the supplier to do the work needed to resolve the TBXs, and this extra work would be reflected in the supplier's proposal. (*A common error in contracting is to fail to include in the Statement of Work (SOW) or SA provisions concerning these kinds of activities. The result is often expensive contract changes.*)

The above-mentioned example reflects a simple use of an TBX to show that the confidence level in the stated value is low. This communicates to the project team that there is additional work required to resolve the TBX. This additional future work represents technical debt; the later the resolution of the TBXs in the development lifecycle, the more technical debt interest is accrued (driving up costs). The effort for resolution of the associated technical debt is often referred to as "TBX Management," as it provides an indication of the effort required to resolve the various TBXs in the resulting requirement set or design.

Not all TBXs are bad. If analysis is needed to fully develop the value, showing a nominal value within brackets, "[ ]" with an TBR shows that it is currently a placeholder. While it is important to reduce technical debt to prevent rework later, it is also important to resist the temptation to set a questionable value that overly constrains the resulting design just to remove the TBR.

TBXs can be managed in multiple ways using common PM techniques.

- One way often used is to control these within the integrated set of needs or set of design input requirements to assign a unique identifier for each TBX (for example, TBD1 or TBR3) and a summary report showing all TBXs and associated unknowns. Often the unique identifier may be the identifier for the need or requirement statement appended to the letters TBD x.x.x or

TBR x.x.x where "xxx" is the number of the need or requirement to which the TBD or TBR applies.

- To aid in the management of TBXs, it is useful to include attributes such as *A26—Stability*—and *A27—Responsible Person* defined in Chapter 15. This will allow reports to be generated from the data set for all needs that have the stability attribute set and who the responsible person is for resolving the TBX.
- In a document-centric practice of SE, it is common to include the TBDs and TBRs in an appendix of the needs or design input requirement documents with sufficient information to track and manage the resolution of the TBXs.
- If multiple sets of needs or requirements exist, the work to resolve the TBXs may become complex and involve several resources and actions. Of particular concern is when a value in one need or requirement statement has a dependency on values in one or more other need or requirement statements. If this is the case, a comprehensive TBX tracking mechanism would be helpful, such as using the project's toolset to connect all TBXs to a common database, where tracking data (such as forward work, assignee, and closure plan) is defined and managed. This could be linked (traced) directly to the needs and design input requirements they impact. The toolset would also allow dependent TBXs to be linked to help ensure consistency as the values are defined and agreed-to.

Keeping track of the TBXs allows for awareness of the maturity of the needs and design input requirements throughout the lifecycle and is a valuable metric to evaluate completion and maturity of the set of needs and set of design input requirements. Some PM and SE management applications enable generation of a report which shows the needs and design input requirements that contain unresolved TBXs, along with associated status and resolution plans.

Capturing TBDs and TBRs contained in a set of needs and set of design input requirement statements that have dependencies within the project's toolset is critical for development of complex systems to help ensure they have the characteristics defined in the GtWR *C3—Unambiguous, C4—Complete*; *C6—Feasible*, *C7—Verifiable*, and *C8—Correct.*

14.2.5 Manage Change

NRVVM involves managing all changes to needs, design input requirements, design output specifications, system verification artifacts, and system validation artifacts baselines over lifecycle. The project must define change management processes at the beginning of the project in their various project plans. Usually, changes are managed within an organization's existing CM process as communicated in the project's CMP.

Note: When referring to needs and design input requirements changes, the reference is to the need expressions and requirement expressions, which include the need statement or requirement statement along with the set of attributes. Managing change applies to changes for both the need statements and requirement statements as well as all associated attributes.

Given that drivers and constraints are imposed on the project, the project must monitor and control changes to NRVV data closely as part of its CM processes. To do this, the project or its organization must have a process in place enabling the project to be notified of these potential changes, assess their impacts, and communicate those impacts to those that have CM authority over the project's drivers and constraints. *Refer to Section 14.2.5.4 for approaches to change impact analysis.*

Not all changes have an adverse impact; for example, a requested change that clarifies a requirement can reduce ambiguity in design, or a requested change that simplifies interfaces may reduce complexity in design and integration. A regulation or standard may change, but specific requirements within the standard or regulation applicable to the SOI under development may not change. A key challenge is for the organization to have a process that will notify projects when a standard or regulation changes so the project team can determine whether the change impacts their SOI. As new needs and requirements are identified, they are assessed for impacts to the project and product and presented to stakeholders for approval using the needs and requirements "bucket" discussed in Sections 4.6.4.4 and 6.2.6.2.

All change requests must be captured and placed under configuration control. In addition to the change requests, the impact analysis and any other documents created during the change request process should be captured. The status and disposition of change requests should be readily available to the relevant stakeholders and monitored through change logs. Most RMTs include the capability to maintain a history of all changes made within the tool.

All changes should be subjected to a review and approval cycle to ensure the resulting changed needs and requirements have these characteristics and that traceability is maintained to ensure that the impacts of any proposed changes are fully assessed.

The CCB may choose to allow change requests under a prescribed dollar amount to approved by the developer's project manager and/or the customer's project manager with change requests over the threshold approved by the CCB. Those thresholds usually are determined by considering impact on risk, cost, schedule, or deliverables and should be defined in the project's PMP, SEMP, and CMP.

When managing changes, the CCB needs to consider the distribution of information related to the decisions made during the change process. The CCB needs to communicate the needs, requirements, and specifications of change decisions to the affected organizations in a timely manner. During a CCB meeting to assess and approve a change, actions to update artifacts and work products need to be included as part of the change package. These actions should be tracked to ensure that affected artifacts and work products are updated in a timely manner to help maintain consistency and correctness. A major challenge is establishing and maintaining an ASoT.

For those changes that are approved, the project will need to formulate change requests for all artifacts affected by that change that were identified during change impact assessment. The project team should also ensure that the approved changes are communicated in a timely manner to all relevant stakeholders.

Note: Beware of changes that are fast tracked through the CM processes. Often fast-tracked changes do not involve the proper degree of change impact analysis that is used to identify possible impacts due to the change. As a result, the fast-tracked change could be implemented with unintended negative consequences. Another type of change to avoid is deferred changes. These are changes that are approved, but their implementation is deferred to a later time, for example, after a key milestone or moved into a different budget cycle. When the deferred change was approved it was approved for the system configuration at that time. The danger is that, depending on the deferred time, other changes could have been approved and implemented, resulting in a different system configuration. When the deferred change is implemented, it again could have unintended negative consequences because of the changed configuration.

Need and requirement changes in later lifecycle stages are more likely to cause significant impacts on cost and schedule as they could lead to possible changes in the design and resulting rework to parts of the system that have already been built or coded. The possibility of these impacts must be a major consideration when doing change impact assessment. Because of this, it is imperative that all changes be thoroughly evaluated to determine the impacts on the cost, schedule, risk, architecture, design, interfaces, lifecycle concepts, higher and lower-level needs and requirements, system verification artifacts, and system validation artifacts. See Section 14.2.5.4 for a more detailed discussion on change impact analysis. *See Section 14.2.5.3 for a more detailed discussion on managing change across the lifecycle.*

14.2.5.1 Control Needs and Requirements Creep Needs and requirements instability is a leading cause of change, and NRDM monitoring and controlling activities ensure that requested changes are processed through the project's change control process.

"Scope creep," "needs creep," and "requirements creep" are the phrases used to describe the subtle way that sets of needs and sets of design input requirements tend to grow incrementally during development, resulting in an SOI that is more expensive and complex than originally intended.

Some of this growth may not be warranted, while some of the sets of needs and sets of design input requirements creep involves new needs and design input requirements that were not known when the original sets of needs and requirements were baselined and were not anticipated, during the *Lifecycle Concepts and Needs Definition* and *Design Input Requirements Definition* activities. These new needs and resulting design input requirements are the result of natural evolution, and if a relevant SOI that can be validated to meet the real-world expectations is to be developed, these changes cannot be ignored.

There are several approaches for avoiding or at least minimizing needs and requirements creep:

- Focus on the problem, Mission, Goals, and Objectives (MGOs), measures, and constraints defined at the beginning of the project and determine whether a change is warranted. Often changes, especially later in the development lifecycle, will require rework with associated costs and schedule issues, which will need to be addressed if the change is approved. If the change does not directly address the problem, MGOs, measures, or constraints, why approve the change?
- In the needs elicitation phase, work with the stakeholders to bring out both explicit as well as implicit needs and requirements that might otherwise not be stated. This helps to avoid missing needs and requirements at the beginning of the project.
- Define a set of lifecycle concepts that have been through the lifecycle concepts analysis and maturation activities (discussed in Chapter 4) and agreed-to by the customers, users, and other relevant stakeholders. In addition to operations, address all lifecycle stages of the SOI as well as alternate nominal and off-nominal cases; this helps to avoid missing needs and requirements.
- Establish a configuration control and change management process. This will determine which stakeholders have the authority to submit changes formally to the CCB and the type of changes that will be allowed. Maintaining configuration control is essential to having a control over the project budget and schedule.
- Establish a process for assessing changes and their impacts on the rest of the system and other artifacts within the project's data and information model. Compare this impact with the consequences of not approving the change.
- Determine the feasibility of implementing a change in terms of budget, schedule, technology, and risk. If it cannot be accommodated within the established resource margins, then the change most likely should be disapproved. If the change does not "fit" within needs and requirements "bucket" discussed in Sections 4.6.4.4 and 6.2.6.3, then do not approve the change.
- Determine criticality. If a change does not address the ability to meet a critical need or requirement, a requirement dealing with safety or security, or a mandatory regulation or standard, then do not approve it.

One of the authors attended a briefing of a project that was getting an award for its success in developing a system that met the stakeholder needs within cost and schedule constraints. When the project manager was asked what was key to the project's success, the project manager said that he had extremely strict rules for change management. His philosophy was that if it is not broken, is safe, or is not breaking a law—then leave it alone! It did not take long for members of the project team to limit any proposed changes to those that met these criteria. While this seems drastic, it worked.

14.2.5.2 Why Needs and Requirements Change It is useful to understand common reasons why needs and requirements change so this knowledge can be used to help avoid, or at least minimize, the number of changes as well as avoid unnecessary needs and requirements creep.

Reasons needs may change:

- *The Lifecycle Concept and Needs Definition activities addressed in Chapter 4 were either not done or done incompletely*: This can result in defective, incorrect, inconsistent, and missing needs as well as needs that are not feasible. The later in the development lifecycle before discovering these problems, the greater the impact on cost and schedule.
- *Some stakeholders may not initially know what they need*: They say: "Show me, and I will tell you if that is what I need." The project team may have developed something, like a graphical user interface (GUI) interface or a report a stakeholder said they needed, but when delivered, they say: "No, that is not what I need, that's not what I meant; what I really meant was ....". This is one reason prototyping and modeling are encouraged to enable stakeholder evaluation of proposed concepts during the lifecycle concepts analysis and maturation activities; this allows quick resolution of issues early in the lifecycle before the integrated set of needs are baselined and transformed into the set of design input requirements.
- *Changing expectations*: Stakeholder real-world expectations tend to change over time if not managed. If the project team does not involve the stakeholders at each phase of development and setting expectations to let them know what to expect, other people could be influencing them and changing their expectations. To help manage changing expectations and the resulting needs and requirements creep, it is critical to involve the stakeholders to set their expectations and minimize potential expectation changes.
- *Changing technology*: Technology is changing at an increasing rate, and often the stakeholders expect the latest and greatest. For SOIs with a long development time, the available technologies can change, and the project will be challenged to keep up. The stakeholders may have not asked for some feature or level of performance initially because the technology was not mature, but now that technology is available and mature, the stakeholders may demand the new features or performance be included and enabled by those technologies.
- *Changing drivers and constraints*: One source of change is changes to drivers and constraints. These changes could be due to a change in schedule or budget, a change to how the SOI needs to interact with an existing external system, a change in the operational environment, or a change to a standard or regulation the project must provide objective evidence of compliance. Failing to implement changes to drivers and constraints could result in issues during integration, compliance failure, and failed system validation.
- *New and changing stakeholders*: Some stakeholders may not exist at the beginning of the project. Over time, stakeholders leave and are replaced with new stakeholders. Each of these stakeholders could see the problem and desired solution (end state) differently. Because of this, they may want to change or add new needs and requirements.

- *Changing needs*: Sometimes the "problem" that the SOI is to address and the MGOs and measures change the operational environment changes, enabling systems change as well as other external systems in which the SOI will interact. When this happens, the whole project needs to be rethought. All the deliverables that have been prepared to date that were based on meeting the originally stated problem statement, MGOs, and needs could be obsolete. When this happens, the project team will have to redo a lot of the upfront activities and artifacts.
- *Missing or ignored stakeholders*: There may be key stakeholders that were not identified or were ignored at the beginning of the project during elicitation. Stakeholders represent needs and requirements. Missing or ignored stakeholders results in missing needs and requirements. For example, the safety or security office stakeholders were not involved at the beginning of the project during elicitation. As a result, specific safety and security needs and requirements were not defined at the beginning of the project. Later, during certification or qualification activities, safety and security representatives will not certify or qualify the system for use because the SOI did not meet their needs and requirements.
- *Overly optimistic budget or schedule*: Promising something that cannot be delivered with the available resources, in the time allotted, or without establishing feasibility before the project is formulated drives a change in approach for the project and is a source of change of the needs. It is human nature to be overly optimistic.
- *All product lifecycle stages not addressed*: If a key product lifecycle was missed (test, system verification, system validation, transportation, storage, transition, operations, upgrades, maintenance, or disposal), then the lifecycle concepts analysis and maturation process effort will result in missing needs. Unique needs to address a missing lifecycle concept will need to be added.
- *Unknowns not addressed*: At the time of baselining, there were unknowns not yet resolved resulting in need statements containing TBDs and TBRs, driving changes to the needs. See Section 14.2.4 for a detailed discussion on managing unknowns.

Reasons why requirements may change:

- *The Design Input Requirement Definition activities addressed in Chapter 6 were either not done or done incompletely*: This can result in defective, incorrect, inconsistent, and missing requirements as well as requirements that are not feasible. The later in the development lifecycle before discovering these problems, the greater the impact on cost and schedule. Given a poorly-formed set of design input requirements, developers will find major problems and issues resulting in change.
- *Failure to define and mature lifecycle concepts and define an integrated set of needs prior to defining the set of design input requirements as was addressed in Chapter 4*: A common problem is the definition of design input requirements with little or no underlying analysis of the lifecycle concepts and needs from which the requirements are transformed, resulting in poorly-formed requirements.
- *Changing needs*: Changing needs due to the issues discussed previously is often perceived as the major reason for change and requirements creep. If the needs change, then the resulting design input requirements transformed from those needs will change.
- *Changing drivers and constraints*: One source of change is changes to drivers and constraints. These changes could be due to a change in schedule or budget, a change to how the SOI needs to interact with an existing external system, a change in the operational environment, or a change to a standard or regulation the project must provide objective evidence of compliance. Failing to implement changes to drivers and constraints could result in issues during integration compliance failure and failed system validation.

- *Ignored requirements*: Another thing that could have happened is that the developers could have ignored some of the original requirements. Either they did not understand the requirements, the requirements were too complex, or the developer just chose to ignore the requirements. As a result, the developers designed the SOI without reference to one of more of the requirements, resulting in failed system verification and system validation.
- *Immature technology:* Projects sometimes base their ability to meet the needs based on a technology that is not mature enough to use. Doing so adds risk to the project as well as the potential for change if the technology does not deliver the needed functionality, performance, or quality.
- *Failure to establish feasibility*: A key characteristic of individual needs and requirements and sets of needs and requirements is feasibility. Failing to establish feasibility before baseline (or before a contract is issued) can result in expensive changes to cost and schedule and, in some cases, a cause to cancel the project.
- *Interfaces not defined*: A key interface was overlooked. Without addressing the interface, the product cannot be integrated into the macro system of which it is a part.
- *Unknowns not addressed*: At the time of baselining, there were unknowns not yet resolved resulting in requirement statements containing TBDs and TBRs, driving changes to the requirements. *See Section* 14.2.4 *for a detailed discussion on managing unknowns.*

The concepts and activities in this Manual are designed to help avoid the above-mentioned issues, resulting in more stable needs and requirements that have less "churn" than if these concepts were not followed.

14.2.5.3 Change Management Across the Lifecycle Change management is of critical importance across all lifecycle activities. The architecture, allocation, and traceability discussed in Chapter 6 result in a three-dimensional web of relationships both vertically between levels and horizontally across the lifecycle. This web is represented within the SOI's federated/integrated data and information model. For today's increasingly complex systems, it is difficult to understand the effects of change without such a model.

Gentry Lee from the Jet Propulsion Laboratory notes: "*A good systems engineer knows the partial derivative of everything in respect to everything*." This means that a change that occurs anywhere in an SOI's integrated/federated data and information model could have an impact on other data and information in that model. With all the PM and SE data and information developed across the lifecycle included in this model and linked together, it is much easier for the project team to assess a change in any data item to determine whether the change could have impacts on other data items.

A change to a stakeholder's needs could impact a lifecycle concept, which could impact multiple needs, which could impact multiple design input requirements, which could impact the design and implementing the design output specifications, and which would impact production, system integration, system verification, and system validation activities and artifacts. Any of these changes could impact cost and schedule or the ability of the system to address the problem or opportunity which the system is supposed to address.

What are the impacts on system verification and validation planning when a level 3 design input requirement changes? Does that change impact the subsystem at level 2 from which its parent requirement(s) were allocated and associated system verification artifacts? Related/dependent peer requirements and associated system verification artifacts at the same level?

How will level 4 system element requirements and associated system verification artifacts be impacted by a change in the allocated level 3 parent requirement? How will a design input requirement change impact the architecture and design? What is the ripple effect of that one change? How will the change impact the design output specifications? Production? Will software need to be recoded or hardware rebuilt? What will be the total cost of the change? How will the change impact the schedule?

> For example, consider a large government program where requirements were defined seven levels deep across multiple projects and the architectures within each project that made up the program.
>
> Due to changes in mission and funding (primarily due to politics), several key changes were mandated at the top levels of management. At the program level, it seemed that only four or five requirements for the overall system architecture would be changed.
>
> What is the big deal? Because of the ripple effect, thousands of requirements at the lower levels were impacted along with design implementations of those requirements. The associated system verification and validation artifacts for those requirements were impacted as well.
>
> As a result, the cost and schedule impacts associated with "just a few" top-level program requirements changes were too great, and the program was canceled.

For today's increasingly complex, software-intensive systems, it is difficult, if not impossible, for a human brain to comprehend all the relationships and impacts across the lifecycle due to change. Because of this, it is critical that projects have the SE tools necessary to develop the integrated/federated data and information model, enabling them to manage these relationships and access impacts of changes across all lifecycle activities and associated artifacts.

For highly regulated systems, such as medical devices or nuclear-based systems, being able to manage the relationships and assess and record the impacts of changes, no matter which level of architecture or where in the lifecycle, are mandatory for certification, qualification, and approval for use.

Because of these possible impacts of a change, it is critical that the developing organization's CM office assess the impacts of changes across the entire web of data and associated artifacts, especially all the artifacts that are part of the project's system verification and validation programs and associated artifacts.

14.2.5.4 Change Impact Analysis Effective change management requires a process that assesses the impact of proposed changes prior to their approval and implementation. For CM to perform this function, a baseline configuration for each artifact under CM should be recorded, and resources and tools used to assess impacts to the baseline.

Change analysis typically involves performing an impact analysis to evaluate the proposed change in relation to how it will affect other needs, requirements, design, risk, the system, the project (if change is approved), as well as consequences if not approved. The systems engineer, project manager, other key stakeholders, and subject matter experts participate in the CM process to assess the impact of the change, including cost, schedule, performance, security, safety, quality, and compliance.

When performing change impact analysis, key considerations include:

- How the change may affect other needs, requirements, design, design output speculations, manufacturing, system verification, system validation, the integrated system, the project, and the program.
- How the change may affect the risks identified in Chapter 4 both if implemented and if not implemented.
- Resources, time, and schedule impacts that are required to incorporate the change.
- Project team members across the lifecycle that could be affected by the change.
- Effects of the change on requirement allocations and budgeting.

- Impact of the change to development teams and suppliers of lower-level subsystems and system elements: Will the change require a change to be made to a supplier contract?
- Schedule and cost implications of the change.
- Given the timing of the change, how the work in progress will be affected.
- When the change is implemented, will the resulting system continue to adequately address the problem and MGOs that were approved by the stakeholders?
- How will the change affect security and safety?
- How will the change affect compliance with standards and regulations?
- How will the change affect risk mitigation?

A key benefit of completing an impact analysis is that it allows for proposed changes within the project to be considered in an integrated fashion, thereby reducing risk, which often arises from changes being made without consideration as to the possible impacts.

Typical data, information, and methods used to analyze the change impact are shown below. Ideally, this information is maintained within the project toolset.

- *Budgeting of resources, performance, and quality. Refer to Section 6.4.5.*
- *Margins and Reserves.* A list of key margins for the system and PM reserves and the status of the margins and reserves. For example, the propellant performance margin will provide the necessary propellant available versus the propellant necessary to complete the mission or budget reserves accounting for unknowns and unknown unknowns during both development and operations. Changes should be assessed for their impact on margins and reserves. A threats list is normally used to identify the costs associated with all the risks for the project. Project reserves are used to mitigate the appropriate risk. Analyses of the reserves available versus the needs identified by the threats list assist in the prioritization for reserve use. *Refer to Sections 4.3.7.2.4, 6.2.1.3.6, and 6.4.6.*
- *Stakeholder Register.* Managed by the project to ensure that the appropriate subject matter experts and other applicable stakeholders are aware of the change and are the changes and communicating impacts that could result from implementing or not implementing the change. All changes need to be provided to the appropriate individuals with the needed experience and knowledge to ensure that the change has had all impacts identified and assessed across the lifecycle. *Refer to Section 4.3.3 for more information on identifying stakeholders.*
- *Risk Identification and Management.* Changes can affect the consequences and likelihood of identified risks or can introduce new risks to the project. *Refer to Section 4.3.7 for a more detailed discussion concerning the identification, assessment, and handling of risk and the INCOSE SE HB Risk Management Process.*
- *Interface Audit.* Changes to external systems in which the SOI interacts or changes to the SOI that affect the interaction of the SOI with an external system are a major risk to the project. In addition, missing interface requirements and/or definitions concerning an interaction represent key risk to the project. Use of an interface audit, as defined in Section 6.2.3.6, will help identify and address these issues.
- *Allocation, Budgeting, and Traceability.* Changes to any requirement need to be assessed in terms of allocation, budgeting, and traceability, as discussed in Chapter 6 and below in Sections 14.2.6 and 14.2.7.
- *Cost and Schedule.* All changes must be assessed in terms of impacts on cost and schedule.

14.2.6 Manage and Control of Flow Down, Allocation, and Budgeting

A key part of managing and controlling the needs and requirements includes traceability between needs and implementing requirements, parent and child requirements, as well as traceability between the design input requirements and design output specifications.

As discussed in Section 6.4.3 during the *Design Input Requirements Definition*, based on analysis of the design input requirements and the SOI whose requirements are being allocated, the INCOSE SE HB *Architecture Definition Process* decomposes the system into subsystems and system elements, resulting in the next level of the architecture. During this analysis, the project team determines what "role," if any, each subsystem and system element at the next level of the architecture has in the implementation of the design input requirement(s) being allocated.

For each of the allocated requirements, the receiving subsystems and system elements define child requirements that are necessary and sufficient to meet the intent of the allocated parent requirement. Each of these child requirements is traced to their allocated parent requirements. "Necessary and sufficient" are addressed both during the definition process as well as part of requirement verification and requirement validation prior to baselining each set of design input requirements. A major part of requirements management is managing the allocations and traceability to 1) ensure each requirement is allocated to the next level until no further elaboration is necessary and 2) once baselined, the changes to the requirements at any level are assessed in terms of the impacts of those changes on allocation and traceability.

As part of allocation, performance, form, and quality requirements are apportioned (budgeted) to each system, subsystem, and system element that has a role in the realization of the budgeted value. Budgets need to be managed and controlled at the integrated system level. A critical concept associated with requirements and budgeting is that the budgeted quantities result in requirements that have a dependency—a change in one will result in the need to change another.

Because of these dependencies, establishing traceability between the child requirements and their allocated parent, and between peer requirements that have a dependency, is critical. Identifying and managing these lower-order dependencies is exceedingly difficult to do unless the project team is using a federated/integrated data and information system model.

Managing these budgets can be challenging in that, initially, they are often estimations with a minimum of analysis, especially for a document-centric approach to SE. The budgets can be very dynamic as the design matures for each subsystem and system element. Some subsystems or system elements may need less than what was budgeted, and others more. These changes have a ripple effect on all dependent budgets up and down the levels of the system architecture. Because of this, it is critical to manage all budgets within the integrated system architecture from the top using a single integrated model of the SOI.

Budgets are established as limits within which a quantity is managed. Given there is uncertainty with the budgets, there is inherent risk to the project being able to stay within the allocated budgeted values. One way to help manage those risks is the use of margins and reserves (Section 6.4.6). Another key role of NRVV management is the management of the margins and reserves.

The uncertainty concerning budgeted quantities decreases as the design matures. A common management role is to "release" reserves as they are needed and to tighten the margins. For example, for spacecraft mass, the project may define a reserve of 30% when the integrated set of needs is baselined for the SOI, but once allocated baseline of all the sets of design input requirements for each part of the SOI architecture is complete, this reserve can be reduced to 20%. As the design matures, more margin can be released at PDR and the rest released at CDR (this is often referred to a maturity margin management).

14.2.7 Combine Allocation and Traceability to Manage Requirements

Another key NRDM activity is managing traceability. Because allocation is closely tied to traceability, it is useful to combine the concepts from a management perspective. Combining the concepts of allocation and traceability provides a powerful method to manage the design input requirements, especially across levels and across subsystems and system elements within a specific level.

A major advantage of RMTs is their ability to produce reports based on filtering the information contained within the requirement attributes and other information. These reports can identify issues within and between requirement sets using the allocation and traceability information that will need to be addressed and corrected.

There are two common issues concerning allocation:

1. Requirements are not allocated.
2. Requirements are not allocated correctly.

For requirements that are not allocated: Unless the project has determined that no further elaboration of the requirements for a subsystem or system element is needed and they are ready to make a buy, make, or reuse decision, all requirements must be allocated to the subsystems or system elements at the next level of the physical architecture that has a role in meeting the allocated requirement. For an SOI that requires further elaboration, the project team can use the RMT to generate a report that lists all design input requirements for the SOI that are not allocated to a lower-level subsystem or system element. Ideally, the report will be "null"; if it is not, the unallocated requirements will need to be allocated. If not allocated, the implementing child requirements will not have been defined and further allocated, resulting in missing requirements for the system elements at lower levels of the architecture. A missing requirement at the system level of the architecture could result in missing requirements for the subsystems and system elements at the lower levels of the architecture.

For requirements that are not allocated correctly: It is one thing to ensure all requirements have been allocated, but another to assess whether the allocations are correct, either to a wrong subsystem or system element or failing to be allocated to all applicable subsystems and system elements that have a role in realizing the allocated parent requirement. RMT reports can indicate that requirements have been allocated, but the RMT has no way to automatically determine whether allocations are correct—that must be done by members of the project team. *Note that traceability occurs naturally (as it should) after requirements are decomposed and allocated. Note: "Traceability" without allocation is merely an unfounded assertion.*

Unfortunately, the concept of allocation is not as well understood as traceability. There are several common misconceptions and bad practices concerning allocation and traceability:

- *Some feel they can use the traceability matrices to assess allocation.* The thought process is that if a parent has child requirements defined for a subsystem or system element, it is a safe assumption that the parent must have been allocated to that subsystem or system element. However, people make mistakes during the allocation process, and that assumption is not always valid. Allocations can be incorrect or incomplete. *Ideally, allocation is both a top-down and bottom-up activity.* The system architect and owner of the parent requirement make an initial allocation, and the owners of the receiving subsystems or system elements confirm the validity of the allocation. The likelihood of incorrect allocations is less when the allocations are based on the architectural, analytical, and behavioral models developed by the project team.
- *Missing information in the traceability matrix.* Another issue is that when the traceability matrices are developed, rather than including the requirement text (or at least a summary or title of the requirement) in the matrix (such as the example shown in Table 6.2), only requirement

numbers are used. With only requirement numbers, it is hard to assess the correctness of the allocations just seeing that the allocated parents have some child requirements without seeing the wording of the parent and child requirements and their rationale.

- *Not all tool vendors understand the real meaning of allocation and thus do not always implement the concept correctly within their tools.* One way to test this is to ask the tool vendors if they can generate a report that lists all child requirements that trace to a parent requirement that was NOT allocated to the subsystem or system element in which the child requirements exist. Ideally, they can generate this report AND the results in the report are "null." If not null, the project team will have to fix the discrepancy. In most cases, it will be a failure to have allocated the parent requirement to that subsystem or system element. Currently, few RMTs can generate such a report.

Common defects that can be identified by combining the concepts of allocation and traceability:

- *Parent requirements with no child requirements.* Unless no further elaboration is needed, all parent requirements must have implementing child requirements defined for each subsystem or system element to which the parent requirement was allocated. As stated earlier, managing the traceability of large sets of requirements in a trace matrix manually is difficult. This is another major advantage and reason for using models and RMTs. The models will help avoid missing child requirements, and RMTs can help identify cases when a parent has no child requirements.

  Use the RMTs to generate a report that lists all allocated parent requirements that do not trace to one or more implementing child design input requirement(s). Again, the report should be "null." If not, the project team must define child requirements that are necessary and sufficient such that when implemented, the intent of the allocated parent requirement will be achieved.
- *Needs with no implementing design input requirements.* All needs must have implementing design input requirements defined for each system, subsystem, or system element to which the need applies. Models will help avoid needs that were not transformed into a design input requirement, RMTs can help identify cases when a need has no implementing design input requirements.

  Use the RMTs to generate a report that lists all needs that do not trace to one or more implementing design input requirements. Again, the report should be "null." If not, the project team must go through the transformation process defined in Chapter 6 resulting in one or more design input requirements that are necessary and sufficient such that when implemented, the intent of the need will be achieved (requirement validation).
- *Orphan needs that do not trace to a source.* As shown in Figure 4.12, needs are transformed or derived from multiple sources. All needs must trace to a source. Models will help avoid needs that do not trace to a source; RMTs can help identify cases when a source has no implementing need.

  Use the RMTs to generate a report that lists all needs that do not trace to one or more sources. Again, the report should be "null." If not, the project team must go through the *Lifecycle Concept and Needs Definition* activities defined in Chapter 4 to determine the source and establish traceability. Without this traceability, it will be difficult to validate each need to ensure it communicates the intent of the source from which it was derived as well as assess the possible impacts of proposed changes.
- *Orphan requirements that do not trace to a need, parent, or a source.* All design input requirements must trace to a need, source, or parent. If a design input requirement does not trace to a need, parent, or source, it should be assumed either that the traceability process is flawed and should be redone or that the requirement is "gold plating" and should be removed from the set.

As stated earlier, traceability helps establish that the set of design input requirements has the GtWR characteristic *C1—Necessary*. If the requirement cannot be traced to a need, source, or parent, why is the requirement in the set? In addition to traceability, attribute *A1—Rationale* also helps establish whether a requirement is needed; if the writer cannot provide rationale, why is the requirement in the set?

The project team can use the RMT to provide a report for all design input requirements that do not trace to a need, parent, or source. Again, the result should be "null." If not null, the project team must determine whether each requirement without a need, parent, or source is needed, and if so, trace the requirement to the correct need, parent, or source and provide rationale for why the requirement is needed. One common reason for requirements not having a need, parent, or source is a failure of the writer to trace the requirement to its need, parent, or source when the requirement was added to the set.

Another reason for lack of trace could be the need, parent, or source is missing. The project team members responsible for the lower-level requirements know that, based on experience with similar subsystems or system elements, the requirement needs to be included in the set even if there was no need to transform it from, no source to derive it from, or no parent requirement allocated to the subsystem or system element.

A major reason for no need is that the project team did not define an integrated set of needs for the system, subsystem, or system element before defining the set of design input requirements as advocated in this Manual. No matter the reason, it is important to perform an analysis to determine if there is a missing need, source, or parent requirement in the applicable higher-level system and add it if the parent should have been included. If not included, the result could be that that parent requirement was not allocated to all the applicable lower-level subsystems or system elements, or a need or source was not identified that should have been, resulting in missing requirements. Once the missing need, source, or parent has been added, the project team will need to allocate the parent to the next level of subsystems and system elements to which it applies or derive the appropriate requirements that address the intent of the added need or source.

- *Needs, sources, or requirements with incorrect or missing implementing child requirements.* Again, it is one thing to ensure all needs, sources, and parent requirements have implementing child requirements, but another to assess whether the child requirements are correct or if there is a missing implementing child requirement. RMTs can indicate that needs, sources, and requirements have implementing child requirements, but the RMT has no way to automatically determine whether the child requirements are correct or there are missing child requirements—that must be done by members of the project team. Models can help the project team avoid incorrect or missing child requirements. A common reason for incorrect child requirements is the result of incorrect traceability—when the child requirement was added to the set, the writer failed to trace it to the proper need, source, or parent within the RMT. A common reason for missing child requirements is a failure to use models as an analysis tool to identify the child requirements. Incorrect or missing child requirements can be hard to identify.
- *Requirements with an incorrect parent or source.* Again, it is one thing to ensure all child requirements trace to need, source, or parent, but another to assess whether the need, source, or parent it traces to is correct. RMTs can indicate that requirements trace to a need, source, or parent, but the RMTs have no way to automatically determine whether the need, source, or parent is correct—that must be done by members of the project team. Models can help the project team avoid incorrect needs, sources, or parents. A common reason for incorrect need, source, or parent is the result of incorrect traceability—when the requirement was added to the set, the

writer failed to trace it to the proper need, source, or parent within the RMT. Traces to incorrect needs, sources, or parents can be hard to identify.

- *Sets of child requirements are not necessary and sufficient to implement the parent requirement, need, or source from which it was transformed/derived.* The sets of child requirements for a given need, source, or parent must form a necessary and sufficient set such that, when implemented, the intent of the need, source, or parent will be met. "Necessary" implies the requirement is needed to meet the intent of the parent requirement—or need—if not necessary, why is it in the set? "Sufficient" implies the set of child requirements, when implemented will result in the intent of the need, source, or parent being met—if there is a missing child requirement, the set will not be sufficient.

  Again, it is one thing to ensure all parent requirements have child requirements or a need or source has implementing design input requirements but another to assess the sets of child requirements or design input requirements are necessary and sufficient. RMTs can indicate that parent requirements have child requirements or that the needs or sources have implementing design input requirements, but the RMTs have no way to automatically determine whether they are necessary and sufficient, which must be done by members of the project team. Addressing the question of necessary and sufficient is difficult to do unless the project team uses models to aid in the assessment.

Addressing the issues discussed above is difficult and can take a considerable amount of time and resources, but the effort is necessary to define and manage the sets of design input requirements for an SOI that have the characteristics defined in the GtWR for well-formed sets of design input requirements.

Using a data-centric approach, as advocated in this Manual can reduce the occurrence of these issues as well as make it much easier to identify and correct these issues. *Failing to spend the time and resources to do these assessments results in an accumulation of technical debt that will have to be paid back later at an exceedingly high-interest rate.*

Note: Modern RMTs allow the project to define a traceability and dependency model within the tool and define rules concerning required traceability. The tool will then enforce these rules, helping to minimize traceability issues discussed in Chapter 6.

This activity helps establish each design input requirement has the GtWR characteristics *C1—Necessary—and C8—Correct.* It also helps establish the set of design input requirements has the GtWR characteristics *C10—Complete, C11—Consistent, and C14—Able to be Validated.*

14.2.8 Manage and Control Interfaces

Note: This section is an elaboration of the INCOSE SE HB Section 3.2.4, Interface Management.

Managing interfaces cuts across all SOI lifecycle processes and activities. Given that the behavior of a system is a function of the interaction of its parts, it is critical that the project team identify and define each of the interactions across interface boundaries between all subsystems and system elements that make up the SOI as well as interactions with external systems including users and operators. *Refer to Section 6.2.3 for definition of interfaces and interactions.*

Failure to identify and manage interface boundaries and interactions across those boundaries is a significant risk to the project, especially during system integration, system verification, system validation, and operations. Failing to do so will result in costly and time-consuming rework. Each interaction across an interface boundary must be accessed in terms of stability, documentation, threats, and risk. The SOI is particularly vulnerable to undesirable things happening at and across the interface boundaries, especially when interfacing with external systems over which they may have little or no control. Identifying risk associated with interface boundaries and interactions across those boundaries is key to exposing potential risk to the project.

For this reason, it is critical that the project team define how they will manage interfaces in their SEMP starting at the beginning of the project as a *distinct project function* defining lifecycle concepts and needs definition for how the project will ensure the SOI will work safely and securely with the external systems with which it must interact in the intended operational environment AND is protected from outside threats across those interface boundaries.

Managing interfaces includes the oversight and management of the identification of interface boundaries, identification of interactions across those boundaries, defining and agreeing on definitions for each interaction, identifying the risk associated with each interaction, and defining interface requirements for each interaction. The identification of interface boundaries is a key function of the *Lifecycle and Needs Definition* activities, the INCOSE SE HB *Architectural Definition Process*, and concurrent *Design Input Requirements Definition* activities. During design, the interface definitions evolve, and the resultant definitions are recorded within interface definition type documents such as an ICD or IDD, which contain definitions dealing with each the characteristics of the "thing involved in the interaction" that is crossing each interface boundary, the media involved, and what each SOI looks like at the boundary.

Managing interfaces is a major activity across the lifecycle. Many of the issues found during system integration, system verification, and system validation involve an interface. Interface issues are also common during the integration of the SOI into the macro system it is a part of. Once the SOI has been successfully integrated into the macro system, it is a part of interfaces that must continue to be managed and maintained. When the system is a part of an System of Systems (SoS), managing the interactions with other systems within the SoS is of critical importance.

As a crosscutting activity starting at the beginning of the project, managing the interfaces across the lifecycle:

- Highlights underlying critical issues much earlier in the project than would otherwise be revealed that could impact the project's budget, schedule, and system performance.
- Clarifies the dependencies an SOI has on other systems (enabling systems) and dependencies other systems have on the SOI.
- Helps ensure compatibility between the SOI and those external systems with which it interacts.

Managing interfaces includes the following key functions:

- Facilitating cooperation and agreements with other stakeholders.
- Defining roles and responsibilities.
- Enabling open communication concerning issues.
- Establishing timing for providing interface information, problem resolution, and agreeing on definitions for the interactions across interface boundaries early in the project.
- Accessing and managing risk as part of the INCOSE SE HB *Risk Management Process,* avoiding potential impacts, especially during integration and operations.
- Establishing baselines for interface requirements, interface definitions, architecture, and design artifacts.
- Ongoing management and CM of the interface requirements and definitions, as well as any associated artifacts (such as ICDs, IDDs, or repositories) as part of the CM activities defined in the project CMP.

14.2.8.1 Best Practices Associated with Managing Interfaces Best practices for defining and managing interactions across interface boundaries and recording interface definitions and requirements include:

1. Define how the project team will manage interfaces in their SEMP and NRDM Plan.
2. Assign responsibility for the identification of all interface boundaries, definition of the interactions across those boundaries, and management of artifacts associated with interface management. It is a customary practice to assign one or more individuals to manage interfaces throughout the system lifecycle.
   a. Involve all interface stakeholders—they are the ones with the knowledge, and they are the ones the project team needs to work with to make sure all interfaces are addressed and associated artifacts developed.
   b. Include all interactions in the integrated set of system needs and transform those needs into interface requirements to be included within the SOI's set of system requirements.
   c. Ensure all interactions have been defined appropriately for the lifecycle stage, recorded, and baselined, and all interface artifacts are put under configuration control.
   d. For developing systems, ensure a process is in place to define, agree, and manage how the developing systems will interact with external systems across the interface boundary.
   e. For systems being developed by suppliers, ensure there is a plan in place concerning how interfaces will be managed within the contract, especially for cases where the interface definitions evolve with the design.
   f. Make all interface documentation available to developers as well as to those responsible for the external systems the SOI needs to interact.
   g. Provide traceability between interface requirement pairs, to parent requirements, and to the common interaction definitions.
   h. Track/monitor changes to all interface artifacts. A change to an interface definition can impact both sides of the interface as well as the parent requirement. Failing to manage interfaces is a major reason for issues during system integration and for systems failing system verification and system validation.
3. Effective communication is a vital part of interface management. Many projects incorporate the use of an Interface Control Working Group (ICWG), which includes members responsible for each of the interfacing subsystems and system elements as well as external systems and enabling systems. The use of ICWGs formalizes and enhances collaboration within project teams or between project teams and external organizations. The use of ICWGs is an effective approach that helps to ensure adequate consideration of all aspects of the interfaces. An ICWG is a critical function when managing an SoS.
4. Identify external interfaces early as part of the definition of drivers and constraints and lifecycle concepts analysis and maturation. Diagrams and models should be used to aid in the identification of all internal interactions between subsystems and system elements. Because of their importance, all interfaces must be identified to ensure compatibility with other systems, define the system's boundaries, and manage risk associated with interactions across those boundaries.
5. Obtain copies of all interface documentation for existing systems. If the interface definitions are not documented, work with the owners of the existing systems to document them. Ensure their current configuration is defined in sufficient detail so that the developing SOI can be designed to successfully interact with the existing systems across each interface boundary.

6. Plan for system verification of all interface requirements, system integration, and assessing and validating the behavior of the integrated system as a function of the interactions across the interface boundaries.

   System verification of all interface requirements can be expensive. To do system verification against an interface requirement, often there will be a need for special equipment to assess the interactions across the interface boundary.

   In support of system integration, system verification, and system validation activities, the system on the other side of the interface may not be available, or verification that the interactions are as required may be a prerequisite to integrate the SOI with the other system until both systems have competed system verification of all interface requirements. To complete system verification, simulators or emulators (software and/or hardware) may be needed. These must be budgeted for, developed, and go through their own system verification and validation in accordance with a schedule consistent with the SOI's system integration, system verification, and system validation schedule.
7. Throughout the lifecycle, bidirectional traceability must be maintained between the interface requirements, definition of interactions across interface boundaries, architecture elements, system analysis results, design artifacts, and system verification and validation artifacts.
8. For today's increasingly complex, software-intensive systems, it is critical that the project team uses a data-centric approach to address all interactions across interface boundaries and development of the associated artifacts. From a completeness and correctness perspective, the project team must use diagrams and models to identify all functional and physical interface boundaries and interactions across those boundaries with both internal and external systems.
9. As part of the interface definition, many projects find the need or benefit to apply interface standards. In some cases, such as for plug-and-play elements or interfaces across open system architectures, it is necessary to strictly apply interface standards to ensure the necessary interoperation with systems the project team does not control. Examples of these standards include Internet Protocol (IP) standards and Modular Open Systems Architecture (MOSA) standards. Interface standards can also be beneficial for systems that are likely to have emergent requirements by enabling the evolution of capabilities through the use of standard interface definitions that allow new system elements to be added.
10. A major issue concerning interface definition is that when an SOI is contracted out to a supplier and the SOI interacts with other supplier-developed SOIs. Often the contracts are issued prior to design, and thus, the design definitions of what the SOIs look like at the interface boundary and the media involved in the interaction have not yet been defined. In addition, it is common for the suppliers to have little insight into the workings of other supplier-developed SOIs with which they interact and how changes to those SOIs could affect the interactions and performance of their SOI or changes to their SOI could affect other SOIs. In these cases, it is important that the customer clearly addresses in the contract how each supplier will support, participate in, and comply with the interface management activities during interface definition, design, system integration, system verification, and system validation.

This activity helps establish the set of design input requirements has the GtWR characteristics *C10—Complete and C11—Consistent, C14—Able to be Validated.*

14.2.9 Manage and Control System Verification and Validation Activities

Key management and control activities include the planning and preparation for, and oversight of, system verification and system validation activities across the lifecycle; and assessing changes to the integrated sets of needs and sets of design input requirements as to their impact on system verification and system validation. Planning and preparation begin during the project's lifecycle concepts analysis and maturation activities concerning the concepts for how the project plans to do system verification and validation and developing and implementing MIVV Plan and individual SIVV plans.

A key part of planning and preparation is ensuring that the system validation attributes for each need in the integrated sets of needs are defined when the need expressions are formulated, as well as the system verification attributes for each design input requirement are defined when the requirement expressions are formulated. This applies to each subsystem and system element within the SOI physical architecture as well as to the integrated system. These attributes include defining the *Success Criteria*, *Strategy*, *Method*, level of integration when system verification or system validation will occur, and the *Responsible Organization* for system verification or system validation (discussed in Chapters 10 and 11).

This planning and preparation are critical as system verification and validation activities represent a significant portion of the project's budget and schedule, as well as the fact that acceptance of the SOI is dependent on successful system verification and system validation activities.

Oversight of the system verification and validation activities includes oversight and management of:

- The implementation of the provisions defined within the MIVV Plan and each SIVV Plan.
- The development of information for each system verification and validation *Instance*.
- The definition of system verification and validation *Activities*.
- The definition of the system verification and validation *Procedure Requirements*.
- The development, review, and approval of the system verification and validation *Procedures*.
- Monitoring and reporting on the status of the performance of these *Procedures*.
- Reporting the results of the execution of the Procedures via the *Execution Records*.
- The combination of the *Execution Records* into *Approval Packages*.
- Submission of the *Approval Packages* to the *Approval Authority*.
- Tracking the status of the *Approval Authority* acceptance, certification, and qualification of the system elements, subsystems, and the integrated system.

To ensure the project's system verification and validation programs are managed effectively to meet schedule milestones and budget, metrics should be developed to track and manage the planning and performance of the system verification and validation programs across all lifecycle stages. To do this, some organizations may develop an in-house application to manage and control their system verification and system validation activities, while others may choose to procure an existing application from a vendor.

Examples of metrics include:

- The number of needs and design input requirements that have the system verification and validation attributes defined concerning *Success Criteria*, *Strategy*, *Method*, level, and *Responsible Organization* as they are written.
- The number of system verification and validation *Instances* (needs, design input requirements, and design output specifications) that have been defined.
- The number of *Instances* that have been assigned to an *Activity*.
- The number of *Activities* that have been defined.
- The number of *Activities* that have *Procedures* developed for their implementation.
- The number of *Procedures*.

- The number of system verification and validation *Events* that have been scheduled.
- The number of *Procedures* that are in work, completed, successful, unsuccessful, and approved.
- The number of *Execution Records* that have been completed.
- The number of *Approval Packages* being developed and their status (in work, completed, approved, and ready for submission to the *Approval Authority*).
- The number of *Approval Packages* approved or disapproved by the *Approval Authority*.
- The percentage of *Instances* that have been closed (status and outcomes).
- The burn-down plan for system verification and validation *Instances* closures (against needs, design input requirements, and design output specifications).
- The number of nonconformances/compliances and associated variances.

System verification and system validation metrics can be generated by the project's toolset to produce reports or dashboards that communicate the status of the project's system verification and validation programs to the project team, upper-level management, and customers. Metrics need to be kept current to provide the project team and customers the latest insight into the status of the project's system verification and system validation program. The closure of hundreds (thousands?) of system verification and validation *Instances* can involve a significant percentage of a project's resources, budget, and schedule.

These metrics can be generated from various sources, including information from the project's budgeting and scheduling tools, the attributes defined for the individual needs and design input requirements expressions, as well as any special metrics contained in the SOI's integrated/federated data and information model developed to help manage and control the project's system verification and validation program.

This activity helps establish each design input requirement has the GtWR characteristic *C7—Verifiable*. It also helps establish the set of design input requirements has the GtWR characteristic *C14—Able to be Validated.*

15

ATTRIBUTES FOR NEEDS AND REQUIREMENTS

Attributes [65] are included as part of the need and requirement expressions to define and capture key information to aid in the definition, verification, validation, management, and reuse of not only the needs and design input requirements but also the management of the System of Interest (SOI) across all lifecycle activities. The attributes discussed in this section apply to needs and requirements, no matter the level at which they exist.

A list of attributes is provided that can be associated with each need or requirement statement. The need or requirement statement plus its attributes results in a need or requirement expression as defined in Chapter 2.

Note: Many of the attributes listed are useful for both managing needs as well as requirements. Others may be more useful as applied to only requirements.

This list is not exhaustive and is not meant to be prescriptive or proscriptive in any way. It is not the intention that an organization should include all these attributes when defining needs or requirements expressions. The purpose of this Chapter is to present a set of attributes that have been and are being used by various organizations. In some cases, an organization may define other attributes not included in this list of attributes. The important thing is that the project team define and enforce an attribute scheme that is specific to their domain, product line, culture, and processes. The operational-level stakeholders within the organization must agree on a set of attributes that will be used for all levels of needs and requirements and implemented within the project team's toolset and integrated dataset.

Although it is unlikely an organization would include all the attributes listed herein, there is a subset of these attributes that is proposed to represent a minimum set to be defined for each need or requirement. This minimum set supports the mandatory characteristics of well-formed need and requirement statements and supports the maintenance of the need and requirement statements, especially Configuration Management (CM) and tracking the progress of the system design, system

INCOSE Needs and Requirements Manual: Needs, Requirements, Verification, Validation Across the Lifecycle,
First Edition. Louis S. Wheatcraft, Michael J. Ryan, and Tami Edner Katz.

verification, and system validation activities throughout the product lifecycle. In the discussion below, the attributes in that minimum set are annotated with an asterisk ("*").

Note that the information contained within the attributes is meant to be informative in a contractual sense (not normative); because of this, the word "shall" must not be used in any of the text contained in attributes to help ensure the amplifying information is not inadvertently interpreted to be an additional requirement.

The attributes are organized within the following five broad categories:

- Attributes to help define need and requirement statements and understand their intent.
- Attributes associated with system verification or system validation.
- Attributes to help manage the needs or requirements across the lifecycle.
- Attributes to show applicability and enable reuse of the needs and requirements.
- Attributes to aid in product line management.

Note: In the following definitions, the term SOI is used. The SOI is the entity to which the need or requirement statement applies. From a software perspective, the entity could be an application, module, feature, or component.

15.1 ATTRIBUTES TO HELP DEFINE NEEDS AND REQUIREMENTS AND THEIR INTENT

This set of attributes is used by both the author of the needs and requirements and anyone who will be reviewing, implementing, verifying, validating, or managing the needs and requirements. The attributes in this set help ensure the needs or requirements have the characteristics of well-formed need or requirement statements defined in the Guide for Writing Requirements (GtWR) as well as help understand the intent and reason for including the need or requirement.

A1—Rationale*

Rationale states the reason for the need or requirement's existence. Rationale defines why the need or requirement should be included and other information relevant to better understand the reason for and intent of the need or requirement. Rationale can also be used to record assumptions that were made when writing the need or requirement statement, the source of numbers, and what design effort, if any, drove the requirement. If the need or requirement includes a constraint or a design implementation, the rationale explains the reason for including that need or requirement in the set. This attribute, along with the attributes of *A2—Trace to Parent*, and *A3—Trace to Source*, helps to support the claim that the need or requirement has the GtWR characteristics *C1—Necessary, C3—Unambiguous, and C13—Comprehensible.*

A2—Trace to Parent*

Requirements at one level are implemented at the next level of the system architecture via allocation (see *A5—Allocation/budgeting*). A child requirement is one that is an elaboration, derivation, or decomposition of an allocated parent. The achievement of each of the child requirements is necessary to the achievement of the parent requirement. Each of the child requirements must be traced to its parent requirement. When managing requirements within the project toolset, when a trace is established from the child to the parent, a bidirectional trace is also established between the parent and that child requirement.

The project toolset should support the concept of traceability and a separate attribute may not be needed as the applications in the toolset provide the ability to link entities together as an inherent capability of the application. However, for those using common office applications to document their requirements, trace to parent is accomplished via this attribute.

This attribute, along with attributes *A1—Rationale* and *A3—Trace to Source*, helps ensure the need or requirement has the GtWR characteristics *C1—Necessary* and *C10—Complete*. This attribute may also be classified under attributes used to manage needs and requirements. *Refer to Chapter 6 for a more detailed discussion on traceability.*

A3—Trace to Source*

Design input requirements result from a transformation of one or more needs, so all requirements must trace to need(s) from which they were transformed. Each need and requirement must be able to be traced to a source. For requirements, this is different from tracing to a parent requirement because it identifies where the requirement came from and/or how the requirement content was determined (rather which specific requirement is its parent). For example, for a requirement derived from a need that invoked a standard or regulation, the requirement could trace to a parent requirement, a need, and a requirement in a standard or regulation from which it was derived.

Each need must trace to a source to establish where the need was derived from. Examples of sources include mission, goals, objectives, measures, constraints, system concepts, user stories, use cases, models, analysis, documents, regulations, standards, risk derived from a trade study, interviews with a stakeholder, a specific stakeholder, minutes of stakeholder workshop, or Engineering Change Request (ECR). Sources could also be a functional area within an enterprise or business unit (for example, marketing, safety, security, compliance, quality, engineering, manufacturing, shipping, installation, operations, or maintenance).

Maintaining this trace is key to being able to show compliance to higher-level organizational requirements, customer requirements, regulations, and standards. This trace is also important in terms of change management. If something related to a source changes, then the need or requirement traced to that source may also need to be changed. For example, if the analysis used to generate a number in a requirement is updated as the design matures, the requirement must also be updated. *This attribute*, along with *A1—Rationale* and *A2—Trace to Parent*, helps to ensure the need or requirement has the GtWR characteristic, *C1—Necessary*.

A4—States and Modes

The state or mode of the SOI to which a need or requirement applies. Some systems have various states and modes, each having a separate set of requirements that apply to the specific state or mode. If the system and associated needs and requirements are structured in this way, this attribute enables needs and requirements to be assigned to the applicable state(s) or mode(s).

Rather than using this attribute, some organizations choose to include the state or mode as a condition within the need or requirement statement. A state could also represent a precondition for the execution of some action. That precondition could be included in this attribute or could be included in the need or requirement text. Refer to the GtWR for more information concerning requirement templates and the inclusion of conditional clauses within a requirement statement.

The advantage of using this attribute is that it enables the project team to search for and display all needs or requirements to which a given state or mode applies. It also avoids having to repeat a requirement with a different condition for each state and mode to which it applies.

This attribute helps to ensure the GtWR characteristic *C10—Complete* is met.

A5—Allocation/Budgeting*

Subsystems or system elements at the next level to which the requirement was allocated/budgeted. As discussed in Chapter 6, requirements at one level are allocated or budgeted to parts of the architecture for implementation. There are two types of allocation. The first type, allocation (budgeting) of performance, resources, or quality, involves allocating a quantity at one level to the next level, where the quantity is budgeted between subsystems and system elements at the next level (such as performance, power, mass, reliability, accuracy, precision, and quality). The second type of allocation is allocation of responsibility, where requirements at one level are assigned to subsystems and system elements at the next level for implementation.

For green field systems, when allocation is performed, there are no requirements at the next level, so allocation is linking a requirement at one level to subsystems or system elements at the next level of the architecture.

The project toolset should support the concepts of allocation and budgeting. A separate attribute may not be needed if the applications in the toolset provide the ability to link entities together as an inherent capability of the application. However, for those who are using common office applications to document their requirements, allocation is accomplished via this attribute.

This attribute helps to ensure a requirement will be addressed at the next level of the architecture.

15.2 ATTRIBUTES ASSOCIATED WITH SYSTEM VERIFICATION AND SYSTEM VALIDATION

This set of attributes is used to help track and manage the system verification and validation activities and status to provide objective evidence that the designed and built SOI meets its set of design input requirements (system verification) or its integrated set of needs (system validation). Each need or requirement represents a system verification or validation *Instance*. This set of attributes can be used for defining each *Instance* as discussed in Chapter 10 as well as provide metrics that management can use to track the status of the project's system verification and validation activities across the lifecycle. Each *Instance* is included within a system validation *Activity* that will be conducted by a system verification or validation *Procedure*. For projects that have a separate application for managing their system verification and validation programs, these attributes can be links or pointers to the information in that application database that would be included in these attributes. *Refer to Chapter 10 concerning defining the information that is included within these attributes.*

A6—System Verification or System Validation *Success Criteria**

The criteria which must be shown to have been met by the *Strategy* and *Method* to provide objective evidence that the SOI has successfully met the need or requirement. Identifying the *Success Criteria* helps to ensure the need or requirement has the GtWR characteristics *C3—Unambiguous*, C6—*Feasible*, *C7—Verifiable*, and *C8— Correct*.

A7—System Verification or System Validation *Strategy**

Strategy to be used to provide a high-level description of the approach to be used to verify the SOI meets a requirement or validate the SOI meets a need as defined by the *Success Criteria* in accordance with an agreed-to *Method* used to obtain the data that provides objective evidence that the system meets a need or requirement.

A8—System Verification or System Validation *Method**

The *Method* to be used (Test, Demonstration, Inspection, or Analysis) to provide objective evidence used to determine the designed and built SOI meets the requirement (system verification) or need (system validation) as defined by the *Success Criteria* with the required degree of confidence. Identifying the *Method* helps ensure the need or requirement has the GtWR characteristics *C3—Unambiguous*, C6—*Feasible*, *C7—Verifiable*, and *C8—Correct*.

A9—System Verification or System Validation *Responsible Organization**

The organization that is responsible for system verification and validation for a need or requirement or sets of needs and sets of design input requirements for the SOI. The responsible organization could be the customer, developing organization, or if, external to the organization, the supplier. Refer to Chapter 13 for a detailed discussion on supplier-developed SOIs concerning system verification and system validation.

A10—System Verification or System Validation Level

Architecture level at which the verification or system validation *Activity* and implementing *Procedure* will be performed to provide objective evidence that the SOI meets the requirement (verification) or need (validation) during system integration. Possible values could include: <system>, <subsystem>, <assembly>, <component>, <hardware>, <software>, or <integration>. The actual name of the level will be organization or domain specific based on the product breakdown structure or architecture description document defined in the project's Master Integration, Verification, and Validation (MIVV) plan or System Integration, Verification, and Validation (SIVV) plans.

A11—System Verification or System Validation Phase

Lifecycle phase in which a system element or subsystem verification or system validation *Activity* and implementing *Procedure* will be performed to provide objective evidence that the system element or subsystem meets a requirement or need. Possible values could include design verification and design validation, production verification, system integration, system verification, system validation, proof testing, certification, qualification, or acceptance. The actual name of the phase can be organization or domain specific as defined in the project's MIVV plan or SIVV plans.

A12—Condition of Use

Description of the operational conditions of use in which the need or requirement applies. Rather than using this attribute, some organizations prefer to include the *condition of use* as part of the need or requirement statement; others state the conditions of use as separate requirement to define the operation environment in which the system will be used. In some cases, condition of use involves a specific trigger event or state of the system. Some organizations prefer to include the trigger or state as part of the need or requirement statement. Understanding the *condition of use* is key to proper implementation in design as well as conducting system verification and validation activities.

Refer to the GtWR for more information concerning requirement templates and the inclusion of conditional clauses within a requirement statement.

The advantage of using this attribute is that it enables the project team to search for and display all needs or requirements to which a given *condition of use* applies.

Identifying the condition of use helps ensure the need or requirement has the GtWR characteristics *C4—Complete*, and *C8—Correct*.

A13—System Verification or System Validation Results

The results of each system verification or validation *Activity* and implementing *Procedure* are communicated within an *Execution Record*. This attribute traces each *Instance* to the associated *Activity* in which it is included and the results as communicated in the *Execution Record* as discussed in Chapters 10 and 11. Possible values could include *complete—successful*, *complete—unsuccessful*, *passed*, or *failed*. This attribute enables reports to be generated or status to be displayed on a dashboard, for example, percent of requirements for the SOI whose system verification *Activity* has been successfully completed, percent of needs for the SOI whose system validation *Activity* has been successfully completed, or the number or percentage of system verification *Activities* that were not successful.

A14—System Verification or System Validation Status

Indicates the status of the system verification or validation *Activities* and implementing *Procedure*, including sign-off/approval of the *Procedure* and resulting *Execution Record* stating whether the SOI has been verified that it meets the design input requirement or validated that it meets the need. Possible values could include not started, in work, [percentage] complete, complete, sign-off in work, or *Execution Record* complete. This attribute enables reports to be generated or status to be displayed on a dashboard, for example, percentage of requirements whose *Execution Record* is complete and included in the system verification or system validation *Approval Package. Refer to Chapters 10 and 11.*

15.3 ATTRIBUTES TO HELP MANAGE THE NEEDS AND REQUIREMENTS

This set of attributes is used to manage and maintain the needs and design input requirements and sets of needs and sets of design input requirements across the lifecycle. Areas of prime importance to management, in addition to the above-mentioned attributes dealing with system verification and validation, include managing risk, managing change, ensuring consistency and completeness, and having insight into the status of needs and requirements definition and implementation across the lifecycle.

Managing risk is a key role of management as well as making sure needs and requirements are tracked across the lifecycle, especially if they are high priority or critical/essential to project success. The inclusion of attributes dealing with these topics allows reports to be generated or information to be displayed on a dashboard by applications within the project's toolset, which will help the project team keep the project on track and manage changes (*refer to Chapter 16*).

A15—Unique Identifier*

A unique identifier, which can be either a number or mixture of characters and numbers used to refer to the specific need or requirement. It can be a separate identifier or automatically assigned by applications within the project's toolset. This identifier is used once and never reused. A unique identifier is also used to link needs and requirements in support of the flow down of requirements (allocation and budgeting) and traceability and to establish peer-to-peer relationships. Some organizations include unique identifier codes that relate to the SOI or level of architecture to which the need or requirement applies, e.g., [SOI]SNxxxx or [SOI]SRxxxx. *Organizations should never use a document paragraph number as a requirement unique identifier.*

A16—Unique Name

A unique name or title for the need or requirement that reflects the main thought the need or requirement is addressing. This is useful to provide a short title to allow the requirements to be viewed in a document form, as a tree structure, or in graphical form. Defining a unique name was common in requirements documents for use as section titles but is not used as often when the sets of requirements are maintained and managed within a database. Caution should be used when using this attribute, as discussed in R25 in the GtWR in which it states, "Avoid relying on headings to support explanation or understanding of the need or requirement."

A17—Originator/Author*

The person defining and entering a need or requirement expressions into the Requirement Management Tool (RMT). When entering a need or requirement expression into the project's database, many of the applications will automatically log the name of the person entering the requirement statement and associated attributes; in this case, the application automatically fills in this attribute.

A18—Date Requirement Entered

The date the need or requirement was entered into the project's toolset. When entering a need or requirement into the project's database, many of the applications will automatically log the date and time of entry; in this case, the application automatically includes this attribute.

A19—Owner*

The person or organizational element that maintains the need or requirement, who has the right to say something about this need or requirement, approves changes to the need or requirement and reports the status of the need or requirement. The same person could be both the Owner and the Source (of the need or requirement), but it is recommended that Owner and Source are two different attributes. The Owner maintains both the need or requirement statements and all associated attributes. *Note: Some prefer to add the position/title or role of the Owner (for example, Manager, SE, or I&T engineer). Especially in today's world, where turnover of staff is common, the author may be gone and forgotten before the project is complete and that name becomes meaningless.*

A20—Stakeholders

List of key stakeholders (name, position/title, or role) that have a stake in the implementation of the need or requirement and who will be involved in the review and approval of the need or requirement as well as any proposed changes. Within the project's toolset, this list of stakeholders is those that will be automatically notified when either the need or requirement statement is entered into the data set or any associated change to a need or requirement statement, or attribute is proposed. This attribute enables collaboration within the RMT.

A21—CM Authority

The organizational entity that has CM authority over the need or requirement expressions and has the authority to baseline the needs and requirements and approve proposed changes to a need or set of needs or changes to a requirement or set of requirements of which it is a part. Depending on the level of the architecture, organizations may have different levels of CM authority.

A22—Change Proposed

An indication that there is one or more proposed changes to the need or requirement that the owner or other stakeholders need to act on. Specific changes will normally be managed as unique entities within the RMT by the organization's CM authority defined above, such that their status can also be tracked within the RMT.

A23—Version Number

An indication of the version of the need or requirement. This is to make sure that the correct version is being implemented as well as to provide an indication of the volatility of the need or requirement. A need or requirement that has a lot of changes could indicate a problem or risk to the project.

The version number may be automatically assigned by an application within the project toolset whenever a change has been approved. A major feature of most RMTs is the ability to store all versions of a need or requirement along with each change associated with a need or requirement and the status of that change. Once the change has been approved, the need or requirement will be updated, and a new version number assigned.

A24—Approval Date

Date the current version of the need or requirement was approved. The approval date is dependent on Attributes A28 - Need or Requirement Verification Status and A 29 - Need or Requirement Validation Status.

A25—Date of Last Change

Date the need or requirement was last changed, i.e., the date of the current version. When entering a change to a need or requirement, many of the applications will automatically log the date of the change automatically filling in this attribute.

A26—Stability/Volatility

An indication of the likelihood that the need or requirement will change. Some needs and requirements, when first proposed, will have numbers that are best guesses at the time, or the actual value is not readily available when the requirement has been entered into the project database. The number may not be agreed to by all stakeholders or there may be an issue on the achievability of the need or requirement. Some needs or requirements will have a to-be-determined (TBD), to-be-computed (TBC), or to-be-resolved (TBR) in the text. Stability/Volatility is related to a high- or medium-risk need or requirement (see *A36—Risk of Implementation* attribute below for more information). Possible values could include stable, likely to change, incomplete, or unresolved TBX (TBR, TBS, TBC).

TBXs represent risk to the project and should be treated as separate entities as they must be managed and resolved before the need or requirement can be implemented by design. The longer they remain unresolved, the more technical debt the project is accumulating. The RMT should provide the means to manage the TBXs as discussed in Chapter 14. This attribute could be a link to the TBX module within or external to the RMT, where the TBXs and other risks are tracked and managed.

This attribute also enables reports to be generated concerning the GtWR characteristics *C6—Feasible and C12—Feasible.*

Refer to Section 14.4 for a more detailed discussion on TBX management and managing unknowns.

A27—Responsible Person

Person (name, position/title, or role) responsible for ensuring that the need or requirement is implemented in the design as well as verifying that the SOI met the requirement or validating that the SOI met the need. This is different from need or requirement owner defined previously. The Responsible Person maintains attribute *A31—Status (of implementation).*

A28—Need or Requirement Verification Status*

An indication that a need or requirement has been verified. As defined in Chapter 2, need or requirement verification is the process of ensuring the need or requirement meets the rules and characteristics (necessary, singular, conforming, appropriate, correct, unambiguous, complete, feasible, and verifiable) for well-formed need or requirement statements as defined in the GtWR or a comparable guide or checklist developed by the organization. Possible values include not started, in work, complete, or verified. Once the need or requirement has been verified as defined in Chapter 5 (for needs) or Chapter 7 (for requirements), it can be approved. The date of approval is included in Attribute A24—Approval Date.

A29—Need or Requirement Validation Status*

An indication the need or requirement has been validated. Requirement validation is the confirmation the requirement is an agreed-to transformation that clearly communicates the intent of needs in a language understood by the designers and those responsible for design and system verification. Need validation is the confirmation the need is an agreed-to transformation that clearly communicates the Mission, Goals, and Objectives (MGOs), measures, and lifecycle concepts in a language understood by the requirement writers, designers, and those responsible for design and system validation. Need validation and requirement validation activities need to involve the customers and users and all other stakeholders of the system being developed. Possible values include not started, in work, complete, or validated. Once the need or requirement has been verified as defined in Chapter 5 (for needs) or Chapter 7 (for requirements), it can be approved. The date of approval is included in Attribute A24—Approval Date.

A30—Status of the Need or Requirement

Maturity of the need or requirement. Possible values include draft, in development, ready for review, in review and approved, disapproved, or deleted. In general, the set of needs or requirements would not be baselined until all the individual need or requirement verification and validation status attributes have been set to indicate that verification and validation of the need or requirement statement is complete.

A status of "approved" should only be allowed if the above-mentioned need or requirement verification attribute (A28) and validation attribute (A29) both indicate the need or requirement has been both verified and validated, and the requirement expression is stable; *A36—Risk (of implementation)* is at an acceptable level and has the organization's mandatory attributes defined. Even if the need or requirement has been disapproved or deleted, it is useful to keep its history within the database for future consideration. The date of approval is included in Attribute A24, Approval Date.

A31—Status of Implementation

For a need, an indicator of the status of transformation of a need into one or more requirements. Possible values include transformation complete, transformation in work, and no plans to transform the need into one or more requirements. When the transformation is complete, all resulting requirements must trace to the need(s) from which it was transformed. See also *A3—Trace to Source*. All needs should have at least one implementing requirement. If the developing organization has no plans for transforming the need into a requirement, it must address the rationale for not doing so with the CM authority (*A21—CM Authority*).

For a requirement, an indicator of the status of the implementation or realization of the requirement in the design and design output specifications. Possible values include requirement implemented in design; requirement not feasible; requirement not implemented and no plan to do so; and requirement not implemented but have plan to do so. If implemented and the project toolset allows, this attribute could be a trace between the design definition artifacts, design output specification, and the design input requirement. If using an SE modeling tool, this attribute could indicate the requirement has been implemented in the model or linked to a portion of the model.

This attribute also enables reports to show the GtWR characteristics *C6—Feasible and C12—Feasible.*

Table 15.1 shows an example of how one organization records and reports the status of implementation of design input requirements.

TABLE 15.1 Example Implementation Status Recording and Reporting.

Status	Meaning
Waived	Implementation waived; No plans to implement
Pending	Implementation planned and is expected but has not occurred yet.
Under review	The decision to implement is under review.
Successful implemented	Implemented in the design successfully.
Unsuccessful implementation	Implementation attempted by failed because of xxxx—waiver (or deviation) in work.
Partial implementation	Implementation attempted but only partially successful. Deviation in work.

Note: If there is no plan to implement a requirement, a change, waiver, deviation, or exception would have to be submitted and approved by the CM authority (A21—CM Authority).

A32—Trace to Interface Definition

As discussed in Chapter 6, the interactions between two systems across an interface boundary are described in interface definitions, which are often contained in a document (or other similar electronic representation), which has a title such as an Interface Control Document (ICD), an Interface Agreement Document (IAD), an Internal Interface Definition Document (IIDD), or an Interface Definition Document (IDD). In software, the interface definition could be in a data dictionary, an Application Programming Interface (API), or Software Development Kit (SDK).

The interface requirements contained in each of the interacting system's requirement sets will include a reference to where the interaction is defined. This attribute provides a link tracing the interface requirements to where the interaction referenced in the requirements are defined.

This attribute facilitates reports out of the SOI's integrated database so that, for any interface definition, it is obvious which interface requirements invoke that definition. This aids in change control. If any interface requirement changes, the definition may need to be changed. If the definition changes, interface requirements that are linked to that definition may need to be assessed or changed as well (or at least the owners of the interface requirement need to be made aware that the definition has changed).

This attribute also enables reports to show the requirement has the GtWR characteristics *C1—Necessary, C4—Completeness, C7—Verifiable,* and *C11—Consistent.* If no requirement points to an interface definition or only one requirement points to the definition, there may be a missing interface requirement. This attribute supports the interface audit activity discussed in Chapter 6.

A33—Trace to Dependent Peer Requirements*

This attribute links requirements that are related to, or dependent on, each other (other than parent–child) at the same level. Peer requirements may be related for several reasons, such as they may be codependent (a change to one could require a change to the other), bundled (to implement this requirement, this other peer requirement must also be implemented), or a complementary interface requirement of another system to which the system interacts.

This trace may also result from the allocation and budgeting process. If a resource was allocated from one level and budgeted to multiple system elements at the next level, the values of the resulting child requirements are dependent—a change to one would result in a change in one or more of the other child requirements. Because of this dependency, all child requirements having a common parent would need to be linked together. (See *A5—Allocation/Budgeting.*) See Chapter 6 for more information on traceability, allocation, budgeting, and interface requirements.

This attribute enables reports to show individual requirements have the GtWR characteristic *C1—Necessary* and for a set of requirements, *C10—Complete* and *C11—Consistent.* For most RMTs or other SE tools, having a trace from one codependent requirement to another results in an automatic bidirectional trace such that if there is a proposed change to one, the tool can notify the applicable stakeholders and the impact of that change on the other(s) dependent requirements and their parent can be assessed.

A34—Priority*

Priority is an indication of how important the need or requirement is to the stakeholders. It may not be a critical/essential requirement—that is, one the system must possess, or it will not work at all (see *A35 – Criticality or Essentiality*)—but maybe something the stakeholder(s) hold dear in that it adds value from their perspective. Priority may be characterized in terms of a number (1, 2, 3) or label (high, medium, low). The reason for classifying a need or requirement as high priority should be communicated in the rationale attribute (*A1 – Rationale*). Assuming each child requirement is necessary, their priority will be inherited from their parent requirement. In the case of needs, all requirements transformed from that need would inherit the priority of the need.

High-priority needs and requirements must always be met for the stakeholder expectations to be met; lower-priority needs and requirements may be traded off when conflicts occur or when there are budget or schedule issues resulting in a de-scope effort. If there is a need to trade off (modify or delete) a need or requirement, a change request must be submitted to the organization that has CM authority for the set of needs and requirements (*A21—CM Authority*).

A35—Criticality or Essentiality*

A critical or essential need or requirement is one that the system must achieve for the system to function at all—that is, without it, the system will not be able to achieve its primary purpose or intended use in the intended operational environment when operated by its intended users. A critical or essential requirement would also be included in the system's key performance requirement set. Criticality/essentiality is most often characterized in terms of yes or no. Not all needs or requirements are critical or essential. The reason for classifying a need or requirement as critical or essential should be communicated in the rationale attribute (*A1—Rationale*).

Assuming each child requirement is necessary, their criticality or essentiality will be inherited from their parent (requirement or need from which the requirement was transformed). Critical or essential needs and requirements must always be met. Unlike high-priority needs or requirements, critical or essential requirements may not be traded off when there are budget or schedule issues. If a critical need or requirement cannot be met, the feasibility of the project will have to be re-examined. To prevent this, the authors advocate that projects identify at least one feasible set of lifecycle concepts prior to baselining the set of needs and transforming those needs into design input requirements.

While all critical/essential needs and requirements will also have a high priority, it is possible to have a noncritical yet high-priority need or requirement.

A36—Risk of Implementation*

A value assigned to each need or requirement indicating the risk of the need or requirement not being met.

Risk of implementation may be characterized in terms of a number (1, 2, 3) or a label (high, medium, low). Some organizations communicate risk as a cross-product of likelihood and impact (a high likelihood and a high impact would be represented in a risk matrix as 1×1, while a low likelihood and high impact would be 3×1). Some organizations use a 3×3 risk matrix; others may use a more granular 5×5 matrix. An unstable requirement is also a risk factor (*see A26—Stability*).

Risk can also address feasibility/attainability in terms of technology, schedule, cost, or politics. If the technology needed to meet the requirement is new with a low maturity (low Technology Readiness Level [TRL]), the risk is higher than if using a more mature technology that has been used successfully in other similar projects (use and environment). The requirement can be high risk if the cost and time to develop a technology is outside what has been planned for the project. Risk may also be inherited from a parent requirement or need.

Needs or requirements that are high risk will also be tracked more closely, especially if they are high priority or are critical. These will be treated as Technical Performance Measures (TPMs) and monitored closely by the project team.

Needs or requirements that are at risk include those that fail to have the characteristics defined in the GFWR. Failing to have these characteristics can result in the need or requirement not being implemented (will fail system verification) and needs not being realized (will fail system validation).

This attribute also enables reports to show the need or requirement has the GtWR characteristic *C6—Feasible* and the set of needs or requirements has the GtWR characteristic *C11—Feasible.*

A37—Risk Mitigation*

This attribute indicates whether a need or requirement is part of a risk mitigation action.

As part of lifecycle concepts analysis and maturation activities, projects should assess management, development, production, system integration, system verification, system validation, compliance, and operational risk, as discussed in Chapter 4. For a risk allocated to the system under development, the management of the risk mitigation is needed throughout all lifecycle activities. To do this, lifecycle concepts for mitigating the risk are defined, and the associated needs to implement these concepts are defined. Design input requirements are transformed from these needs and allocated to lower levels of the architecture for implementation. Validation attributes are defined for those needs and verification attributes are also defined for those requirements.

To manage the risk mitigation across the lifecycle and to ensure the risk allocated to the system under development has been mitigated, the inclusion of this attribute is needed. This attribute, along with trace to parent, trace to source, allocation, priority, and criticality/essentiality, allows the risk mitigation to be managed throughout the SOI lifecycle.

The risk (to be mitigated) may be characterized in terms of a value of yes or no. The actual risk will be communicated as a cross-product of likelihood and impact and included in the organization's risk-tracking database. Having this attribute will allow the status of the mitigation to be tracked.

This attribute, along with *A2—Trace to Parent*, *A3—Trace to Source*, *A5—Allocation*, *A34—Priority*, and *A35—Critically or Essentiality*, allows the risk mitigation to be managed throughout the development lifecycle.

A38—Key Driving Need or Requirement

A Key Driving Need (KDN) or Key Driving Requirement (KDR) is a need or requirement that, to implement (design, manufacturing, verification, and validation), can have a significant impact on cost or schedule. A KDN or KDR can be of any priority or criticality/essentiality. Knowing the impact that a KDN or KDR has on the budget and schedule allows better management of sets of needs and sets of requirements.

If a project is over budget and late, the project team may decide to de-scope. As part of their de-scope efforts, they may want to consider deleting a low-priority, noncritical, high-risk KDN or KDR. If there is a need to delete or waive a KDN or KDR need or requirement, a change request must be submitted to the organization that has CM authority for the set of needs and requirements (*A21—CM Authority*).

This attribute also enables reports to show the need or requirement has the GtWR characteristics *C6—Feasible and the sets of needs and requirements have the GtWR characteristic C11—Feasible.*

A39—Additional Comments

Generic comment field that can be used to document possible issues with the need or requirement, any conflicts, status of negotiations, actions, design notes, and implementation notes, for example. Further, evaluation and prototyping of the system concept may have identified important guidelines for the implementation of the need or requirement. This information may be useful as the need or requirement is being reviewed and serves as a place to document information not addressed by other attributes.

A40—Type/Category

Each organization will define types or categories to which a need or requirement fits based on how they wish to organize and manage their sets of needs and requirements. The Type/Category field is most useful because it allows the database to be viewed by a large number of designers and stakeholders for a wide range of uses. For example, maintainers could review the database by asking to be shown all the maintenance needs and requirements, engineers developing test plans could extract all corrosion-control needs and requirements, and so on. See also the GtWR rules *R40—Related Requirements together* and *R41—Structured.*

Examples of Type/Category of needs and requirements include:

- Function: Functional/Performance.
- Operational (fit): Interactions with external systems—input, output, external interfaces, operational environmental, facility, ergonomic, compatibility with existing systems, logistics, users, training, installation, transportation, storage, safety, and security.
- Physical characteristics (form).
- Quality (-ilities): reliability, availability, maintainability, accessibility, transportability, quality provisions, and growth capacity.
- Compliance:
  - Standards and regulations—policy and regulatory.
  - Constraints—imposed on the project and the project must show compliance.
  - Business rules—a rule imposed by the enterprise or business unit.
  - Business requirements—a requirement imposed by the enterprise or business unit.

Refer to Chapters 4 and 6 for a more detailed discussion of this approach to organizing needs and requirements.

This attribute also enables reports to show the sets of needs or requirements have the GtWR characteristic *C10—Complete.*

15.4 ATTRIBUTES TO SHOW APPLICABILITY AND ENABLE REUSE

These attributes may be used by an organization that has a family of similar product lines to identify the applicability of the need or requirement and reuse of needs or requirements that may apply to a new product being developed or the upgrade of an existing product.

The use of these attributes and the terminology used within the text for each are very domain and product line dependent.

A41—Applicability

Can be used to indicate the increment, build, release, or model to which a need or requirement applies.

A42—Region

Region where the product will be marketed, sold, and used to which a need or requirement applies.

A43—Country

Countries where the product will be marketed, sold, and used a need or requirement applies to.

A44—State/Province

State(s) or province(s) within a country or region where the product will be marketed, sold, and used to which a need or requirement applies.

A45—Market Segment

Segment of the market that will be using the product to which a need or requirement applies.

A46—Business Unit

A specific business unit within the enterprise that produces the product to which a need or requirement applies.

15.5 ATTRIBUTES TO AID IN PRODUCT LINE MANAGEMENT

When developing a product line or a specific product within a product line, different versions of a given product are referred to as "variants." Within the sets of needs and requirements there will be needs and requirements that will have different applicability and values depending on the variant of the product they apply.

Depending on the various product variants, there may be similar needs and requirements defined, with slightly different values depending on the product variant. Thus, there are also needs and requirements variants as well.

The project's toolset (RMT and modeling applications) should allow users to define and manage needs and requirements of a product variant—that is, to identify needs and requirements that are common to all product variants as well as needs and requirements that are unique to a specific variant.

Individual product variants may have common requirements that exist for all variants as well as unique needs and requirements that exist only for that variant. For example, needs and requirements dealing with quality and compliance may be common to all variants. Some functions may be common to all variants but have variations in performance.

Other functions may be unique to a specific variant. For example, a coffee maker product line may have one variant that is a smaller model for individual consumption, another variant that is a larger model designed for a larger household, and an even larger variant for use in a business. The project team will ensure the common product line features are maintained in each variant as well as address the unique needs of the user community that is reflected in the individual variants.

For the definition of needs and requirements within a product line, the needs include both the overall product line needs as well as needs associated with each variant.

In addition to the previous attributes concerning reuse, the following attributes can be used to manage needs and requirements that apply to a specific variant within the product line. Using these attributes, the project toolset could be used to form specific sets of needs and requirements for each product variant. Each set would include both the needs or requirements common to each variant as well as the needs or requirements specific to a given variant.

A47—Product Line

A specific brand or product line within a given business unit to which a need or requirement applies.

A48—Product Line Common Needs and Requirements

An indication that a need or requirement applies to all variants within the product line.

A49—Product Line Variant Needs and Requirements

A list of variants within a product line. For needs and requirements that apply to a specific variant, this attribute would be set for that specific variant.

Using this approach, each need or requirement variant would be entered into the sets of needs and requirements, and the above-mentioned attributes would be defined for that product line and variants of the products within that product line. Once done, the user will be able to view only the needs and requirements that apply to a specific variant. For a case when there are four different performance values set for a given function, there would be four versions of the functional requirement in the dataset. Which one applies to a given variant would be determined by how the above-mentioned attributes are defined.

Using this approach, there is an issue with maintaining correctness and consistency among dependent needs and requirements and child requirements of a variant parent requirement.

It may be that selecting a specific variant of a functional/performance requirement would mean that any other requirements where a dependency exists would have to have their variants selected as well, and the project team would need to ensure they are consistent with the selected variant.

This consistency would need to be accessed with other peer requirements at the same level either within the system, subsystem, or system element or peer requirements in other subsystems and system elements. For example, different variants of a functional/performance requirement may have different power or cooling requirements.

In the case of a parent variant requirement, its set of child requirements would need to be selected as well. Selecting a variant requirement that has child requirements defined for the subsystems and system elements at the next level of the architecture would mean that all the child requirements would need to be selected automatically within the RMT. If at higher levels of the architecture, there could be a chain of multiple lower levels of requirements effected by that variant. The child requirements may be allocated to lower-level subsystems and system elements, each with its own set of child requirements.

Unless the relationships of all the needs and requirements that have variants are established and managed within the RMT, it will be difficult to maintain correctness and consistency across all the sets of needs and requirements at each level of the system architecture for each variant.

While this section deals with attributes as applied to needs and requirements, a project could also define attributes for subsystems and system elements within the SOI physical architecture. Then, the above-mentioned attributes dealing with variants could be set for these subsystems and system elements. Selecting a specific system element would result in all the requirements for that system element being selected. If that system element also has variants, the individual needs and requirements for that specific variant of the variant system, subsystem, or system element could be set.

The result is that once these attributes have been set, the user would be able to display needs and requirements that apply only to that variant product and the subsystems and system elements within that variant's architecture.

15.6 GUIDANCE FOR USING ATTRIBUTES

The information communicated within the attributes must be managed to the same extent as the needs and requirement expressions are managed. Changes to some attributes may have a direct impact on how a need or requirement is addressed in design and the *Strategy* and *Method* used to obtain objective evidence the SOI meets the need or requirement during system verification and system validation. For example, a requirement that was originally of low priority becoming high priority.

The reason for using attributes is to help the project team better manage the project. Given that needs, requirements, verification, and validation are the common threads that tie each of the development lifecycle process activities and artifacts together, having insight into these activities and artifacts is necessary to manage the project effectively.

Attributes help support the SE activities, enabling the information contained within the attributes to be structured for ease of viewing, filtering, sorting, reporting, and statusing; enabling the needs or requirements to be sorted or selected for further action, and enabling management to have insight into the needs and design input requirements definition as well as the development activities of the SOI across its lifecycle.

A major feature of applications within the project toolset is the reports that can be generated to give insight into the progress being made across all lifecycle stages. Managers can define specific dashboards and reports that target various aspects of their project based on the attributes. Some applications within the project toolset allow dashboard displays to be defined that include graphics based on the selected attributes. This is only possible if the project includes in the toolset schema the attributes describing the information needed to produce these reports and dashboards at the beginning of the project.

Knowing the priority helps to plan better project activities. Knowing the risk of implementation allows mitigation of that risk. Combining implementation risk and priority or criticality/essentiality can provide management with valuable information they can use to manage the project's needs and requirements and associated risks. Identifying needs and requirements that are both high-priority (or high-criticality/essentiality) and high implementation risk allows management to focus on the most important issues.

Again, the project team can create a 3×3 matrix similar to the risk matrix, but in this case, one axis is risk, and the other axis is priority or criticality/essentiality. This allows management to know which high-priority or critical needs and requirements are high-risk and their status. If these attributes are not documented for each need and associated design input requirement(s), how will management have this knowledge?

Knowing the criticality/essentiality of a requirement or the fact a need is a KDN or a requirement is a KDR allows the project team to plan accordingly. When under schedule or budget pressure, a KDN or KDR that is low priority or low criticality but high risk, may be a candidate for deletion if a project must be de-scoped.

Activities associated with system verification and validation are key cost and schedule drivers on any project. Knowing the priority, criticality/essentiality, and risk of a need or requirement is valuable information when defining the *Success Criteria* and selecting the system *Strategy* and *Method* as discussed in Chapter 10. Knowing the *Strategy* and *Method* facilitates the development of a more realistic budget and schedule. Knowing the status of a design input requirement and status of the system verification and validation activities allow the project team to manage the budget and schedule more effectively. With this knowledge, actions can be taken when there are indications of problems that could lead to budget overruns and schedule slips before they become real problems.

A major source of budget and schedule issues is change. Attributes allow managers to assess the change and the potential impacts of that change. Attributes that link artifacts from one systems engineering lifecycle to another aid in change management and assessing the impacts of change across all lifecycle activities.

A word of caution; as with the use of all information, a "lean" approach should be taken when deciding which attributes will be used. Do not include a specific attribute unless the project team, their management, or their customer has asked for that attribute and will be using that attribute in some manner to manage the project and sets of needs and requirements. While attributes are all potentially useful, too many should not be defined because of the time and effort needed to define and maintain them. The attributes the project team selects should "add value."

Attributes are used to manage projects more effectively. To use attributes to do so, they need to be accurate and current. As the old saying goes, "garbage in; garbage out." Another applicable saying is "A time saving tool should not take more time to maintain than the time it saves."

16

DESIRABLE FEATURES OF AN SE TOOLSET

16.1 CHOOSING AN APPROPRIATE TOOLSET

As discussed in Chapter 3, the Information-based Needs and Requirements Definition and Management (I-NRDM) data-centric approach advocated in this Manual requires the project team to have a toolset that allows the creation of an integrated/federated data and information model of the System of Interest (SOI) being developed.

Individual tools tend to focus on specific needs and types of work products. Organizations at the strategic and operational levels need to perform trade studies to see if the expense of purchasing a specific application or toolset, maintaining the licenses, training people to use the tool(s), maintaining the tool(s), maintaining needs, requirements, models, and other work products and their underlying data and information developed by the tool(s) are going to provide a positive Return on Investment (ROI), reduce time to market, improve the approval process for highly regulated systems, reduce the number of product defects, reduce the amount of rework, or reduce warranty work and recalls.

Given today's increasingly complex, software-intensive systems, a toolset that includes requirement management tools (RMTs), diagramming tools, modeling tools, budgeting tools, and scheduling tools is needed. Ideally, these tools can share data allowing Systems Engineering (SE) and Project Management (PM) artifacts to be linked, helping ensure consistency among artifacts and work products, establishing an authoritative source of truth (ASoT), as well as providing management insight into the status and progress of a project across the lifecycle.

There are SE tools whose sole purpose is to establish traceability between:

- Data, information, and artifacts developed and managed in an SE tool (such as needs and requirements in an RMT),
- Data, information, and artifacts developed and managed in other dedicated tools for specific activities such as risk and test (verification and validation) management,

INCOSE Needs and Requirements Manual: Needs, Requirements, Verification, Validation Across the Lifecycle,
First Edition. Louis S. Wheatcraft, Michael J. Ryan, and Tami Edner Katz.

- Models using a language-based modeling tool, and
- Design artifacts to the actual physical configuration items.

Interrelationships of these various SE tools are shown in Figure 16.1. For example, a requirement managed in an RMT is the same requirement that exists in a modeling tool; the two representations are linked via traceability between the tools using some type of data integration tool.

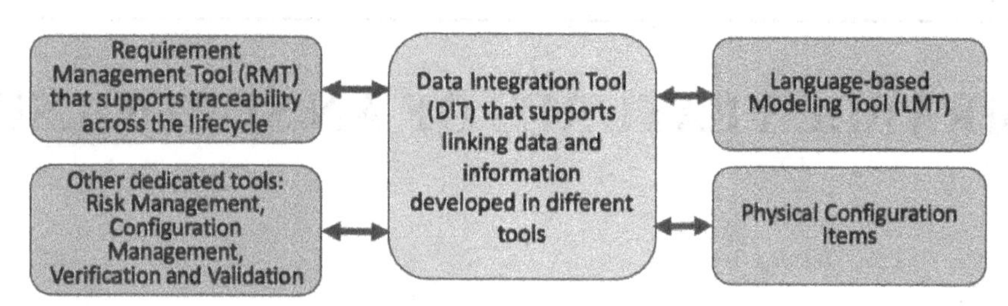

FIGURE 16.1 Establishing Traceability Between Data and Information From Different Tools.

A major challenge for the future, as discussed in International Council on System Engineering (INCOSE) Vision 2035 [24], is "Federation across different domain specific tools, and integration of data are becoming a focus for enabling collaboration and analysis, but many obstacles remain." For the interactions shown in Figure 16.1, tool vendors must address this challenge.

To develop systems more effectively, the toolset needs to be tailored to an organization's needs, as evidenced by statements such as the following in NASA's NPR 7123.1, NASA Systems Engineering Processes and Requirements [10] "*… technical teams and individuals should use the appropriate and available sets of tools and methods to accomplish required common technical process activities. This would include the use of modeling and simulation as applicable to the product-line phase, location of the Work Breakdown Structure (WBS) model in the system structure, and the applicable phase exit criteria.*"

Going beyond needs and requirements, there are tools that can support the entire system lifecycle including budgeting, scheduling, defining, designing, building/coding, verifying, validating, and sustaining engineering activities. These tools are used to collect, link, visualize, analyze, manage, and communicate data and information across all system lifecycle stages. These more robust tools allow the organization to produce various views of the system under development and create and maintain the various work products and artifacts (needs, requirements, design artifacts, system verification artifacts, system validation artifacts, documents, databases, reports, diagrams, drawings, models, budgets, schedules, WBS, and Product Breakdown Structure [PBS]) and their underlying data and information needed to manage the system development efforts more effectively.

A key issue with tools that claim to support the entire lifecycle of a product is that they lack the robustness of tools that focus on a specific capability. For example, while an RMT is good at supporting the definition and management of sets of needs and requirements across the lifecycle, these tools often lack the capability to include diagramming and language-based models to support the underlying analysis from which the needs and requirements are derived. Similarly, tools that are good at supporting diagramming and development of language-based models often lack key features of RMTs to adequately support the definition and management of sets of needs and requirements.

These tools are like any other software application … one size does not fit all. The project toolset that is best for an organization is the toolset that meets the organization's needs and requirements management, PM, SE, and modeling needs. When selecting a toolset, consider the outcomes resulting from the use of the tools and the ROI resulting from these outcomes.

What capabilities are needed from a project toolset depends on the product line, its complexity, green-field versus brown-field products, issues the organization is having and wants to address, and the workforce knowledge and experience. Organizations need to understand what data and information best meet their needs and which set of tools they need to develop and manage this data and information.

Before embarking on a project toolset evaluation and selection initiative, work with business and operations management stakeholders, project teams, engineering staff, and other key stakeholders to determine what the organization needs to help better manage the development of the systems in their domain. What features and functionality are needed in a project toolset so the projects can effectively and efficiently manage their needs, requirements, design, design output specifications, system integration, system verification, and system validation throughout the lifecycle?

Specifically, choose the toolset that supports the level of data-centricity the project has decided to strive for. Select tools that support PM and SE from a data-centric perspective using the SOI's integrated or federated, shareable sets of data. *Refer to INCOSE RWG whitepaper* [6] *"Integrated Data as a Foundation of Systems Engineering."*

It is advisable to choose tools that support the generation and management of multiple lifecycle artifacts and work products and their underlying data and information, especially tools that fully support interoperability standards for compatible tools, schemas, and databases.

The perfect case would be to procure a single SE tool that "does it all," i.e., the one tool would result in the project having integrated or federated, shareable sets of data by default. That would help to ensure all data and information are current and consistent across all system lifecycle stages. Again, vendors are in the process of developing tools that will support the development and shareability of PM and SE data and information across all lifecycle stages. However, it is doubtful that a single tool will be developed that can "do it all" with the features and capabilities needed by all project team members.

16.2 DESIRABLE FEATURES OF AN SE TOOLSET

This section identifies key features to be considered when selecting a set of tools to procure for inclusion within the project toolset. For consistency, it is recommended that this selection be made at the strategic and operational levels of the organization. Doing so results in consistency between projects within the organization and enables the reuse of data and information, as well as the personnel that have experience in the use of these tools. For example, for a given enterprise, a common set of standards and regulations may apply across multiple product lines, or to manage a product line, there is a benefit to being able to reuse data and information, including needs and requirements rather than each new project beginning from nothing. With a common toolset, the requirements in these standards and regulations can be imported once into a "library" and shared across all projects, allowing the associated data and information to be reused resulting in a shorter time to market and increased profitability.

Including the features in this section will enable the organizations to achieve the capabilities they need to effectively implement their PM and SE processes from a data-centric perspective.

The order of the list does not in any way imply priority. Priority of these features and functions and the "importance-weighing factor" for each are left up to the evaluating organization based on its unique product development and management needs.

Currently, it is unlikely that any one tool will include everything in this list, as many vendors tailor their tool to a specific client base or a specific system lifecycle stage and set of artifacts and work products. However, it would be preferable to minimize the number of different applications in the organization's toolset tailored to their specific domain, product line, and processes consistent with the level of data centricity they are moving toward.

The INCOSE Tools Integration and Model Lifecycle Management (TIMLM) WG has developed an SE Tools Database (SETDB) that is available to INCOSE members. This database includes a listing of SE tools, with specific capabilities and features defined. The SETDB is interactive, allowing users to specify the capabilities and features they need and to set up priority/weighting factor of each to allow them to select and rank SE tools available to meet the needs of their organization.

16.2.1 Functionality

This section addresses the recommended capabilities and features of a toolset.

Needs and Requirements Quality Does the toolset include the capability to support best practices concerning the quality of need and requirement statements? For example, does the toolset enforce needs and requirements standards such as those defined in the *INCOSE Guide to Writing Requirements (GtWR)*? This includes the ability of the toolset to help authors define well-formed need and requirement statements (spelling, grammar, unambiguous terms, requirement statement structure, and consistency as examples) and to assess the quality of a set of needs and requirements based on the organization's standards for writing needs and requirements.

Grammatical Structure of Needs and Requirements Rather than treating need and requirement statements as an indivisible entity, does the toolset allow text-based need and requirement statements to be managed as a grammatical structure? Within this structure, the requirement statements refer to various entities: architectural entities, functions, features, conditions, states, modes, events, transitions, interfaces, and units of measure. To relate those references to the corresponding entities in the overall system model involves getting inside the grammar of a need and requirement statement. This capability will allow needs and requirements to be an integral part of the overall system model. This capability is also key to ensuring the terminology used within the need and requirement statements is consistent with the project-wide ontology.

Allocation, Budgeting, and Traceability Does the toolset support the key concepts of allocation, budgeting, and traceability between not just requirements but all work products and their underlying data—no matter the level or lifecycle stage as discussed in Chapters 6 and 14? If developing functional, analytical, and behavioral models, this capability allows needs and requirements to be linked to the applicable parts of the architecture within the model. From a budgeting perspective, the toolset should allow the budgeting of performance, physical characteristics, and quality from one level of the physical architecture to another and establish relationships between the resulting child requirements and with their allocated parent.

Interface Management Does the toolset support the documentation of interface definitions (such as Interface Control Documents [ICDs] or Interface Definition Documents [IDDs]) and the corresponding interface requirements that are linked to those definitions? Can the toolset be configured to link an interface requirement from one system, subsystem, or system element of the SOI architecture to the corresponding interface requirement for another system, subsystem, or system element with which the first element interacts with? Can the toolset be configured to notify owners of a proposed change to a complementary interface requirement(s) or their definition? Will the toolset enable users to perform the interface audit described in Chapter 6? *Note: This topic deals with not only internal interfaces but also interfaces between the SOI under development and external systems it is required to interact with.*)

Dependencies Between Artifacts and Work Products and Their Underlying Data and Information Does the toolset allow users to link dependent requirements and other artifacts and work products across all system lifecycle process activities and their underlying data and information to each other? (This is important when a change to one artifact or work product could necessitate a change in another artifact or work product along with the underlying data and information.) The dependent artifact or work product may be part of the SOI under development or another system, subsystem, or system element. Does the toolset allow users to do consistency assessments between dependent requirements and artifacts and work products? Can the toolset be configured to notify owners of dependent artifacts and work products when they are defined or a change is made to one of the dependent artifacts and work products? *Note: This feature is supported by the traceability feature.*

Impact Assessment Does the toolset allow the user to assess the impact of a change vertically among subsystems and system elements at different levels of the architecture as well as horizontally across all work products from all system lifecycle activities, helping the user to understand the impact of a change to other artifacts and work products and the project's delivery schedule, cost, and quality? Does the toolset allow the user to do a change impact assessment to other artifacts and work products generated in other system lifecycle activities whose underlying data resides in a separate database? For example, what are the impacts of a requirement change on design? A change to an analytical or behavioral model on a need or requirement linked to that model? A change to a dependent requirement in a different system, subsystem, or system element? A change to a requirement on system verification planning? A change to a need on system validation planning?

Ontology Does the toolset allow users to define an ontology for a specific SOI as well as include the capability to expand its inherent ontology to define a project-wide ontology? An ontology includes the formal naming and definition of a set of terms, entities, data types, and properties, as well as defining the relationships between these terms, entities, and data types that are fundamental to a domain. Defining an ontology is critical to ensuring consistency and allowing the sharing of data and information across system lifecycle process activities as well as reusability within the enterprise. *Note: Most tools provide the capability to define an ontology for the data and information managed within the tool, but often the ontology is proprietary. From a shareability perspective, what is needed is for all tools in the project's toolset to have an ontology that is consistent with the enterprise-wide ontology.*

Schema Does the project's toolset include the capability to define a master project schema for its integrated or federated, shareable sets of data consistent with the enterprise-wide schema? The schema is a description, in a formal language, of the database structure that defines the objects in the databases, relationships between those objects, shows how real-world entities are modeled in the database, and integrity constraints that ensure compatibility between parts of the schema. Do the tools in the toolset conform to standards for development of a common schema (such as Open Services Lifecycle Collaboration [OSLC])? SE tools in the project's toolset need to ensure their schemas are consistent with the project's master schema defined for the SOI's integrated or federated, shareable sets of data to ensure compatibility and consistency of the data, allowing the data to be shared among the various tools within the project's toolset. (*See also interoperability and tool integration later in the list.*)

Embedded Objects Does the SE toolset allow the user to embed objects with various electronic formats (tables, pictures, drawings, diagrams, RTF files, word processing documents, spreadsheet documents, test procedures, and drawings) that can be linked to other work products?

Diagrams and Drawings Does the SE toolset support the development and management of diagrams and drawings as electronic files that can be linked to other work products and their underlying data independently from an analytical modeling tool?

Modeling Does the toolset support the development and use of functional, architectural, analytical, behavioral, and physical language-based models? Does the toolset allow the development and documentation of use cases, functional flow block diagrams, states and modes diagrams, timing diagrams, external interface diagrams, requirement expressions, and other types of models needed by the project and store the underlying data and information representing these activities and diagrams in a database consistent with a project's master schema? Will the toolset allow the user to develop high-fidelity models that support simulations—if that capability is needed by the organization? Does the toolset allow the creation of an extensible data model that can be easily constrained by a rule set; an extensible Application Programming Interface (API) to allow incorporation of custom data creation and manipulation utilities; and a rich, natural language query engine?

Reusability What features do the tools within the project toolset have that will enable the reuse of work products and their underlying data and information for similar projects or projects involved in updating an existing product? Can the work products and underlying data for one version of a product be duplicated and used as the basis for the next version? Does the toolset support the development of a library of common needs, requirements, standards, and regulations applicable to types of systems developed by the enterprise? With such a library, each project can draw applicable needs and requirements from the library rather than having to recreate and enter these requirements separately for each project.

Product-Line Management In addition to reusability, what features does the SE toolset have that support product-line management? Does the tool allow branching of work products; for example, for a class of systems, the same root requirement can be branched to multiple versions of the root requirement? Does the toolset support the definition of variant systems, subsystem, and system elements, as well as variant needs and requirements?

16.2.2 Tool Attributes

This section addresses toolset features that enable tailoring to support the organization's specific needs.

Tailorable Can the SE toolset be tailored within the organization based on a project's needs, for example, complexity, team knowledge, development methodology, product line, size, processes, timeframe, customer requirements, ontology, and schema?

Configuration/Customization Does the toolset include the ability to configure and/or customize the toolset to the customer's domain, product line, culture, and organizational processes? Does the toolset allow individual users to configure their interface based on their unique role and use of the SE toolset with minimum help from the vendor? *Note: Configure refers to the ability to configure the tool to meet user needs without changing the code. Customize involves changes to the tool code to provide new or tailored features needed by the customer. Having a tool that can be configured to meet an organization's needs is cheaper than paying a vendor to customize the tool to meet those needs.*

Learnability/Usability Do tools within the project toolset have a user interface that is intuitive, user-friendly, and easy to use with a short learning curve? How much training is necessary and available? Does the tool require users to go through significant training to learn a specific language to use the tool? Are online documentation and help functionality included? Does the toolset provide methods allowing the user to navigate between various artifacts and work products and visualizations such as needs, design input requirements, documents, CM information, reports, diagrams, models, design artifacts, or design output specifications?

Security Does the toolset provide security of the data and information in terms of access (at multiple levels, within levels, and different user classes), protection (from loss), and integrity? Does the toolset support the security standards applicable to the organization's domain and product type(s) and data governance policy?—Information security includes the principles of confidentiality, integrity, and availability (CIA). Data and information must be protected from unauthorized use and disclosure, must be able to be trusted (accurate, consistent, of high quality, and managed), and accessible to authorized personnel—and not accessible by non-authorized personnel. Tools within the project's toolset must support the enterprise data governance and security policies and requirements for both internal data storage and cloud storage.

Accessibility (Devices/Location) Does the toolset allow users to access data securely via desktops, laptops, and portable devices (tablets and smartphones) both inside and outside the organization's firewalls? Are the sets of data created by the toolset accessible by another organization's (customer, vendor, or supplier) toolset based on user class and respective access privileges?

Online Versus Offline Modes Does the toolset require the user to be connected to the server continuously (online) to use the toolset, or does the toolset allow offline work to be accomplished with synchronization after going back online?

Concurrent Access How many concurrent users does the SE toolset allow to work within the same area? What happens when more than one user wants to edit the same work products and underlying data? For some complex systems, there may be over a hundred users modifying various work products in the SOI's integrated or federated, shareable sets of data simultaneously.

Performance What is the maximum wait time between user actions? How does the toolset minimize performance impacts as the number and complexity of work products increase, as well as when the number of concurrent users increase?

Collaboration Does the toolset support real-time collaboration among the users within the tool across all lifecycle process activities? Does the toolset allow users to collaborate no matter where their workplace is located? Globally? Does the toolset allow external organizations (customers/vendors/suppliers) to collaborate with the project team?

Tool Integration Does the SE toolset allow the integration of needs and requirements developed in an RMT to be linked to entities defined and managed in external diagramming and modeling tools? For example, can needs and requirements developed and managed within an RMT be linked to need and requirement entities within System Modeling Language (SysML) or other types of models developed and managed in another application?

Interoperability—Data Sharing Does the toolset allow the sharing of data among tools within the project toolset as well as with external organizations involved with the project (ReqIF compliant, for example) as well as with other word processing and spreadsheet application-supported formats? How easy is it for information to be transferred into and out of the toolset to support the organization's processes and people? Does the toolset provide a standardized interface for importing or exporting data from/to other applications, rather than requiring specialized scripting, to achieve a transfer/interaction? Can the tools perform extraction, transformation, and loading (ETL) of data created by other toolsets external to the SOI's integrated or federated, shareable sets of data so the data and information in the external tool's database can be imported into the SOI's integrated or federated, shareable sets of data and used by the project? How well can the individual tools be integrated with each other, i.e., can one tool access and manipulate the data and information created by a different tool? Do the tools in the toolset conform to a common data exchange standard (for example, AP239, AP233 XML)? *When a tool vendor says their tool conforms with a standard, does it conform partially or fully?* Do the tools in the toolset allow data exchange between tools seamlessly, with minimal and straightforward data model mapping required on the part of the user? How well does the degree of integration of the tools in the toolset meet the needs of the organization?

Sharing of Data with External Organizations Does the SE toolset allow the project to identify and securely share specific sets of data with external organizations, for example, customers or suppliers? Does the toolset include an industry-standard import/export utility? Is the tool partially or fully compliant with that standard?

Storage Location Does the toolset require artifacts and work products and their underlying data to be stored in the "cloud" provided by the tool vendor or stored in-house on the organization's server(s) or cloud-based storage? While "in the cloud" aids in collaboration, security and protection of sensitive, classified, or IP information is a concern when storing data in the cloud. Accessibility and protection (from loss) are also key considerations for any storage location. Is the storage location consistent with the enterprise data governance policy?

Scalability/Extendibility Will the toolset be able to support the development and management of the volume of work products consistent with the size and complexity of the systems the organization develops? If the enterprise is procuring the toolset, will the toolset be able to support the number of projects within the enterprise given the size and complexity of the systems developed within the enterprise or an increase in the number of projects using the toolset?

Archive/Backup/Long Term Availability Does the toolset provide the capability to archive and backup the data and information consistent with enterprise data governance policy? Do the formats used provide long-term availability as storage and retrieval technologies evolve or a tool changes, or applications within the toolset change? (Organizations *should avoid using a backup/archive format that is proprietary and may no longer be accessible if the tool vendor goes out of business.)*

16.2.3 Management and Reporting

This section addresses toolset features that support project management.

Attributes Does the toolset allow the user to define and manage attributes for work products? For example, for needs and requirements, does the SE tool allow the user to define attributes needed to help manage their needs and requirements as discussed in Chapter 15?

Measures Does the toolset allow the enterprise and project to define specific measures that will allow managers and systems engineers to monitor and assess progress, identify issues, and ensure the system being developed will meet the stakeholder needs? Several key measures are commonly used that reflect overall customer/user satisfaction (for example, performance, safety, reliability, availability, maintainability, and workload requirements: Key Performance Indicators (KPIs), Key Performance Parameters (KPPs), Measures of Performance [MOPs], Measures of Effectiveness [MOEs], and Leading Indicators [LIs].)

Reports Does the toolset include a robust, well-documented report capability that allows users to create unique reports (using the attributes and measures defined previously) as well as customize standard reports provided with the tool? Does the toolset allow reports to be exported in multiple formats that are readable by common office applications? At the beginning of the quest for a toolset, one of the first things a project should do is identify the number and type of reports the tool needs to support, their content, and their required form. That will drive the schema of the data, meta-data, measures, and attributes to be included in the SOI's integrated or federated database.

Metrics/Dashboards Closely related to a report feature, does the toolset provide the capability to do "data mining" and analytics of the measures and information in the attributes to enable the display of and report historical and trend data? It is often not enough to just know what percent of the requirements for the system have been successfully verified. Often it would be useful to see if the trend to completion of system verification and validation activities is on the right pace, is slowing down, or speeding up. If slowing down, the project may not be able to complete all the system verification and validation activities in time for the customer acceptance review. Management needs to know this in order to take corrective action before it becomes a problem.

Notifications Can the applications within the toolset send notifications to applicable stakeholders via email or texting concerning changes to data and information contained within artifacts and work products? Can the notifications be sent from one user to another user (or group of users) concerning actions, comments, and questions? Can the SE toolset send notifications to the appropriate users when a specific measure is predicted to or has exceeded a pre-specified threshold?

Project Management Work Products Does the project toolset allow various PM work products to be managed within the SE toolset or at least linked to artifacts within the SE toolset? This includes budget, schedule, and risk management work products, as well as development and management of a WBS. Can these work products and underlying data and information be linked to parts of the PBS and other SE artifacts and their underlying data? For example, can the WBS be linked to a PBS?

Lifecycle Support Does the project toolset support system development across all system lifecycle process activities: needs and requirements definition and management, gate reviews, architecture, design, manufacturing, coding, system integration, system verification, system validation, and sustaining engineering? For system verification and system validation, does the toolset allow the linkage of needs and requirements to their system verification and validation planning artifacts, *Procedures*, *Execution Records*, and *Approval Packages*?

Workflow Does the toolset provide the ability to define and support the organization's PM and SE process workflow within the tools (for example, for needs and requirements, does the SE toolset allow the project to track the needs and requirement's status: draft, review, approve, baseline; design, test, code/manufacture, system verification, and system validation)? Can the SE toolset allow the creation, management, and execution of PM and SE processes, procedures, and work instructions within the toolset?

Configuration Management (CM) Does the project toolset provide robust CM of all system lifecycle artifacts and work products and underlying data and information, including change, version, and baseline control? Does the toolset allow the user to access the change history of any artifact or work product? If artifacts and work products are developed and maintained within the database, does the toolset allow configuration control of the database (versus the various reports/visualizations representing the data and information in the database)?

16.2.4 General Considerations

This section addresses general considerations that are important to consider when selecting tools.

Price Is the toolset affordable in relation to the number and size of the projects, the number of needs and requirements and sets of needs and requirement, modeling activities and resulting artifacts, system verification and validation artifacts, and number of concurrent users? Concerning affordability, is there a single upfront application fee, individual license fee (if a license fee, one-time or yearly)? Are the licenses fixed or floating? Does the price include initial setup, installation, configuration or customization or is that extra? Is ongoing technical support included or extra? Is training included or extra? Would it be more cost-effective to spend more on a single tool that has most of the above features or multiple tools to give the organization all the features needed to manage system development across all lifecycle stages?

Cost of Infrastructure to Support the Use of the SE Toolset What are the IT requirements to host and deploy the toolset? What specialty skills are required to operate, extend, tailor, and maintain the applications within the toolset?

Vendor/Product Maturity How long have applications in the toolset been on the market? How long have the vendors been in business? Is the vendor stable? Will the product likely be supported for the foreseeable future?

User Feedback and Satisfaction In today's social media-driven world, access is provided to actual user comments concerning the applications within the toolset, the vendors, ease of use, reliability, technical support, and other factors of concern to users. Do not get overwhelmed by the hype and sales pitch from the vendor. See what the actual users have to say about the applications being considered for inclusion in the organization's toolset.

APPENDIX A: REFERENCES

The following references were consulted when developing this Manual. When possible, the source is referenced in the text. In some cases, concepts and information from several sources were used as a basis of the text but were not directly attributable to any single source.

1. INCOSE-TP-2003-002-05 (2023). *Systems Engineering Handbook*, 5e. Wiley.
2. SEBoK (2024). Guide to the Systems Engineering Body of Knowledge (SEBoK). https://sebokwiki.org/wiki/Guide_to_the_Systems_Engineering_Body_of_Knowledge_(SEBoK).
3. INCOSE-TP-2021-003-01 (2022). Guide to Needs and Requirements, Prepared by the Requirements Working Group. INCOSE.
4. INCOSE-TP-2021-004-01 (2022). Guide to Verification and Validation, Prepared by the Requirements Working Group. INCOSE.
5. INCOSE-TP-2010-006-04 (2023). Guide to Writing Requirements, Prepared by the Requirements Working Group. INCOSE.
6. INCOSE-TP-2018-001-01 (2018). INCOSE Integrated Data as a Foundation of Systems Engineering, Prepared by the Requirements Working Group. INCOSE.
7. ISO/IEC/IEEE 15288 (2023). *Systems and Software Engineering — System Lifecycle Processes*, New York: IEEE.
8. ISO/IEC/IEEE 29148 (2018). *Systems and Software Engineering — Lifecycle Processes — Requirements Engineering*, 2e. New York: IEEE.
9. NASA NPR 7123.1D (2023). NASA SE Processes and Requirements.
10. NASA (2016). NASA Systems Engineering Handbook, NASA SP-2016-6105 Rev2.
11. Dick, J., Wheatcraft, L., Long, D. et al. (2017). Integrating requirement expressions with system models. *Paper Presented at ASEC 2017*, UK.
12. Ryan, M., Wheatcraft, L., Dick, J., and Zinni, R. (2014). An improved taxonomy for definitions associated with a requirement expression. *Systems Engineering/Test and Evaluation Conference SETE2014*, pp. 28–30, Adelaide.

INCOSE Needs and Requirements Manual: Needs, Requirements, Verification, Validation Across the Lifecycle,
First Edition. Louis S. Wheatcraft, Michael J. Ryan, and Tami Edner Katz.

13. Ryan, M.J. (2013). An improved taxonomy for major needs and requirements artefacts. *INCOSE International Symposium IS2013.*
14. Ryan, M. and Wheatcraft, L. (2017). On the use of the terms verification and validation. *INCOSE International Symposium IS2017*, July 2017.
15. EIA-632 (2003). Processes for Engineering a System.
16. Armstrong, J.R. (2007). The continued evolution of validation: issues and answers. *INCOSE 2007 – 17th Annual International Symposium Proceedings.*
17. Drues, M. (2020). Why design validation is more than testing: how to validate your validation. Green light Guru webinar series on YouTube.
18. Hoehne, O. (2017). I do not need requirements—I know what I'm doing usability as a critical human factor in requirements management. *27th INCOSE International Symposium, IS2017*, Adelaide.
19. Wheatcraft, L. and Ryan, M. (2018). Communicating requirements – effectively! *Paper Presented at INCOSE IS.*
20. Letouzey, J.L. and Declan Whelan, D. (2016). *Introduction to Technical Debt.* Agile Alliance.
21. Wheatcraft, L., Ryan, M., and Svensson, C. (2017). 2017-1 Integrated data as the foundation of systems engineering. *Presented at INCOSE IS 2017*, Adelaide, Australia.
22. Wheatcraft, L., Ryan, M., Dick, J., and Llorens, J. (2019). Information-based requirement development and management. *Paper Presented at INCOSE IS.*
23. Wheatcraft, L., Ryan, M., Dick, J., and Llorens, J. (2019). The need for an information-based approach to requirement development and management. *Paper and Poster Presentation at INCOSE IS.*
24. INCOSE (2021). *Systems Engineering Vision 2035.* Engineering Solutions for a Better World.
25. Ryan, M. (2017). *Requirements Practice.* Canberra, Australia: Argos Press.
26. Llorens, J., Fraga, A., Alonso, L., and Fuentes, J. (2015). Ontology-assisted systems engineering process with focus in the requirements engineering process. In: *Complex Systems Design & Management* (ed. F. Boulanger, D. Krob, G. Morel, and J.C. Roussel), 149–161. Cham: Springer.
27. The ReUse Company (TRC-1) (2018). Knowledge Centric Systems Engineering. Web page.
28. Wheatcraft, L. (2019). The role of requirements in an MBSE world. *SyEN Newsletter* (June).
29. Object Management Group What is SysML. http://www.omgsysml.org/index.htm (accessed 20 May 2024).
30. Carson, R.S. (2004). Requirements completeness. *Proceedings of INCOSE International Symposium* **14** (1): 930–944.
31. NASA, CXP 70024 (2008). Human systems integration requirements, Rev B, 3/3/2008, Appendix J, allocation matrix.
32. NASA, CXP 72208 (2009). Suit element requirements document, Rev c.3, 9/22/2009, Appendix C, mission applicability, Appendix D, subsystem allocation matrix, and Appendix F, HSIR allocation matrix.
33. Katz, T. (2020). When to constrain the design? *Application of Design Standards on a New Development Program, Presented at INCOSE IS2020.*
34. FDA (2024). USC Title 21, Part 820, Quality System Regulation. As of April 8, 2024.
35. GAO-20-48G (2020). Technology Readiness Assessment Guide.
36. Bilbro, J.W. (2007). Systematic assessment of the program/project impacts of technological advancement and insertion, revision A. JB Consulting International.
37. Wheatcraft, L. (2011). Triple your chances of project success – risk and requirements. *Paper Presented at INCOSE IS 2011*, June 2011, Denver, Colorado and NASA's PM Challenge 2012, February 2012, Orlando, Florida.
38. FDA (2018). Web page: the most common FDA audit findings from 2017. Posted January 11, 2018.
39. D'Souza, A., Wheatcraft, L.S., Katz, T. et al. (2024). Traceability–a vision for now and tomorrow. *Presented at INCOSE IS2024.*
40. Wheatcraft, L. (2016). Resource Margins and Reserves, August 2016.

41. Carson, R.S. and Noel, R.A. (2018). Formal requirements verification and validation. *Proceedings of INCOSE International Symposium* **28** (1): 805–818.
42. Carson, R.S. (2004). Integrating failure mode and effects with the system requirements analysis. *Proceedings of INCOSE International Symposium* **14** (1): 1785–1796.
43. Micouin, P. (2014). *Model Based Systems Engineering: Fundamentals and Methods*. Wiley.
44. Carson, R.S. (1995). A set theory model for anomaly handling in system requirements analysis. *Proceedings of NCOSE International Symposium* **5** (1): 557–562.
45. Carson, R.S. (1996). Designing for failure: anomaly identification and treatment in system requirements analysis. *Proceedings of INCOSE International Symposium* **6** (1): 785–792.
46. Ross, R., McEvilley, M., and Winstead, M. (2022). *Engineering Trustworthy Secure Systems*. Gaithersburg, MD: National Institute of Standards and Technology, NIST Special Publication (SP) NIST SP 800-160v1r1. doi: https://doi.org/10.6028/NIST.SP.800-160v1r1.
47. INCOSE (2023). *Human Systems Integration (HSI): A Primer*, vol. **1** R1.2, September 2023. ISBN: 978-1-937076-12-2.
48. Carson, R.S. (2019). Using MBSE to develop requirements. *Presentation to INCOSE Requirements Working Group* (24 October 2019).
49. Carson, R.S. (2020). Using architecture and MBSE to develop validated requirements. *Presented at INCOSE Western States Regional Conference,* Seattle, WA (September).
50. Roques, P. (2015). How modeling can be useful to better define and trace requirements. *Requirements Engineering Magazine*.
51. Wikipedia. OSI model. https://en.wikipedia.org/wiki/OSI_model (accessed 20 May 2024).
52. Dick, J.A. (2004). The systems engineering sandwich: combining requirements, models and design. *Proceedings of INCOSE 20014*, pp. 1401–1414. Toulouse, FR: Wiley.
53. NASA (2016). *Expanded Guidance for Systems Engineering*, vol. 1 and 2. Washington, DC: National Aeronautics and Space Administration.
54. Wheatcraft, L. (2012). Thinking ahead to verification and validation. *Presented at NASA's PM Challenge 2012*, February 2012, Orlando, Florida.
55. Charette, R.N. (2021). How software is eating the car. *IEEE Spectrum Posted 2021-06-07*.
56. Scukanec, S.J. and VanGaasbeek, J.R. (2010). A day in the life of a verification requirement. *Presented at INCOSE IS 2010*.
57. ANSI/EIA-649-B-2011 (2011). *EIA Standard – National Consensus Standard for Configuration Management*. Arlington, VA: Electronic Industries Association.
58. MIL-HDBK-61A(SE) (2001). *Military Handbook—Configuration Management Guidance*. Washington, DC: United States of America Department of Defense.
59. MIL-STD-481B (1988). *Military Standard Configuration Control – Engineering Changes (Short Form), Deviations and Waivers*. Washington, DC: US Department of Defense.
60. ISO 10007 (2017). *Quality Management Systems—Guidelines for Configuration Management*.
61. MIL-STD-973 (1992). *Military Standard—Configuration Management*. Washington, DC: United States of America Department of Defense.
62. Wilson, B. (2009). Integrated testing: a necessity, not just an option. *ITEA Journal of Test & Evaluation* **30**: 375–380.
63. Katz, T. (2019). Evaluation of OTS hardware assemblies for use in risk averse, cost constrained space-based systems. *29th Annual INCOSE International Symposium (IS2019)*.
64. Smith, R.W. (1989). *The Space Telescope: A Study of NASA, Science, Technology and Politics*. Cambridge: Cambridge University Press and Hubble Space Telescope. https://en.m.wikipedia.org/wiki/Hubble_Space_Telescope#.
65. Wheatcraft, L., Ryan, M., and Dick, J. (2016). On the use of attributes to manage requirements. *Systems Engineering Journal* **19** (5): 448–458.

APPENDIX B: ACRONYMS AND ABBREVIATIONS

AD^2	Advancement Degree of Difficulty
AFD	Activity Flow Diagram
AFOTEC	Air Force Operational Test and Evaluation Center
AI	Artificial Intelligence
AIAA	American Institute of Aeronautics and Astronautics
ANSI	American National Standards Institute
API	Application Programming Interface
ASME	American Society of Mechanical Engineers
ASoT	Authoritative Source of Truth
BOL	Beginning of Life
BOM	Beginning of Mission
CAD	Computer-aided Design
CAPA	Corrective and Preventive Action
CCB	Change Control Board
CCB	Configuration Control Board
CDR	Critical Design Review
CDRL	Contract Deliverables Requirements List
CF	Completion Form
CFR	Code of Federal Regulations
CI	Configuration Item
CIA	Confidentiality, Integrity, and Availability
CM	Configuration Management
CMMI	Capability Maturity Model Integration
CMP	Configuration Management Plan

INCOSE Needs and Requirements Manual: Needs, Requirements, Verification, Validation Across the Lifecycle, First Edition. Louis S. Wheatcraft, Michael J. Ryan, and Tami Edner Katz.

CNR	Concepts, Needs, and Requirements
ConOps	Concept of Operations
COTS	Commercial Off-the-Shelf
CR	Concept Review
CR	Cost Reimbursement
DFD	Data Flow Diagram
DFI	Development Flight Instrumentation
DHF	Design History File
DHR	Device History Record
DMR	Device Master Record
DoD	Department of Defense
DoE	Department of Energy
DT&E	Developmental Test and Evaluation
ECR	Engineering Change Request
EIA	Electronic Industries Alliance
EMC	Electromagnetic compatibility
EMI	Electromagnetic interference
EOL	End of Life
EOM	End of Mission
ERD	Entity Relationship Diagram
ETL	Extraction, Transformation, and Loading
FAA	Federal Aviation Administration
FAT	Factory Acceptance Test
FCA	Functional Configuration Audit
FDA	Food and Drug Administration
FFBD	Functional Flow Block Diagram
FMEA	Failure Mode and Effects Analysis
FP	Fixed Price
FQT	Formal Qualification Test
FSoT	Federated Source of Truth
GAO	Government Accountability Office
GtNR	Guide to Needs and Requirements
GtVV	Guide to Verification and Validation
GtWR	Guide to Writing Requirements
HB	Handbook
HMI	Human Machine Interface
HSE	Human Systems Engineering
HSI	Human Systems Integration
HVAC	Heating, Ventilation, Air Conditioning
ICWG	Interface Control Working Group
I-NRDM	Information-based Needs and Requirements Definition and Management
I&T	Integration and Test
IAD	Interface Agreement Document
ICD	Interface Control Document
IDA	Interface Definition Agreement
IDD	Interface Definition Document
IEC	International Electrotechnical Commission
IEEE	Institute of Electrical and Electronics Engineers

IFMEA	Interface Failure Mode and Effects Analysis
IFU	Instructions for Use
IIDD	Internal Interface Definition Document.
INCOSE	International Council on System Engineering
IPO	Input, Process, Output
IRD	Interface Requirement Document
IRR	Integration Readiness Review
ISO	International Organization for Standardization
IT	Information Technology
KDN	Key Driving Need
KDR	Key Driving Requirement
KPI	Key Performance Indicator
KPP	Key Performance Parameter
LI	Leading Indicators
LIR	Lid Installation Robot
LIS	Lid Installation System
LOE	Level of Effort
MBD	Model-Based Design
MBSE	Model-Based Systems Engineering
MCR	Mission Concept Review
MGO	Mission, Goals, and Objectives
MIVV	Master Integration, Verification, and Validation
MMS	Mission Monitoring System
MOE	Measures of Effectiveness
MOP	Measures of Performance
MOS	Measures of Suitability
MOTS	Modified Off-the-Shelf
MRL	Manufacturing Readiness Level
MTBF	Mean Time Before Failure
MTTF	Mean Time to Failure
MTTR	Mean Time to Repair
NASA	National Aeronautics and Space Administration
NGO	Need, Goals, Objectives
NLP	Natural Language Processing
NPR	NASA Procedural Requirements
NRDM	Needs and Requirements Definition and Management
NRM	Needs and Requirements Manual
NRVV	Needs, Requirements, Verification, and Validation
NRVVM	Needs, Requirements, Verification, and Validation Management
OpsCon	Operational Concept
OSI	Open Systems Interconnection
OSLC	Open Services for Lifecycle Collaboration
OTS	Off-the-Shelf
OT&E	Operational Test and Evaluation
PA	Purchase Agreement
PAD	Project Authorization Document
PB	Performance Based
PBC	Performance-Based Contract

PBS	Product Breakdown Structure
PCA	Physical Configuration Audit
PDD	Process Definition Documents
PDR	Preliminary Design Review
PM	Project Management
PM&I WG	Project Management and Integration Working Group
PMI	Project Management Institute
PMP	Program/Project Management Plan
PO	Purchase Order
POC	Point of Contact
PRD	Program/Project Requirements Document
psi	Pounds per Square Inch
PTR	Post-test Review
PWS	Performance Work Statement
QA	Quality Assurance
QC	Quality Control
QM	Quality Management
QMS	Quality Management System
RFI	Request for Information
RFP	Request for Proposal
RM	Requirement Management
RMT	Requirement Management Tool
ROI	Return on Investment
RVCM	Requirements Verification Compliance Matrix
RTF	Rich Text Format
RVM	Requirements Verification Matrix
RWG	Requirements Working Group
SA	Supplier Agreement
SAD	System Architecture Diagrams
SAT	Site Acceptance Test
SBP	Strategic Business Plan
SCL	SE Capability Level
SCMP	Stakeholder Communications Management Plan
SCR	System Concept Review
SDK	Software Development Kit
SDR	System Design Review
SE	Systems Engineering
SE HB	Systems Engineering Handbook
SE&I WG	Systems Engineering and Integration Working Group
SEBoK	Systems Engineering Body of Knowledge
SEI	Software Engineering Institute
SEMP	System Engineering Management Plan
SIPOC	Source, Input, Process, Output, Customer
SIVV	System Integration, Verification, and Validation
SME	Subject Matter Expert
SOI	System of Interest
SOO	Statement of Objectives
SOP	Standard Operating Procedures

SoS	System of Systems
SOW	Statement of Work
SR	Scope Review
SRD	Software Requirements Document
SRD	Systems Requirement Document
SRR	System Requirements Review
SRS	Software Requirement Specification
StND	Stakeholder Needs Document
StRD	Stakeholders Requirements Document
StRS	Stakeholder Requirements Specification
STL	"Standard Triangle Language" or "Standard Tessellation Language"
SVaCM	System Validation Compliance Matrix
SVaM	System Validation Matrix
SVCM	System Verification Compliance Matrix
SVM	System Verification Matrix
SySML	System Modeling Language
TBC	To Be Computed
TBD	To Be Determined
TBR	To Be Resolved
TBS	To Be Supplied
TBX	Generic version of TBD, TBR, TBS, TBC
TDP	Technical Data Package
TMP	Technology Maturity Plan
TO	Task Order
TRA	Technology Readiness Assessment
TRL	Technology Readiness Level
TRR	Test Readiness Review
UFMEA	Use or User Failure Mode and Effects Analysis
URD	Users Requirements Document
USB	Universal Serial Bus
USC	United States Code
UL	Underwriters Laboratory
VIP	Vested, Influential, Participate
VOX	Voices of all Stakeholders
WBS	Work Breakdown Structure
WG	Working Group
WI	Work Instruction
XML	Extensible Markup Language

APPENDIX C: GLOSSARY

Note: This Appendix is a compilation of the major terms and concepts communicated in the context in which they are used within this Manual, and as such, the intent is to not replace but to complement and elaborate classic definitions found in other texts and standards.

Acceptance: An activity or series of activities conducted before the transition of the SOI from supplier to customer against predefined and agreed-to "necessary for acceptance" criteria such that the customer can determine whether the realized SOI is suitable to change ownership from supplier to customer. Acceptance can include system verification, system validation, certification, and qualification activities done by the customer, or a review of supplier system verification, system validation, certification, and qualification results that contain objective evidence that the acceptance criteria was met as agreed to in a contract. It is critical that system verification and validation *Success Criteria*, *Method*, and *Strategy* are stated for each need and requirement that clearly define what is *necessary for acceptance*.

Accuracy: A measure of how close an average of measures is to the baseline value expressed as a percentage. Using a target as an example, accuracy is the average of the distance each arrow or bullet hits in relation to the bullseye. *See also Precision.*

Allocated Baseline: The SOI physical architecture and resulting family of sets of needs and sets of requirements for the SOI and each subsystem, and

INCOSE Needs and Requirements Manual: Needs, Requirements, Verification, Validation Across the Lifecycle, First Edition. Louis S. Wheatcraft, Michael J. Ryan, and Tami Edner Katz.

system element or software component within the SOI architecture along with interface definition documentation for each interface boundary within the integrated SOI and interface boundaries across which the SOI interacts with an external system. *See also functional and product baseline.*

Allocation: The process by which the design input requirements defined for an entity at one level of the architecture are assigned (flow down) to the entities at the next lower level of the architecture that have a role in the implementation of the allocated requirement.

Allocation Matrix: A table or matrix visualization showing the allocation and budgeting of requirements for an entity at one level of the architecture to the entities at the next lower level of the architecture that have a role in the implementation of the allocated requirements.

Analysis (as a method of verification or validation): A *Method* that uses established mathematical models, simulations, algorithms, charts, graphs, circuit diagrams, or other scientific principles and procedures to provide objective evidence that the design input requirements or needs were met. Analysis also includes engineering assessments and logic usage based on previously recorded evidence obtained during design verification, design validation, early system verification, early system validation, and production verification activities, as well as results of lower-level subsystem or system element system verification or system validation. There are several subcategories of Analysis, including Analogy or Similarity, Certification, Model or Simulation, Engineering Assessment, and Sampling.

Approval Authority: Any individual, group of individuals, or organization that has the authority to accept and approve an SOI for use in its operational environment by its intended users. An *Approval Authority* could be a combination of an internal or external customer, a regulatory agency, or a third-party certification organization. It is the *Approval Authority* that 1) *decides* what *constitutes necessary for acceptance* and 2) *determines* what *is necessary for acceptance.*

Assumption: A proposition that is taken for granted as if it were known to be true, whether it is true. An assumption must not be considered to be true until there is objective evidence that it is true.

Authoritative Source of Truth: The official state or baseline version of data and information for a project. *See also Federated Source of Truth.*

Attribute: Additional information associated with an entity that is used to aid in its definition and management. Attributes are included as part of the needs and design input requirements expressions to define and capture key information to aid in the definition, verification, validation, management, and reuse of not only the needs and design input requirements but also the management of the SOI across all lifecycle activities.

Bidirectional Traceability: The ability to establish a two-way link between entities such that each has knowledge of the other.

Brownfield System: Legacy or heritage system where there is an existing predecessor system that can be evolved or transformed into the desired system.

Budgeting: The allocation of some quantity such as resource production or utilization, performance, quality, or some physical attribute. *See also Allocation.*

Certification: The result of an audit or a written assurance that the realized SOI has been developed and can perform its assigned functions, in accordance with legal or workmanship, design, and construction standards (such as for an aircraft, consumer product, or medical device). The development reviews, system verification, and system validation results form the basis for certification.

Commercial Off-the-Shelf (COTS): Existing elements in the marketplace that can be considered for purchase and use within the SOI architecture when making a Buy, Make, or Reuse decision for a system element. *See also Modified Off-the-Shelf and Off-the-Shelf.*

Compliance Matrix: A matrix/table visualization of the system verification or validation information that lists the needs or requirements, the system verification or validation *Procedure*, system verification or validation result in terms of compliance (pass/fail) of the system verification or validation *Success Criteria* for each system verification or validation *Instance*, and a summary of the objective evidence that was used to determine compliance or a pointer to where the evidence is located.

Compliance Risk: Risk concerning the ability to show compliance with applicable standards and regulations.

Compliance Needs and Requirements: A category of needs and requirements that address conformance with design and construction standards and regulations.

Concept: A textual or graphic representation that concisely expresses how an entity can fulfill the problem, threat or opportunity, mission, goals, objectives, and measures it was defined to address. The concept demonstrates how the entity provides a business capability in terms of people, processes, and products within constraints and acceptable risk. Concepts can also be defined concerning specific capabilities the stakeholders expected the entity to address, such as security, safety, resilience, and sustainability.

Confidence Level: The degree of confidence in the correctness, repeatability, and trustworthiness of the results of the data from a system verification and validation *Activity* obtained during the performance of the system verification and validation *Procedure* implementing that *Activity*. The Confidence Level is a measure on a scale of 0–100 percentage points on how confident that subsequent system verification or validation actions will yield the same result or how much confidence there is concerning the results of system verification or validation by one of the *Methods* of system verification or validation.

Cost Reimbursement (CR): A type of contract where the customer pays the supplier for allowed expenses, plus an additional amount to allow for a profit. (Sometimes referred to as cost-plus).

Criticality: Addresses needs and requirements from the perspective that their realization is essential in terms of the system's ability to fulfill its primary purpose in the operational environment when operated by its intended users. *See also Priority.*

Customer: The organizations or persons (internal or external) requesting or procuring a work product, and/or will be the recipient of the work product when delivered. Customers are key stakeholders that exist at multiple levels of an organization and may be internal or external to the enterprise.

Customer requirements: Customer-supplied requirements transformed from the customer needs. For example, a set of customer-supplied system requirements provided to a supplier within a contract. Customer-supplied requirements could apply to the SOI or to the organization developing the SOI. If they apply to the organization developing the SOI, they would be included in a SOW, PO, or SA. *See also Stakeholder Requirements and User Requirements.*

Demonstration (as a method of system verification or validation): A *Method* that involves the operation of an SOI to provide objective evidence that the required functions were accomplished under specific scenarios and operational environments. Demonstration is the manipulation of a system, subsystem, or system element, as it is intended to be used to verify that the results are as expected as defined by the *Success Criteria.* Demonstration consists of a *qualitative* determination made through observation, with or without the use of special test equipment or instrumentation, to verify characteristics such as human engineering features, services, access features, and physical accommodation (compared with Test, which uses a *quantitative* approach).

Design Controls: Work instructions and processes containing requirements on the organization involved in the transformation of work products developed within one lifecycle stage to work products developed within another lifecycle stage. The design controls are part of a larger Quality Management System (QMS). *See also Transformation.*

Design Input Requirements: The *technical* design-to set of system requirements for the SOI that were transformed from the baselined integrated set of needs and are inputs to the architecture and design definition processes. The set of design input requirements communicates the system perspective concerning what the system must do to meet the integrated set of needs and is what the design and SOI are verified against. Their focus is on the "what," not "how." To help distinguish requirements from needs, the requirements are written as "shall," statements (or other agreed-to term) to make clear that these are mandatory requirements that must be met and against which the SOI will be verified.

Design Output Specifications: The set of design definition records, artifacts, reports, strategy/approach as well as system design characteristics and design descriptions described in a form suitable for implementation.

Design outputs are also often referred to as end-item specifications or the Technical Data Package (TDP), which include parts lists, drawings, wiring diagrams, plumbing diagrams, labeling diagrams, requirements, logic diagrams, algorithms, computer-aided design (CAD) files, or Standard Tessellation Language (STL) files (for 3D printing), and represent criteria to which the SOI will be manufactured or coded; as such, they can be thought of as the build-to/code-to requirements no matter their form.

Design Validation: Confirmation that the design, as communicated in the set of design output specifications, will result in a system that meets its intended purpose in its operational environment when operated by the intended users as defined by the integrated set of needs and does not enable unintended users to negatively impact the intended use of the system.

Design Verification: Confirmation that: (1) the design reflects the set of design input requirements, (2) the set of design output specifications clearly implements the intent of the design as communicated by the set of design input requirements, and (3) the design meets the rules and characteristics defined for the organization's processes, guidelines, and requirements for design as communicated within the design controls.

Development Risk: Risk concerning problems that can occur due to a failure to follow PM and SE lifecycle process activity best practices, resulting in work products and artifacts that are not complete, correct, consistent, and feasible. Development risk also concerns the feasibility of performance and quality based on the maturity of critical technologies.

Deviation: A form of Request for Variance defined as "A specific written authorization, granted prior to the manufacture of an item, to depart from a particular requirement(s) of an item's current approved configuration documentation for a specific number of units or a specified period of time." [61] *See also Waivers and Request for Variance.*

Document Tree: A hierarchical view and grouping of artifacts and work products associated with each entity within a PBS.

Drivers and Constraints: Things outside the project's control that drive and constrain the lifecycle concepts analysis and maturation activities, as well as the solution space available to the project team. Compliance is mandatory—failing to show compliance will result in the system failing system validation, certification, acceptance, and approval for use.

Early Validation: Any validation activity (needs, requirements, or design) that occurs prior to system validation.

Early System Validation: System validation activities that occur concurrently with design validation prior to formal system validation.

Early System Verification: System verification activities that occur concurrently with design verification prior to formal system verification.

Elicitation: Engaging with the stakeholders to discover and understand what is needed, what processes exist, how stakeholders expect to interact with SOI, what happens over the SOI's lifecycle (good and bad), what modes, states, and transitions the SOI might undergo or experience during use (such as nominal, alternate-nominal, and off-nominal operations as well as misuse and loss scenarios). Misuse addresses cases where unintended users could misuse the SOI in unintended ways and could result in losses of some kind (for example, loss of the SOI, loss of the intended use of the SOI, loss of information, loss of money, and even loss of life).

Enabling system: Systems external to the SOI needed to "enable" certain lifecycle activities or enable the SOI to operate. Enabling systems include support systems and services across the SOI lifecycle, for example, development, production, integration, system verification, system validation, deployment, training, storage, transportation, operations, maintenance, and disposal. They could consist of laboratories, test equipment, simulators, test fixtures, power supplies, clean rooms, transportation, storage, and integration facilities, to name a few.

Entity: A single item to which a concept, need, or requirement applies: an organization, business unit, project, supplier, service, procedure, SOI—hardware (system, subsystem, system element), SOI—software (application, package, module, and feature), product, process, or stakeholder class (user, operator, tester, maintainer, as examples). There are three general types of entities—physical or software entities such as the engineered systems to be developed; process entities such as procedures or work instructions; and business or human entities such as business units, users, customers, developers, suppliers, and other stakeholders.

Factory Acceptance Test (FAT): The FAT series of activities are conducted at the developing organization's facility and witnessed by customer representatives. FAT activities are designed, to the extent possible given it is not in the operational environment, to provide objective evidence that can be used to determine that the SOI meets the customer's needs (system validation) but in a more controlled (and often less expensive) test environment. While the FAT activities are more focused on system validation against the integrated set of needs, a subset of the system-level requirements may also undergo final system verification during the FAT.

Feasibility: The ability of the SOI to meet its needs and requirements within cost, schedule, technology, legal, and ethical constraints.

Federated Source of Truth (FSoT): Some larger organizations with multiple systems are using the concept of "federated source of truth" instead of an ASoT at the Operational level discussed in Chapter 2. In this context, while each individual system may have its own ASoT, there may be inconsistencies between systems. An FSoT enables the organization to ensure all systems have access to the same data and information, which helps to eliminate inconsistencies and errors that can arise when different systems use different data and information sources. With an FSoT, data and information are indexed, retrieved, and presented in a user interface layer where they are created and can be accessed by all stakeholders. This approach is used to ensure that the information used by different systems is consistent and current.

Fit (Operational) Needs and Requirements: A category of needs and requirements dealing with functions that concern secondary or enabling capabilities, functions, and interactions between the SOI and external systems needed for the system to accomplish its primary functions. This includes functions concerning the ability of the system to interface with, interact with, connect to, operate within, and become an integral part of the macrosystem it is a part. Fit includes human system interactions and interfaces as well as both the induced and natural environments (conditions of operations, transportation, storage, and maintenance).

Fixed Price Contracting: Type of contract where the supplier is paid a fixed amount, no matter the supplier expenses to provide a system. Fixed-price contracts are used to control costs and risks to the customer. The customer has no oversight of the supplier during the development of the SOI to be supplied.

Form (Physical Characteristics) Needs and Requirements: A category of needs and requirements that address the shape, size, dimensions, mass, weight, and other observable parameters and characteristics that uniquely distinguish a system. For software, form needs and requirements could address programming language, lines of code, and memory requirements as examples.

Formal Qualification Test (FQT): A series of formal activities that involve the project manager and/or the customer that are designed to provide objective evidence that can be used to determine compliance (verification) with all contractual requirements. The FQT activities may involve an environment and users/operators that are different than that expected during operations. This environment and test inputs are meant to stimulate the system in a particular way to verify the system meets each requirement. This is not a validation test.

Function: A task, action, or activity that must be performed to achieve a desired outcome.

Functional baseline: The system-level baselined lifecycle concepts, integrated set of needs, and set of design input requirements for an SOI along with interface definition documentation for all interface boundaries across which the SOI will interact. *See also Allocated and Product Baseline.*

Function/Performance Needs and Requirements: A category of needs and requirements that communicate the primary functions and associated performance that the SOI needs to perform in terms of its intended use. The functions address the capabilities and features the stakeholders expect the SOI to have; performance addresses how well, how many, and how fast type attributes of the function.

Functional/Performance Requirement: A requirement stating a task, action, or activity that the SOI must do to meet the need it was transformed from or an allocated parent requirement. Functional requirements include a performance measure such as how well, how many, and how fast a function needs to be performed.

Goals (as part of MGOs): Upper-level needs that form the second level of the hierarchy of the integrated set of needs. Goals are elaborated from the mission statement communicating what needs to be achieved that will result in achieving the mission. *See also Mission Statement and Objectives.*

Goals (communicated within a set of needs or set of requirements): Statements that communicate a customer or stakeholder desire for some feature or capability that, while not critical, adds some value, as long as it is low risk and does not impact critical functions nor drive cost nor impact schedule. Often communicated as "would like" or "should" statements. These goals are not requirements, are not binding, and are not subject to design and system verification or validation.

Gold Plating: The act of adding needs or requirements for features that are not necessary to address the problem or opportunity the SOI is addressing and meet the stated MGOs and measures.

Greenfield System: A new system is being developed where there is not an adequate predecessor system.

Horizontal Traceability: Forward and backward traceability between entities across the SOI lifecycle (from lifecycle concepts to retirement).

Induced Environment: Part of the operational environment including temperature, vibrations, mechanical loads, electrical emissions, and acoustics created (induced) from external systems, as well as temperature, vibrations, mechanical loads, electrical emissions, and acoustics resulting from the operation of the SOI. Induced environments from existing systems that the SOI has no control over are constraints within which the SOI must operate. Induced environments for existing systems or for the SOI under development may be negotiated and thus are part of the needs and requirements definition processes. Once defined, the expectation is that the SOI will be able to operate within the induced operational environments.

Information-based Needs and Requirements Definition and Management (I-NRDM): A practical implementation of PM and SE from a data-centric perspective as defined in the INCOSE RWG whitepaper *Integrated Data as a Foundation of Systems Engineering*: "SE, from a data-centric perspective, involves the formalized application of shareable sets of data to represent the SE work products and underlying data and information generated to support concept maturation, needs and requirements development, design, analysis, integration, system verification and validation activities throughout the system lifecycle, from conceptual design to retirement."

Inspection (as a method of verification or validation): A *Method* that is nondestructive and typically includes the use of sight, hearing, smell, touch, and taste; simple physical manipulation; or mechanical and electrical gauging and measurement. Inspection could be inspection of the end item system (code or physical attributes) or artifacts that represent the SOI such as drawings, parts listings, or part specifications.

Integrated Set of Needs: A representation of the integrated and baselined set of needs that were transformed from the set of lifecycle concepts and other sources for the SOI. The integrated set of needs communicates an integrated stakeholder's perspective of the SOI as viewed externally concerning their real-world expectations of what they need the SOI to do and what the SOI design input requirements, design, and SOI are to be validated against.

Interface: An interface is a boundary where, or across which, two or more systems interact in accordance with an agree-to interface definition.

Interface Control Document (ICD): [Also referred to as an Interface Definition Document (IDD)]. A document that contains a common location or repository to record an agreement concerning information defining interactions across an interface boundary. These definitions are communicated as statements of fact and not requirement (shall) statements. The form could be text, figures, drawings, tables, or other forms most suited to communicating the definition. The ICD could contain definitions that represent the as-designed or as-coded or built definition of the interactions across an interface boundary or be evolutionary first stating as design inputs the characteristics of what is crossing the interface boundary and then including the design outputs concerning the media involved in the interaction and what each system looks like at the interface boundary. An ICD could be one-sided (owned by an existing system) or multi-sided (owned by an integrating organization controlling the interactions between two or more systems within the system architecture).

Interface Definition: A common agreement concerning a specific interaction across an interface boundary between the SOI and another system commonly recorded in an ICD, IDD, or similar type interface definition document or location in a database.

Interface Requirement: A functional requirement that involves a defined interaction of a system across an interface boundary with another system or entity. The interaction is communicated as a verb/object pair. An interface requirement includes a reference that defines the interaction.

Interface Risk: Risk associated with interface boundaries in terms of stability, threats, or failures that could impact the operation of the SOI.

Level of Effort: An approach to contracting that is a type of CR contract where the supplier is paid for a specified level of effort (time and materials) over a stated period on work that is stated in general terms. The customer has both insight and oversight of the supplier activities.

Levels: Levels or layers refer to a hierarchy in which SE artifacts are defined and managed. Levels could refer to levels of an organization or levels of architecture. The term level is also used in terms of "levels of detail" and "levels of abstraction."

Lifecycle Concepts: Lifecycle concepts (plural) refer to the multiple concepts across the lifecycle for how the organization (and stakeholders within an organization) expects to manage, acquire, define, develop, build/code, integrate, verify, validate, transition, install, operate, support, maintain, and retire an entity. Lifecycle concepts can be communicated in various forms including textual (for example, OpsCon or ConOps), graphical representations (for example, diagrams and models), and/or electronic (databases). There are multiple lifecycle concepts that apply to each entity, so it is useful to develop a necessary and sufficient set of lifecycle concepts from which an integrated set of needs can be derived.

Margins (development and technical): The difference between the estimated budgeted value and the actual value at the end of development when the system is delivered.

Management Reserve: The portion of the available quantity held back or kept "in reserve" by management or the quantity owner during development and not made available through allocation.

Management Risk: Risk concerning budget, schedule, resources, and upper-level management support of the project throughout the system life.

Master Integration, Verification, and Validation (MIVV) Plan: A planning document that expands and implements the project's design verification, design validation, system integration, system verification, and system validation philosophy/concepts as defined in the PMP and SEMP. The MIVV Plan defines the detailed processes and approach that will be used for design verification, design validation, system integration, system verification, and system validation. The information in the MIVV Plan is used to manage and plan design verification, design validation, system verification, and validation activities for all system, subsystem, and system elements that are part of the SOI physical architecture. The MIVV Plan further defines the design verification, design validation, system integration, system verification, and system validation concepts and schedules contained in the PMP and SEMP. *See also System Integration, Verification, and Validation (SIVV) Plan.*

Measures: Data and information that will be used to both validate the objectives against and manage system development across the lifecycle. When defining objectives, the project team must define what measures the SOI will be validated against. Measures are referred to by various terms: Measures of Suitability (MOS), Measures of Effectiveness (MOEs), Measures of Performance (MOPs), Key Performance Parameters (KPPs), Technical Performance Measures (TPMs), Leading Indicators (LIs), mission success criteria, primary science objectives, secondary science objectives, acceptance criteria, as examples.

Mission Statement: The top tier of the hierarchy of needs based on the analysis of a problem, threat, or opportunity that the project was formed to address. The mission statement defines the "why"—why does the project exist? and the "what"—what does the organization's strategic- and operational-level stakeholders or the customer need the system to accomplish (what is the expected outcome) that will address the defined problem statement?

Modified Off-the-Shelf (MOTS): Existing OTS or COTS elements that can be considered for use, after modification, within the SOI architecture when making a Buy, Make, or Reuse decision for a system element. *See also Off-the-Shelf and Commercial Off-the-Shelf.*

Natural Environment: Part of the operational environment (for example, temperature, humidity, and pollutants) that exists naturally—part of nature that the SOI has no control over. As such, the natural operational environment is a constraint within which the SOI must operate.

Needs: Well-formed textual statements of expectations for an *entity* stated in a structured, natural language from the perspective of what the stakeholders need the entity to do, in a specific operational environment, communicated at a level of abstraction appropriate to the level at which the entity exists.

Need Expression: Includes a *need statement* plus a set of associated *attributes*.

Need Set: A well-formed set of agreed-to need expressions for an entity and its external interfaces. *See also Integrated Set of Needs.*

Need Statement: The result of a formal transformation of one or more sources or *lifecycle concepts* into an agreed-to expectation for an *entity* to perform some function or possess some quality within specified constraints with acceptable risk. Need statements are written in a structured, natural language and have the characteristics defined in the GtWR.

Need Validation: Confirmation that the needs clearly communicate the intent of the agreed-to lifecycle concepts, constraints, and stakeholder real-world expectations from which they were transformed in a language understood by the requirement writers. Confirmation that the *integrated set of needs* correctly and completely captures what the stakeholders need and expect the system to do in the context of its intended use in the operational environment when operated by its intended users.

Need Verification: Confirmation that the *need statements* and *integrated set of needs* meet the rules and characteristics defined for writing well-formed needs and sets of needs in accordance with the organization's standards, guidelines, rules, and checklists.

Objective Evidence: Information based on tangible data and information based on facts that can be proven to be true through test, analysis, demonstration, inspection, measurement, observation, and other such *Methods* of verification and validation.

Objectives: Upper-level needs that form the third level of the hierarchy of the integrated set of needs. Objectives are elaborated from the goals providing more details concerning what must be done to meet the goals that will result in the mission to be achieved, i.e., what the project team and the SOI need to achieve so the SOI can fulfill its intended purpose (mission) in its operational environment when operated by its intended users.

Off-the-Shelf (OTS): Existing OTS elements that can be considered for use within the SOI architecture when deciding to make, buy, or reuse a system, subsystem, or system element. These could be existing elements within the developing organization, elements that can be purchased in the marketplace (COTS), or OTS or COTS elements that will be modified before use (MOTS). *See also Modified Off-the-Shelf and Commercial Off-the-Shelf.*

Open Systems Interconnection (OSI) Model: A conceptual model that characterizes and standardizes the communication functions of a telecommunication or computing system without regard to its underlying internal structure and technology. Its goal is the interoperability of diverse communication systems with standard communication protocols. The model partitions the flow of data in a communication system into seven abstraction layers, from the physical implementation of transmitting bits across a communications medium to the highest-level representation of data of a distributed application (https://en.wikipedia.org/wiki/OSI_model).

Operational capability: Planned availability of additional resources during surge or peak consumption periods as well as enabling the system to succeed even when the unexpected happens. Operational capability also provides additional resources when operational conditions or the environment are different from those assumed when the system resource requirements were generated, and the system was not designed to accommodate.

Operational Environment: The environment in which the SOI will be operated. The operational environment includes the natural environment, induced environments, conditions for use the SOI is intended to be used. The operational environment also includes external systems the SOI must interact with across the interface boundaries (external interfaces).

Operational Margin: The difference between what is required during operations and what is actually provided in addition to what is required.

Operational Risk: Risk associated with the ability of the system to fulfill its intended use when operated by its intended users, in its intended environment, and risk associated with unintended users negatively impacting the intended use of the system or using the system in an unintended way resulting in loss.

Performance-Based Contracting: An approach to contracting where the focus is on well-defined measures and outcomes rather than how the supplier does the work to produce the required outcomes. With this method, the customer may be provided some insight into the supplier activities as defined in the SOW, but no oversight (control) of those activities.

Problem statement: A short statement developed at the beginning of a project that defines the problem, threat, or opportunity for which the project team is to address.

Precision: A function of the distribution of measurements taken, i.e., closeness of agreement between independent, repeated measures obtained from the same sample under specific conditions. Precision is represented by the standard deviation (in units of the test) or coefficient of variation (in units of percent). Using a target as an example, precision deals with how "tight" the cluster of arrows or bullets is—are the arrows or bullets all over the target versus they are in a tight grouping? *See also Accuracy.*

Priority: Relative importance of a need or requirement from the perspective of the stakeholders. *See also Criticality.*

Product: The SOI that will be delivered to a customer or customers.

Product (System) Baseline: The baselined sets of design output specifications for the subsystems and system elements that are part of the product (system) physical architecture (the realization of the allocated baseline). *See also Allocated and Functional Baseline.*

Product Breakdown Structure (PBS): A hierarchical view of the physical architecture showing and naming actual physical parts (hardware, mechanical, and software) subsystems and system elements that are part of the SOI architecture. Similar to the concept of a WBS.

Production Risk: Risk concerning the ability to manufacture or code the SOI.

Production Verification: Verifying that the system was "built right" as defined by the design output specifications, as well as verifying that it was "built correctly" per the organizational design controls (guidelines, best practices, and requirement) concerning the equipment and processes used to manufacture and code the system element were followed.

Post-Test Review (PTR): A PTR often concludes each test phase (FQT, FAT, and SAT) of the acceptance process. At the PTR, the result of each *Procedure* is presented along with a list of any discrepancies and non-conformances.

Qualification: Requires that all system verification, system validation, and required certification actions have been successfully performed. The qualification process demonstrates that each subsystem and system element that is part of the system physical architecture and the integrated system meets the applicable needs, requirements, and design output specifications, including performance and safety margins.

Quality Management System (QMS): A set of organizational process and activities that are defined by a collection of policies, processes, documented procedures, and records. The QMS defines how a company will achieve the creation and delivery of the quality products and services it provides to its customers. Included within the QMS are Design Controls for how the organization will implement the QMS for each lifecycle stage of activities (https://advisera.com/9001academy/knowledgebase/quality-management-system-what-is-it). *See also Design Controls.*

Quality needs and requirements: A category of needs and requirements that address fitness for use often referred to as the "-ilities." For example, reliability, testability, operability, availability, maintainability, operability, supportability, manufacturability, and interoperability.

Rationale: An attribute included within each need and requirement expression that communicates the reason for the need or requirement's existence. Rationale defines why the need or requirement should be included and other information relevant to better understand the reason for and intent of the need or requirement.

Readiness for Use: A determination of readiness for use by an *Approval Authority* based on an analysis of the system verification, system validation, certification, and qualification results.

Request for Variance: A nonconformance with baselined needs, requirements, or design output specifications. Sometimes referred to as *deviations* or *waivers* [57–59] or *concessions* [60]. *See also Deviations and Waivers.*

Requirement (project): Requirements on a project or organizational elements within the enterprise that will be recorded in a project management plan and other plans, procedures, and work instructions.

Requirement (supplier): Requirements on a supplier, vendor, or contractor that will be recorded in Statements of Work (SOW), Purchase Order (PO), or Supplier Agreements (SA).

Requirement Expression: Includes a *requirement statement* and a set of associated *attributes*.

Requirement set: A well-formed set of agreed-to *requirement expressions* for an entity and its external interfaces.

Requirement Statement: The result of a formal transformation of one or more sources, needs, or higher-level requirements into an agreed-to obligation for an entity to perform some function or possess some quality within specified constraints with acceptable risk.

Requirement Validation: Confirmation that the requirements clearly communicate the intent of the needs, parent requirements, and other sources from which they were transformed, in a language understandable by the design and manufacturing/coding teams.

Requirement Verification: Confirmation that the requirement statements and sets of requirements meet the rules and characteristics defined for writing well-formed requirements and sets of requirements in accordance with the organization's standards, guidelines, rules, and checklists such as the INCOSE GtWR.

Risk Assessment: The evaluation, analysis, and estimation or quantification (likelihood versus consequence) for each identified risk, hazard, or threat.

Risk Mitigation: Actions concerning how the risk associated with a hazard or associated threat could be eliminated, the likelihood reduced, and/or the severity of the impacts/consequences be reduced.

Risk Treatment/Handling: The planning, tracking, and controlling of a risk hazard or threat across the lifecycle.

SE Vee Model: The SE Vee model is a visualization to help communicate the SE processes over time commonly used to show the iterative and recursive nature of SE as a system is decomposed into a hierarchical architecture of individual subsystems and system elements, realized by design, built, and coded, and integrated into the system. The left side of the SE Vee shows a top-down series of activities where the definition of lifecycle concepts, integrated sets of needs, sets of design input requirements, architecture, and design take place. Correspondingly, the right side of the SE Vee shows a bottom-up series of activities where the physical system elements and subsystems that make up the system physical architecture are verified against their respective requirements and validated against their respective needs that were defined during the SE activities on the left side of the SE Vee.

SIPOC Diagram: SIPOC stands for **S**ource-**I**nput-**P**rocess (function/activity)-**O**utput-**C**ustomer. For each function or activity, its inputs and outputs are identified. For each input, the source is listed, and for each output, the customer or destination of the output is listed.

Site Acceptance Test (SAT): A system validation series of *Activities* and associated *Procedures* whose successful completion results in *Approval Authority* acceptance of the SOI for use or release to the public. After the SOI is delivered to the customer's location or location of intended use, the SAT activities are executed in the operational environment with the intended users to validate system operability and to perform a final system verification of the interfaces with external systems. The SAT activities will often execute operational scenarios—usually "day in the life" type scenarios based on use cases or an operational scenario that exercises critical functionality and performance of the system. Depending on the contract type, this is equivalent to the final DT&E event.

Stakeholders: Any individual or organization who may be affected by the SOI, who will participate in the development of the SOI, who is able to influence the definition and development of the SOI, or with whom the project team will interact across the SOI lifecycle. Stakeholders are the primary source of needs and requirements for an SOI. There are stakeholders internal to the developing organization as well as external stakeholders including customers and user/operators.

Stakeholder Needs: A user-oriented view of the system as viewed externally: what stakeholders need from the system, what they need the system to do, how they plan to interact with the system, and what interactions the system has with its external environment.

Stakeholder Register: List of stakeholders involved in some way with the SOI across its life along with key information for each stakeholder.

Stakeholder Requirements: Requirements transformed from the stakeholder needs communicating a black-box, external view of the SOI. Stakeholder Requirements could apply to the SOI or to the organization developing the SOI. If they apply to the SOI, they would be inputs to the *Lifecycle Concepts and Needs Definition* activities. If they apply to the organization developing the SOI, they would be included in a project management plan, SEMP, or other plan. Those requirements that apply to an external organization developing the SOI would be included in an SOW, PO, or SA. Stakeholder requirements include customer-supplied and owned requirements. *See also Customer Requirements and User Requirements.*

System: An entity that can be decomposed through elaboration that involves analysis, allocation, and budgeting into lower-level subsystems and system elements defined within the SOI architecture.

System (Product) Baseline: The baselined sets of design output specifications for the subsystems, system elements, and software components that are part of the SOI (product) physical architecture. *See also Allocated and Functional Baseline.*

System Element: An entity within the SOI architecture where the project has made a buy, make, or reuse determination and no further decomposition by the project team is deemed necessary for successful system element design and realization.

System of Interest (SOI): The system, subsystem, or system element that will be developed, verified, validated, and delivered to either an internal or an external customer. The SOI could be a product, system, or system element within the end item architecture.

System Integration, Verification, and Validation (SIVV) Plan: The SIVV Plans are used to implement the SOI project team's MIVV Plan design verification, design validation, system integration, system verification, and system validation concepts and activities for a given system, subsystem, system element, or software component within the SOI architecture. The SIVV Plan contains a detailed identification of the activities to be performed as part of design verification, design validation, production verification, system verification, and system validation of a specific subsystem, system element, or software component that makes up the integrated system. Resources including test equipment, facilities, and personnel are addressed. A detailed schedule consistent with the design verification and design validation, production verification, system integration, system verification, and system validation schedule defined in the MIVV Plan is included.

System Integration, Verification, and Validation Risk: Risks associated with the project's ability to successfully integrate the parts that make up the SOI, successfully verify the SOI meets its requirements, and validate the SOI meets its needs.

System Requirement: *See Design Input Requirement.*

System Validation: Confirmation that the designed, built, and verified SOI will result in or has resulted in an SOI that meets its intended purpose in its operational environment when operated by its intended users and does not enable unintended users to negatively impact the intended use of the system as defined by its integrated set of needs.

System Verification: Confirmation that the designed and built or coded SOI: (1) has been produced by an acceptable transformation of design inputs into design outputs; (2) meets its set of design input requirements and set of design output specifications; (3) no error/defect/fault has been introduced at the time of any transformation; and (4) was produced per the requirements, rules, and characteristics defined by the organization's best practices and guidelines defined in the organization's QMS design controls.

System Verification or System Validation *Activity*: A system verification or validation *Activity* that addresses one or more system verification or system validation *Instances*. The *Activity* will be realized via a system verification or validation *Procedure*. Some may refer an *Activity* as a Test Case.

System Verification or System Validation *Approval Package*: A collection of *Execution Records* and other information to be submitted to the *Approval Authority* that contain objective evidence that the SOI has been successfully verified or validated.

System Verification or System Validation *Event*: A scheduled *Event* within the project Master Schedule concerning when a system verification or validation *Procedure* will be performed.

System Verification or System Validation *Execution Record*: A record that states the status and outcome of a system verification or validation *Procedure* for an *Activity* in terms of whether the results provide objective evidence that the SOI met the *Success Criteria* for each *Instance* within that *Activity.*

System Verification or System Validation *Instance*: A unique entity within a system model for system verification and validation used to manage the system verification or system validation actions associated with a given need or design input requirement. There is one system verification or system validation *Instance* for each need and design input requirement.

System Verification or System Validation Matrix: A matrix/table visualization of the system verification and validation attributes and other planning information associated with each need and design input requirement used to help both plan and manage system verification and validation *Activities*. Each row of the Matrix represents a system verification or system validation *Instance*.

System Verification or System Validation *Method*: The *Method* to be used to verify the SOI meets a requirement or is validated to meet a need. *Methods* include Test, Demonstration, Inspection, and Analysis.

System Verification or System Validation *Procedure*: A sequence of steps or actions (often in the form of a document) that result in the collection of data that provide objective evidence that the system meets the *Success Criteria* defined for each *Instance* within a given *Activity* in accordance with the set of *Procedure Requirements* and other key information gathered for the *Activity* the *Procedure* applies.

System Verification and Validation *Procedure Requirements*: A set of requirement statements on the responsible organization or personnel involved in performing a *Procedure* concerning the details involved in implementing the *Strategy*, *Method, and Success Criteria* defined for each *Instance* that will result in objective evidence a need or design input requirement has been satisfied by the SOI for each *Instance* included in the *Activity* addressed by the *Procedure.* Some may refer to Procedure Requirements as Verification or Validation Requirements or Test Requirements.

System Verification and System Validation *Strategy*: An overview of the methodology to be used for a verification or validation *Activity* that will result in the data needed to provide objective evidence that the SOI has met the defined system verification or validation *Success Criteria.*

System Verification and System Validation *Success Criteria*: A set of criteria included within a need or requirement expression that must be shown to have been met, with some level of confidence, before an *Approval Authority* will qualify, certify, or accept the SOI to be used as intended in the operational environment when operated by its intended users. The *Success Criteria* represent the expected outcome of a system verification or system validation *Procedure* for each *Activity* and *Instance* within that *Activity* included within the *Procedure*.

Technical Debt: Technical debt [20] is a metaphor coined by Ward Cunningham, coauthor of the *Manifesto for Agile Software Development*, to describe what occurs when a project team uses a quick, short-term solution that will require additional development work later to meet the needs of the stakeholders. From a project perspective, technical debt refers to the eventual consequences of poor SE and PM practices.

Technology Maturation Plan (TMP): A project's plan and concept for maturing critical technologies needed to be used within the SOI to meet stakeholders' needs and requirements for capabilities, performance, quality, and compliance. This plan addresses what is needed in terms of resources, budget, and time to advance the TRL of the critical technology to enable its use within the SOI.

Technology Readiness Assessment (TRA): An assessment of the maturity and associated level of risk (TRL) associated with a given technology being considered for use in the SOI.

Technology Readiness Level (TRL): A measure addressing the maturity and associated level of risk associated with a given technology being considered for use in the SOI. TRLs are expressed on a scale of 1 to 9. The lower the number, the less mature the technology is for that specific use and operational environment, and the higher the risk to the project.

Test (as a method of system verification or validation): A *Method* that leverages data taken from one or more actions to provide objective evidence that the SOI satisfies a need or requirement. The *Method* Test uses controlled, predefined sets of inputs, data, or stimuli to exercise the SOI. Using special test equipment or instrumentation to obtain the required data, results are compared against expected outcomes as defined by the *Success Criteria*

Test Readiness Review (TRR): A formal gate review that must be passed prior to the execution of a *Procedure* or family of *Procedures*. At the TRR, an overall assessment of readiness of the system, enabling systems, *Procedures*, and personnel is conducted. A TRR will typically also address system configuration (including any simulators or emulators), risk, and other information pertinent to the performance of a *Procedure*.

Test Requirements: Requirements on at test to be conducted. *See also Procedure Requirements*.

Textual Need and Requirement Statements: Needs and requirements expressed using a structured, natural language having the characteristics defined in the GtWR resulting from following the activities discussed in this Manual and defined by a set of rules such as those defined within the INCOSE GtWR. Textual needs and requirements can be expressed in an RMT as well as a language-based modeling application.

Traceability: The ability to establish an association or relationship between two or more entities and to track entities from their origin to the activities and deliverables that satisfy them, as well as assess the effects on artifacts across the lifecycle when change occurs.

Traceability Record: A representation of the traceability links established within the project's integrated dataset between requirements, needs, and other artifacts. The traceability record includes directionality and type of links established to describe unique relationships between needs and requirements and other artifacts.

Traceability Relationship Meta-model: A model that is established at the beginning of a project defining the artifacts to be managed within the project's toolset and the traceability and relationships between those artifacts. Many modern tools enable the project to define rules within the model to ensure the model is followed and to flag discrepancies when the rules are not followed.

Trace Matrix: A table visualization of the traceability record showing relationships in a matrix form that can be used to manage sets of needs and sets of design input requirements throughout the system lifecycle.

Transformation: A formal process used when transforming an SE artifact developed in one lifecycle phase to SE artifacts for another lifecycle stage. For example, moving from lifecycle concepts to the needs that communicate what the stakeholders need from the SOI and then defining the design input requirements that communicate what the SOI must do to meet those needs. The transformation process is defined in work instructions and processes containing requirements on the organization involved in the transformation. *See also Design Controls.*

Qualification Criteria: A set of criteria defined in a regulation or standard that must be achieved that provides objective evidence that the system complies with the requirements within that regulation or standard. Qualification criteria often include safety margins for system performance in case where normal operating performance values are exceeded due to a failure or unexpected changes in the operational environment.

Unidirectional traceability: The ability to establish a one-way trace from one entity to another, where the receiving entity has no knowledge of the source entity.

User: Stakeholders that directly interact with the product or system being developed throughout its life.

User Needs: A subset of stakeholder needs. *See also Stakeholder Needs.*

User Requirements: A subset of stakeholder requirements. *See also Stakeholder Requirements.*

Vertical Traceability: Is most often referred to in the context of levels of organization or architectural levels of the system or product under development. Various levels exist from a hierarchical architecture view of an SOI. Higher-level needs and requirements are allocated to the SOI, and bidirectional vertical traceability is established as these requirements are defined.

Validation (generic): Confirmation and provision of objective evidence that an engineering element will result, or has resulted in, a system that meets its intended purpose in its operational environment when operated by its intended users.

Verification (generic): Confirmation and provision of objective evidence that an engineering element: (1) has been produced by an acceptable transformation, (2) meets its requirements, and (3) was produced per the rules and characteristics defined for the organization's best practices and guidelines defined in the organization's QMS design controls.

Waiver: A form of Request for Variance defined as "A written authorization to accept an item, which during manufacture, or after having been submitted for Government inspection or acceptance, is found to depart from specified requirements, but nevertheless is considered suitable for use 'as is' or after repair by an approved method." [61] *See also Deviations and Request for Variance.*

Work Breakdown Structure: A hierarchical view of the tasks and activities to be performed by a project within a given budget and schedule. *See also Product Breakdown Structure.*

Zero-Based Approach: An approach where all needs and requirements are thoroughly evaluated by the project for not only feasibility but also applicability and value regarding the agreed to MGOs, measures, drivers and constraints, risk mitigation, and lifecycle concepts, and thereby "earn" their way into the *integrated set of needs* and set of *design input requirements.*

APPENDIX D: COMMENT FORM

<table>
<tr><td colspan="5">Name of submitter (first name and last name):</td><td colspan="3"></td></tr>
<tr><td colspan="5">Date submitted:</td><td colspan="3"></td></tr>
<tr><td colspan="5">Contact information (email address):</td><td colspan="3"></td></tr>
<tr><td colspan="5">Type of submission (individual/group):</td><td colspan="3"></td></tr>
<tr><td colspan="5">Group name and number of contributors (if applicable):</td><td colspan="3"></td></tr>
<tr><td>Comment sequence number</td><td>Commenter name</td><td>Category (TH, TL, E, G)</td><td>Section number</td><td>Specific reference (for example, paragraph, line, Figure no., and Table no.)</td><td>Issue, comment, and rationale (Rationale must make comment clearly evident and supportable)</td><td>Proposed Changed/ New Text

-- MANDATORY ENTRY (Must be substantial to increase the odds of acceptance)</td><td>Importance Rating (R = Required, I = Important, T = Think About for future version)</td></tr>
<tr><td></td><td></td><td></td><td></td><td></td><td></td><td></td><td></td></tr>
<tr><td></td><td></td><td></td><td></td><td></td><td></td><td></td><td></td></tr>
<tr><td></td><td></td><td></td><td></td><td></td><td></td><td></td><td></td></tr>
<tr><td></td><td></td><td></td><td></td><td></td><td></td><td></td><td></td></tr>
</table>

Submit comments to the Requirements Working Group leaders at: requirements-leaders@incose.net

If this fails, comments may be sent to info@incose.org (the INCOSE main office), which can relay your comments to the working group leaders, if so, requested in the comment cover page.

INCOSE Needs and Requirements Manual: Needs, Requirements, Verification, Validation Across the Lifecycle,
First Edition. Louis S. Wheatcraft, Michael J. Ryan, and Tami Edner Katz.

INDEX

INCOSE Needs and Requirements Manual: Needs, Requirements, Verification, Validation Across the Lifecycle, First Edition. Louis S. Wheatcraft, Michael J. Ryan, and Tami Edner Katz.